Springer Optimization and Its Applications

Volume 96

Aims and Scope
Optimization has been expanding in all directions at an astonishing rate during the last few decades. New algorithmic and theoretical techniques have been developed, the diffusion into other disciplines has proceeded at a rapid pace, and our knowledge of all aspects of the field has grown even more profound. At the same time, one of the most striking trends in optimization is the constantly increasing emphasis on the interdisciplinary nature of the field. Optimization has been a basic tool in all areas of applied mathematics, engineering, medicine, economics, and other sciences.

The series *Springer Optimization and Its Applications* publishes undergraduate and graduate textbooks, monographs and state-of-the-art expository work that focus on algorithms for solving optimization problems and also study applications involving such problems. Some of the topics covered include nonlinear optimization (convex and nonconvex), network flow problems, stochastic optimization, optimal control, discrete optimization, multi-objective programming, description of software packages, approximation techniques and heuristic approaches.

More information about this series at http://www.springer.com/series/7393

Themistocles M. Rassias
Editor

Handbook of Functional Equations

Stability Theory

Springer

Editor
Themistocles M. Rassias
Department of Mathematics
National Technical University of Athens
Athens, Greece

ISSN 1931-6828 ISSN 1931-6836 (electronic)
ISBN 978-1-4939-5309-7 ISBN 978-1-4939-1286-5 (eBook)
DOI 10.1007/978-1-4939-1286-5
Springer New York Heidelberg Dordrecht London

Mathematics Subject Classification (2010): 39B05, 39B22, 39B52, 39B62, 39B82, 40A05, 41A30, 54C60, 54C65.

Printed on acid-free paper

Springer is part of Springer Science + Business Media (www.springer.com)

Preface

Handbook of Functional Equations: Stability Theory consists of 17 chapters written by eminent scientists from the international mathematical community, who present important research works in the field of mathematical analysis and related subjects, particularly in the Ulam stability theory of functional equations. These works provide an insight in a large domain of research with emphasis to the discussion of several theories, methods and problems in approximation theory, influenced by the seminal work of the well-known mathematician and physicist Stanislaw Ulam (1909–1984). Emphasis is given to one of his fundamental problems concerning approximate homomorphisms.

The chapters of this book focus mainly on both old and recent developments on the equation of homomorphism for square symmetric groupoids, the linear and polynomial functional equations in a single variable, the Drygas functional equation on amenable semigroups, monomial functional equation, the Cauchy–Jensen type mappings, differential equations and differential operators, operational equations and inclusions, generalized module left higher derivations, selections of set-valued mappings, D'Alembert's functional equation, characterizations of information measures, functional equations in restricted domains, as well as generalized functional stability and fixed point theory. It is a pleasure to express our deepest thanks to all the mathematicians who, through their works, participated in this publication. I would like to thank Dr. Michael Batsyn for his invaluable help during the preparation of this book. I would also wish to acknowledge the superb assistance that the staff of Springer has provided for the publication of this work.

Athens, Greece Themistocles M. Rassias

Contents

Contributors

Marcin Adam Institute of Mathematics, Silesian University of Technology, Gliwice, Poland

Z. Afshari Department of Mathematics and Computer Sciences, Sabzevar, Iran

Anna Bahyrycz Department of Mathematics, Pedagogical University, Kraków, Poland

Janusz Brzdęk Department of Mathematics, Pedagogical University, Kraków, Poland

Liviu Cădariu Department of Mathematics, Politehnica University of Timişoara, Timişoara, Romania

Jeongwook Chang Department of Mathematics Education, Dankook University, Yongin, Republic of Korea

Jaeyoung Chung Department of Mathematics, Kunsan National University, Kunsan, Republic of Korea

Stefan Czerwik Institute of Mathematics, Silesian University of Technology, Gliwice, Poland

Elhoucien Elqorachi Department of Mathematics, Faculty of Sciences, University Ibn Zohr, Agadir, Morocco

Gian Luigi Forti Dipartimento di Matematica, Università degli Studi di Milano, Milano, Italy

M. E. Gordji Department of Mathematics, Semnan University, Semnan, Iran

Eszter Gselmann Department of Analysis, Institute of Mathematics, University of Debrecen, Debrecen, Hungary

Kil-Woung Jun Department of Mathematics, Chungnam National University, Daejeon, Korea

H. Khodaei Department of Mathematics, Malayer University, Malayer, Iran

Hark-Mahn Kim Department of Mathematics, Chungnam National University, Daejeon, Korea

Krzysztof Król Institute of Mathematics, Silesian University of Technology, Gliwice, Poland

Jung Rye Lee Department of Mathematics, Daejin University, Daejin, Korea

Gyula Maksa Department of Analysis, Institute of Mathematics, University of Debrecen, Debrecen, Hungary

Youssef Manar Department of Mathematics, Faculty of Sciences, University Ibn Zohr, Agadir, Morocco

Choonkil Park Research Institute for Natural Sciences, Hanyang University, Hanyang, Korea

Magdalena Piszczek Department of Mathematics, Pedagogical University, Kraków, Poland

Dorian Popa Technical University of Cluj-Napoca, Department of Mathematics, Cluj-Napoca, Romania

Ioan Raşa Technical University of Cluj-Napoca, Department of Mathematics, Cluj-Napoca, Romania

Themistocles M. Rassias Department of Mathematics, National Technical University of Athens, Zografou Campus, Athens, Greece

A. Roukbi Department of Mathematics, Ibn Tofail University, Kenitra, Morocco

Ioan A. Rus Department of Mathematics, Babeş-Bolyai University, Cluj-Napoca, Romania

T. L. Shateri Department of Mathematics and Computer Sciences, Sabzevar, Iran

Dong Yun Shin Department of Mathematics, University of Seoul, Seoul, Korea

Eunyoung Son Department of Mathematics, Chungnam National University, Daejeon, Korea

D. Zeglami Department of Mathematics, E.N.S.A.M., Moulay Ismail University, Al Mansour, Meknes, Morocco

On Some Functional Equations

Marcin Adam, Stefan Czerwik and Krzysztof Król

Subject Classifications: 39B22, 39B52, 39B82, 40A05, 41A30.

Abstract This chapter consists of three parts. In the first part we consider so-called Adomian's polynomials and present the proof of the convergence of the sequence of such polynomials to the solution of the equation. The second part is devoted to present several approximation methods for finding solutions of so-called Kordylewski–Kuczma functional equation. Finally, in the last one we present a stability result in the sense of Ulam–Hyers–Rassias for generalized quadratic functional equation on topological spaces.

Keywords Adomian's polynomials · Decomposition method · Convergence of Adomian's iterations · Approximate solutions of functional equations · Generalized quadratic functional equation · Stability

1 On the Convergence of Adomian's Method

1.1 Introduction

G. Adomian in several papers (see e.g. [6–9]) developed a numerical technique using special kinds of polynomials (called Adomian polynomials) for solving non-linear functional equations. In this method the solution is given by a series having terms which are Adomian's polynomials. Unfortunately, the problems of convergence are not satisfactorily solved.

M. Adam (✉) · S. Czerwik · K. Król
Institute of Mathematics, Silesian University of Technology,
Kaszubska 23, 44-100 Gliwice, Poland
e-mail: marcin.adam@polsl.pl

S. Czerwik
e-mail: stefan.czerwik@polsl.pl

K. Król
e-mail: krolk@poczta.onet.pl

T. M. Rassias (ed.), *Handbook of Functional Equations*,
Springer Optimization and Its Applications 96, DOI 10.1007/978-1-4939-1286-5_1

Yves Cherruault [10] presents some results on convergence of Adomian's method. But the result (and the proof) presented in this paper contains some incorrectness. In this chapter we present a similar result and the complete proof.

Throughout this chapter we denote the sets of nonnegative integers, positive integers, and real numbers by $\mathbb{N}_0$, $\mathbb{N}$, and $\mathbb{R}$ respectively.

1.2 The Decomposition Method

Let us consider the following functional equation:

$$y - N(y) = f, \tag{1}$$

where $N : X \to X$ is a given function on a Banach space X and $f \in X$. Assume that (1) has exactly one solution $y \in X$ for every $f \in X$.

The Adomian method is the following one. Let

$$y = \sum_{n=0}^{\infty} y_n, \tag{2}$$

and assume that

$$N(y) = \sum_{n=0}^{\infty} A_n, \tag{3}$$

where A_n are so-called Adomian's polynomials obtained from the relation:

$$\text{for} \quad z = \sum_{n=0}^{\infty} \lambda^n y_n, \quad N\left(\sum_{n=0}^{\infty} \lambda^n y_n\right) = \sum_{n=0}^{\infty} \lambda^n A_n,$$

where λ is a parameter. Then we have (under suitable assumptions on N)

$$n!A_n = \frac{d^n}{d\lambda^n}\left[N\left(\sum_{n=0}^{\infty} \lambda^n y_n\right)\right]_{\lambda=0}, \quad n \in \mathbb{N}_0. \tag{4}$$

In general, one can verify that A_n depends only on $y_0, \ldots, y_n$.

First, we neglect the problem of convergence of the series involved into the method. Then from (1), (2), and (3) we get

$$\sum_{n=0}^{\infty} y_n - \sum_{n=0}^{\infty} A_n = f,$$

and therefore

$$y_0 = f,$$

$$y_1 = A_0,$$

$$\vdots$$

$$y_n = A_{n-1}, \qquad n \in \mathbb{N}.$$

Consequently, we can determine all term of the series (2).

In [10] the following modification is presented. For a convergent series

$$y = \sum_{n=0}^{\infty} y_n,$$

we define

$$N(y) = \sum_{n=0}^{\infty} A_n(y_0, \ldots, y_n),$$

where A_n's are given by (4). Let

$$U_n = \sum_{i=0}^{n} y_i,$$

then

$$N_n(U_n) = \sum_{i=0}^{n} A_i(y_0, \ldots, y_i).$$

Therefore the Adomian method is equivalent to finding the sequence

$$S_n = y_1 + \ldots + y_n$$

given by the formula

$$S_{n+1} = N_n(y_0 + S_n),\ S_0 = 0,\ n \in \mathbb{N}_0 \tag{5}$$

(for more details, see [10]).

The sequence defined by

$$S_{n+1} = N(y_0 + S_n),\ S_0 = 0,\ n \in \mathbb{N}_0,$$

one can associate with the following equation:

$$N(y_0 + S) = S. \tag{6}$$

Under suitable assumptions we shall prove in the next part of the chapter that the sequence given by (5) will converge to S, the solution of (6) (the proof presented in [10] contains some incorrectness).

1.3 Convergence Result

We shall prove the following result concerning the convergence of the approximations to the solution of the equation.

Theorem 1 *Let X be a Banach space and let $N, N_n : X \to X$, $n \in \mathbb{N}_0$, be functions such that*

$$\|N(x) - N(y)\| \le \delta\|x - y\|, \qquad x, y \in X, \tag{7}$$

where

$$0 \le \delta < 1, \tag{8}$$

and there exist constants $L_n \ge 0$, $n \in \mathbb{N}_0$, such that

$$\|(N_n - N)(x)\| \le L_n\|x\|, \quad x \in X,\ n \in \mathbb{N}_0, \tag{9}$$

where

$$L_n \to 0 \quad as \quad n \to \infty. \tag{10}$$

Then the sequence $\{S_n\}$ given by

$$S_{n+1} = N_n(y_0 + S_n),\ S_0 = 0,\ n \in \mathbb{N}_0, \tag{11}$$

where $y_0 \in X$ is arbitrarily fixed, converges to S, the unique solution of the equation

$$N(y_0 + S) = S. \tag{12}$$

Proof Let us note that, in view of (7) and (8), from the well-known Banach fixed point theorem for the strict contraction in complete metric space the Eq. (12) has exactly one solution $S \in X$.

Take $\varepsilon > 0$ such that $\varepsilon + \delta < 1$. Then from (10) it follows that there exists a number $n_0 \in \mathbb{N}$ such that $L_n < \frac{\varepsilon}{4}$ for $n \ge n_0$. Assume that M_0 is the number satisfying the conditions

$$\|S_{n_0}\| \le \frac{M_0}{4}, \quad \|S\| \le \frac{M_0}{4}, \quad \|y_0\| \le M_0. \tag{13}$$

Now we shall prove the inequality

$$\|S_n - S\| \le \frac{M_0}{2}, \qquad n \ge n_0. \tag{14}$$

For $n = n_0$ we have

$$\|S_{n_0} - S\| \le \|S_{n_0}\| + \|S\| \le \frac{M_0}{4} + \frac{M_0}{4} = \frac{M_0}{2}.$$

Let us assume that for $n = m \ge n_0$ we have

$$\|S_m - S\| \le \frac{M_0}{2}. \tag{15}$$

From (7), (9), and (11) we get

$$\begin{aligned}\|S_{n+1}-S\| &= \|N_n(y_0+S_n)-N(y_0+S)\|\\ &= \|N_n(y_0+S_n)-N(y_0+S_n)+N(y_0+S_n)-N(y_0+S)\|\\ &\le \|(N_n-N)(y_0+S_n)\|+\delta\|S_n-S\|\\ &\le L_n\|y_0+S_n\|+\delta\|S_n-S\|\\ &\le L_n(\|y_0\|+\|S_n\|)+\delta\|S_n-S\|.\end{aligned}$$

Therefore

$$\|S_{n+1}-S\| \le L_n(\|y_0\|+\|S_n\|)+\delta\|S_n-S\| \tag{16}$$

for all $n \in \mathbb{N}$. From (15) one obtains

$$\|S_m\| \le \|S_m-S\|+\|S\| \le \frac{M_0}{2}+\frac{M_0}{4} \le M_0$$

and consequently, by (16) for $n=m$ we get

$$\|S_{m+1}-S\| \le L_m(M_0+M_0)+\delta\frac{M_0}{2} \le (\varepsilon+\delta)\frac{M_0}{2} \le \frac{M_0}{2},$$

i.e.,

$$\|S_{m+1}-S\| \le \frac{M_0}{2},$$

which shows by the induction principle that (14) holds true. Consequently, from (14) there exists

$$0 \le \limsup_{n\to\infty}\|S_n-S\| = q < \infty.$$

To prove our theorem it is enough to verify that $q=0$. In fact, (14) implies

$$\|S_n\| \le M_0, \qquad n \ge n_0,$$

and consequently by (16) we obtain for $n \ge n_0$

$$\|S_{n+1}-S\| \le 2L_nM_0+\delta\|S_n-S\|.$$

Hence, on account of (10), we get

$$\begin{aligned}q &= \limsup_{n\to\infty}\|S_{n+1}-S\|\\ &\le \limsup_{n\to\infty}(2L_nM_0)+\limsup_{n\to\infty}(\delta\|S_n-S\|)\\ &\le \delta\limsup_{n\to\infty}\|S_n-S\|\\ &\le \delta q,\end{aligned}$$

i.e.,

$$q \leq \delta q,$$

which implies by (8) the equality $q = 0$. This means that there exists the limit

$$0 \leq \lim_{n\to\infty} \|S_n - S\| \leq \limsup_{n\to\infty} \|S_n - S\| = 0$$

and the proof is completed. □

2 Approximation Methods for Solving Functional Equations

2.1 The Collocation Method

In the first and in the second part of this section we contemplate two boundary element methods. The first will be the collocation method. We can read about this method in [29]. Furthermore, we can read about the linear functional equation and the nonlinear functional equation in [27].

We consider the linear functional equation

$$y[f(x)] = g(x)y(x) + F(x), \tag{17}$$

where functions f, g, F are given and y is unknown function.

In this section we consider continuous solutions of the Eq. 17 in the interval $[a, b]$ and taking values in $\mathbb{R}$.

The class of functions defined in an interval I and taking values in $\mathbb{R}$ will be denoted by $\mathbf{Y[I]}$. Let $\mathbf{R}^{\mathbf{0}}_{\xi}\mathbf{[I]}$ be the class of continuous and strictly increasing functions in I fulfilling the conditions for $\xi \in \overline{I}$:

1. $(f(x) - x)(\xi - x) > 0, \quad x \in I,\ x \neq \xi,$
2. $(f(x) - \xi)(\xi - x) < 0, \quad x \in I,\ x \neq \xi.$

Theorem 2 [27] *Assume that $f \in \mathbf{R}^{\mathbf{0}}_{\xi}\mathbf{[I]}$, where $\xi \in I$. Let functions $g, F \in \mathbf{Y[I]}$ be continuous in I and $g(x) \neq 0$ for $x \in I$, $x \neq \xi$. Further, let condition $|g(\xi)| > 1$ be fulfilled. Then Eq.* (17) *has a unique continuous solution $y \in \mathbf{Y[I]}$ in I. This solution is given by the formula*

$$y[x] = -\sum_{n=0}^{\infty} \frac{F[f^n(x)]}{G_{n+1}(x)}, \quad x \in I, \tag{18}$$

where

$$G_n(x) = \prod_{i=0}^{n-1} g[f^i(x)], \qquad x \in I,\ n \in \mathbb{N}.$$

In the sequel we present the following result.

Theorem 3 *Let functions* $g, F : [a,b] \to \mathbb{R}$, $f : [a,b] \to [a,b]$ *fulfill the assumptions of Theorem 2 for* $P = [a,b]$, *where* $a < b$, $a, b \in \mathbb{R}$. *Let*

$$S := \{x_i : x_i \in [a,b],\ x_i \neq x_j \text{ for } i \neq j,\ i,j = 1,\ldots,n\}.$$

Assume that $\Phi_j : [a,b] \to \mathbb{R}$, $j = 1,\ldots,n$, *are given and linearly independent functions on the interval* $[a,b]$. *Then the solution of the* Eq. (17) *on the interval* $[a,b]$ *is approximated by the function*

$$y_n(x) = \sum_{j=1}^{n} p_j \Phi_j(x), \qquad x \in [a,b], \tag{19}$$

where coefficients p_j, $j = 1,\ldots,n$, *are solutions of the equations*

$$\sum_{j=1}^{n} p_j \Psi_j(x_i) = F(x_i), \qquad i = 1,\ldots,n, \tag{20}$$

where $\Psi_j(x_i) := \Phi_j[f(x_i)] - g(x_i)\Phi_j(x_i)$ *and* $x_i \in S$, $i, j = 1,\ldots,n$.

Proof Define the following error function

$$R[y(x)] := y[f(x)] - g(x)y(x) - F(x).$$

We calculate value $R[y_n(x_i)]$ for $x_i \in S$:

$$\begin{aligned} R[y_n(x_i)] &= y_n[f(x_i)] - g(x_i)y_n(x_i) - F(x_i) \\ &= \sum_{j=1}^{n} p_j \Phi_j[f(x_i)] - g(x_i)\sum_{j=1}^{n} p_j \Phi_j(x_i) - F(x_i) \\ &= \sum_{j=1}^{n} p_j \Big[\Phi_j[f(x_i)] - g(x_i)\Phi_j(x_i)\Big] - F(x_i). \end{aligned}$$

In the collocation method we calculate unknown parameters p_i by finding place, where error function $R[y(x)]$ takes value equal to 0 on the set S, that is

$$R[y_n(x_i)] = 0, \qquad i = 1,\ldots,n.$$

From the above conditions we get the system of Eq. (20). This completes the proof. □

In this section we use the same examples to show solution accuracy.

Example 1 Find approximate solution of the equation

$$y\Big(\frac{1}{2}x\Big) = (2x+4)y(x) - 4x^2 - 7x, \qquad x \in [-1,1], \tag{21}$$

using the collocation method.

Solution 1 In this case we take the following knots: $x_1 = -1$, $x_2 = 0$, $x_3 = 1$ from the interval $[-1, 1]$. From the Eq. (21) we get $f(x) = \frac{1}{2}x$, $g(x) = 2x + 4$, $F(x) = -4x^2 - 7x$. These functions fulfill the assumptions of Theorem 2. We take the following linearly independent functions:

$$\Phi_1(x) = 1, \qquad \Phi_2(x) = x, \qquad \Phi_3(x) = x^2.$$

We calculate the functions $\Psi_j(x)$ for $j = 1, 2, 3$:

$$\Psi_1(x) = \Phi_1[f(x)] - g(x)\Phi_1(x) = -2x - 3,$$

$$\Psi_2(x) = \Phi_2[f(x)] - g(x)\Phi_2(x) = -2x^2 - \frac{7}{2}x,$$

$$\Psi_3(x) = \Phi_3[f(x)] - g(x)\Phi_3(x) = -2x^3 - \frac{15}{4}x^2.$$

By putting the knots to above functions and from (20) we get the following system of equations

$$\begin{cases} -p_1 + \frac{3}{2}p_2 - \frac{7}{4}p_3 = 3, \\ -3p_1 + 0p_2 + 0p_3 = 0, \\ -5p_1 - \frac{11}{2}p_2 - \frac{23}{4}p_3 = -11, \end{cases}$$

to which the only solution is

$$\begin{cases} p_1 = 0, \\ p_2 = 2, \\ p_3 = 0. \end{cases}$$

Hence we get approximate solution of the Eq. (21) in the form

$$y_3(x) = p_1\Phi_1(x) + p_2\Phi_2(x) + p_3\Phi_3(x) = 2x.$$

This is the accurate solution of the Eq. (21).

Next we apply this method to nonlinear functional equation. We consider the nonlinear functional equation

$$y[f(x)] = g(x, y(x)), \tag{22}$$

where functions f, g are given and y is unknown function.

Let Ω be a region (an open, simply connected set) on the real plane and let I be a real interval. We introduce the sets (see [27])

$$\Omega_x = \{y : (x, y) \in \Omega\}.$$

Let g be a function defined in Ω. We denote by Γ_x the set of values of $g(x, y)$ for $y \in \Omega$.

We assume the following hypotheses (see [27]).

Conjecture 1 For every $x \in I$ the set Ω_x is an open interval (possibly infinite).

Conjecture 2 For every $x \in I$ we have $\Gamma_x = \Omega_{f(x)}$.

Conjecture 3 The function $g(x, y)$ is defined and continuous in the strip

$$\{(x, y) : \ x \in I, \ y \in \Omega_x\}.$$

Conjecture 4 For every fixed $x \in I$ the function $g_x(y) \overset{df}{=} g(x, y)$ is invertible in Ω_x.

Theorem 4 [27] *Let* $f \in \mathbf{R}^0_\xi[\mathbf{I}]$, *where* ξ *is an endpoint of* I, $\xi \notin I$, *and let conjectures 1, 2, 3, and 4be fulfilled. Then Eq. (22) has in* I *a continuous solution* y *depending on an arbitrary function, given in the interval* $[x_0, f(x_0)]$.

Now we consider the following example.

Example 2 Find the approximate solution of the equation

$$y\Big(\frac{1}{2}x\Big) = y^3(x) - 8x^3 + x, \qquad x \in [-1, 1], \tag{23}$$

using the collocation method.

Solution 2 From the Eq. (23) we get $f(x) = \frac{1}{2}x$, and $g(x, y) = -8x^3 + x + y^3$. These functions fulfill the assumptions of Theorem 4 for $\Omega = (-1, 1) \times \mathbb{R}$ and $\Omega_x = \Omega_{f(x)} = \Gamma_x = \mathbb{R}$. In this case we take the following knots: $x_1 = -1$, $x_2 = 0$, $x_3 = 1$ from the interval $[-1, 1]$. We take the following linear independent functions:

$$\Phi_1(x) = 1, \qquad \Phi_2(x) = x, \qquad \Phi_3(x) = x^2.$$

We calculate the error function

$$R[y(x)] = y\Big(\frac{1}{2}x\Big) - y^3(x) + 8x^3 - x.$$

Hence for

$$y_3(x_i) = \sum_{j=1}^{3} p_j \Phi_j(x_i), \quad i = 1, 2, 3,$$

we get

$$R[y_3(x_i)] = \sum_{j=1}^{3} p_j \Phi_j\Big(\frac{1}{2}x_i\Big) - \left[\sum_{j=1}^{3} p_j \Phi_j(x_i)\right]^3 + 8x_i^3 - x_i \quad i = 1, 2, 3. \tag{24}$$

In the collocation method $R[y_3(x_i)] = 0$ for $i = 1, 2, 3$. From (24) we get for $i = 1, 2, 3$

$$p_1 - p_1^3 + \Big(\frac{p_2}{2} - 3p_1^2 p_2\Big) x_i + \Big(\frac{p_3}{4} - 3p_1 p_2^2 - 3p_1^2 p_3\Big) x_i^2 + \big(-p_2^3 - 6p_1 p_2 p_3\big) x_i^3$$

$$+\left(-3p_2^2p_3-3p_1p_3^2\right)x_i^4+\left(-3p_2p_3^2\right)x_i^5-p_3^3x_i^6=-8x_i^3+x_i.$$

By putting the knots to above equations we get the following nonlinear system of equations

$$\begin{cases} p_1-p_1^3-\frac{p_2}{2}+3p_1^2p_2+\frac{p_3}{4}-3p_1p_2^2-3p_1^2p_3+p_2^3+6p_1p_2p_3-3p_2^2p_3\\ -3p_1p_3^2+3p_2p_3^2-p_3^3=7,\\ p_1-p_1^3=0,\\ p_1-p_1^3+\frac{p_2}{2}-3p_1^2p_2+\frac{p_3}{4}-3p_1p_2^2-3p_1^2p_3-p_2^3-6p_1p_2p_3-3p_2^2p_3\\ -3p_1p_3^2-3p_2p_3^2-p_3^3=-7, \end{cases}$$

to which the solution is

$$\begin{cases} p_1=0,\\ p_2=2,\\ p_3=0. \end{cases}$$

Hence we get approximate solution of the Eq. (23) in the form

$$y_3(x)=p_1\Phi_1(x)+p_2\Phi_2(x)+p_3\Phi_3(x)=2x.$$

This is the solution of the Eq. (23).

2.2 *The Method of Moments*

One can read about this method one can read in [29].

Theorem 5 *Let functions $g, F:[a,b]\to\mathbb{R}$, $f:[a,b]\to[a,b]$ fulfill the assumptions of Theorem 2 for $P=[a,b]$, where $a<b$, $a,b\in\mathbb{R}$. Assume that $\Phi_j:[a,b]\to\mathbb{R}$, $j=1,\dots,n$, are given and linearly independent functions on the interval $[a,b]$. Then the solution of the* Eq. (17) *on the interval $[a,b]$ is approximated by the function*

$$y_n(x)=F(x)+\sum_{j=1}^{n}p_j\Phi_j(x),\qquad x\in[a,b],\tag{25}$$

where coefficients p_j, $j=1,\dots,n$, are solutions of the equations

$$\sum_{j=1}^{n}p_j\left[\int_a^b\Psi_j(x)\Phi_i(x)\,dx\right]=\int_a^b\Big((1+g(x))F(x)-F[f(x)]\Big)\Phi_i(x)\,dx\tag{26}$$

for $i=1,\dots,n$ and $\Psi_j(x):=\Phi_j[f(x)]-g(x)\Phi_j(x)$, $x\in[a,b]$.

Proof Define the following error function

$$R[y(x)] := y[f(x)] - g(x)y(x) - F(x), \quad x \in [a, b].$$

We calculate the value $R[y_n(x)]$ for $x \in [a, b]$:

$$\begin{aligned}
R[y_n(x)] &= y_n[f(x)] - g(x)y_n(x) - F(x) \\
&= F[f(x)] + \sum_{j=1}^{n} p_j \Phi_j[f(x)] - g(x)\left[F(x) + \sum_{j=1}^{n} p_j \Phi_j(x)\right] - F(x) \\
&= \sum_{j=1}^{n} p_j \Big[\Phi_j[f(x)] - g(x)\Phi_j(x)\Big] + F[f(x)] - (1 + g(x))F(x) \\
&= \sum_{j=1}^{n} p_j \Psi_j(x) + F[f(x)] - (1 + g(x))F(x). \qquad (27)
\end{aligned}$$

In the method of moments we require the error function to be orthogonal to functions $\Phi_i(x)$, $i = 1, \ldots, n$. It means that it must be satisfied by the following conditions

$$\int_a^b R[y_n(x)]\Phi_i(x)\,dx = 0, \qquad i = 1, \ldots, n. \tag{28}$$

From (27) and (28) we get the system of equations

$$\int_a^b \sum_{j=1}^{n} p_j \Psi_j(x)\Phi_i(x)dx - \int_a^b (1+g(x))F(x)\Phi_i(x)dx + \int_a^b F[f(x)]\Phi_i(x)dx = 0.$$

After transformations we get the system of Eq. (26). This completes the proof. □

The system of Eq. (26) takes the following matrix form

$$\mathbf{C} \cdot \mathbf{p} = \mathbf{F}, \tag{29}$$

where

$$\begin{aligned}
&\mathbf{C} = \{C_{ij}\}_{i,j=1,\ldots,n}, \; C_{ij} = \int_a^b \Psi_j(x)\Phi_i(x)\,dx, \\
&\mathbf{p} = [p_1, \ldots, p_n]^T, \\
&\mathbf{F} = [F_1, \ldots, F_n]^T, \; F_i = \int_a^b [(1 + g(x))F(x) - F[f(x)]]\Phi_i(x)\,dx.
\end{aligned}$$

Remark 1 It would be interesting to find conditions when the system of Eq. (29) has a solution.

Example 3 Find the approximate solution of the equation

$$y\Big(\frac{1}{2}x\Big) = (2x + 4)y(x) - 4x^2 - 7x, \qquad x \in [-1, 1], \tag{30}$$

using the method of moments.

Solution 3 From the Eq. (30) we get $f(x) = \frac{1}{2}x$, $g(x) = 2x+4$, $F(x) = -4x^2-7x$. These functions fulfill the assumptions of Theorem 2. We take the following linearly independent functions:

$$\Phi_1(x) = 1, \qquad \Phi_2(x) = x, \qquad \Phi_3(x) = x^2.$$

We calculate the functions $\Psi_j(x)$ for $j = 1, 2, 3$:

$$\Psi_1(x) = \Phi_1[f(x)] - g(x)\Phi_1(x) = -2x - 3,$$

$$\Psi_2(x) = \Phi_2[f(x)] - g(x)\Phi_2(x) = -2x^2 - \frac{7}{2}x,$$

$$\Psi_3(x) = \Phi_3[f(x)] - g(x)\Phi_3(x) = -2x^3 - \frac{15}{4}x^2.$$

Furthermore, we get the following elements of matrix **C**:

$$C_{11} = \int_{-1}^{1} \Psi_1(x)\Phi_1(x)\,dx = \int_{-1}^{1} (-2x - 3)\,dx = -6,$$

$$C_{12} = \int_{-1}^{1} \Psi_2(x)\Phi_1(x)\,dx = \int_{-1}^{1} (-2x^2 - \frac{7}{2}x)\,dx = -\frac{4}{3},$$

$$C_{13} = \int_{-1}^{1} \Psi_3(x)\Phi_1(x)\,dx = \int_{-1}^{1} (-2x^3 - \frac{15}{4}x^2)\,dx = -\frac{5}{2},$$

$$C_{21} = \int_{-1}^{1} \Psi_1(x)\Phi_2(x)\,dx = \int_{-1}^{1} (-2x - 3)x\,dx = -\frac{4}{3},$$

$$C_{22} = \int_{-1}^{1} \Psi_2(x)\Phi_2(x)\,dx = \int_{-1}^{1} (-2x^2 - \frac{7}{2}x)x\,dx = -\frac{7}{3},$$

$$C_{23} = \int_{-1}^{1} \Psi_3(x)\Phi_2(x)\,dx = \int_{-1}^{1} (-2x^3 - \frac{15}{4}x^2)x\,dx = -\frac{4}{5},$$

$$C_{31} = \int_{-1}^{1} \Psi_1(x)\Phi_3(x)\,dx = \int_{-1}^{1} (-2x - 3)x^2\,dx = -2,$$

$$C_{32} = \int_{-1}^{1} \Psi_2(x)\Phi_3(x)\,dx = \int_{-1}^{1} (-2x^2 - \frac{7}{2}x)x^2\,dx = -\frac{4}{5},$$

$$C_{33} = \int_{-1}^{1} \Psi_3(x)\Phi_3(x)\,dx = \int_{-1}^{1} (-2x^3 - \frac{15}{4}x^2)x^2\,dx = -\frac{3}{2}$$

and elements of the column **F**:

$$F_1 = \int_{-1}^{1} [(2x + 5)(-4x^2 - 7x) + 4\left(\frac{1}{2}x\right)^2 + 7\left(\frac{1}{2}x\right)]\,dx = -22,$$

$$F_2 = \int_{-1}^{1} [(2x + 5)(-4x^2 - 7x) + 4\left(\frac{1}{2}x\right)^2 + 7\left(\frac{1}{2}x\right)]x\,dx = -\frac{121}{5},$$

$$F_3 = \int_{-1}^{1} [(2x+5)(-4x^2-7x) + 4\left(\frac{1}{2}x\right)^2 + 7\left(\frac{1}{2}x\right)]x^2\,dx = -\frac{66}{5}.$$

Finally, we get the following system of equations

$$\begin{cases} -6p_1 - \frac{4}{3}p_2 - \frac{5}{2}p_3 = -22, \\ -\frac{4}{3}p_1 - \frac{7}{3}p_2 - \frac{4}{5}p_3 = -\frac{121}{5}, \\ -2p_1 - \frac{4}{5}p_2 - \frac{3}{2}p_3 = -\frac{66}{5}, \end{cases}$$

to which the only solution is

$$\begin{cases} p_1 = 0, \\ p_2 = 9, \\ p_3 = 4. \end{cases}$$

Hence we get the approximate solution of the Eq. (30) in the form

$$y_3(x) = p_1\Phi_1(x) + p_2\Phi_2(x) + p_3\Phi_3(x) + F(x) = 9x + 4x^2 - 4x^2 - 7x = 2x.$$

Moreover, this is the solution of the Eq. (30).

Next we apply this method to nonlinear functional equation.

Example 4 Find the approximate solution of the equation

$$y\left(\frac{1}{2}x\right) = y^3(x) - 8x^3 + x, \qquad x \in [-1,1], \tag{31}$$

using the method of moments.

Solution 4 From the Eq. (31) we get $f(x) = \frac{1}{2}x$ and $g(x,y) = -8x^3 + x + y^3$. This functions fulfill the assumptions of Theorem 4 for $\Omega = (-1,1) \times \mathbb{R}$ and $\Omega_x = \Omega_{f(x)} = \Gamma_x = \mathbb{R}$. From (31) we obtain $F(x) = -8x^3 + x$. We take the following linearly independent functions:

$$\Phi_1(x) = 1, \qquad \Phi_2(x) = x, \qquad \Phi_3(x) = x^2, \qquad \Phi_4(x) = x^3.$$

We calculate the error function

$$R[y(x)] = y\left(\frac{1}{2}x\right) - y^3(x) + 8x^3 - x.$$

Hence for

$$y_4(x) = F(x) + \sum_{j=1}^{4} p_j\Phi_j(x), \quad x \in [-1,1],$$

we get

$$R[y_4(x)] = F\left(\frac{1}{2}x\right) + \sum_{j=1}^{4} p_j\Phi_j\left(\frac{1}{2}x\right) - \left[F(x) + \sum_{j=1}^{4} p_j\Phi_j(x)\right]^3 + 8x^3 - x \tag{32}$$

for $x \in [-1, 1]$. In the method of moments it must be satisfied by the conditions

$$\int_{-1}^{1} R[y_4(x)]\Phi_i(x)\,dx = 0, \qquad i = 1, 2, 3, 4. \tag{33}$$

From (33) for $\Phi_1(x) = 1$ we get

$$\begin{aligned}\int_{-1}^{1} & (512 - 192p_4 + 24p_4^2 - p_4^3)x^9 + (-192p_3 + 48p_3p_4 - 3p_3p_4^2)x^8\\
&+ (-192 - 192p_2 + 24p_3^2 + 48p_4 + 48p_2p_4 - 3p_3^2p_4 - 3p_4^2 - 3p_2p_4^2)x^7\\
&+ (-192p_1 + 48p_3 + 48p_2p_3 - p_3^3 + 48p_1p_3 - 6p_3p_4 - 6p_2p_3p_4 - 3p_1p_4^2)x^6\\
&+ (24 + 48p_2 + 24p_2^2 + 48p_1p_3 - 3p_3^2 - 3p_2p_3^2 - 3p_3 - 6p_2p_4 - 3p_2^2p_4\\
&- 6p_1p_3p_4)x^5\\
&+ (48p_1 + 48p_1p_2 - 3p_3 - 6p_2p_3 - 3p_2^2p_3 - 3p_1p_2^2 - 6p_1p_4 - 6p_1p_2p_4)x^4\\
&+ (6 + 24p_1^2 - 3p_2 - 3p_2^2 - p_2^3 - 6p_1p_3 - 6p_1p_2p_3 + \frac{p_4}{8} - 3p_1^2p_4)x^3\\
&+ (-3p_1 - 6p_1p_2 - 3p_1p_2^2 + \frac{p_3}{4} - 3p_1^2p_3)x^2\\
&+ (-\frac{1}{2} - 3p_1^2 + \frac{p_2}{2} - 3p_1^2p_2)x + (p_1 - p_1^3)\,dx = 0.\end{aligned}$$

From (33) for $\Phi_2(x) = x$ we obtain

$$\begin{aligned}\int_{-1}^{1} & (512 - 192p_4 + 24p_4^2 - p_4^3)x^{10} + (-192p_3 + 48p_3p_4 - 3p_3p_4^2)x^9\\
&+ (-192 - 192p_2 + 24p_3^2 + 48p_4 + 48p_2p_4 - 3p_3^2p_4 - 3p_4^2 - 3p_2p_4^2)x^8\\
&+ (-192p_1 + 48p_3 + 48p_2p_3 - p_3^3 + 48p_1p_3 - 6p_3p_4 - 6p_2p_3p_4 - 3p_1p_4^2)x^7\\
&+ (24 + 48p_2 + 24p_2^2 + 48p_1p_3 - 3p_3^2 - 3p_2p_3^2 - 3p_3 - 6p_2p_4 - 3p_2^2p_4\\
&- 6p_1p_3p_4)x^6\\
&+ (48p_1 + 48p_1p_2 - 3p_3 - 6p_2p_3 - 3p_2^2p_3 - 3p_1p_2^2 - 6p_1p_4 - 6p_1p_2p_4)x^5\\
&+ (6 + 24p_1^2 - 3p_2 - 3p_2^2 - p_2^3 - 6p_1p_3 - 6p_1p_2p_3 + \frac{p_4}{8} - 3p_1^2p_4)x^4\\
&+ (-3p_1 - 6p_1p_2 - 3p_1p_2^2 + \frac{p_3}{4} - 3p_1^2p_3)x^3\\
&+ (-\frac{1}{2} - 3p_1^2 + \frac{p_2}{2} - 3p_1^2p_2)x^2 + (p_1 - p_1^3)x\,dx = 0.\end{aligned}$$

From (33) for $\Phi_3(x) = x^2$ we get

$$\int_{-1}^{1} (512 - 192p_4 + 24p_4^2 - p_4^3)x^{11} + (-192p_3 + 48p_3p_4 - 3p_3p_4^2)x^{10}$$

$$+(-192-192p_2+24p_3^2+48p_4+48p_2p_4-3p_3^2p_4-3p_4^2-3p_2p_4^2)x^9$$
$$+(-192p_1+48p_3+48p_2p_3-p_3^3+48p_1p_3-6p_3p_4-6p_2p_3p_4-3p_1p_4^2)x^8$$
$$+(24+48p_2+24p_2^2+48p_1p_3-3p_3^2-3p_2p_3^2-3p_3-6p_2p_4-3p_2^2p_4$$
$$-6p_1p_3p_4)x^7$$
$$+(48p_1+48p_1p_2-3p_3-6p_2p_3-3p_2^2p_3-3p_1p_2^2-6p_1p_4-6p_1p_2p_4)x^6$$
$$+(6+24p_1^2-3p_2-3p_2^2-p_2^3-6p_1p_3-6p_1p_2p_3+\frac{p_4}{8}-3p_1^2p_4)x^5$$
$$+(-3p_1-6p_1p_2-3p_1p_2^2+\frac{p_3}{4}-3p_1^2p_3)x^4$$
$$+(-\frac{1}{2}-3p_1^2+\frac{p_2}{2}-3p_1^2p_2)x^3+(p_1-p_1^3)x^2\,dx=0.$$

From (33) for $\Phi_4(x)=x^3$ we obtain

$$\int_{-1}^{1}(512-192p_4+24p_4^2-p_4^3)x^{12}+(-192p_3+48p_3p_4-3p_3p_4^2)x^{11}$$
$$+(-192-192p_2+24p_3^2+48p_4+48p_2p_4-3p_3^2p_4-3p_4^2-3p_2p_4^2)x^{10}$$
$$+(-192p_1+48p_3+48p_2p_3-p_3^3+48p_1p_3-6p_3p_4-6p_2p_3p_4-3p_1p_4^2)x^9$$
$$+(24+48p_2+24p_2^2+48p_1p_3-3p_3^2-3p_2p_3^2-3p_3-6p_2p_4-3p_2^2p_4$$
$$-6p_1p_3p_4)x^8$$
$$+(48p_1+48p_1p_2-3p_3-6p_2p_3-3p_2^2p_3-3p_1p_2^2-6p_1p_4-6p_1p_2p_4)x^7$$
$$+(6+24p_1^2-3p_2-3p_2^2-p_2^3-6p_1p_3-6p_1p_2p_3+\frac{p_4}{8}-3p_1^2p_4)x^6$$
$$+(-3p_1-6p_1p_2-3p_1p_2^2+\frac{p_3}{4}-3p_1^2p_3)x^5$$
$$+(-\frac{1}{2}-3p_1^2+\frac{p_2}{2}-3p_1^2p_2)x^4+(p_1-p_1^3)x^3\,dx=0.$$

Finally, we get the following nonlinear system of equations

$$\begin{cases} -\frac{1248}{35}p_1 - 2p_1^3 + \frac{76}{5}p_1p_2 - 2p_1p_2^2 - \frac{2099}{70}p_3 - 2p_1^2p_3 + \frac{396}{35}p_2p_3 - \frac{6}{5}p_2^2p_3 \\ -\frac{6}{5}p_1p_3^2 - \frac{2}{7}p_3^3 + \frac{396}{35}p_1p_4 - \frac{12}{5}p_1p_2p_4 + \frac{188}{21}p_3p_4 - \frac{12}{7}p_2p_3p_4 - \frac{6}{7}p_1p_4^2 \\ -\frac{2}{3}p_3p_4^2 = 0, \\ \frac{38}{5}p_1^2 - \frac{3131}{105}p_2 - 2p_1^2p_2 + \frac{198}{35}p_2^2 - \frac{2}{5}p_2^3 + \frac{396}{35}p_1p_3 - \frac{12}{5}p_1p_2p_3 + \frac{94}{21}p_3^2 \\ -\frac{6}{7}p_2p_3^2 - \frac{115729}{4620}p_4 - \frac{6}{5}p_1^2p_4 + \frac{188}{21}p_2p_4 - \frac{6}{7}p_2^2p_4 - \frac{12}{7}p_1p_3p_4 - \frac{2}{3}p_3^2p_4 \\ +\frac{122}{33}p_4^2 - \frac{2}{3}p_2p_4^2 - \frac{2}{11}p_4^3 = -\frac{22849}{385}, \\ -\frac{1032}{35}p_1 - \frac{2}{3}p_1^3 + \frac{396}{35}p_1p_2 - \frac{6}{5}p_1p_2^2 - \frac{57749}{2310}p_3 - \frac{6}{5}p_1^2p_3 + \frac{188}{21}p_2p_3 - \frac{6}{7}p_2^2p_3 \\ -\frac{6}{7}p_1p_3^2 - \frac{2}{9}p_3^3 + \frac{188}{21}p_1p_4 - \frac{12}{7}p_1p_2p_4 + \frac{244}{33}p_3p_4 - \frac{4}{3}p_2p_3p_4 - \frac{2}{3}p_1p_4^2 \\ -\frac{6}{11}p_3p_4^2 = 0, \\ \frac{198}{35}p_1^2 - \frac{28759}{1155}p_2 - \frac{6}{5}p_1^2p_2 + \frac{94}{21}p_2^2 - \frac{2}{7}p_2^3 + \frac{188}{21}p_1p_3 - \frac{12}{7}p_1p_2p_3 + \frac{122}{33}p_3^2 \\ -\frac{2}{3}p_2p_3^2 - \frac{257563}{12012}p_4 - \frac{6}{7}p_1^2p_4 + \frac{244}{33}p_2p_4 - \frac{2}{3}p_2^2p_4 - \frac{4}{3}p_1p_3p_4 - \frac{6}{11}p_3^2p_4 \\ +\frac{450}{143}p_4^2 - \frac{6}{11}p_2p_4^2 - \frac{2}{13}p_4^3 = -\frac{761377}{15015}, \end{cases}$$

to which the solution is

$$\begin{cases} p_1 = 0, \\ p_2 = 1, \\ p_3 = 0, \\ p_4 = 8. \end{cases}$$

Hence we get the approximate solution of the Eq. (31) in the form

$$\begin{aligned} y_4(x) &= p_1\Phi_1(x) + p_2\Phi_2(x) + p_3\Phi_3(x) + p_4\Phi_4(x) + F(x) \\ &= 0 + 1x + 0x^2 + 8x^3 - 8x^3 + x = 2x. \end{aligned}$$

As we know this is the solution of the Eq. (31).

2.3 *The Least Squares Method*

In this section we apply the method presented in [25], to nonlinear functional equation. We present this method on the following example.

Example 5 Find approximate solution of the equation

$$y\left(\frac{1}{2}x\right) = y^3(x) - 8x^3 + x, \qquad x \in [-1, 1], \tag{34}$$

using the least squares method.

Solution 5 From the Eq. (34) we get $f(x) = \frac{1}{2}x$ and $g(x, y) = -8x^3 + x + y^3$. These functions fulfill the assumptions of Theorem 4 for $\Omega = (-1, 1) \times \mathbb{R}$ and $\Omega_x = \Omega_{f(x)} = \Gamma_x = \mathbb{R}$. We take the following linearly independent functions:

$$\Phi_1(x) = 1, \qquad \Phi_2(x) = x.$$

We calculate the error function

$$R[y(x)] = y\Big(\frac{1}{2}x\Big) - y^3(x) + 8x^3 - x.$$

In the least squares method we tend to minimize the following expression

$$I(\mathbf{p}) = I(p_1, p_2) = \int_{-1}^{1} (R[y_2(x)])^2 \, dx,$$

where $y_2(x) = p_1\Phi_1(x) + p_2\Phi_2(x)$. Hence we get

$$I(\mathbf{p}) = \int_{-1}^{1} \left(\sum_{j=1}^{2} p_j\Phi_j\Big(\frac{1}{2}x\Big) - \left[\sum_{j=1}^{2} p_j\Phi_j(x)\right]^3 + 8x^3 - x \right)^2 dx.$$

Thus coefficients p_i must satisfy the equations

$$\frac{\partial I(\mathbf{p})}{\partial p_i} = 0, \qquad i = 1, 2.$$

Therefore we get for $i = 1, 2$:

$$\int_{-1}^{1} \left(\sum_{j=1}^{2} p_j\Phi_j\Big(\frac{1}{2}x\Big) - \left[\sum_{j=1}^{2} p_j\Phi_j(x)\right]^3 + 8x^3 - x \right) \cdot \left(\Phi_i\Big(\frac{1}{2}x\Big) - 3\left[\sum_{j=1}^{2} p_j\Phi_j(x)\right]^2 \cdot \Phi_i(x) \right) dx = 0. \tag{35}$$

From (35) for $i = 1$ we have

$$\int_{-1}^{1} \left(p_1 + p_2\left(\frac{1}{2}x\right) - (p_1 + p_2x)^3 + 8x^3 - x \right) \cdot \left(1 - 3\,(p_1 + p_2x)^2\right) dx = 0,$$

and for $i = 2$ one gets

$$\int_{-1}^{1} \left(p_1 + p_2\left(\frac{1}{2}x\right) - (p_1 + p_2x)^3 + 8x^3 - x \right) \cdot \left(\frac{1}{2}x - 3\,(p_1 + p_2x)^2 \cdot x\right) dx = 0.$$

Finally, we obtain the following nonlinear system of equations

$$\begin{cases} 2p_1 - 8p_1^3 + 6p_1^5 - \frac{76}{5}p_1p_2 - 6p_1p_2^2 + 20p_1^3p_2^2 + 6p_1p_2^4 = 0, \\ -\frac{38}{5}p_1^2 + \frac{1}{6}p_2 - 6p_1^2p_2 + 10p_1^4p_2 - \frac{198}{35}p_2^2 - \frac{4}{5}p_2^3 + 12p_1^2p_2^3 + \frac{6}{7}p_2^5 = -\frac{19}{15}, \end{cases}$$

to which the solution is

$$\begin{cases} p_1 = 0, \\ p_2 = 2. \end{cases}$$

Hence we get the approximate solution of the Eq. (34) in the form

$$y_2(x) = p_1\Phi_1(x) + p_2\Phi_2(x) = 2x.$$

Moreover, this is the solution of the Eq. (34).

2.4 The Adomian Decomposition Method

We can read about this method in [1, 2, 4, 5–9]. In this section we apply the decomposition method, presented in [2, 26], to nonlinear functional equation. Let us consider the following functional equation

$$y - N(y) = \mathbf{f}, \tag{36}$$

where $N : X \to X$ is a given function on a Banach space X and $\mathbf{f} \in X$.

Theorem 6 [2] Let X be a Banach algebra and let $N : X \to X$ be the class of C^∞. Then

$$A_n = \sum_{\alpha_1+\ldots+\alpha_n=n} c_{\alpha_1,\ldots,\alpha_n}(N(y_0))^{n+1-\alpha_1} \circ (N'(y_0))^{\alpha_1-\alpha_2} \\ \ldots (N^{(n-1)}(y_0))^{\alpha_{n-1}-\alpha_n} \circ (N^{(n)}(y_0))^{\alpha_n} \tag{37}$$

and $A_0 = N(y_0)$, where $y_0 \in X$ and

$$c_{\alpha_1,\ldots,\alpha_n} \\ = \frac{n!}{(\alpha_1-\alpha_2)!\ldots(\alpha_{n-1}-\alpha_n)!\alpha_n!(1!)^{\alpha_1-\alpha_2}\ldots(n-1)!^{\alpha_{n-1}-\alpha_n}(n!)^{\alpha_n}(n+1-\alpha_1)!} \tag{38}$$

for $n \in \mathbb{N}$.

From (38) we get the table of coefficients $c_{\alpha_1,\ldots,\alpha_n}$ for $n = 1, 2$:

n	$c_{\alpha_1,\ldots,\alpha_n}$	value
1	c_1	1
2	c_{11}	$\frac{1}{2}$
	c_{20}	1

Theorem 7 [2] *If N is C^∞ and satisfies $\|N^{(n)}(y_0)\| \leq M < 1$ for any $n \in \mathbb{N}$, then the decompositional series $\sum_{n=0}^{\infty} y_n$ is absolutely convergent and we have*

$$\|y_{n+1}\| = \|A_n\| \leq M^{n+1} n^{\sqrt{n}} e^{\pi\sqrt{\frac{2}{3}n}}. \tag{39}$$

We present this method on the following example.

Example 6 Find the approximate solution of the equation

$$y\left(\frac{1}{2}x\right) = \frac{1}{12}y^3(x) - \frac{2}{3}x^3 + x, \qquad x \in [-1, 1], \tag{40}$$

using the decomposition method.

Solution 6 From the Eq. (40) we get $f(x) = \frac{1}{2}x$ and $g(x, y) = -\frac{2}{3}x^3 + x + \frac{1}{12}y^3$. These functions fulfill the assumptions of Theorem 4 for $\Omega = (-1, 1) \times \mathbb{R}$ and $\Omega_x = \Omega_{f(x)} = \Gamma_x = \mathbb{R}$.

In this example we consider a Banach algebra $(X, \|\cdot\|)$, where $X = C([-1, 1])$, $\|x\| = \sup_{t\in[-1,1]} |x(t)|$ and $N : C([-1, 1]) \to C([-1, 1])$, $N \in C^\infty(X, Y)$ for $Y = C([-1, 1])$. From (40) we get $N(y) = \frac{1}{12}y^3$ and $\mathbf{f}(x) = -\frac{2}{3}x^3 + x$. In this method we have

$$y_0\left(\frac{1}{2}x\right) = \mathbf{f} = -\frac{2}{3}x^3 + x,$$

$$y_1\left(\frac{1}{2}x\right) = A_0,$$

$$\vdots$$

$$y_n\left(\frac{1}{2}x\right) = A_{n-1}, \qquad n \in \mathbb{N}.$$

By differentiating $N(y)$ we get $N'(y) = \frac{1}{4}y^2$, $N''(y) = \frac{1}{2}y$, $N'''(y) = \frac{1}{2}$ and $N^{(4)}(y) = 0$. From (37) and applying the table of coefficients we have

$$y_0(x) = 2x - \frac{16}{3}x^3,$$

$$y_1\left(\frac{1}{2}x\right) = A_0 = N(y_0) = \frac{1}{12}y_0^3 = \frac{1}{12}\left(2x - \frac{16}{3}x^3\right)^3$$

$$= \frac{2}{3}x^3 - \frac{16}{3}x^5 + \frac{128}{9}x^7 - \frac{1024}{81}x^9,$$

$$y_1(x) = \frac{16}{3}x^3 - \frac{512}{3}x^5 + \frac{16384}{9}x^7 - \frac{524288}{81}x^9,$$

$$y_2\left(\frac{1}{2}x\right) = A_1 = c_1 N(y_0) \cdot N'(y_0) = \frac{1}{4}\left(2x - \frac{16}{3}x^3\right)^2$$

$$\cdot\left(\frac{16}{3}x^3 - \frac{512}{3}x^5 + \frac{16384}{9}x^7 - \frac{524288}{81}x^9\right)$$

$$= \frac{16}{3}x^5 - \frac{1792}{9}x^7 + \frac{74752}{27}x^9 - \frac{1409024}{81}x^{11} + \frac{11534336}{243}x^{13} - \frac{33554432}{729}x^{15},$$

$$y_2(x) = \frac{512}{3}x^5 - \frac{229376}{9}x^7 + \frac{38273024}{27}x^9 - \frac{2885681152}{81}x^{11} + \frac{94489280512}{243}x^{13} - \frac{1099511627776}{729}x^{15},$$

$$y_3\left(\frac{1}{2}x\right) = A_2 = c_{11}(N(y_0))^2 \cdot N''(y_0) + c_{20}N(y_0)\cdot(N'(y_0))^2$$

$$= \frac{1}{2}\left(\frac{16}{3}x^3 - \frac{512}{3}x^5 + \frac{16384}{9}x^7 - \frac{524288}{81}x^9\right)^2 \cdot \frac{1}{2}\left(2x - \frac{16}{3}x^3\right) + 1\left(\frac{16}{3}x^3 - \frac{512}{3}x^5 + \frac{16384}{9}x^7 - \frac{524288}{81}x^9\right)\left(\frac{1}{4}\left(2x - \frac{16}{3}x^3\right)^2\right)^2$$

$$= \frac{176}{9}x^7 - \frac{31744}{27}x^9 + \frac{825344}{27}x^{11} - \frac{107773952}{243}x^{13} + \frac{2800943104}{729}x^{15} - \frac{14266925056}{729}x^{17} + \frac{347422588928}{6561}x^{19} - \frac{1105954078720}{19683}x^{21},$$

$$y_3(x) = \frac{22528}{9}x^7 - \frac{16252928}{27}x^9 + \frac{1690304512}{27}x^{11} - \frac{882884214784}{243}x^{13} + \frac{91781303631872}{729}x^{15} - \frac{1869994400940032}{729}x^{17} + \frac{182149494303883264}{6561}x^{19} - \frac{2319353808095805440}{19683}x^{21},$$

In the decomposition method we consider the solution having the series form

$$y = \sum_{n=0}^{\infty} y_n. \tag{41}$$

In our case we get the following partial sum of series (41)

$$\mathbf{y_3}(x) := y_0(x) + y_1(x) + y_2(x) + y_3(x)$$

$$= 2x - \frac{63488}{3}x^7 + \frac{65536000}{81}x^9 + \frac{2185232384}{81}x^{11} - \frac{262798311424}{81}x^{13} + \frac{90681792004096}{729}x^{15} - \frac{1869994400940032}{729}x^{17} + \frac{182149494303883264}{6561}x^{19} - \frac{2319353808095805440}{19683}x^{21}.$$

Next we use Theorem 7 to prove the convergence of this series. First we verify the assumption of this theorem. From above calculations we have

$$\begin{aligned}
\|N(y_0)\| &\leq \frac{\sqrt{2}}{162},\\
\|N'(y_0)\| &\leq \frac{1}{18},\\
\|N''(y_0)\| &\leq \frac{\sqrt{2}}{6},\\
\|N'''(y_0)\| &\leq \frac{1}{2},\\
\|N^{(n)}(y_0)\| &\leq 0, \qquad n = 4, 5, \ldots
\end{aligned}$$

Hence we get

$$\|N^{(n)}(y_0)\| \leq \frac{1}{2} < 1, \qquad n \in \mathbb{N}_0.$$

From Theorem 7 we guess that the series (41) is absolutely convergent to the solution of (40).

If we look into the partial sums of the series we discover that most expressions cancel out in this series. From (41) and Theorem 4 we get that the Eq. (40) has continuous solution. One can check that $y(x) = 2x$ is the solution of the Eq. (40).

3 Stability of the Generalized Quadratic Functional Equation on Topological Spaces

3.1 Introduction

In the theory of functional equations the problem of the stability has its origin in the following question, posed by S. Ulam [33] in 1940, concerning the stability of group homomorphisms.

Let G_1 be a group and let G_2 be a metric group with a metric $d(\cdot,\cdot)$. Given $\varepsilon > 0$, does there exist a $\delta > 0$ such that if a function $f : G_1 \to G_2$ satisfies the inequality

$$d[f(xy), f(x)f(y)] < \delta \quad \text{for all } x, y \in G_1,$$

then there exists a homomorphism $a : G_1 \to G_2$ with

$$d[f(x), a(x)] < \varepsilon \quad \text{for all } x \in G_1?$$

In the next year, D. H. Hyers [19] gave a partial affirmative answer to the question of Ulam in the context of Banach spaces. That was the first significant breakthrough and a step toward more solutions in this area. Since then, a large number of papers have been published in connection with various generalizations of Ulam's problem and Hyers' theorem. For more information concerning the stability problems of functional equations the reader is referred to the next monographs [13, 14, 21, 22] and papers, e.g. [11, 12, 16, 20, 31].

Let $(S, +)$ be a commutative semigroup and throughout this part of the chapter let X be a sequentially complete locally convex linear topological Hausdorff space. We recall a few results concerning the stability of the Cauchy and Pexider functional equations on topological spaces.

In [32] L. Székelyhidi and in [17] Z. Gajda have proved that if $f : S \to X$ is a function for which the Cauchy difference

$$Cf(x, y) := f(x + y) - f(x) - f(y), \quad x, y \in S$$

is bounded on S, then there exists an additive function $a : S \to X$ such that $f - a$ is bounded on S. In such a case we say that the Cauchy functional equation

$$f(x + y) = f(x) + f(y), \quad x, y \in S$$

is stable in the sense of Hyers and Ulam.

For an arbitrary set $A \subset X$, we denote by $conv\ A$ the convex hull of A, by $cl\ A$ the closure of A, and by $seq\ cl\ A$ the sequential closure of A. In [30] K. Nikodem has proved, assuming additionally that S is a semigroup with zero, that if for arbitrary functions $f, g, h : S \to X$ for which the Pexider difference $P(f, g, h)$ satisfies the condition

$$P(f, g, h)(x, y) := f(x + y) - g(x) - h(y) \in V, \quad x, y \in S,$$

where V is a bounded, convex, and symmetric with respect to zero subset of X, then there exist functions $F, G, H : S \to X$ satisfying the Pexider functional equation

$$F(x + y) = G(x) + H(y), \quad x, y \in S$$

such that $f(x) - F(x) \in 3\ seq\ cl\ V$, $g(x) - G(x) \in 4\ seq\ cl\ V$, $h(x) - H(x) \in 4\ seq\ cl\ V$ for all $x \in S$.

Motivated by this result, E. Głowacki and Z. Kominek [18] have proved the stability of the Pexider functional equation without the assumption that S contains the zero element. Moreover, in [24] Z. Kominek generalized this result, and similarly to the above Nikodem's theorem, he gave the appropriate bounds of functions $f - F$, $g - G$, and $h - H$.

In [3] M. Adam and S. Czerwik have proved the stability of the generalized quadratic functional equation. This result reads as follows.

Theorem 8 *Let G be an Abelian 2-divisible group and let $B \subset X$ be a nonempty bounded set. If functions $f, g : G \to X$ satisfy*

$$f(x + y) + f(x - y) - g(x) - g(y) \in B, \quad x, y \in G,$$

then there exists exactly one quadratic function $Q : G \to X$ such that

$$Q(x) + f(0) - f(x) \in \frac{2}{3}\ seq\ cl\ conv\,(B - B), \quad x \in G, \tag{42}$$

$$2Q(x) + g(0) - g(x) \in \frac{2}{3}\, seq\ cl\ conv\ (B - B), \quad x \in G. \tag{43}$$

Moreover, the function Q is given by the formulae

$$Q(x) = \lim_{n\to\infty} \frac{f(2^n x)}{2^{2n}} = \frac{1}{2} \lim_{n\to\infty} \frac{g(2^n x)}{2^{2n}}, \quad x \in G \tag{44}$$

and the convergence is uniform on G.

In the present part of the chapter we determine the general solution of the functional equation

$$f(x + 3y) + g(x - y) = h(x - 3y) + k(x + y), \quad x, y \in G, \tag{45}$$

which is a generalized version of the functional equation

$$f(x + 3y) + 3\, f(x - y) = f(x - 3y) + 3\, f(x + y), \quad x, y \in G. \tag{46}$$

The general solution of the above equation is of the form $f = Q+A+c$, where Q is a quadratic mapping, A is an additive one and c is an arbitrary constant. It is also worth noting that this equation is equivalent to the functional equation $\Delta^3_{2y} f(x - 3y) = 0$, where Δf is the difference operator defined by $\Delta_h f(x) = f(x+h) - f(x)$ and $\Delta^3 f$ denotes its third iteration. Therefore a solution of the above equation is a polynomial of degree at most two (see, e.g., [28]). It is a classical result in the theory of functional equations. Moreover, we will prove the Hyers–Ulam stability of the above functional equation on topological spaces.

Now we give some auxiliary results. Given sets $A, B \subset X$ and a number $k \in \mathbb{R}$, we define the well-known operations

$$A + B := \{x \in X\colon\ x = a + b,\ a \in A,\ b \in B\},$$

$$kA := \{x \in X\colon\ x = ka,\ a \in A\}.$$

One can prove (see, e.g., [15]) the following lemmas.

Lemma 1 *If $A, B \subset X$ and $0 \le \alpha \le \beta$, then*

$$\alpha A \subset \beta\ conv[A \cup \{0\}],$$

$$conv\ A + conv\ B = conv(A + B).$$

Lemma 2 *For any sets $A, B \subset X$ and numbers $\alpha, \beta \in \mathbb{R}$ we have*

$$\alpha(A + B) = \alpha A + \alpha B,$$

$$(\alpha + \beta)A \subset \alpha A + \beta A.$$

Moreover, if A is a convex set and $\alpha, \beta \ge 0$, then

$$\alpha A + \beta A = (\alpha + \beta)A.$$

Let us recall that a set $A \subset X$ is said to be bounded iff for every neighbourhood U of zero there exists a number $\alpha > 0$ such that $\alpha A \subset U$.

Lemma 3 *If $A, B \subset X$ are bounded sets, then*

$$A \cup B, \quad A + B, \quad conv\, A$$

are also bounded subsets of X.

3.2 Stability

In the following theorem we prove the stability of the functional Eq. (46) on topological spaces.

Theorem 9 *Let G be an Abelian group uniquely divisible by 2 and 3, and assume that B is nonempty and bounded subset of X. If $f : G \to X$ satisfies the condition*

$$f(x+3y) + 3f(x-y) - f(x-3y) - 3f(x+y) \in B, \quad x, y \in G, \tag{47}$$

then there exist a unique quadratic function $Q : G \to X$ and a unique additive function $A : G \to X$ such that

$$Q(x) + A(x) + f(0) - f(x) \in \frac{1}{24}\, seq\ cl\ conv\,(11B - 3B), \quad x \in G,$$

$$Q(x) + f(0) - \frac{f(x) + f(-x)}{2} \in \frac{1}{3}\, seq\ cl\ conv\, B, \quad x \in G,$$

$$A(x) - \frac{f(x) - f(-x)}{2} \in \frac{1}{8}\, seq\ cl\ conv\,(B - B), \quad x \in G.$$

Moreover, the functions Q and A are given by the formulae

$$Q(x) = \lim_{n\to\infty} \frac{f(2^n x) + f(-2^n x) - 2f(0)}{2 \cdot 2^{2n}}, \quad x \in G,$$

$$A(x) = \lim_{n\to\infty} \frac{f(3^n x) - f(-3^n x)}{2 \cdot 3^n}, \quad x \in G.$$

Proof Putting $x = y = 0$ in (47) we have $0 \in B$. Let us define a function $f_e : G \to X$ by the formula

$$f_e(x) := \frac{f(x) + f(-x)}{2} - f(0), \quad x \in G.$$

Clearly, f_e is even and $f_e(0) = 0$. Since f satisfies (47), then

$$f_e(x+3y) + 3f_e(x-y) - f_e(x-3y) - 3f_e(x+y) \in conv\, B, \quad x, y \in G. \tag{48}$$

Substitution $y = x$ in (48) leads to

$$f_e(4x) - 4f_e(2x) \in conv\ B, \quad x \in G,$$

i.e.,

$$\frac{1}{2^2} f_e(2x) - f_e(x) \in \frac{1}{4}\ conv\ B, \quad x \in G. \tag{49}$$

By a standard way (see, e.g., [3]) one can inductively check that

$$\frac{1}{2^{2n}} f_e(2^n x) - f_e(x) \in \frac{1}{3}\left(1 - \frac{1}{2^{2n}}\right) conv\ B, \quad n \in \mathbb{N},\ x \in G. \tag{50}$$

Define

$$Q_n(x) := \frac{1}{2^{2n}} f_e(2^n x), \quad n \in \mathbb{N},\ x \in G. \tag{51}$$

For all $m, n \in \mathbb{N}$ and $x \in G$, we have by (50)

$$Q_{m+n}(x) - Q_n(x) = \frac{1}{2^{2(m+n)}} f_e\left(2^{m+n} x\right) - \frac{1}{2^{2n}} f_e(2^n x)$$

$$= \frac{1}{2^{2n}}\left[\frac{1}{2^{2\ m}} f_e(2^m \cdot 2^n x) - f_e(2^n x)\right] \in \frac{1}{2^{2n}} \cdot \frac{1}{3}\left(1 - \frac{1}{2^{2\ m}}\right) conv\ B.$$

From the boundedness of the set $conv\ B$ (see Lemma 3) we have that $(Q_n)_{n\in\mathbb{N}}$ is a Cauchy sequence of elements of X uniformly convergent on G by the completeness of X. Therefore we can define a function $Q : G \to X$ by the following formula

$$Q(x) := \lim_{n\to\infty} Q_n(x), \quad x \in G.$$

Obviously, Q is even and $Q(0) = 0$. Taking the limit in (50) as $n \to \infty$ we obtain

$$Q(x) - f_e(x) \in \frac{1}{3}\ seq\ cl\ conv\ B, \quad x \in G. \tag{52}$$

Substituting $2^n x$, $2^n y$ instead of x and y in (48), respectively, and dividing both sides of the resulting expression by 2^{2n} we get

$$\frac{1}{2^{2n}} f_e(2^n(x + 3y)) + 3 \cdot \frac{1}{2^{2n}} f_e(2^n(x - y))$$

$$-\frac{1}{2^{2n}} f_e(2^n(x - 3y)) - 3 \cdot \frac{1}{2^{2n}} f_e(2^n(x + y)) \in \frac{1}{2^{2n}}\ conv\ B, \quad x, y \in G.$$

Taking the limit in the above expression as $n \to \infty$ we conclude that

$$Q(x + 3y) + 3Q(x - y) - Q(x - 3y) - 3Q(x + y) = 0, \quad x, y \in G.$$

It is well known (see, e.g., [23]) that if Q is even then the above functional equation is equivalent to the original quadratic functional equation

$$Q(x + y) + Q(x - y) = 2Q(x) + 2Q(y), \quad x, y \in G.$$

Therefore, Q is quadratic.

To prove the uniqueness of the solution assume that there exists another quadratic function $Q_1 : G \to X$ satisfying the condition (52). Then we have

$$Q_1(x) - Q(x) = Q_1(x) - f_e(x) - [Q(x) - f_e(x)] \in \frac{2}{3}\, seq\ cl\ conv\ B,$$

i.e.,

$$Q_1(x) - Q(x) \in \frac{2}{3}\, seq\ cl\ conv\ B, \quad x \in G.$$

Replacing x in the above condition by $2^n x$ and dividing both sides by 2^{2n}, and then taking the limit in the resulting expression, we obtain $Q_1(x) - Q(x) = 0$, i.e., $Q_1 = Q$.

According to (52) we get

$$Q(x) + f(0) - \frac{f(x) + f(-x)}{2} \in \frac{1}{3}\, seq\ cl\ conv\ B, \quad x \in G. \tag{53}$$

Moreover

$$Q(x) = \lim_{n\to\infty} \frac{f(2^n x) + f(-2^n x) - 2f(0)}{2 \cdot 2^{2n}}, \quad x \in G.$$

Now we consider the second case where a function $f_o : G \to X$ is defined by the formula

$$f_o(x) := \frac{f(x) - f(-x)}{2}, \quad x \in G.$$

Clearly, f_0 is odd and $f_o(0) = 0$. Since f satisfies (47), then

$$f_o(x + 3y) + 3f_o(x - y) - f_o(x - 3y) - 3f_o(x + y) \in \frac{1}{2}\, conv\ (B - B), x, y \in G. \tag{54}$$

Taking $x = 0$ in (54) we see that

$$f_o(3x) - 3f_o(x) \in \frac{1}{4}\, conv\ (B - B), \quad x \in G,$$

hence

$$\frac{1}{3} f_o(3x) - f_o(x) \in \frac{1}{12}\, conv\ (B - B), \quad x \in G. \tag{55}$$

Proceeding similarly as in the previous case we can prove by induction the following formula

$$\frac{1}{3^n} f_o(3^n x) - f_o(x) \in \frac{1}{8}\left(1 - \frac{1}{3^n}\right) conv\ (B - B), \quad n \in \mathbb{N},\ x \in G. \tag{56}$$

Moreover, $\left(\frac{f_o(3^n x)}{3^n}\right)_{n\in\mathbb{N}}$ is a Cauchy sequence and thus we can define a function $A : G \to X$ by the following formula

$$A(x) := \lim_{n\to\infty} \frac{f_o(3^n x)}{3^n}, \quad x \in G.$$

Obviously, A is odd and $A(0) = 0$. Taking the limit in (56) as $n \to \infty$ we obtain

$$A(x) - f_o(x) \in \frac{1}{8}\, seq\ cl\ conv\,(B - B), \quad x \in G. \tag{57}$$

It follows from (54) that

$$A(x + 3y) + 3A(x - y) - A(x - 3y) - 3A(x + y) = 0, \quad x, y \in G.$$

It is also well known (see, e.g., [23]) that if A is odd then the above functional equation is equivalent to the Cauchy functional equation. Therefore, A is additive. Similarly as before, we can prove the uniqueness of the solution. By (57) we get

$$A(x) - \frac{f(x) - f(-x)}{2} \in \frac{1}{8}\, seq\ cl\ conv\,(B - B), \quad x \in G. \tag{58}$$

Moreover

$$A(x) = \lim_{n\to\infty} \frac{f(3^n x) - f(-3^n x)}{2 \cdot 3^n}, \quad x \in G.$$

Finally, from (53) and (58) we obtain

$$Q(x) + A(x) + f(0) - f(x) \in \frac{1}{24}\, seq\ cl\ conv\,(11B - 3B), \quad x \in G,$$

which completes the proof. □

In the following theorem we determine the general solution of the functional Eq. (45) without assuming any regularity condition on the unknown functions f, g, h, k.

Theorem 10 *Let G_1 be an Abelian group divisible by 2 and 3, and let G_2 be an Abelian group uniquely divisible by 2 and 3. Suppose that functions $f, g, h, k : G_1 \to G_2$ satisfy the following functional equation*

$$f(x + 3y) + g(x - y) = h(x - 3y) + k(x + y), \quad x, y \in G_1. \tag{59}$$

Then there exist a quadratic function $Q : G_1 \to G_2$, additive functions $E, F : G_1 \to G_2$ and constants $C_1, C_2, C_3, C_4 \in G_2$ such that $C_1 + C_2 = C_3 + C_4$ and

$$\begin{aligned} f(x) &= Q(x) + E(x) + F(x) + C_1, \\ g(x) &= 3Q(x) + 3E(x) - F(x) + C_2, \\ h(x) &= Q(x) + E(x) - F(x) + C_3, \\ k(x) &= 3Q(x) + 3E(x) + F(x) + C_4 \end{aligned}$$

for all $x \in G_1$.

Proof Since the group G_2 is uniquely divisible by 2 (i.e., $2G = G$), then we may split f into its even and odd parts $f^+, f^- : G_1 \to G_2$ by the formulae

$$f^+(x) := \frac{f(x) + f(-x)}{2}, \qquad f^-(x) := \frac{f(x) - f(-x)}{2}, \quad x \in G_1.$$

Clearly, f^+ is even, f^- is odd, and $f = f^+ + f^-$. Similarly we define the functions $g^+, g^-, h^+, h^-, k^+, k^-$. Obviously $f^-(0) = g^-(0) = h^-(0) = k^-(0) = 0$. Since f, g, h, k satisfy (59), then

$$f^+(x + 3y) + g^+(x - y) = h^+(x - 3y) + k^+(x + y), \quad x, y \in G_1, \tag{60}$$

$$f^-(x + 3y) + g^-(x - y) = h^-(x - 3y) + k^-(x + y), \quad x, y \in G_1. \tag{61}$$

Let us denote $C_1 := f^+(0)$, $C_2 := g^+(0)$, $C_3 := h^+(0)$, $C_4 := k^+(0)$. For $x = y = 0$ in (60) we have $C_1 + C_2 = C_3 + C_4$. Define

$$f_0(x) := f^+(x) - C_1,$$

$$g_0(x) := g^+(x) - C_2,$$

$$h_0(x) := h^+(x) - C_3,$$

$$k_0(x) := k^+(x) - C_4$$

for all $x \in G_1$. Obviously, the functions f_0, g_0, h_0, k_0 are also even and $f_0(0) = g_0(0) = h_0(0) = k_0(0) = 0$. Moreover

$$f_0(x + 3y) + g_0(x - y) = h_0(x - 3y) + k_0(x + y), \quad x, y \in G_1. \tag{62}$$

Setting, successively, $x = 0$, $y = 0$, $y = x$, $y = -x$, $x = 3y$ and $x = 5y$ in (62) and applying the fact that f_0, g_0, h_0, k_0 are even, we get

$$f_0(3x) + g_0(x) = h_0(3x) + k_0(x), \tag{63}$$

$$f_0(x) + g_0(x) = h_0(x) + k_0(x), \tag{64}$$

$$f_0(2x) = h_0(x) + k_0(x), \tag{65}$$

$$f_0(x) + g_0(x) = h_0(2x), \tag{66}$$

$$f_0(3x) + g_0(x) = k_0(2x), \tag{67}$$

$$f_0(4x) + g_0(2x) = h_0(x) + k_0(3x) \tag{68}$$

for all $x \in G_1$. Subtracting (65) and (66) from (64) we have

$$f_0(2x) = h_0(2x), \quad x \in G_1,$$

i.e.,

$$f_0(x) = h_0(x), \quad x \in G_1. \tag{69}$$

Comparing (64) and (69) we see that

$$g_0(x) = k_0(x), \quad x \in G_1. \tag{70}$$

Therefore, using (69) and (70) one can check that Eqs. (66), (67), and (68) are now given as follows

$$f_0(x) + g_0(x) = f_0(2x), \tag{71}$$

$$f_0(3x) + g_0(x) = g_0(2x), \tag{72}$$

$$f_0(4x) + g_0(2x) = f_0(x) + g_0(3x) \tag{73}$$

for all $x \in G_1$. Replacing x by $2x$ in (71) we obtain

$$f_0(2x) + g_0(2x) = f_0(4x), \quad x \in G_1. \tag{74}$$

Adding (71), (73), and (74) we have

$$g_0(3x) = 2g_0(2x) + g_0(x), \quad x \in G_1. \tag{75}$$

Replacing x by $3x$ in (71) we get

$$f_0(3x) + g_0(3x) = f_0(6x), \quad x \in G_1. \tag{76}$$

Adding (72) and (75), and using (76) yields

$$f_0(6x) = 3g_0(2x), \quad x \in G_1,$$

whence

$$f_0(3x) = 3g_0(x), \quad x \in G_1. \tag{77}$$

Combining (72) and (77) we arrive at

$$g_0(2x) = 4g_0(x), \quad x \in G_1. \tag{78}$$

Multiplying both sides of (78) by 2 and adding to the resulting Eq. (75) we obtain

$$g_0(3x) = 9g_0(x), \quad x \in G_1. \tag{79}$$

Comparing (77) and (79) we see that

$$f_0(3x) = 3g_0(x) = \frac{1}{3}g_0(3x), \quad x \in G_1,$$

i.e.,

$$3f_0(x) = g_0(x), \quad x \in G_1. \tag{80}$$

Finally, from (69), (70), and (80) we get

$$3f_0(x) = g_0(x) = 3h_0(x) = k_0(x), \quad x \in G_1. \tag{81}$$

Let $Q := f_0$. Therefore from (62) and (81) we have

$$Q(x+3y) + 3Q(x-y) = Q(x-3y) + 3Q(x+y), \quad x, y \in G_1.$$

Since Q is even and satisfies the above equation, then Q is a quadratic function. Moreover

$$\begin{aligned} f^+(x) &= Q(x) + C_1, \\ g^+(x) &= 3Q(x) + C_2, \\ h^+(x) &= Q(x) + C_3, \\ k^+(x) &= 3Q(x) + C_4 \end{aligned}$$

for all $x \in G_1$.

Now we consider the second case. Replacing y by $-y$ in (61) we obtain

$$f^-(x-3y) + g^-(x+y) = h^-(x+3y) + k^-(x-y), \quad x, y \in G_1. \tag{82}$$

Thus, subtracting (82) from (61) we see that

$$\begin{aligned} &f^-(x+3y) + h^-(x+3y) + g^-(x-y) + k^-(x-y) \\ &= f^-(x-3y) + h^-(x-3y) + g^-(x+y) + k^-(x+y), \quad x, y \in G_1. \end{aligned} \tag{83}$$

Define

$$\begin{aligned} E_1(x) &:= f^-(x) + h^-(x), \quad x \in G_1, \\ E_2(x) &:= g^-(x) + k^-(x), \quad x \in G_1. \end{aligned}$$

Clearly, these functions are odd and $E_1(0) = E_2(0) = 0$. Then (83) becomes

$$E_1(x+3y) + E_2(x-y) = E_1(x-3y) + E_2(x+y), \quad x, y \in G_1. \tag{84}$$

Setting, successively, $x = 0$, $y = x$, $x = 3y$ and $x = 5y$ in (84) we get

$$E_1(3x) = E_2(x), \tag{85}$$

$$E_1(2x) = -E_1(x) + E_2(x), \tag{86}$$

$$E_1(3x) + E_2(x) = E_2(2x), \tag{87}$$

$$E_1(4x) + E_2(2x) = E_1(x) + E_2(3x) \tag{88}$$

for all $x \in G_1$. From (85) and (87) it easily follows that

$$E_2(2x) = 2E_2(x), \quad x \in G_1. \tag{89}$$

Comparing (87) and (86), and using (89) we have

$$E_1(3x) = E_1(2x) + E_1(x), \quad x \in G_1. \tag{90}$$

Now, replace x by $2x$ in (86) to get

$$E_1(4x) = -E_1(2x) + E_2(2x), \quad x \in G_1. \tag{91}$$

Then subtracting (90) and (91) from (88), and using (85) and (89) we obtain

$$E_2(3x) = 3E_2(x), \quad x \in G_1. \tag{92}$$

Next, using (85) and (92) we have

$$E_1(3x) = E_2(x) = \frac{1}{3}E_2(3x), \quad x \in G_1,$$

i.e.,

$$3E_1(x) = E_2(x), \quad x \in G_1. \tag{93}$$

Substituting twice (93) back into (84) we see that

$$E_1(x+3y) + 3E_1(x-y) = E_1(x-3y) + 3E_1(x+y), \quad x, y \in G_1, \tag{94}$$

$$E_2(x+3y) + 3E_2(x-y) = E_2(x-3y) + 3E_2(x+y), \quad x, y \in G_1. \tag{95}$$

Since E_1 and E_2 are odd, and satisfy the above equations, then these functions are additive.

Adding (61) and (82) we have

$$f^-(x+3y) - h^-(x+3y) + g^-(x-y) - k^-(x-y)$$

$$= -f^-(x-3y) + h^-(x-3y) - g^-(x+y) + k^-(x+y), \quad x, y \in G_1. \tag{96}$$

Define

$$F_1(x) := f^-(x) - h^-(x), \quad x \in G_1,$$

$$F_2(x) := g^-(x) - k^-(x), \quad x \in G_1.$$

Obviously, these functions are also odd and $F_1(0) = F_2(0) = 0$. Then (96) becomes

$$F_1(x+3y) + F_2(x-y) = -F_1(x-3y) - F_2(x+y), \quad x, y \in G_1. \tag{97}$$

Setting $y = 0$ in (97) we get

$$F_1(x) = -F_2(x), \quad x \in G_1. \tag{98}$$

Now, using twice (98) in (97) we see that

$$F_1(x+3y) + F_1(x-3y) = F_1(x+y) + F_1(x-y), \quad x, y \in G_1, \tag{99}$$
$$F_2(x+3y) + F_2(x-3y) = F_2(x+y) + F_2(x-y), \quad x, y \in G_1.$$

Setting, successively, $y = x$ and $x = 3y$ in (99) we obtain

$$F_1(2x) = 2F_1(x), \quad x \in G_1, \tag{100}$$
$$F_1(3x) = F_1(2x) + F_1(x), \quad x \in G_1. \tag{101}$$

Combining (100) and (101) we arrive at

$$F_1(3x) = 3F_1(x), \quad x \in G_1. \tag{102}$$

Now we show that F_1 and F_2 are additive. Interchanging the roles of variables in (99) and applying the oddness of F_1 we deduce that

$$F_1(3x+y) - F_1(3x-y) = F_1(x+y) - F_1(x-y), \quad x, y \in G_1. \tag{103}$$

Replacing y by $3y$ in (103) and using (102) one can obtain

$$3F_1(x+y) - 3F_1(x-y) = F_1(x+3y) - F_1(x-3y), \quad x, y \in G_1,$$

i.e.,

$$F_1(x+3y) + 3F_1(x-y) = F_1(x-3y) + 3F_1(x+y), \quad x, y \in G_1. \tag{104}$$

Since F_1 is odd and satisfies the above equation, then F_1 is additive. From (98) the function F_2 is also additive. Finally

$$E_1(x) = f^-(x) + h^-(x), \tag{105}$$
$$3E_1(x) = g^-(x) + k^-(x), \tag{106}$$
$$F_1(x) = f^-(x) - h^-(x), \tag{107}$$
$$-F_1(x) = g^-(x) - k^-(x) \tag{108}$$

for all $x \in G_1$. Define

$$E(x) := \frac{1}{2}E_1(x), \qquad F(x) := \frac{1}{2}F_1(x), \quad x \in G_1.$$

Then from (105), (107), and (106), (108) we obtain

$$f^-(x) = E(x) + F(x), \quad x \in G_1,$$

$$g^-(x) = 3E(x) - F(x), \quad x \in G_1$$

and

$$h^-(x) = E(x) - F(x), \quad x \in G_1,$$
$$k^-(x) = 3E(x) + F(x), \quad x \in G_1,$$

respectively. Since $f = f^+ + f^-$, then

$$f(x) = Q(x) + E(x) + F(x) + C_1, \quad x \in G_1.$$

Similarly, one can show that

$$g(x) = 3Q(x) + 3E(x) - F(x) + C_2,$$
$$h(x) = Q(x) + E(x) - F(x) + C_3,$$
$$k(x) = 3Q(x) + 3E(x) + F(x) + C_4$$

for all $x \in G_1$, where $C_1 + C_2 = C_3 + C_4$. This completes the proof. □

Using a similar argument to that of the proof in Theorem 10 and applying Theorem 9, we have the following result concerning the stability of (59) on topological spaces.

Theorem 11 *Let G be an Abelian group uniquely divisible by 2 and 3. Suppose that $B \subset X$ is a nonempty bounded set and let functions $f, g, h, k : G \to X$ satisfy the condition*

$$f(x + 3y) + g(x - y) - h(x - 3y) - k(x + y) \in B, \quad x, y \in G. \tag{109}$$

Then there exist exactly one quadratic function $Q : G \to X$ and two unique additive functions $E, F : G \to X$ such that

$$Q(x) + F(x) + G(x) + f(0) - f(x) \in \frac{91}{6}\, seq\ cl\ conv\,(B - B),$$

$$3Q(x) + 3F(x) - G(x) + g(0) - g(x) \in \frac{65}{4}\, seq\ cl\ conv\,(B - B),$$

$$Q(x) + F(x) - G(x) + h(0) - h(x) \in \frac{91}{6}\, seq\ cl\ conv\,(B - B),$$

$$3Q(x) + 3F(x) + G(x) + k(0) - k(x) \in \frac{65}{4}\, seq\ cl\ conv\,(B - B)$$

for all $x \in G$.

References

1. Abbaoui, K., Cherruault, Y.: Convergence of Adomian's method applied to differential equations. Comput Math. Appl. **28**(5), 103–109 (1994)
2. Abbaoui, K., Cherruault, Y.: New ideas for proving convergence of decomposition methods. Comput Math. Appl. **29**(7), 103–108 (1995)
3. Adam, M., Czerwik, S.: On the stability of the quadratic functional equation in topological spaces. Banach J. Math. Anal. **1**(2), 245–251 (2007)
4. Adomian, G.: Stochastic Systems. Academic, New York (1983)
5. Adomian, G.: A review of the decomposition method in applied mathematics. J. Math. Anal. Appl. **135**, 501–544 (1988)
6. Adomian, G.: Nonlinear Stochastic Systems Theory and Applications to Physics. Kluwer, Dordrecht (1989)
7. Adomian, G., Rach, R.: Nonlinear stochastic differential delay equations. J. Math. Anal. Appl. **91**(1), 94–101 (1983)
8. Adomian, G., Rach, R.: On the solution of algebraic equations by the decomposition method. J. Math. Anal. Appl. **105**(1), 141–166 (1985)
9. Adomian, G., Rach, R., Sarafyan, D.: On the solution of equations containing radicals by the decomposition method. J. Math. Anal. Appl. **111**(2), 423–426 (1985)
10. Cherruault, Y.: Convergence of Adomian's method. Kybernetes **18**(2), 31–38 (1989)
11. Czerwik, S.: On the stability of the quadratic mapping in normed spaces. Abh. Math. Sem. Univ. Hambg. **62**, 59–64 (1992)
12. Czerwik, S.: The stability of the quadratic functional equation. In: Rassias, Th.M., Tabor, J. (eds.) Stability of Mappings of Hyers-Ulam Type, pp. 81–91. Hadronic Press, Palm Harbor (1994)
13. Czerwik, S.: Functional Equations and Inequalities in Several Variables. World Scientific, New Jersey (2002)
14. Czerwik, S.: Stability of Functional Equations of Ulam-Hyers-Rassias Type. Hadronic Press, Palm Harbor (2003)
15. Czerwik, S.: On stability of the equation of homogeneous functions on topological spaces. In: Rassias, Th.M., Brzdęk, J. (eds.) Functional Equations in Mathematical Analysis. Springer Optimization and Its Applications, vol. 52 , pp. 87–96. (2012)
16. Forti, G.L.: Hyers-Ulam stability of functional equations in several variables. Aequationes Math. **50**, 143–190 (1995)
17. Gajda, Z.: On stability of Cauchy equation on semigroups. Aequationes Math. **36**, 76–79 (1988)
18. Głowacki, E., Kominek, Z.: On stability of the Pexider equation on semigroups. In: Rassias, Th.M., Tabor, J. (eds.) Stability of Mappings of Hyers-Ulam Type, pp. 111–116. Hadronic Press, Palm Harbor (1994)
19. Hyers, D.H.: On the stability of the linear functional equation. Proc. Nat. Acad. Sci. U S A **27**, 222–224 (1941)
20. Hyers, D.H., Rassias, Th.M.: Approximate homomorphisms. Aequationes Math. **44**, 125–153 (1992)
21. Hyers, D.H., Isac, G., Rassias, Th.M.: Stability of Functional Equations in Several Variables. Birkhäuser Verlag, Basel (1998)
22. Jung, S.-M.: Hyers-Ulam-Rassias Stability of Functional Equations in Mathematical Analysis. Hadronic Press, Palm Harbor (2000)
23. Jun, K.-W., Kim, H.-M., Lee, D.O.: On the Hyers-Ulam-Rassias stability of a modified additive and quadratic functional equation. J. Korea Soc. Math. Educ. Ser. B: Pure Appl. Math. **11**(4), 323–335 (2004)
24. Kominek, Z.: On Hyers-Ulam stability of the Pexider equation. Demonstratio Math. **37**, 373–376 (2004)

25. Król, K.: Application the least squares method to solving the linear functional equation. In: Materiały konferencyjne: Młodzi naukowcy wobec wyzwań współczesnej techniki, pp. 263–270. Politechnika Warszawska (2007)
26. Król, K.: Application of the decomposition metod to solving functional equations. Nonlinear Funct. Anal. Appl. **14**(5), 809–815 (2009)
27. Kuczma, M.: Functional Equations in a Single Variable. Monografie Mat., vol. 46, Polish Scientific Publishers, Warsaw (1968)
28. Kuczma, M.: An Introduction to the Theory of Functional Equations and Inequalities. Cauchy's Equation and Jensen's Inequality, 2nd edn. Birkhäuser , Basel (2009) (edited by A. Gilányi)
29. Michlin, S.G.: Variacionnye metody v matematicheskoi fizike, Gosudarstvennoe izdatelstvo tehniko-teoreticheskoi literatury. Moskva (1957)
30. Nikodem, K.: The stability of the Pexider equation. Annales Math. Silesianae **5**, 91–93 (1991)
31. Rätz, J.: On approximately additive mappings. In: Beckenbach, E.F. (ed.) General Inequalities 2, ISNM, vol. 47 , pp. 233–251. Birkhäuser, Basel (1980)
32. Székelyhidi, L.: Note on Hyers's theorem. C. R. Math. Rep. Acad. Sci. Can. **8**, 127–129 (1986)
33. Ulam, S.M.: Problems in Modern Mathematics. Science Editions. Wiley, New York (1960)

Remarks on Stability of the Equation of Homomorphism for Square Symmetric Groupoids

Anna Bahyrycz and Janusz Brzdęk

Mathematics Subject Classification (2010) *39B22, 39B52, 39B82.*

Abstract Let $(G, \star)$ and $(H, \circ)$ be square symmetric groupoids and $S \subset G$ be nonempty. We present some remarks on stability of the following conditional equation of homomorphism

$$f(x \star y) = f(x) \circ f(y) \qquad x, y \in S, x \star y \in S,$$

in the class of functions mapping S into H. In particular, we consider the situation where $H = \mathbb{R}$ and

$$-\nu(x, y) \leq h(x \star y) - h(x) \circ h(y) \leq \mu(x, y) \qquad x, y \in S, x \star y \in S,$$

with some functions $\mu, \nu : S^2 \to [0, \infty)$.

Keywords Hyers-Ulam stability · Square symmetric groupoids · Homomorphism · Fixed point · Complete metric · Linear equation

1 Introduction

The issue of stability of functional equations has been a very popular subject of investigations for the last nearly 50 years (see, e.g., [6, 11, 23, 25, 31, 32, 35, 36]). The main motivation for it was given by S.M. Ulam (cf. [58]) in 1940 in his talk at the University of Wisconsin, where he presented some unsolved problems and, in particular, the following one.

A. Bahyrycz (✉) · J. Brzdęk
Department of Mathematics, Pedagogical University,
Podchorążych 2, 30-084 Kraków, Poland
e-mail: bah@up.krakow.pl

J. Brzdęk
e-mail: jbrzdek@up.krakow.pl

T. M. Rassias (ed.), *Handbook of Functional Equations*,
Springer Optimization and Its Applications 96, DOI 10.1007/978-1-4939-1286-5_2

Let G_1 be a group and (G_2, d) a metric group. Given $\varepsilon > 0$, does there exist $\delta > 0$ such that if $f : G_1 \to G_2$ satisfies

$$d(f(xy), f(x)f(y)) < \delta \qquad x, y \in G_1,$$

then a homomorphism $T : G_1 \to G_2$ exists with

$$d(f(x), T(x)) < \varepsilon \qquad x, y \in G_1?$$

In 1941 D.H. Hyers [21] published a partial answer to it, which can be stated as follows.
Let E and Y be Banach spaces and $\varepsilon > 0$. Then, for every $g : E \to Y$ with

$$\sup_{x,y \in E} \|g(x+y) - g(x) - g(y)\| \leq \varepsilon,$$

there is a unique solution $f : E \to Y$ of the Cauchy equation

$$f(x+y) = f(x) + f(y) \tag{1}$$

such that

$$\sup_{x \in E} \|g(x) - f(x)\| \leq \varepsilon.$$

We can describe the result of Hyers by simply saying that *the Cauchy functional Eq. (1) is Hyers-Ulam stable (or has the Hyers-Ulam stability)*. At the moment it is well known (see, e.g., [23]) that in the Hyers result it is enough to assume that $(E, +)$ is an amenable semigroup or a square symmetric groupoid (either of those assumptions is fulfilled when $(E, +)$ is a commutative semigroup).

In the next few years Hyers and Ulam published some further stability results for polynomial functions, isometries, and convex functions in [22, 26–28].

We should mention here that now we are aware of an earlier result concerning stability of functional equations that is due to Gy. Pólya and G. Szegö [40, Teil I, Aufgabe 99] (see also [41, Part I, Chap. 3, Problem 99] and [18, p. 125]) and reads as follows ($\mathbb{N}$ stands for the set of positive integers).
For every real sequence $(a_n)_{n \in \mathbb{N}}$ such that

$$\sup_{n,m \in \mathbb{N}} |a_{n+m} - a_n - a_m| \leq 1,$$

there is a real number ω with $\sup_{n \in \mathbb{N}} |a_n - \omega n| \leq 1$. Moreover,

$$\omega = \lim_{n \to \infty} \frac{a_n}{n}.$$

A next significant result for the Cauchy equation was obtained by T. Aoki [1], who proved the subsequent (for another extension of the result of Hyers see [55]).

Theorem 1 *Assume that E_1 and E_2 are two normed spaces, E_2 is complete, $c \geq 0$ and $0 < p < 1$. Let $f : E_1 \to E_2$ be a mapping such that*

$$\|f(x+y) - f(x) - f(y)\| \leq c(\|x\|^p + \|y\|^p) \qquad x, y \in E_1. \tag{2}$$

Then there exists a unique solution $T : E_1 \to E_2$ of (1) *with*

$$\|f(x) - T(x)\| \leq \frac{c\|x\|^p}{1 - 2^{p-1}} \qquad x \in E_1. \tag{3}$$

Unfortunately that result had remained unnoticed by a wider audience for quite a long time. In 1978 it was rediscovered by Th.M. Rassias [47], who came across it while solving a problem communicated to him by Ulam (evidently not familiar with the Aoki outcome, as well); moreover, Th.M. Rassias [47] obtained the linearity of T under the assumption of continuity of f and thus provided an outcome on stability of linear mappings and a method of proving it. In this way and through his numerous further publications (see, e.g., [12, 24, 25, 29, 49–54]) Th.M. Rassias strongly stimulated investigations of that kind of stability (for further information and references see, e.g., [6, 23, 31, 32]). In recognition of this, the results obtained in [47] are quite often referred to as the Hyers-Ulam-Rassias stability of the Cauchy equation (cf., e.g., [16, 31–32]).

In [49] (see also [53, p. 326]) it has been noticed that a result analogous to that of Theorem 1 is also valid for $p < 0$. Next, motivated by the problem raised by Th.M. Rassias (see [48]), Z. Gajda [15] extended the result contained in Theorem 1 for $p > 1$ and gave an example that for $p = 1$ it is not valid (for further such examples see [53]; some complementary result for the case $p = 1$ are given in [17, 56]). Recently, it has been proved in [8] (see also [39]) that, for $p < 0$, each function $f : E_1 \to E_2$ satisfying (2) (but of course only for $x \neq 0$ and $y \neq 0$) must be additive and the completness of E_2 is not necessary then.

All those stability results finally yield the following theorem.

Theorem 2 *Let E_1 and E_2 be two normed spaces, $c \geq 0$ and $p \neq 1$ be fixed real numbers. Let $f : E_1 \to E_2$ be a mapping such that*

$$\|f(x+y) - f(x) - f(y)\| \leq c(\|x\|^p + \|y\|^p) \qquad x, y \in E_1 \setminus \{0\}. \tag{4}$$

If $p \geq 0$ and E_2 is complete, then there exists a unique additive function $T : E_1 \to E_2$ with

$$\|f(x) - T(x)\| \leq \frac{c\|x\|^p}{|2^{p-1} - 1|} \qquad x \in E_1 \setminus \{0\}. \tag{5}$$

If $p < 0$, then f is additive.

It has been proved in [7] that estimation (5) is optimum for $p \geq 0$ in the general case. Further extensions and generalizations of the idea of stability, described above, have been proposed in [4, 5, 16, 20, 45, 46] (for more information we refer, e.g., to [6, 11, 13, 23, 31]), where the authors studied mappings f satisfying the inequality

$$\|f(x+y) - f(x) - f(y)\| \leq \varphi(x, y) \tag{6}$$

for functions φ of some other or more general form than in condition (4). In particular, the function φ of the form

$$\varphi(x, y) \equiv c(\|x\|^p \|y\|^q)$$

has been used in [45, 46], with suitable real c, p, q.

2 An Auxiliary Result

In the sequel we use in the proofs some kind of fixed point approach (for a survey on related results we refer to [10]).The following lemma is very simple, but nevertheless very useful in that ($\mathbb{R}_+$ denotes the set of nonnegative reals).

Lemma 1 *Assume that $\Gamma : \mathbb{R}_+ \to \mathbb{R}_+$ is nondecreasing and*

$$\sum_{n=0}^{\infty} \Gamma^n(t) < \infty \qquad t \in \mathbb{R}_+. \tag{7}$$

Then the following three statements are valid.

(a) $\Gamma(0) = 0$.

(b) *Γ is continuous at* 0 *or, for each $t \in \mathbb{R}_+$, there is $m \in \mathbb{N}$ such that*

$$\Gamma^n(t) = 0 \qquad n > m.$$

(c) *If Γ is subadditive, i.e.,*

$$\Gamma(t+s) \le \Gamma(t) + \Gamma(s) \qquad t, s \in \mathbb{R}_+,$$

then

$$\Gamma\left(\sum_{n=k}^{\infty} \Gamma^n(t)\right) \le \sum_{n=k+1}^{\infty} \Gamma^n(t) \qquad k \in \mathbb{N}_0, t \in \mathbb{R}_+. \tag{8}$$

Proof The proof seems to be a routine by now, but for the convenience of readers we present it here.

Since Γ is nondecreasing, we have

$$\Gamma(0) \le \Gamma^n(0) \qquad n \in \mathbb{N}.$$

Consequently, it is easily seen that (a) is a consequence of (7).

Next, for the proof of (b), assume that there are $t \in \mathbb{R}_+$ and a sequence $(k_n)_{n\in\mathbb{N}} \in \mathbb{N}^{\mathbb{N}}$ such that

$$\lim_{n\to\infty} k_n = \infty$$

and

$$\Gamma^{k_n}(t) > 0.$$

Suppose that Γ is not continuous at 0, i.e., according to (a) there is $d > 0$ with

$$\Gamma(c) > d \qquad c > 0.$$

Then

$$\Gamma^{k_n+1}(t) = \Gamma\left(\Gamma^{k_n}(t)\right) \geq d \qquad n \in \mathbb{N},$$

which is a contradiction to (7). Thus we have proved (b).

Finally, we prove (c). So, fix $t \in \mathbb{R}_+$ and $k \in \mathbb{N}$. First, suppose that Γ is not continuous at 0. Then, according to (b), there is $m \in \mathbb{N}$ such that

$$\Gamma^n(t) = 0 \qquad n > m.$$

Next, by subadditivity of Γ, we get

$$\begin{aligned}\Gamma\left(\sum_{n=k}^{\infty}\Gamma^n(t)\right) &\leq \sum_{n=k}^{m}\Gamma^{n+1}(t) + \Gamma\left(\sum_{n=m+1}^{\infty}\Gamma^n(t)\right)\\ &= \sum_{n=k}^{m}\Gamma^{n+1}(t) + \Gamma(0),\end{aligned}$$

whence, by (a),

$$\Gamma\left(\sum_{n=k}^{\infty}\Gamma^n(t)\right) \leq \sum_{n=k}^{m}\Gamma^{n+1}(t) = \sum_{n=k+1}^{\infty}\Gamma^n(t).$$

It remains to consider the case when Γ is continuous at 0. Then, analogously as above, the subadditivity of Γ yields

$$\Gamma\left(\sum_{n=k}^{\infty}\Gamma^n(t)\right) \leq \sum_{n=k}^{m}\Gamma^{n+1}(t) + \Gamma\left(\sum_{n=m+1}^{\infty}\Gamma^n(t)\right) \tag{9}$$

for $m \in \mathbb{N}$, $m > k$. Clearly, (7) and (a) imply that

$$\lim_{m\to\infty}\Gamma\left(\sum_{n=m+1}^{\infty}\Gamma^n(t)\right) = 0.$$

Hence, letting $m \to \infty$ in (9), we obtain (8). □

We need yet the following simple observation ($\mathbb{R}$ stands for the set of real numbers).

Proposition 1 *Let S be a nonempty set, $\mathcal{T} : \mathbb{R}^S \to \mathbb{R}^S$ and $\gamma_n, \delta_n : S \to \mathbb{R}_+$ for $n \in \mathbb{N}_0$. Assume that*

$$-\delta_{n-1}(t) \le \mathcal{T}^n(\varphi)(t) - \mathcal{T}^{n-1}(\varphi)(t) \le \gamma_{n-1}(t) \qquad t \in S, n \in \mathbb{N}$$

and

$$\Delta(t) := \sum_{n=0}^{\infty} \delta_n(t) < \infty, \qquad \Gamma(t) := \sum_{n=0}^{\infty} \gamma_n(t) < \infty \qquad t \in S.$$

Then, for each $t \in S$, the limit

$$\Phi(t) := \lim_{n\to\infty} \mathcal{T}^n(\varphi)(t)$$

exists and

$$-\Delta(t) \le \Phi(t) - \varphi(t) \le \Gamma(t).$$

Proof Since, for every $k, n \in \mathbb{N}_0 := \mathbb{N} \cup \{0\}$, $k > 0$, $t \in S$, we have

$$\begin{aligned} -\sum_{i=n}^{\infty} \delta_i(t) &\le \sum_{i=1}^{k} \mathcal{T}^{n+i}(\varphi)(t) - \mathcal{T}^{n+i-1}(\varphi)(t) \\ &= \mathcal{T}^{n+k}(\varphi)(t) - \mathcal{T}^n(\varphi)(t) \le \sum_{i=n}^{\infty} \gamma_i(t), \end{aligned}$$

the limit $\Phi(t)$ exists and, with $n = 0$ and $k \to \infty$, we obtain

$$\Delta(t) \le \Phi(t) - \varphi(t) \le \Gamma(t).$$

□

3 Modified Stability on Square Symmetric Groupoids

In the case where f takes values in the set of real numbers $\mathbb{R}$, condition (6) can be written in the form

$$-\varphi(x, y) \le f(x + y) - f(x) - f(y) \le \varphi(x, y) \tag{10}$$

and there arises a natural question whether results analogous to those in Theorem 2 can be obtained also for functions $f : E_1 \to \mathbb{R}$, which satisfy the inequalities

$$-\nu(x, y) \le f(x + y) - f(x) - f(y) \le \mu(x, y), \tag{11}$$

with some suitable functions $\mu, \nu : E_1 \times E_1 \to \mathbb{R}_+$, not necessarily equal. In this section we investigate this issue. Moreover, we do it on a restricted domain and for functions with a domain being a subset of a square symmetric groupoid.

Let us recall that a groupoid $(G, \diamond)$ (i.e., a nonempty set G with a binary operation $\diamond : G \times G \to G$) is square symmetric provided the operation $\diamond$ is square symmetric, i.e.,

$$(x \diamond x) \diamond (y \diamond y) = (x \diamond y) \diamond (x \diamond y) \qquad x, y \in G.$$

Clearly every commutative semigroup is a square symmetric groupoid. Next, let X be a linear space over a field $\mathbb{K}$, $a, b \in \mathbb{K}$, $z \in X$ and define a binary operation $\star : X^2 \to X$ by:

$$x \star y := ax + by + z \qquad x, y \in X.$$

Then it is easy to check that $(X, \star)$ provides a simple example of a square symmetric groupoid. Clearly, $\star$ is commutative only if $a = b$. Moreover, it is easy to check that it is associative only for $a = b = 1$.

If in a nonempty set G we define a binary operation $\star : G^2 \to G$ by one of the following two conditions:

$$x \star y := x \qquad x, y \in G,$$

$$x \star y := y \qquad x, y \in G,$$

then it is square symmetric. Also, it is easily seen that $(\mathbb{R}, \circ)$, with

$$x \circ y := A(x - y) \qquad x, y \in \mathbb{R},$$

with some fixed mapping $A : \mathbb{R} \to \mathbb{R}$, is another very simple example of a square symmetric groupoid.

One more example of a square symmetric operations is described in [18, Theorem 12]. Namely, let $(G, \cdot)$ be a groupoid with a right (or left) unit element and $F : G^2 \to G$ be such that

$$F(z \cdot x, z \cdot y) = z \cdot F(x, y), \qquad F(x \cdot z, y \cdot z) = F(x, y) \cdot z \qquad x, y, z \in G.$$

Then the operation $\star : G^2 \to G$, given by: $x \star y := F(x, y)$ for $x, y \in G$, is square symmetric.

Observe yet that, if $(G, \circ)$ is a square symmetric groupoid, H is a nonempty set and $h : H \to G$ is a bijection, then the operation $\star : H^2 \to H$, given by:

$$x \star y := h^{-1}(h(x) \circ h(y)) \qquad x, y \in H,$$

is also square symmetric.

The square symmetric groupoids have been already considered in several papers investigating the stability of some functional equations (see, e.g., [18, 33, 34, 37, 38, 43, 44, 55, 57]). For a description of square symmetric operations we refer to [14].

Finally, let us mention that $(G, +, d)$ is a complete metric groupoid provided $(G, +)$ is a groupoid, (G, d) is a complete metric space and the operation $+ : G^2 \to G$ is continuous, in both variables simultaneously, with respect to the metric d.

In the sequel, given a groupoid $(G, \diamond)$, we define a mapping $\tau_\diamond : G \to G$ by:

$$\tau_\diamond(x) := x \diamond x.$$

Now we are in a position to prove the next theorem, which generalizes to some extent the main results in [18, 33, 34, 37, 38].

Theorem 3 *Let $(G, \star)$ be a square symmetric groupoid, $\circ : \mathbb{R}^2 \to \mathbb{R}$ be a continuous square symmetric operation (cf. [14]), $S \subset G$ be nonempty, $\mu, \nu : S^2 \to \mathbb{R}_+$, $\tau_\star(S) \subset S$, $\Gamma_n : \mathbb{R} \to \mathbb{R}_+$ be nondecreasing for $n \in \mathbb{N}$, and $h : S \to \mathbb{R}$ satisfy*

$$-\nu(x, y) \le h(x \star y) - h(x) \circ h(y) \le \mu(x, y) \qquad x, y \in S, x \star y \in S. \tag{12}$$

Suppose that one of the following two conditions is valid.

(i) $\tau_\circ$ is bijective and $\sigma := \tau_\circ^{-1}$, $\rho := \tau_\star$ satisfy

$$\sigma^n(x) - \sigma^n(y) \le \Gamma_n(x - y) \qquad x, y \in \mathbb{R}, n \in \mathbb{N}, \tag{13}$$

$$\widehat{\mu}(x) := \sum_{n=0}^{\infty} \Gamma_{n+1}(\mu(\rho^n(x), \rho^n(x))) < \infty \qquad x \in S,$$

$$\widehat{\nu}(x) := \sum_{n=0}^{\infty} \Gamma_{n+1}(\nu(\rho^n(x), \rho^n(x))) < \infty \qquad x \in S,$$

$$\liminf_{n\to\infty} \Gamma_n(\mu(\rho^n(x), \rho^n(y))) = 0 \qquad x, y \in S, x \star y \in S, \tag{14}$$

$$\liminf_{n\to\infty} \Gamma_n(\nu(\rho^n(x), \rho^n(y))) = 0 \qquad x, y \in S, x \star y \in S. \tag{15}$$

(ii) $\tau_\star|_S$ is a bijection onto S, (13), (14), *and* (15) *hold with $\rho := (\tau_\star|_S)^{-1}$ and $\sigma := \tau_\circ$, and*

$$\widehat{\mu}(x) := \nu(\rho(x), \rho(x)) + \sum_{n=1}^{\infty} \Gamma_n(\nu(\rho^{n+1}(x), \rho^{n+1}(x))) < \infty \qquad x \in S,$$

$$\widehat{\nu}(x) := \mu(\rho(x), \rho(x)) + \sum_{n=1}^{\infty} \Gamma_n(\mu(\rho^{n+1}(x), \rho^{n+1}(x))) < \infty \qquad x \in S.$$

Then the limit

$$F(x) := \lim_{n\to\infty} \sigma^n(h(\rho^n(x)))$$

exists for each $x \in S$,

$$F(x \star y) = F(x) \circ F(y) \qquad x, y \in S, x \star y \in S, \tag{16}$$

and

$$-\widehat{\nu}(x) \leq F(x) - h(x) \leq \widehat{\mu}(x) \qquad x \in S. \tag{17}$$

Moreover, if Γ_1 *is subadditive and*

$$\liminf_{n\to\infty} \Gamma_1^n(\widehat{\mu}(\rho^n(x))) = 0 \qquad x \in S, \tag{18}$$

$$\liminf_{n\to\infty} \Gamma_1^n(\widehat{\nu}(\rho^n(x))) = 0 \qquad x \in S, \tag{19}$$

then F *is the unique function mapping* S *into* $\mathbb{R}$ *such that* (16) *and* (17) *are valid.*

Proof Write $\Gamma_0(t) = t$, for $t \in \mathbb{R}$ and define $\Delta_n : \mathbb{R} \to \mathbb{R}$ by

$$\Delta_n(x) := -\Gamma_n(-x) \qquad x \in \mathbb{R}, n \in \mathbb{N}_0.$$

Then Δ_n is nondecreasing for each $n \in \mathbb{N}_0$ and, by (13),

$$\Delta_n(x-y) \leq \sigma^n(x) - \sigma^n(y) \leq \Gamma_n(x-y) \qquad x, y \in \mathbb{R}, n \in \mathbb{N}_0. \tag{20}$$

Let

$$\mathcal{T}(\alpha)(t) := \sigma(\alpha(\rho(t)))$$

for $\alpha \in \mathbb{R}^S$, $t \in S$. Next, (12) with $x = y$ yields

$$-\nu(x,x) \leq h(\tau_\star(x)) - \tau_\circ(h(x)) \leq \mu(x,x) \qquad x \in S. \tag{21}$$

Hence, in the case when (i) holds, by (20), for each $n \in \mathbb{N}_0$ and $z \in S$ we get

$$\begin{aligned}
\mathcal{T}^{n+1}(h)(z) - \mathcal{T}^n(h)(z) &= \sigma^{n+1} \circ h \circ \rho^{n+1}(z) - \sigma^n \circ h \circ \rho^n(z)\\
&= \sigma^{n+1} \circ h \circ \rho^{n+1}(z) - \sigma^{n+1} \circ \tau_\circ \circ h \circ \rho^n(z)\\
&\leq \Gamma_{n+1}(h \circ \rho(\rho^n(z)) - \tau_\circ(h(\rho^n(z))))\\
&\leq \Gamma_{n+1}(\mu(\rho^n(z), \rho^n(z)))
\end{aligned}$$

and

$$\begin{aligned}
\mathcal{T}^{n+1}(h)(z) - \mathcal{T}^n(h)(z) &= \sigma^{n+1} \circ h \circ \rho^{n+1}(z) - \sigma^n \circ h \circ \rho^n(z)\\
&= \sigma^{n+1} \circ h \circ \rho^{n+1}(z) - \sigma^{n+1} \circ \tau_\circ \circ h \circ \rho^n(z)\\
&\geq \Delta_{n+1}(h \circ \rho(\rho^n(z)) - \tau_\circ(h(\rho^n(z))))\\
&\geq \Delta_{n+1}(-\nu(\rho^n(z), \rho^n(z)))\\
&= -\Gamma_{n+1}(\nu(\rho^n(z), \rho^n(z))).
\end{aligned}$$

If (ii) holds, then analogously we obtain

$$\mathcal{T}^{n+1}(h)(z) - \mathcal{T}^n(h)(z) = \sigma^{n+1} \circ h \circ \rho^{n+1}(z) - \sigma^n \circ h \circ \rho^n(z)$$

$$\begin{aligned}
&\leq \Gamma_n(\sigma \circ h(\rho^{n+1}(z)) - h(\rho^n(z)))\\
&= \Gamma_n(\tau_\circ \circ h(\rho^{n+1}(z)) - h \circ \tau_\star(\rho^{n+1}(z)))\\
&\leq \Gamma_n(\nu(\rho^{n+1}(z), \rho^{n+1}(z))),
\end{aligned}$$

$$\begin{aligned}
\mathcal{T}^{n+1}(h)(z) - \mathcal{T}^n(h)(z) &\geq \Delta_n(\sigma \circ h(\rho^{n+1}(z)) - h(\rho^n(z)))\\
&= \Delta_n(\tau_\circ \circ h(\rho^{n+1}(z)) - h \circ \tau_\star(\rho^{n+1}(z)))\\
&\geq \Delta_n(-\mu(\rho^{n+1}(z), \rho^{n+1}(z)))\\
&= -\Gamma_n(\mu(\rho^{n+1}(z), \rho^{n+1}(z)))
\end{aligned}$$

for every $n \in \mathbb{N}_0$ and $z \in S$. So, in view of Proposition 1 with

$$\gamma_n(z) := \begin{cases} \Gamma_{n+1}(\mu(\rho^n(z), \rho^n(z))), & \text{if } (i) \text{ holds;}\\ \Gamma_n(\nu(\rho^{n+1}(z), \rho^{n+1}(z))), & \text{if } (ii) \text{ holds,}\end{cases}$$

and

$$\delta_n(z) := \begin{cases} -\Gamma_{n+1}(\nu(\rho^n(z), \rho^n(z))), & \text{if } (i) \text{ holds;}\\ -\Gamma_n(\mu(\rho^{n+1}(z), \rho^{n+1}(z))), & \text{if } (ii) \text{ holds,}\end{cases}$$

for $z \in S$ and $n \in \mathbb{N}_0$, the limit

$$F(x) := \lim_{n\to\infty} \mathcal{T}^n(h)(x)$$

exists for every $x \in S$ and (17) holds.

Now we show that F satisfies (16). So take $x, y \in S$ with $x \star y \in S$. Then it is easy to check that, for every $n \in \mathbb{N}$,

$$\rho^n(x), \rho^n(y) \in S,$$

$$\rho^n(x \star y) = \rho^n(x) \star \rho^n(y),$$

$$\sigma^n(h(\rho^n(x))) \circ \sigma^n(h(\rho^n(y))) = \sigma^n(h(\rho^n(x)) \circ h(\rho^n(y))),$$

and consequently

$$\begin{aligned}
\sigma^n(h(\rho^n(x \star y))) &- \sigma^n(h(\rho^n(x))) \circ \sigma^n(h(\rho^n(y)))\\
&= \sigma^n(h(\rho^n(x) \star \rho^n(y))) - \sigma^n(h(\rho^n(x)) \circ h(\rho^n(y)))\\
&\leq \Gamma_n(h(\rho^n(x) \star \rho^n(y)) - h(\rho^n(x)) \circ h(\rho^n(y)))\\
&\leq \Gamma_n(\mu(\rho^n(x), \rho^n(y))),
\end{aligned}$$

$$\sigma^n(h(\rho^n(x \star y))) - \sigma^n(h(\rho^n(x))) \circ \sigma^n(h(\rho^n(y)))$$

$$\geq \Delta_n(h(\rho^n(x) \star \rho^n(y)) - h(\rho^n(x)) \circ h(\rho^n(y)))$$
$$\geq \Delta_n(-\nu(\rho^n(x), \rho^n(y))) = -\Gamma_n(\nu(\rho^n(x), \rho^n(y))).$$

Letting $n \to \infty$, in view of (14), (15) and the continuity of $\circ$, we obtain

$$F(x \star y) = F(x) \circ F(y).$$

Finally we prove the statement concerning the uniqueness of F. So, assume that Γ_1 is subadditive and (18), (19) are valid. Let $A : S \to \mathbb{R}$ satisfy

$$A(x \star y) = A(x) \circ A(y) \qquad x, y \in S, x \star y \in S,$$

and

$$-\hat{\nu}(x) \leq A(x) - h(x) \leq \hat{\mu}(x) \qquad x \in S.$$

According to Lemma 1, for every $x \in S$ and $n \in \mathbb{N}$,

$$\begin{aligned}\sigma^n(F(\rho^n(x))) - \sigma^n(A(\rho^n(x))) &\leq \Gamma_1^n(F(\rho^n(x)) - A(\rho^n(x)))\\ &\leq \Gamma_1^n(F(\rho^n(x)) - h(\rho^n(x)))\\ &\quad + \Gamma_1^n(h(\rho^n(x)) - A(\rho^n(x)))\\ &\leq \Gamma_1^n(\widehat{\mu}(\rho^n(x))) + \Gamma_1^n(\widehat{\nu}(\rho^n(x)))\end{aligned}$$

and

$$\begin{aligned}\sigma^n(F(\rho^n(x))) - \sigma^n(A(\rho^n(x))) &\geq \Delta_1^n(F(\rho^n(x)) - A(\rho^n(x)))\\ &\geq \Delta_1^n(F(\rho^n(x)) - h(\rho^n(x))) + \Delta_1^n(h(\rho^n(x)) - A(\rho^n(x)))\\ &\geq \Delta_1^n(-\widehat{\nu}(\rho^n(x))) + \Delta_1^n(-\widehat{\mu}(\rho^n(x)))\\ &= -\Gamma_1^n(\widehat{\nu}(\rho^n(x))) - \Gamma_1^n(\widehat{\mu}(\rho^n(x))).\end{aligned}$$

Consequently, with $n \to \infty$, in view of (18) and (19) we obtain that

$$F(x) = A(x) \qquad x \in S. \qquad \square$$

Below we give a very simple example of possible applications of Theorem 3 with functions μ and ν having some natural forms, mentioned already before.

Corollary 1 *Let X be a normed space, $c_1, c_2, p, q, r \in [0, \infty)$,*

$$(p-1)(q+r-1) > 0,$$

$S \subset X$ be nonempty, $2S = S$, and $h : S \to \mathbb{R}$ satisfy the inequality

$$-c_1\|x\|^q\|y\|^r \leq h(x+y) - h(x) - h(y) \leq c_2(\|x\|^p + \|y\|^p) \tag{22}$$
$$x, y \in S, x + y \in S.$$

Then there exists a unique function $F : S \to \mathbb{R}$ *such that*

$$F(x + y) = F(x) + F(y) \qquad x, y \in S, x + y \in S, \tag{23}$$

and

$$-\frac{c_1\|x\|^{q+r}}{|1 - 2^{q+r-1}|} \le F(x) - h(x) \le \frac{c_2\|x\|^p}{|1 - 2^{p-1}|} \qquad x \in S. \tag{24}$$

Proof If $p < 1$ and $q + r < 1$, then it is enough to use Theorem 3 (i). If $p > 1$ and $q + r > 1$, then we use Theorem 3 (ii). □

4 Some Complementary Results

In this section we show that a fixed point approach, analogous as in the proof of Theorem 3, can be applied to obtain stability results also for functions mapping a square symmetric groupoid $(G, \star)$ into a metric space (Y, d) and satisfying the "usual stability inequality"

$$d(h(x \star y), h(x) \circ h(y)) \le \mu(x, y) \qquad x, y \in S, x \star y \in S,$$

with a nonempty $S \subset G$, some square symmetric operation $\circ : Y^2 \to Y$, and a suitable function $\mu : S^2 \to [0, \infty)$. In the case where $S = G$ and

$$d(x \circ y, z \circ y) \le d(x, z) \qquad x, y, z \in Y, \tag{25}$$

such results can be derived from [18, Theorems 4 and 6]. However, the reasonings and notations in [18] are somewhat involved, while the proof that we present below is direct, quite elementary, on a restricted domain, and without condition (25). We should add here that actually our proof applies some ideas from [18] (and also from [33, 34, 37, 38, 44, 55, 57]).

We start with the following modification of Proposition 1.

Proposition 2 *Let* (X, d) *be a complete metric space,* S *be a nonempty set,* $\varphi \in X^S$, $\mathcal{T} : X^S \to X^S$, $\gamma_n : S \to \mathbb{R}_+$ *for* $n \in \mathbb{N}_0$,

$$d(\mathcal{T}^n(\varphi)(t), \mathcal{T}^{n-1}(\varphi)(t)) \le \gamma_{n-1}(t) \qquad t \in S, n \in \mathbb{N},$$

and

$$h(t) := \sum_{n=0}^{\infty} \gamma_n(t) < \infty \qquad t \in S. \tag{26}$$

Then the limit

$$\Phi(t) := \lim_{n \to \infty} \mathcal{T}^n(\varphi)(t)$$

exists for every $t \in S$ *and*

$$d(\varphi(t), \Phi(t)) \le h(t) \qquad t \in S. \tag{27}$$

Proof For every $k, n \in \mathbb{N} \cup \{0\}$, $k > 0$, $t \in S$, we have

$$d(\mathcal{T}^{n+k}(\varphi)(t), \mathcal{T}^n(\varphi)(t)) \le \sum_{i=1}^{k} d(\mathcal{T}^{n+i}(\varphi)(t), \mathcal{T}^{n+i-1}(\varphi)(t)) \le \sum_{i=n}^{\infty} \gamma_i(t). \tag{28}$$

So, for each $t \in S$, the limit

$$\Phi(t) := \lim_{n\to\infty} \mathcal{T}^n(\varphi)(t)$$

exists and, in view of (28) (with $n = 0$ and $k \to \infty$), we get (27). □

Now we are in a position to prove the main result in this section.

Theorem 4 *Let $(G, \star)$ and $(H, \circ)$ be square symmetric groupoids, d be a complete metric in H, the operation $\circ$ be continuous, $\mu : S^2 \to \mathbb{R}$, $S \subset G$ be nonempty,*

$$\tau_\star(S) \subset S,$$

$\Gamma_n : \mathbb{R}_+ \to \mathbb{R}_+$ be non-decreasing for $n \in \mathbb{N}$, and $h : S \to H$ satisfy the inequality

$$d(h(x \star y), h(x) \circ h(y)) \le \mu(x, y) \qquad x, y \in S, x \star y \in S. \tag{29}$$

Suppose that one of the following two conditions is valid.

(*i*) *$\tau_\circ$ is bijective, and $\sigma := \tau_\circ^{-1}$, $\rho := \tau_\star$ satisfy*

$$d(\sigma^n(x), \sigma^n(y)) \le \Gamma_n(d(x, y)) \qquad x, y \in H, n \in \mathbb{N}, \tag{30}$$

$$\widehat{\mu}(x) := \sum_{n=0}^{\infty} \Gamma_{n+1}(\mu(\rho^n(x), \rho^n(x))) < \infty \qquad x \in S,$$

$$\liminf_{n\to\infty} \Gamma_n(\mu(\rho^n(x), \rho^n(y))) < \infty \qquad x, y \in S, x \star y \in S. \tag{31}$$

(*ii*) *$\tau_\star|_S$ is a bijection onto S, (30) and (31) hold with $\rho := (\tau_\star|_S)^{-1}$ and $\sigma := \tau_\circ$, and*

$$\widehat{\mu}(x) := \mu(\rho(x), \rho(x)) + \sum_{n=1}^{\infty} \Gamma_n(\mu(\rho^{n+1}(x), \rho^{n+1}(x))) < \infty \qquad x \in S\,.$$

Then the limit

$$F(x) := \lim_{n\to\infty} \sigma^n(h(\rho^n(x)))$$

exists for each $x \in S$,

$$F(x \star y) = F(x) \circ F(y) \qquad x, y \in S, x \star y \in S, \tag{32}$$

and

$$d(F(x), h(x)) \le \widehat{\mu}(x) \qquad x \in S. \tag{33}$$

Moreover, if Γ_1 is subadditive and

$$\liminf_{n\to\infty} \Gamma_1^n(\widehat{\mu}(\rho^n(x))) = 0 \qquad x \in S, \tag{34}$$

then F is the unique function mapping S into H such that (32) and (33) are valid.

Proof Note that $\rho(z) \in S$ for $z \in S$. Write

$$\mathcal{T}(\alpha)(t) := \sigma(\alpha(\rho(t)))$$

for $\alpha \in H^S$, $t \in S$. From (29), with $x = y = z$, we deduce that

$$d(h \circ \tau_\star(z), \tau_\circ \circ h(z)) \le \mu(z, z) \qquad z \in S.$$

Consequently, (30) implies that, for each $n \in \mathbb{N}_0$ and $z \in S$,

$$\begin{aligned} d(\mathcal{T}^{n+1}(h)(z), \mathcal{T}^n(h)(z)) &= d(\sigma^{n+1} \circ h \circ \rho^{n+1}(z), \sigma^n \circ h \circ \rho^n(z)) \\ &= d(\sigma^{n+1} \circ h \circ \rho^{n+1}(z), \sigma^{n+1} \circ \tau_\circ \circ h \circ \rho^n(z)) \\ &\le \Gamma_{n+1}(d(h \circ \rho(\rho^n(z)), \tau_\circ(h(\rho^n(z))))) \\ &\le \Gamma_{n+1}(\mu(\rho^n(z), \rho^n(z))) \end{aligned}$$

if (i) holds and, analogously, we get

$$\begin{aligned} d(\mathcal{T}^{n+1}(h)(z), \mathcal{T}^n(h)(z)) &= d(\sigma^{n+1} \circ h \circ \rho^{n+1}(z), \sigma^n \circ h \circ \rho^n(z)) \\ &\le \Gamma_n(d(\sigma \circ h \circ \rho(\rho^n(z)), h(\rho^n(z)))) \\ &= \Gamma_n(d(\tau_\circ \circ h(\rho^{n+1}(z)), h \circ \tau_\star(\rho^{n+1}(z)))) \\ &\le \Gamma_n(\mu(\rho^{n+1}(z), \rho^{n+1}(z))) \end{aligned}$$

if (ii) holds, where $\Gamma_0(t) = t$, for $t \in \mathbb{R}_+$. So, in view of Proposition 2 with

$$\gamma_n(z) := \begin{cases} \Gamma_{n+1}(\mu(\rho^n(z), \rho^n(z))), & \text{if } (i) \text{ holds;} \\ \Gamma_n(\mu(\rho^{n+1}(z), \rho^{n+1}(z))), & \text{if } (ii) \text{ holds,} \end{cases}$$

for $z \in S$ and $n \in \mathbb{N}_0$, the limit $F(x)$ exists for every $x \in S$ and (33) holds.

Now we show that F satisfies (32). So take $x, y \in S$ with $x \star y \in S$. Then it is easy to check that, for every $n \in \mathbb{N}$,

$$\rho^n(x), \rho^n(y) \in S,$$
$$\rho^n(x \star y) = \rho^n(x) \star \rho^n(y),$$
$$\sigma^n(h(\rho^n(x))) \circ \sigma^n(h(\rho^n(y))) = \sigma^n(h(\rho^n(x)) \circ h(\rho^n(y))),$$

and consequently

$$\begin{aligned} d(\sigma^n(h(\rho^n(x \star y))), \sigma^n(h(\rho^n(x))) \circ \sigma^n(h(\rho^n(y)))) \\ &= d(\sigma^n(h(\rho^n(x) \star \rho^n(y))), \sigma^n(h(\rho^n(x)) \circ h(\rho^n(y)))) \\ &\le \Gamma_n(d(h(\rho^n(x) \star \rho^n(y)), h(\rho^n(x)) \circ h(\rho^n(y)))) \\ &\le \Gamma_n(\mu(\rho^n(x), \rho^n(y))). \end{aligned}$$

Letting $n \to \infty$, in view of (31) and the continuity of $\circ$, we obtain

$$F(x \star y) = F(x) \circ F(y).$$

Finally we prove the statement concerning the uniqueness of F. So, assume that $A : S \to H$,

$$A(x \star y) = A(x) \circ A(y) \qquad x, y \in S, x \star y \in S,$$

and

$$d(A(x), h(x)) \le \widehat{\mu}(x) \qquad x \in S.$$

Then

$$\begin{aligned} d(F(x), A(x)) &= d(\sigma^n(F(\rho^n(x))), \sigma^n(A(\rho^n(x)))) \\ &\le \Gamma_1^n(d(F(\rho^n(x)), A(\rho^n(x)))) \\ &\le \Gamma_1^n(d(F(\rho^n(x)), h(\rho^n(x)))) \\ &\quad + \Gamma_1^n(d(h(\rho^n(x)), A(\rho^n(x)))) \\ &\le 2\Gamma_1^n(\widehat{\mu}(\rho^n(x))) \qquad x \in S, n \in \mathbb{N}. \end{aligned}$$

Now, from (34) we derive that

$$F(x) = A(x) \qquad x \in S. \qquad \square$$

Remark 1 Note that in the case when Γ_1 is subadditive, (i) holds and

$$\Gamma_n(t) \le \Gamma_1^n(t) \qquad t \in S, n \in \mathbb{N}, \tag{35}$$

in view of Lemma 1 we have

$$\begin{aligned} \Gamma_1^n(\widehat{\mu}(\rho^n(x))) &\le \Gamma_1^n\left(\sum_{i=0}^{\infty} \Gamma_1^{i+1}(\mu(\rho^{i+n}(x), \rho^{i+n}(x)))\right) \\ &\le \sum_{i=0}^{\infty} \Gamma_1^{i+n+1}(\mu(\rho^{i+n}(x), \rho^{i+n}(x))) \\ &= \sum_{i=n}^{\infty} \Gamma_1^{i+1}(\mu(\rho^{i}(x), \rho^{i}(x))) \qquad x \in S, n \in \mathbb{N}. \end{aligned}$$

Consequently, in such situation, the condition

$$\sum_{i=0}^{\infty} \Gamma_1^{i+1}(\mu(\rho^i(x), \rho^i(x))) < \infty \qquad x \in S$$

implies (34).

Analogously, in the case of (ii), (34) follows from the conditions (35) and

$$\sum_{i=0}^{\infty} \Gamma_1^{i}(\mu(\rho^{i+1}(x), \rho^{i+1}(x))) < \infty \qquad x \in S.$$

Remark 2 Let $(H, \circ)$ be a square symmetric groupoid and $\sigma := \tau_\circ$. Define functions $\widehat{\Gamma}_n : \mathbb{R} \to \mathbb{R}$, for $n \in \mathbb{N}$, by (cf. [18]):

$$\widehat{\Gamma}_n(t) := \sup\{d(\sigma^n(x), \sigma^n(z)) : x, z \in H, d(x, z) \le t\} \qquad t \in [0, \infty), \tag{36}$$
$$\widehat{\Gamma}_n(t) := 0 \qquad t \in (-\infty, 0).$$

It is easily seen that they are nondecreasing, fulfill (30) and, for every family of nondecreasing functions $\Gamma_n : \mathbb{R} \to \mathbb{R}$, with $n \in \mathbb{N}$, satisfying (30), we have

$$\Gamma_n(t) \ge \widehat{\Gamma}_n(t) \qquad t \in \mathbb{R}.$$

Moreover,

$$d(\sigma^n(x), \sigma^n(y)) \le \widehat{\Gamma}_1^{\,n}(d(x, y)) \qquad x, y \in H.$$

There arises a natural question whether it is possible at all that

$$\widehat{\Gamma}_n \ne \widehat{\Gamma}_1^{\,n}$$

for some square symmetric groupoids H and some n. The answer is *yes*. This is the case for instance when H is a multiplicative subgroup of the field of complex numbers $\mathbb{C}$, of the form

$$H = \{1, -1, i, -i\}$$

(where $i^2 = -1$), with $\circ$ being the usual multiplication in $\mathbb{C}$, and

$$d(x, y) := |x - y| \qquad x, y \in H.$$

In fact, then

$$\begin{aligned}\widehat{\Gamma}_1(2) &= \max\{|x^2 - z^2| : x, z \in H, |x - z| \le 2\}\\ &= \max\{|x^2 - z^2| : x, z \in H\} = 2\end{aligned}$$

and consequently, by induction, we get

$$\widehat{\Gamma}_1^{\,n+1}(2) = \widehat{\Gamma}_1^{\,n}(2) = \widehat{\Gamma}_1(2) = 2 \qquad n \in \mathbb{N},$$

but clearly

$$\widehat{\Gamma}_n(2) = \max \{|x^{2^n} - z^{2^n}| : x, z \in H\} = 0 \qquad n \in \mathbb{N}, n > 1,$$

because $\sigma^n(x) = x^{2^n} = 1$ for every $x \in H, n \in \mathbb{N}, n > 1$.

Remark 3 There arises a natural question concerning the optimality of estimations (17) and (33) (cf., e.g., [18]). Unfortunately, this is not the case, which shows Theorem 2 with $p < 0$.

Remark 4 Let $(G, +)$, $(H, +)$ be commutative semigroups, d be a complete metric in H, $g, h : G \to G$ and $A, B : H \to H$ be homomorphisms with

$$g \circ h = h \circ g, \qquad A \circ B = B \circ A, \tag{37}$$

and $\zeta \in G$, $z_0 \in H$. Write

$$x \star y := g(x) + h(y) + \zeta \qquad x, y \in G$$

and

$$u \circ v := A(u) + B(v) + z_0 \qquad u, v \in H.$$

Then it is easy to check that $(G, \star)$ and $(H, \circ)$ are square symmetric groupoids. Therefore from Theorem 4 one can easily derive a stability result for the functional equation

$$f(g(x) + h(y) + \zeta) = A(f(x)) + B(f(y)) + z_0$$

in the class of functions f mapping a nonempty set $S \subset G$ into H. Note that condition (37) is valid for instance when $g = h^n$ and $A = B^m$ with some $m, n \in \mathbb{N}$.

Below we present a simplified case of such result, concerning the general linear functional equation of the form

$$F(\alpha x + \beta y + \gamma) = AF(x) + BF(y) + C \tag{38}$$

for functions F mapping a subset $S \neq \emptyset$ of a linear space X over a field $\mathbb{K}$ into Banach space Y over a field $\mathbb{F} \in \{\mathbb{R}, \mathbb{C}\}$, with some fixed $\alpha, \beta \in \mathbb{K}$, $A, B \in \mathbb{F}$, $\gamma \in X$, and $C \in Y$. Let us mention here that, motivated by a problem formulated by Th.M. Rassias and J. Tabor in [54, pp. 67–68], several authors (see [2, 3, 9, 19, 30, 42]) studied stability of various particular cases of (38).

Corollary 2 *Let $E := A + B \neq 1$, $A \neq -B$, $S \subset X$ be nonempty, $s : X \to X$ be given by*

$$s(x) := (\alpha + \beta)x + \gamma \qquad x \in X,$$

$\varphi : S^2 \to \mathbb{R}$, and $f : S \to Y$ satisfy

$$\|f(\alpha x + \beta y + \gamma) - Af(x) - Bf(y) - C\| \leq \varphi(x, y) \tag{39}$$

for every $x, y \in S$ *with* $\alpha x + \beta y + \gamma \in S$. *Suppose that there is* $\varepsilon \in \{-1, 1\}$ *such that* $s^\varepsilon(S) \subset S$ *and, for every* $x, y \in S$,

$$H(x) := \sum_{i=0}^{\infty} |E|^{-i\varepsilon} \varphi(s^{i\varepsilon}(x), s^{i\varepsilon}(x)) < \infty,$$

$$\liminf_{n\to\infty} |E^{-n\varepsilon} \varphi(s^{n\varepsilon}(x), s^{n\varepsilon}(y))| = 0. \tag{40}$$

Then there exists a unique $F : S \to Y$ *such that (38) holds for every* $x, y \in S$ *with* $\alpha x + \beta y + \gamma \in S$ *and*

$$\|F(x) - f(x)\| \leq H_0(x) \qquad x \in S,$$

where

$$H_0(x) := \begin{cases} \varphi(s^{-1}(x), s^{-1}(x)), & \text{if } E = 0\,; \\ |E|^{-1} H(x), & \text{if } E \neq 0 \text{ and } \varepsilon = 1; \\ H(s^{-1}(x)), & \text{if } E \neq 0 \text{ and } \varepsilon = -1. \end{cases}$$

Proof Note that condition (i) of Theorem 4 holds when $\varepsilon = 1$ and condition (ii) of Theorem 4 is fulfilled when $\varepsilon = -1$. Consequently it is enough to apply Theorem 4. □

Remark 5 Corollary 2 contains actually the main result in [9]. Note that if

$$A + B = 1 \neq \alpha + \beta,$$

then with

$$x = y = \frac{\gamma}{1 - \alpha - \beta}$$

in (38) we get $C = 0$, which means that for $C \neq 0$ Eq. (38) has no solutions $F : X \to Y$. On the other hand (39) holds for every $x, y \in X$ whenever f is constant and

$$\|C\| \leq \inf_{x,y\in X} \varphi(x, y).$$

This means that Corollary 2 is not true when $A + B = 1 \neq \alpha + \beta$ and $S = X$.

Corollary 2 generalizes several already classical results on stability of (1) described in the Introduction. In fact, if we take $\varepsilon = -1$, $V \subset X$ and

$$\varphi(x, y) := L_1\|x\|^p + L_2\|y\|^q + L_3\|x\|^r\|y\|^s \qquad x, y \in V$$

with some $L_1, L_2, L_3 \in \mathbb{R}_+$, $p, q \in (1, \infty)$, and $r, s \in \mathbb{R}$ with $r + s > 1$, then H_0 has the form

$$H_0(x) = \frac{L_1\|x\|^p}{2^p - 2} + \frac{L_2\|x\|^q}{2^q - 2} + \frac{L_3\|x\|^{r+s}}{2^{r+s} - 2} \qquad x \in V.$$

On the other hand, if $\varepsilon = 1$, $V \subset X \setminus \{0\}$ and

$$\varphi(x, y) := \delta + L_1\|x\|^p + L_2\|\|^q + L_3\|x\|^r\|y\|^s \qquad x, y \in V$$

with some $\delta, L_1, L_2, L_3 \in \mathbb{R}_+$, $q, r \in (-\infty, 1)$, and $r, s \in \mathbb{R}$ with $r + s < 1$, then

$$H_0(x) = \delta + \frac{L_1\|x\|^p}{2 - 2^p} + \frac{L_2\|x\|^q}{2 - 2^q} + \frac{L_3\|x\|^{r+s}}{2 - 2^{r+s}} \qquad x \in V.$$

References

1. Aoki, T.: On the stability of the linear transformation in Banach spaces. J. Math. Soc. Jpn. **2**, 64–66 (1950)
2. Badea, C.: On the Hyers-Ulam stability of mappings: The direct method. In: Rassias, Th.M., Tabor, J. (eds.) Stability of Mappings of Hyers-Ulam Type, pp. 3–7. Hadronic, Palm Harbor (1994)
3. Badea, C.: The general linear equation is stable. Nonlinear Funct. Anal. Appl. **10**, 155–164 (2005)
4. Bourgin, D.G.: Approximately isometric and multiplicative transformations on continuous function rings. Duke Math. J. **16**, 385–397 (1949)
5. Bourgin, D.G.: Classes of transformations and bordering transformations. Bull. Am. Math. Soc. **57**, 223–237 (1951)
6. Brillouët-Belluot, N., Brzdęk, J., Ciepliński, K.: On some recent developments in Ulam's type stability. Abstr. Appl. Anal. **2012**, Article ID 716936, 41 p. (2012)
7. Brzdęk, J.: A note on stability of additive mappings. In: Rassias, Th.M. , Tabor, J. (eds.)Stability of Mappings of Hyers-Ulam Type, pp. 19–22. Hadronic, Palm Harbor (1994)
8. Brzdęk, J.: Hyperstability of the Cauchy equation on restricted domains. Acta Math. Hung. **141**, 58–67 (2013)
9. Brzdęk, J., Pietrzyk, A.: A note on stability of the general linear equation. Aequ. Math. **75**, 267–270 (2008)
10. Ciepliński, K.: Applications of fixed point theorems to the Hyers-Ulam stability of functional equations—a survey. Ann. Funct. Anal. **3**, 151–164 (2012)
11. Czerwik, S.: Functional Equations and Inequalities in Several Variables. World Scientific, London (2002)
12. Faĭziev, V.A., Rassias, Th.M., Sahoo, P.K.: The space of (ψ, γ)-additive mappings on semigroups. Trans. Am. Math. Soc. **354**, 4455–4472 (2002)
13. Forti, G.L.: Hyers-Ulam stability of functional equations in several variables. Aequ. Math. **50**, 143–190 (1995)
14. Forti, G.L.: Continuously increasing weakly bisymmetric groupoids and quasi-groups in $\mathbb{R}$. Math. Pannonica **8**, 49–71 (1997)
15. Gajda, Z.: On stability of additive mappings. Int. J. Math. Math. Sci. **14**, 431–434 (1991)
16. Găvruţa, P.: A generalization of the Hyers-Ulam-Rassias stability of approximately additive mappings. J. Math. Anal. Appl. **184**, 431–436 (1994)
17. Ger, R.: On functional inequalities stemming from stability equations. In: Walter, W. (ed.) General Inequalities 6. International Series of Numerical Mathematics, vol. 103, pp. 227–240. Birkhäuser, Basel (1992)
18. Gilányi, A., Kaiser, Z., Páles, Z.: Estimates to the stability of functional equations. Aequ. Math. **73**, 125–143 (2007)
19. Grabiec, A.: The generalized Hyers-Ulam stability of a class of functional equations. Publ. Math. Debr. **48**, 217–235 (1996)
20. Gruber, P.M.: Stability of isometries. Trans. Am. Math. Soc. **245**, 263–277 (1978)
21. Hyers, D.H.: On the stability of the linear functional equation. Proc. Natl. Acad. Sci. U. S. A. **27**, 222–224 (1941)
22. Hyers, D.H.: Transformations with bounded mth differences. Pac. J. Math. **11**, 591–602 (1961)
23. Hyers, D.H., Isac, G., Rassias, Th.M.: Stability of Functional Equations in Several Variables. Birkhäuser, Boston (1998)
24. Hyers, D.H., Isac, G., Rassias, Th.M.: On the asymptoticity aspect of Hyers-Ulam stability of mappings. Proc. Am. Math. Soc. **126**, 425–430 (1998)
25. Hyers, D.H., Rassias, Th.M.: Approximate homomorphisms. Aequ. Math. **44**, 125–153 (1992)
26. Hyers, D.H., Ulam, S.M.: On approximate isometries. Bull. Am. Math. Soc. **51**, 288–292 (1945)

27. Hyers, D.H., Ulam, S.M.: Approximate isometries of the space of continuous functions. Ann. Math. **48**(2), 285–289 (1947)
28. Hyers, D.H., Ulam, S.M.: Approximately convex functions. Proc. Am. Math. Soc. **3**, 821–828 (1952)
29. Isac, G., Rassias, Th.M.: On the Hyers-Ulam stability of ψ-additive mappings. J. Approx. Theory **72**, 131–137 (1993)
30. Jung, S.M.: On modified Hyers-Ulam-Rassias stability of a generalized Cauchy functional equation. Nonlinear Stud. **5**, 59–67 (1998)
31. Jung, S.M.: Hyers-Ulam-Rassias Stability of Functional Equations in Mathematical Analysis. Hadronic, Palm Harbor (2001)
32. Jung, S.M.: Hyers-Ulam-Rassias Stability of Functional Equations in Nonlinear Analysis. Springer Optimization and Its Applications, vol. 48. Springer, New York (2011)
33. Kim, G.H.: On the stability of functional equations with a square-symmetric operation. Math. Inequal. Appl. **4**, 257–266 (2001)
34. Kim, G.H.: Addendum to 'On the stability of functional equations on square-symmetric groupoid'. Nonlinear Anal. **62**, 365–381 (2005)
35. Moszner, Z.: Sur la définitions différentes de la stabilité des équations fonctionnelles. Aequ. Math. **68**, 260–274 (2004)
36. Moszner, Z.: On the stability of functional equations. Aequ. Math. **77**, 33–88 (2009)
37. Páles, Z.: Hyers-Ulam stability of the Cauchy functional equation on square-symmetric groupoids. Publ. Math. Debr. **58**, 651–666 (2001)
38. Páles, Z., Volkmann, P., Luce, R.D.: Hyers-Ulam stability of functional equations with a square-symmetric operation. Proc. Natl. Acad. U. S. A. **95**, 12772–12775 (1998)
39. Piszczek, M.: Remark on hyperstability of the general linear equation. Aequ. Math. (2013). doi:10.1007/s00010-013-0214-x
40. Pólya, Gy., Szegö, G.: Aufgaben und Lehrsätze aus der Analysis I. Springer, Berlin (1925)
41. Pólya, Gy., Szegö, G.: Problems and Theorems in Analysis I. Springer, Berlin (1972)
42. Popa, D.: Hyers-Ulam-Rassias stability of the general linear equation. Nonlinear Funct. Anal. Appl. **4**, 581–588 (2002)
43. Popa, D.: Functional inclusions on square-symmetric groupoids and Hyers-Ulam stability. Math. Inequal. Appl. **7**, 419–428 (2004)
44. Popa, D.: Selections of set-valued maps satisfying functional inclusions on square-symmetric grupoids. In: Rasssias, Th.M., Brzdęk, J. (eds.) Functional Equations in Mathematical Analysis. Springer Optimization and Its Applications, vol. 52, pp. 261–272. Springer, New York (2012)
45. Rassias, J.M.: On approximation of approximately linear mappings by linear mappings. J. Funct. Anal. **46**, 126–130 (1982)
46. Rassias, J.M.: On a new approximation of approximately linear mappings by linear mappings. Discuss. Math. **7**, 193–196 (1985)
47. Rassias, Th.M.: On the stability of the linear mapping in Banach spaces. Proc. Am. Math. Soc. **72**, 297–300 (1978)
48. Rassias, Th.M.: Problem. Aequ. Math. **39**, 309 (1990)
49. Rassias, Th.M.: On a modified Hyers-Ulam sequence. J. Math. Anal. Appl. **158**, 106–113 (1991)
50. Rassias, Th.M.: The problem of S.M. Ulam for approximately multiplicative mappings. J. Math. Anal. Appl. **246**, 352–378 (2000)
51. Rassias, Th.M.: On the stability of functional equations and a problem of Ulam. Acta Appl. Math. **62**, 23–130 (2000)
52. Rassias, Th.M.: On the stability of functional equations in Banach spaces. J. Math. Anal. Appl. **251**, 264–284 (2000)
53. Rassias, Th.M., Šemrl, P.: On the behavior of mappings which do not satisfy Hyers-Ulam stability. Proc. Amer. Math. Soc. **114**, 989–993 (1992)
54. Rassias, Th.M., Tabor, J.: What is left of Hyers-Ulam stability? J. Nat. Geom. **1**, 65–69 (1992)

55. Rätz, J.: On approximately additive mappings. In: Beckenbach, E.F. (ed.) General Inequalities 2, pp. 233–251. Birkhäuser, Basel (1980)
56. Šemrl, P.: The stability of approximately additive mappings. In: Rassias, Th.M. , J. Tabor (eds.) Stability of Mappings of Hyers-Ulam Type, pp. 135–140. Hadronic, Palm Harbor (1994)
57. Tabor, J., Tabor, J.: Stability of the Cauchy functional equation in metric groupoids. Aequ. Math. **76**, 92–104 (2008)
58. Ulam, S.M.: A Collection of Mathematical Problems. Interscience, New York (1960). (Reprinted as: Problems in Modern Mathematics. Wiley, New York (1964))

On Stability of the Linear and Polynomial Functional Equations in Single Variable

Janusz Brzdęk and Magdalena Piszczek

2010 Mathematics Subject Classification: Primary 39B82; Secondary 39B62

Abstract We present a survey of selected recent results of several authors concerning stability of the following polynomial functional equation (in single variable)

$$\varphi(x) = \sum_{i=1}^{m} a_i(x)\varphi(\xi_i(x))^{p(i)} + F(x),$$

in the class of functions φ mapping a nonempty set S into a Banach algebra X over a field $\mathbb{K} \in \{\mathbb{R}, \mathbb{C}\}$, where m is a fixed positive integer, $p(i) \in \mathbb{N}$ for $i = 1, \ldots, m$, and the functions $\xi_i : S \to S$, $F : S \to X$ and $a_i : S \to X$ for $i = 1, \ldots, m$, are given. A particular case of the equation, with $p(i) = 1$ for $i = 1, \ldots, m$, is the very well-known linear equation

$$\varphi(x) = \sum_{i=1}^{m} a_i(x)\varphi(\xi_i(x)) + F(x).$$

Keywords Hyers–Ulam stability · Polynomial functional equation · Linear functional equation · Single variable · Banach space · Characteristic root

1 Introduction

In what follows $\mathbb{N}$, $\mathbb{Z}$, $\mathbb{R}$, and $\mathbb{C}$ denote the sets of positive integers, integers, reals, and complex numbers, respectively; moreover, $\mathbb{R}_+ := [0, \infty)$, and $\mathbb{N}_0 := \mathbb{N} \cup \{0\}$.

J. Brzdęk (✉) · M. Piszczek
Department of Mathematics, Pedagogical University,
Podchorążych 2, 30-084 Kraków, Poland
e-mail: jbrzdek@up.krakow.pl

M. Piszczek
e-mail: magdap@up.krakow.pl

T. M. Rassias (ed.), *Handbook of Functional Equations*,
Springer Optimization and Its Applications 96, DOI 10.1007/978-1-4939-1286-5_3

The issue of stability of functional equations has been a very popular subject of investigations for more than 50 years. The first known result on it is due to Gy. Pólya and G. Szegö [54] and reads as follows.

For every real sequence $(a_n)_{n\in\mathbb{N}}$ *with*

$$\sup_{n,m\in\mathbb{N}} |a_{n+m} - a_n - a_m| \leq 1,$$

there is a real number ω *such that*

$$\sup_{n\in\mathbb{N}} |a_n - \omega n| \leq 1.$$

Moreover,

$$\omega = \lim_{n\to\infty} \frac{a_n}{n}.$$

But the main motivation for investigation of that subject was given by S. M. Ulam, who in 1940 in his talk at the University of Wisconsin discussed a number of unsolved problems. The following question concerning the stability of homomorphism was among them. *Let* G_1 *be a group and* (G_2, d) *a metric group. Given* $\varepsilon > 0$*, does there exist* $\delta > 0$ *such that if* $f : G_1 \to G_2$ *satisfies*

$$d(f(xy), f(x)f(y)) < \delta$$

for all $x, y \in G_1$*, then a homomorphism* $T : G_1 \to G_2$ *exists with*

$$d(f(x), T(x)) < \varepsilon$$

for all $x, y \in G_1$*?*

The first answer to it was published in 1941 by D. H. Hyers [40]. The subsequent theorem contains an extension of it.

Theorem 1 *Let* E_1 *and* E_2 *be two normed spaces,* $c \geq 0$ *and* $p \in \mathbb{R} \setminus \{1\}$*. Assume that* $f : E_1 \to E_2$ *satisfies the inequality*

$$\|f(x+y) - f(x) - f(y)\| \leq c(\|x\|^p + \|y\|^p), \qquad x, y \in E_1 \setminus \{0\}.$$

If E_2 *is complete and* $p \geq 0$*, then there is a unique* $T : E_1 \to E_2$ *that is additive (i.e.,* $T(x+y) = T(x) + T(y)$ *for* $x, y \in E_1$*) and fulfills*

$$\|f(x) - T(x)\| \leq \frac{c}{|2^{p-1} - 1|} \|x\|^p, \qquad x \in E_1 \setminus \{0\}. \tag{1}$$

If $p < 0$*, then* f *is additive.*

It contains the results of Hyers [40] ($p = 0$), Aoki [2] and Rassias [59] ($p \in (0, 1)$), Gajda [38] ($1 < p$), and Brzdęk [11] ($p < 0$).

From [38] it follows that an analogous result is not true for $p = 1$ (see [41–43] for more details). Moreover, it has been proved in [10] that estimation (1) is optimum.

Results similar to Theorem 1 have been proved for numerous other functional equations. Also, the theorem has been generalized and extended in various directions. For more detailed information we refer to [3, 7, 39, 41–43, 48, 60, 62].

We can introduce the following general definition of the notion of stability that corresponds to the outcomes collected in Theorem 1 (for some comments on various possible definitions of stability we refer to [51–53]).

Definition 1 Let $n \in \mathbb{N}$, A be a nonempty set, (X, d) be a metric space, $\mathcal{C} \subset {\mathbb{R}_+}^{A^n}$ be nonempty, $\mathcal{T}$ be a function mapping $\mathcal{C}$ into ${\mathbb{R}_+}^A$, and $\mathcal{F}_1, \mathcal{F}_2$ be functions mapping nonempty $\mathcal{D} \subset X^A$ into X^{A^n}. We say that the equation

$$\mathcal{F}_1\varphi(x_1, \ldots, x_n) = \mathcal{F}_2\varphi(x_1, \ldots, x_n) \tag{2}$$

is $\mathcal{T}$ – stable provided for every $\varepsilon \in \mathcal{C}$ and $\varphi_0 \in \mathcal{D}$ with

$$d(\mathcal{F}_1\varphi_0(x_1, \ldots, x_n), \mathcal{F}_2\varphi_0(x_1, \ldots, x_n)) \leq \varepsilon(x_1, \ldots, x_n)$$
$$x_1, \ldots, x_n \in A,$$

there is a solution $\varphi \in \mathcal{D}$ of (2) such that

$$d(\varphi(x), \varphi_0(x)) \leq \mathcal{T}\varepsilon(x), \qquad x \in A.$$

Let us mention that given two nonempty sets, by A^B we denote, as usual, the family of all functions mapping B into A.

2 Stability of Zeros of Polynomials

That notion of stability of functional equations, described above, inspired numerous authors to investigate stability of other mathematical objects, in a similar manner (see, e.g., [7, 35, 41–43]).

For instance Li and Hua [49] started to study stability of the solutions of the following polynomial equation

$$x^n + \alpha x + \beta = 0, \tag{3}$$

with $x \in [-1, 1]$, where α and β are fixed real numbers and n is a positive integer. They have proved the following theorem.

Theorem 2 *Assume that* $|\alpha| > n$ *and*

$$|\beta| < |\alpha| - 1.$$

Then there exists a real constant $K > 0$*, such that for each* $\varepsilon > 0$ *and* $y \in [-1, 1]$ *with*

$$|y^n + \alpha y + \beta| \leq \varepsilon,$$

there is a solution $v \in [-1, 1]$ *of* Eq. (3) *such that*

$$|y - v| \leq K\varepsilon.$$

They have asked if an analogous property is true for more general polynomials of the form

$$a_n z^n + a_{n-1} z^{n-1} + \ldots + a_1 z + a_0 = 0.$$

In this way they have inspired authors of the papers [6, 44]. For example, the following result has been proved in [6].

Theorem 3 *Let* $\varepsilon > 0$ *and* $a_0, \ldots, a_n \in \mathbb{R}$ *be such that*

$$|a_0| < |a_1| - (|a_2| + |a_3| + \ldots + |a_n|),$$

$$|a_1| > 2|a_2| + 3|a_3| + \ldots + (n-1)|a_{n-1}| + n|a_n|.$$

If $y \in [-1, 1]$ *fulfills the inequality*

$$|a_n y^n + a_{n-1} y^{n-1} + \ldots + a_1 y + a_0| \leq \varepsilon,$$

then there is $z \in [-1, 1]$ *with*

$$a_n z^n + a_{n-1} z^{n-1} + \ldots + a_1 z + a_0 = 0$$

and

$$|y - z| \leq \lambda\varepsilon,$$

where

$$\lambda := \frac{2|a_2| + 3|a_3| + \ldots + (n-1)|a_{n-1}| + n|a_n|}{|a_1|} < 1.$$

S.-M. Jung [44] has proved the subsequent theorem.

Theorem 4 *Let* $\mathbb{K} \in \{\mathbb{R}, \mathbb{C}\}$, $n \in \mathbb{N}$, $a_0, a_1, \ldots, a_n \in \mathbb{K}$, $r > 0$ *and*

$$B_r = \{\omega \in \mathbb{K} : |\omega| \leq r\}.$$

Assume that

$$|a_1| > \sum_{i=2}^{n} i r^{i-1} |a_i|,$$

$$|a_0| \leq \sum_{i=2}^{n} (i-1) r^i |a_i|.$$

If $\varepsilon > 0$ *and* $z \in B_r$ *fulfill the inequality*

$$|a_n z^n + a_{n-1} z^{n-1} + \ldots + a_1 z + a_0| \leq \varepsilon,$$

then there is $z_0 \in B_r$ *such that*

$$a_n z_0^n + a_{n-1} z_0^{n-1} + \ldots + a_1 z_0 + a_0 = 0$$

and

$$|z - z_0| \leq \frac{\varepsilon}{(1-\lambda)|a_1|},$$

where

$$\lambda := \frac{1}{|a_1|} \sum_{i=2}^{n} i r^{i-1} |a_i| < 1.$$

Further generalization of those two theorems have been obtained in [16], where stability of the following functional equation

$$f(x) + \sum_{j=1}^{m} a_j(x) f(\xi_j(x))^{p(j)} = G(x), \tag{4}$$

has been studied in the class of functions f mapping a nonempty set S into a commutative Banach algebra X over a field $\mathbb{K} \in \{\mathbb{R}, \mathbb{C}\}$, with the unit element denoted by e, where $m \in \mathbb{N}, a_1, \dots, a_m \in X^S$, $p : \{1, \dots, m\} \to \mathbb{N}$, $G \in X^S$ and $\xi_1, \dots, \xi_m \in S^S$. We write $f(y)^0 = e$ and

$$f(y)^k := (f(y))^k, \qquad k \in \mathbb{N}.$$

Note that the linear functional equation (in single variable)

$$f(x) + \sum_{j=1}^{m} a_j(x) f(\xi_j(x)) = G(x) \tag{5}$$

is a particular case of Eq. (4) (when $p(i) = 1$ for $i = 1, \dots, m$). It is very well known and its stability has already been studied in several papers, under various additional assumptions. For more information on its solutions we refer to [46, 47].

In this chapter we present a survey of those stability results concerning Eqs. (4) and (5), published by various authors.

For examples of other stability results for functional equations in single variable see for instance to [1, 4, 5, 7–9, 12, 13, 18, 22, 27–30, 36, 37, 45, 65–69]. For information on polynomials and their solutions we refer to [50, 61].

3 Stability of the Linear Equation: The General Case

In what follows we assume that S is a nonempty set, $\mathbb{K} \in \{\mathbb{R}, \mathbb{C}\}$, $m \in \mathbb{N}$, and $\xi_1, \dots, \xi_m \in S^S$, unless explicitly stated otherwise.

We start our survey with the following general result that can be easily deduced from [22, Corollary 4].

Theorem 5 *Let X be a commutative Banach algebra over a field $\mathbb{K}$, $a_1, \ldots, a_m \in X^S$, $\varepsilon : S \to \mathbb{R}_+$, $\phi : S \to X$,*

$$q(x) := \sum_{i=1}^{m} \|a_i(x)\| < 1,$$

$$\varepsilon(\xi_i(x))\varepsilon(x)$$

and

$$q(\xi_i(x)) \leq q(x) \tag{6}$$

for $x \in S$, $i = 1, \ldots, m$. Assume that

$$\Big\| \phi(x) + \sum_{i=1}^{m} a_i(x)\phi(\xi_i(x)) - G(x) \Big\| \leq \varepsilon(x), \qquad x \in S.$$

Then, for each $x \in S$, the limit

$$f(x) := \lim_{n\to\infty} \mathcal{T}^n \phi(x)$$

exists and the function $f : S \to X$, defined in this way, is the unique solution to Eq. (5) *such that*

$$\|\phi(x) - f(x)\| \leq \frac{\varepsilon(x)}{1 - q(x)}, \qquad x \in S,$$

where $\mathcal{T} : X^S \to X^S$ is given by:

$$\mathcal{T}\, g(x) := G(x) - \sum_{i=1}^{m} a_i(x) g(\xi_i(x)), \qquad g \in X^S,\ x \in S.$$

Clearly, assumption (6) is fulfilled when

$$\|a_i(\xi_i(x))\| \leq \|a_i(x)\|, \qquad x \in S, i = 1, \ldots, m;$$

this is the case, e.g., when the functions $a_1, \ldots, a_m$ are constant.

In the case $m = 1$ Eq. (5) takes the form

$$\varphi(x) + a_1(x)\varphi(\xi_1(x)) = G(x). \tag{7}$$

If ξ_1 is bijective, then it can be rewritten in the form

$$\varphi(\xi(x)) = a(x)\varphi(x) + F(x) \tag{8}$$

with $\xi := \xi_1^{-1}$,

$$a(x) := -a_1(\xi(x)), \qquad x \in S,$$

and

$$F(x) := G(\xi(x)), \qquad x \in S.$$

Also, if a_1 takes only the scalar values and $0 \notin a_1(S)$, then (7) can be written as (8) with

$$a(x) := -\frac{1}{a_1(x)}, \qquad x \in S,$$

and

$$F(x) := \frac{G(x)}{a_1(x)}, \qquad x \in S.$$

Stability of (8) has been investigated in [5, 8, 24, 56–58, 63] (for some related results see, e.g., [8, 9, 29–34, 64–69]); it seems that the most general result has been provided in [24, Lemma 1] and it is presented below. As usual, for each $p \in \mathbb{N}_0$, we write ξ^p for the p-th iterate of ξ, i.e.,

$$\xi^0(x) = x, \qquad x \in S,$$

and

$$\xi^{p+1}(x) = \xi(\xi^p(x)), \qquad p \in \mathbb{N}_0, x \in S,$$

and, only if ξ is bijective,

$$\xi^{-p} = (\xi^{-1})^p,$$

where ξ^{-1} denotes the function inverse to ξ.

From now on we assume that X is a Banach space over $\mathbb{K}$, $F \in X^S$ and $\xi \in S^S$, unless and explicitly stated otherwise.

Theorem 6 *Let* $\varepsilon_0 : S \to \mathbb{R}_+$, $a : S \to \mathbb{K}$,

$$S' := \{x \in S : a(\xi^p(x)) \neq 0 \text{ for } p \in \mathbb{N}_0\},$$

$$\varepsilon'(x) := \sum_{k=0}^{\infty} \frac{\varepsilon_0(\xi^k(x))}{\prod_{p=0}^{k} |a(\xi^p(x))|} < \infty, \qquad x \in S',$$

and $\varphi_s : S \to X$ *be a function satisfying the inequality*

$$\|\varphi_s(\xi(x)) - a(x)\varphi_s(x) - F(x)\| \leq \varepsilon_0(x), \qquad x \in S. \tag{9}$$

Suppose that the function

$$\xi_0 := \xi|_{S \setminus S'} \tag{10}$$

(i.e., the restriction of ξ *to the set* $S\setminus S'$*) is injective and*

$$\xi(S\setminus S') \subset S\setminus S', \qquad a(S\setminus S') \subset \{0\}.$$

Then the limit

$$\varphi'(x) := \lim_{n\to\infty}\left[\frac{\varphi_s(\xi^n(x))}{\prod_{j=0}^{n-1} a(\xi^j(x))} - \sum_{k=0}^{n-1}\frac{F(\xi^k(x))}{\prod_{j=0}^{k} a(\xi^j(x))}\right]$$

exists for every $x \in S'$ *and the function* $\varphi : S \to X$*, given by:*

$$\varphi(x) := \begin{cases} \varphi'(x), & \text{if } x \in S'; \\ F(\xi_0^{-1}(x)), & \text{if } x \in \xi(S)\setminus S'; \\ \varphi_s(x) + u(x), & \text{if } x \in S\setminus[S'\cup\xi(S)], \end{cases}$$

with any $u : S \to X$ *such that*

$$\|u(x)\| \le \varepsilon_0(x), \qquad x \in S,$$

is a solution of functional Eq. (8) *with*

$$\|\varphi_s(x) - \varphi(x)\| \le \varepsilon'(x), \qquad x \in S, \tag{11}$$

where

$$\varepsilon'(x) := \begin{cases} \varepsilon_0(\xi_0^{-1}(x)), & \text{if } x \in \xi(S)\setminus S'; \\ \varepsilon_0(x), & \text{if } x \in S\setminus[S'\cup\xi(S)]. \end{cases}$$

Moreover, φ *is the unique solution of* (8) *that satisfies* (11) *if and only if*

$$S = S' \cup \xi(S).$$

To simplify the statements, in Theorem 6 it is assumed that assumption (10) is fulfilled by every function $\xi : S \to S$ when the set $S\setminus S'$ is empty. Note that in the case $S\setminus S' = \emptyset$, Theorem 6 takes the following much simpler form, which is actually [63, Theorem 2.1].

Theorem 7 *Let* $\varepsilon_0 : S \to \mathbb{R}_+$, $a : S \to \mathbb{K}\setminus\{0\}$,

$$\varepsilon'(x) := \sum_{k=0}^{\infty}\frac{\varepsilon_0(\xi^k(x))}{\prod_{p=0}^{k}|a(\xi^p(x))|} < \infty, \qquad x \in S,$$

and $\varphi_s : S \to X$ *be a function satisfying inequality* (9). *Then the limit*

$$\varphi(x) := \lim_{n\to\infty}\left[\frac{\varphi_s(\xi^n(x))}{\prod_{j=0}^{n-1} a(\xi^j(x))} - \sum_{k=0}^{n-1}\frac{F(\xi^k(x))}{\prod_{j=0}^{k} a(\xi^j(x))}\right]$$

exists for every $x \in S$ and the function $\varphi : S \to X$, defined in this way, is the unique solution of functional Eq. (8) *that satisfies inequality* (11).

The next result has been stated in [24, Corollary 1].

Theorem 8 *Let $a : S \to \mathbb{K}$, $\varepsilon_0 : S \to \mathbb{R}_+$, $\varphi_s : S \to X$ satisfy* (8), *ξ be bijective,*

$$S'' := \{x \in S : a(\xi^{-p}(x)) \neq 0 \text{ for } p \in \mathbb{N}\},$$

$\xi(S'') \subset S''$, $a(S \setminus S'') \subset \{0\}$, and

$$\varepsilon''(x) := \sum_{k=1}^{\infty} \varepsilon_0(\xi^{-k}(x)) \prod_{p=1}^{k-1} |a(\xi^{-p}(x))| < \infty, \qquad x \in S''.$$

Then, for every $x \in S''$, the limit

$$\varphi''(x) := \lim_{n\to\infty} \left[\varphi_s(\xi^{-n}(x)) \prod_{j=1}^{n} a(\xi^{-j}(x)) + \sum_{k=1}^{n} F(\xi^{-k}(x)) \prod_{j=1}^{k-1} a(\xi^{-j}(x)) \right]$$

exists and the function $\varphi : S \to X$, given by

$$\varphi(x) := \begin{cases} \varphi''(x), & \text{if } x \in S''; \\ F(\xi^{-1}(x)), & \text{if } x \in S \setminus S'', \end{cases}$$

is the unique solution of Eq. (8) *such that*

$$\|\varphi_s(x) - \varphi(x)\| \le \varepsilon''(x), \qquad x \in S,$$

where

$$\varepsilon''(x) = \varepsilon_0(\xi^{-1}(x)), \qquad x \in S \setminus S''.$$

For some remarks and examples complementing the above results see [26, pp. 96, 97].

Let us yet present one more simple result from [24, Lemma 2] (a function h mapping S into a nonempty set P is ξ-invariant provided $h(\xi(x)) = h(x)$ for $x \in S$).

Theorem 9 *Assume that ξ is bijective, $\varepsilon_0 : S \to \mathbb{R}_+$ and $a : S \to \mathbb{K}$ are ξ-invariant,*

$$\overline{S} := \{x \in S : |a(x)| \neq 1\},$$

and $\varphi_s : S \to X$ satisfies (9). *Then there exists a unique solution $\varphi : \overline{S} \to X$ of* Eq. (8) *such that*

$$\|\varphi_s(x) - \varphi(x)\| \le \frac{\varepsilon_0(x)}{|1 - |a(x)||}, \qquad x \in \overline{S}.$$

It follows from [24, Remark 7.7] that, in the statement of Theorem 9, in some situations φ cannot be extended to a solution of (8) that maps S into X.

In several cases it can be proved that the assumptions, that appear in the theorems containing the stability results, are necessary. So, one could guess that in the case when some of them are not fulfilled, we should be able to obtain a kind of nonstability outcomes. It is true, but the point is that in general it is very difficult to give a (reasonably simple) general definition of nonstability; for examples of such definitions we refer to [17, 20, 23–25]. If we base on Definition 1, then such nonstability notion should refer to the operator $\mathcal{T}$ and it seems that we should speak of $\mathcal{T}$-nonstability. Below we give an example of such nonstability result for $m = 1$, given in [21, Theorem 1], and the reader will easily identify the suitable operator $\mathcal{T}$.

Theorem 10 *Assume that $(\overline{a}_n)_{n\in\mathbb{N}_0}$ is a sequence in $\mathbb{K}\setminus\{0\}$, $(b_n)_{n\in\mathbb{N}_0}$ is a sequence in X and $(\varepsilon_n)_{n\in\mathbb{N}_0}$ is a sequence of positive real numbers such that*

$$\lim_{n\to\infty}\frac{\varepsilon_n|\overline{a}_{n+1}|}{\varepsilon_{n+1}} = 1.$$

Then there exists a sequence $(x_n)_{n\in\mathbb{N}_0}$ in X satisfying

$$\|x_{n+1} - \overline{a}_n x_n - b_n\| \le \varepsilon_n, \qquad n \in \mathbb{N}_0,$$

and such that, for every sequence $(y_n)_{n\in\mathbb{N}_0}$ in X, given by

$$y_{n+1} = \overline{a}_n y_n + b_n, \qquad n \in \mathbb{N}_0,$$

we have

$$\lim_{n\to\infty}\frac{\|x_n - y_n\|}{\varepsilon_{n-1}} = \infty.$$

For further examples of nonstability results we refer to [17, 20, 23–25]. At the end of the next section we give examples of nonstability results for $m > 1$.

4 Stability of the Linear Equation: Iterative Case

In this section we focus on a special *iterative* case of (5), when there is a function $\xi : S \to S$ such that

$$\xi_j := \xi^j, \qquad j = 1,\ldots,m.$$

Then (5) takes the form

$$f(x) + \sum_{j=1}^{m} a_j(x) f(\xi^j(x)) = G(x). \tag{12}$$

If ξ is bijective, then it can be rewritten in the form

$$f(\eta^m(x)) = \sum_{j=1}^{m} b_j(x) f(\eta^{m-j}(x)) + F(x) \tag{13}$$

(analogously as in the previous section by replacing x by $\xi^{-m}(x)$) with $\eta := \xi^{-1}$ and

$$b_i(x) := -a_i(\eta^m(x)), \qquad F(x) := G(\eta^m(x)), \qquad x \in S, i = 1, \dots, m.$$

Also, if a_m takes only the scalar values and $0 \notin a_m(S)$, then (12) can be written in the form of (13) with $\eta := \xi$ and

$$b_i(x) := -\frac{a_{m-i}(x)}{a_m(x)}, \qquad F(x) := \frac{G(x)}{a_m(x)}, \qquad x \in S, i = 1, \dots, m-1.$$

In what follows we use the following hypothesis concerning the roots of the equation

$$z^m - \sum_{j=1}^{m} b_j(x) z^{m-j} = 0, \tag{14}$$

which (for $x \in S$) is the characteristic equation of functional Eq. (13). The hypothesis reads as follows.

($\mathcal{H}$) $\eta : S \to S, b_1, \dots, b_m : S \to \mathbb{K}\, F : S \to X$ and functions $r_1, \dots, r_m : S \to \mathbb{C}$ satisfy the following condition

$$\prod_{i=1}^{m} (z - r_i(x)) = z^m - \sum_{j=1}^{m} b_j(x) z^{m-j}, \qquad x \in S, z \in \mathbb{C}.$$

It is easily seen that ($\mathcal{H}$) means that $r_1(x), \dots, r_m(x) \in \mathbb{C}$ are the complex roots of Eq. (14) for every $x \in S$. Moreover, the functions $r_1, \dots, r_m$ are not unique, but for every $x \in S$ the sequence

$$(r_1(x), \dots, r_m(x))$$

is uniquely determined up to a permutation. Clearly,

$$0 \notin b_m(S) \quad \text{if and only if} \quad 0 \notin r_j(S) \quad \text{for} j = 1, \dots, m.$$

As before, we say that that a function $\varphi : S \to X$ is f-invariant provided

$$\varphi(f(x)) = \varphi(x), \qquad x \in S.$$

Note, that under the assumption that ($\mathcal{H}$) holds, $b_1, \dots, b_m$ are f-invariant if and only if $r_1, \dots, r_m$ can be chosen f-invariant (see [24, Remark 3]).

To simplify some statements we write

$$\prod_{p=1}^{0} \lambda(h^p(x)) := 1$$

for every $h : S \to S$, $\lambda : S \to \mathbb{K}$, $x \in S$. Moreover, we assume that the restriction to the empty set of any function is injective.

Now we are in a position to present [24, Theorem 1] (see also [24, Remark 7]), which reads as follows.

Theorem 11 *Let* $\varepsilon_0 : S \to \mathbb{R}_+$, $(\mathcal{H})$ *be valid,* $\varphi_s : S \to X$,

$$\left\|\varphi_s(\eta^m(x)) - \sum_{i=1}^{m} b_i(x)\varphi_s(\eta^{m-i}(x)) - F(x)\right\| \le \varepsilon_0(x), \qquad x \in S, \tag{15}$$

r_j *be* η*-invariant for* $j > 1$, $(i_1, \ldots, i_m) \in \{-1, 1\}^m$. *Write*

$$s_j := \frac{1}{2}(1 - i_j), \qquad j = 1, \ldots, m,$$

$$S_1 := \{x \in S : r_1(\eta^{i_1 p}(x)) \neq 0 \, for \; p \in \mathbb{N}_0\}.$$

Assume that, for each $j \in \{1, \ldots, m\}$, *one of the following three conditions holds:*

1° $i_j = 1$ *for* $j = 1, \ldots, m$ *and* $0 \notin b_m(S)$;
2° $i_j = 1$ *for* $j = 1, \ldots, m$, η *is injective,* $\eta(S \setminus S_1) \subset S \setminus S_1$, $r_1(S \setminus S_1) \subset \{0\}$;
3° η *is bijective,* $\eta(S_1) \subset S_1$, *and* $r_1(S \setminus S_1) \subset \{0\}$.

Further, suppose that

$$\varepsilon_1(x) := \sum_{k=s_1}^{\infty} \varepsilon_0(\eta^{i_1 k}(x)) \prod_{p=s_1}^{k-s_1} |r_1(\eta^{i_1 p}(x))|^{-i_1} < \infty, \qquad x \in S_1,$$

$$\varepsilon_j(x) := \sum_{k=s_j}^{\infty} \varepsilon_{j-1}(\eta^{i_j k}(x))|r_j(x)|^{-i_j(k+i_j)} < \infty, \qquad x \in S_j, j \in \{2, \ldots, m\},$$

where

$$S_j := \{x \in S : r_j(x) \neq 0\}, \qquad j > 1,$$

and, in the case $S \setminus S_j \neq \emptyset$,

$$\varepsilon_j(x) := \begin{cases} \varepsilon_{j-1}(\eta^{-1}(x)), & if \, x \in \eta(S) \setminus S_j; \\ \varepsilon_{j-1}(x), & if \, x \in S \setminus [S_j \cup \eta(S)], \end{cases}$$

for $x \in S \setminus S_j$, $j \in \{1, \ldots, m\}$. *Then* Eq. (13) *has a solution* $\varphi : S \to X$ *with*

$$\|\varphi_s(x) - \varphi(x)\| \le \varepsilon_m(x), \qquad x \in S.$$

Moreover, if r_1 *is* η*-invariant and*

$$S \setminus S_j \subset \eta(S \setminus S_j), \qquad j = 1, \ldots, m,$$

then for each η*-invariant function* $h : S \to \mathbb{R}$ Eq. (13) *has at most one solution* $\varphi : S \to X$ *such that*

$$\|\varphi_s(x) - \varphi(x)\| \le h(x)\varepsilon_m(x), \qquad x \in S.$$

A simplified version of Theorem 11, with constant coefficient functions b_j, can be found in [19].

If we assume that the functions $\varepsilon_0, b_1, \dots, b_m$ are η-invariant and η is bijective, then we obtain the following result, which is much simpler than Theorem 11 (see [24, Theorem 2]).

Theorem 12 *Suppose that hypothesis* $(\mathcal{H})$ *holds,* η *is bijective,* $\varepsilon_0 : S \to \mathbb{R}_+$ *and* $b_1, \dots, b_m$ *are* η*-invariant,*

$$\widetilde{S} := \{x \in S : |r_j(x)| \neq 1 \, for j = 1, \dots, m\},$$

and a function $\varphi_s : S \to X$ *is an* ε_0*-solution of* Eq. (13) *that is* (15) *holds. Then there is a unique solution* $\varphi : \widetilde{S} \to X$ *of* (13) *such that*

$$\|\varphi_s(x) - \varphi(x)\| \leq \frac{\varepsilon_0(x)}{|(1 - |r_1(x)|) \cdot \ldots \cdot (1 - |r_m(x)|)|}, \qquad x \in \widetilde{S}. \tag{16}$$

Moreover, for each η*-invariant function* $\varepsilon : \widetilde{S} \to \mathbb{R}$, φ *is the unique solution of* (13) *such that*

$$\|\varphi_s(x) - \varphi(x)\| \leq \varepsilon(x), \qquad x \in \widetilde{S}.$$

It follows from [26, Remark 7.13] that, in the case $\mathbb{K} = \mathbb{R}$ and

$$r_j(S) \subset [0, \infty), \qquad j = 1, \dots, m,$$

estimation (16) in Theorem 12 is the best possible in the general situation. But in some other situations we can get sometimes much better estimations than (16), as for instance in [14, Theorem 3.1] (cf.[14, p. 3]), which is stated for $m = 2$, $F(x) \equiv 0$ and ε_0, b_1 and b_2 being constant functions; it reads as follows.

Theorem 13 *Let* $\eta : S \to S$, $b_1, b_2 \in \mathbb{K}$, $b_2 \neq 0$, $\bar{\varepsilon} > 0$ *and* $g : S \to X$ *satisfy the inequality*

$$\sup_{x \in S} \|g(\eta^2(x)) - b_1 g(\eta(x)) - b_2 g(x)\| \leq \bar{\varepsilon} \qquad x \in S.$$

Suppose that one of the following three conditions is valid:

(*i*) $|s_i| < 1$ *for* $i = 1, 2$ *and* $s_1 \neq s_2$;
(*ii*) $|s_i| \neq 1$ *for* $i = 1, 2$ *and* η *is bijective;*
(*iii*) (*ii*) *holds and* $s_1 \neq s_2$,

where s_1 *and* s_2 *denote the complex roots of the equation*

$$b_2 z^2 + b_1 z - 1 = 0.$$

Then there exists a solution $f : S \to X$ *of the equation*

$$f(\eta^2(x)) = b_1 f(\eta(x)) + b_2 f(x), \qquad x \in S \tag{17}$$

such that

$$\sup_{x\in S}\|g(x)-f(x)\| \le M\varepsilon,$$

where

$$M=\begin{cases}\min\{M_1,M_2\}, & \textit{if } (i) \textit{ or } (iii) \textit{ holds;}\\ M_2, & \textit{if } (ii) \textit{ holds}\end{cases}$$

and

$$M_1 := \frac{1}{|s_1-s_2|}\left(\frac{|s_1|}{||s_1|-1|}+\frac{|s_2|}{||s_2|-1|}\right),$$

$$M_2 := \frac{1}{|(|s_1|-1)(|s_2|-1)|}.$$

Moreover, if $|s_i|<1$ *for* $i=1,2$*, then there exists exactly one solution* $f:S\to X$ *of* Eq. (17) *such that*

$$\sup_{x\in S}\|g(x)-f(x)\|<\infty.$$

Related and even more general (to some extent) results for Eq. (17) can be derived from [15, Theorem 2.1].

Now, let us recall [24, Definition 3] (cf. [26, Definition 7.3]).

Definition 2 Eq. (13) is said to be strongly Hyers–Ulam stable (in the class of functions $\psi:S\to X$) provided there exists $\alpha\in\mathbb{R}$ such that, for every $\delta>0$ and for every $\psi:S\to X$ satisfying

$$\sup_{x\in S}\left\|\psi(\eta^m(x))-\sum_{i=1}^{m}b_i(x)\psi(\eta^{m-i}(x))-F(x)\right\|\le\delta,$$

there exists a solution $\varphi:S\to X$ of (13) with

$$\sup_{x\in S}\|\varphi(x)-\psi(x)\|\le\alpha\delta.$$

In [26, Corollary 7.4] the following result is stated.

Theorem 14 *Suppose that hypothesis* $(\mathcal{H})$ *is valid,* η *is bijective and*

$$\inf_{x\in S}|1-|r_j(x)||>0,\qquad j=1,\dots,m. \tag{18}$$

Then, in the case where $b_1,\dots,b_m$ *are* η*-invariant,* Eq. (13) *is strongly Hyers–Ulam stable.*

From [26, Example 7.5] it follows (see [26, Remark 7.14]) that assumption (18) is necessary in the theorem above.

In the special case when the functions $b_1,\dots,b_m$ are constant, Eq. (13) becomes the following functional equation:

$$\varphi(\eta^m(x))=\sum_{i=1}^{m}b_i\varphi(\eta^{m-i}(x))+F(x) \tag{19}$$

with given fixed $b_1, \dots, b_m \in \mathbb{K}$. Then Theorems 11 and 12 obtain much simpler forms described in [24, Corollaries 3 and 4]. They show in particular that Eq. (19) is strongly Hyers–Ulam stable under the assumption that its characteristic equation

$$r^m - \sum_{i=1}^{m} b_i r^{m-i} = 0 \tag{20}$$

has no roots of module one. The assumption is necessary (see [26, Examples 7.6 and 7.7]).

Clearly, a simple particular case of functional Eq. (13), with S being either the set of nonnegative integers $\mathbb{N}_0$ or the set of integers $\mathbb{Z}$, is the difference equation

$$y_{n+m} = \sum_{i=1}^{m} b_i(n) y_{n+m-i} + d_n, \qquad n \in S, \tag{21}$$

for sequences $(y_n)_{n\in S}$ in X, where $(d_n)_{n\in S}$ is a fixed sequence in X; namely Eq. (13) becomes difference Eq. (21) with

$$f(n) = n + 1, \quad y_n := f(n) = f(\eta^n(0)), \quad d_n := F(n), \qquad n \in S.$$

Stability and nonstability results for such difference equations can be found in [20], with constant functions b_i. Let us recall here a nonstability outcome from [20, Theorem 4].

Theorem 15 *Let $T \in \{\mathbb{N}_0, \mathbb{Z}\}$, $b_1, \dots, b_m \in \mathbb{K}$ and $r_1, \dots, r_m$ denote all the complex roots of* Eq. (20). *Assume that $|r_j| = 1$ for some $j \in \{1, \dots, m\}$. Then, for any $\delta > 0$, there exists a sequence $(y_n)_{n\in T}$ in X, satisfying the inequality*

$$\Big\| y_{n+m} - \sum_{i=1}^{m} b_i y_{n+m-i} - d_n \Big\| \le \delta, \qquad n \in T,$$

such that

$$\sup_{n\in T} \|y_n - x_n\| = \infty$$

for every sequence $(x_n)_{n\in T}$ in X, fulfilling the recurrence

$$x_{n+m} = \sum_{i=1}^{m} b_i x_{n+m-i} + d_n, \qquad n \in T. \tag{22}$$

Moreover, if $r_1, \dots, r_m \in \mathbb{K}$ or there is a bounded sequence $(x_n)_{n\in T}$ in X fulfilling (22), *then $(y_n)_{n\in T}$ can be chosen unbounded.*

The next theorem provides one more nonstability result from [26, Theorem 7.4].

Theorem 16 *Suppose that* $\eta \in S^S$, $F \in X^S$, $b_1, \dots, b_m \in \mathbb{K}$, Eq. (19) *has a solution in the class of functions mapping* S *into* X, *characteristic* Eq. (20) *has a complex root of module* 1, *and there exists* $x_0 \in S$ *such that*

$$\eta^k(x_0) \neq \eta^n(x_0)\,, \qquad k, n \in \mathbb{N}_0, k \neq n,$$

and

$$\eta(S \setminus S_0) \subset S \setminus S_0,$$

where

$$S_0 := \{\eta^n(x_0) : n \in \mathbb{N}_0\}.$$

Then, for each $\delta > 0$, *there is a function* $\psi : S \to X$, *satisfying the inequality*

$$\sup_{x \in S} \Big\| \psi(\eta^m(x)) - \sum_{i=1}^{m} b_i \psi(\eta^{m-i}(x)) - F(x) \Big\| \leq \delta,$$

such that

$$\sup_{x \in S} \|\psi(x) - \varphi(x)\| = \infty$$

for arbitrary solution $\varphi : S \to X$ *of* Eq. (19).

Moreover, if all the roots of characteristic Eq. (20) *are in* $\mathbb{K}$, *then* ψ *can be chosen unbounded.*

A similar, but more general result has been obtained in [23, Theorem 1]. Below we present next two nonstability outcomes from [23, Theorems 2 and 3]. As before, $\eta \in S^S$, $F \in X^S$ and, in the second theorem (see [23, Remark 1]), $d_1, \dots, d_{m-1}$ are the unique complex numbers such that

$$b_1 = r_1 + d_1\,, \qquad b_m = -r_1 d_{m-1}$$

and, in the case $m > 2$,

$$b_j = r_1 d_{j-1} + d_j\,, \qquad j = 2, \dots, m-1.$$

Theorem 17 *Let* $b_1, \dots, b_m \in \mathbb{K}$, $m > 1$, $S_0 \subset S$ *be nonempty*, $\eta(S_0) \subset S_0$,

$$\sup_{x \in S_0} \|F(x)\| < \infty,$$

$$\sum_{j=1}^{m} b_j = 1,$$

and

$$\lim_{n \to \infty} \Big\| \sum_{k=0}^{n} F(\eta^k(x_0)) \Big\| = \infty$$

for some $x_0 \in S_0$. *Then* Eq. (19) *is nonstable on* S_0, *that is there is a function* $\psi : S \to X$ *such that*

$$\sup_{x \in S_0} \left\| \psi(\eta^m(x)) - \sum_{i=1}^{m} b_i \psi(\eta^{m-i}(x)) - F(x) \right\| < \infty$$

and

$$\sup_{x \in S_0} \|\psi(x) - \varphi(x)\| = \infty$$

for arbitrary solution $\varphi : S \to X$ *of* Eq. (19).

Theorem 18 *Let* $b_1, \dots, b_m \in \mathbb{K}$, $m > 1$, $S_0 \subset S$ *be nonempty. Suppose that* Eq. (20) *have a root* $r_1 \in K$, *there is a function* $\psi_0 : S \to X$ *such that*

$$\sup_{x \in S_0} \|\psi_0(\eta(x)) - r_1 \psi_0(x) - F(x)\| < \infty$$

and the equation

$$\psi_1(\eta(x)) = r_1 \psi(x) + F(x)$$

has no solutions $\psi_1 : S \to X$ *with*

$$\sup_{x \in S_0} \|\psi_0(x) - \psi_1(x)\| < \infty.$$

Further, assume that the equation

$$\gamma(\eta^{m-1}(x)) = \sum_{i=1}^{m-1} d_i \gamma(\eta^{m-i}(x)) + \psi_0(x)$$

is nonstable on S_0 *(in the sense described in Theorem* 17*) or has a solution* $\gamma : S \to X$. *Then Eq.* (19) *is nonstable on* S_0.

5 Set-Valued Case

In this part we present two theorems that contain results on selections of set-valued maps satisfying linear inclusions, which can be derived from Theorems 1 and 2 in [55]. They are closely related to the issue of stability of the corresponding functional equations.

Let K be a nonempty set and (Y, d) be a metric space. We will denote by $n(Y)$ the family of all nonempty subsets of Y. The nonnegative real number

$$\delta(A) := \sup\{d(x, y) : x, y \in A\}$$

is said to be the diameter of a nonempty set $A \subset Y$. For $F : K \to n(Y)$ we denote by $\mathrm{cl}F$ the multifunction defined by

$$(\mathrm{cl}F)(x) := \mathrm{cl}F(x), \qquad x \in K.$$

Each function $f : K \to Y$ such that

$$f(x) \in F(x), \qquad x \in K,$$

is said to be a selection of the multifunction F.

The theorems read as follows.

Theorem 19 *Let $F : K \to n(Y)$, $m \in \mathbb{N}$, $a_1, \ldots, a_m : K \to \mathbb{R}$, $\xi_1, \ldots, \xi_m : K \to K$ and*

$$\liminf_{n\to\infty} \sum_{i_1=1}^{k} |a_{i_1}(x)| \sum_{i_2=1}^{k} |a_{i_2} \circ \xi_{i_1}(x)| \ldots \sum_{i_n=1}^{k} |a_{i_n} \circ \xi_{i_{n-1}} \circ \ldots \circ \xi_{i_1}(x)|$$
$$\times \delta(F \circ \xi_{i_n} \circ \ldots \circ \xi_{i_1}(x)) = 0, \qquad x \in K.$$

(a) *If Y is complete and*

$$\sum_{i=1}^{m} a_i(x) F(\xi_i(x)) \subset F(x), \qquad x \in K,$$

then there exists a unique selection $f : K \to Y$ of the multifunction $\mathrm{cl}F$ such that

$$\sum_{i=1}^{m} a_i(x) f(\xi_i(x)) = f(x), \qquad x \in K.$$

(b) *If*

$$F(x) \subset \sum_{i=1}^{m} a_i(x) F(\xi_i(x)), \qquad x \in K,$$

then F is a single-valued function and

$$\sum_{i=1}^{m} a_i(x) F(\xi_i(x)) = F(x), \qquad x \in K.$$

Theorem 20 *Let $m \in \mathbb{N}$, $a_1, \ldots, a_m : K \to \mathbb{R}$, $\xi_1, \ldots, \xi_m : K \to K$, $F, G : K \to n(Y)$,*

$$k(x) := \delta(F(x) + G(x))$$
$$+ \sum_{l=1}^{\infty} \sum_{i_1=1}^{m} |a_{i_1}(x)| \sum_{i_2=1}^{m} |a_{i_2} \circ \xi_{i_1}(x)| \ldots \sum_{i_l=1}^{m} |a_{i_l} \circ \xi_{i_{l-1}} \circ \ldots \circ \xi_{i_1}(x)|$$

$$\times\, \delta(F \circ \xi_{i_l} \circ \ldots \circ \xi_{i_1}(x) + G \circ \xi_{i_l} \circ \ldots \circ \xi_{i_1}(x)) < \infty,$$

$$\sum_{i=1}^{m} a_i(x) F(\xi_i(x)) \subset F(x) + G(x),$$

and $0 \in G(x)$ *for all* $x \in K$. *Then there exists a unique function* $f : K \to Y$ *such that, for each* $x \in K$,

$$\sum_{i=1}^{m} a_i(x) f(\xi_i(x)) = f(x),$$

$$\sup_{y \in F(x)} d(f(x), y) \le k(x).$$

6 Stability of the Polynomial Equation

We end this chapter with a result proved in [16] and concerning stability functional Eq. (4), i.e., the equation

$$f(x) + \sum_{j=1}^{m} a_j(x) f(\xi_j(x))^{p(j)} = G(x).$$

In this section, as at the end of Sect. 2, X denotes a Banach commutative algebra over $\mathbb{K}, m \in \mathbb{N}, a_1, \ldots, a_m : S \to X, p : \{1, \ldots, m\} \to \mathbb{N}, G \in X^S$ and $\xi_1, \ldots, \xi_m \in S^S$.

In what follows, $r > 0$ is a fixed real number and

$$\mathcal{B}_r := \{f \in X^S : \|f(x)\| \le r \text{ for } x \in S\}.$$

To simplify statements of the main results we define operators $\mathcal{L} : X^S \to X^S$ and $\Psi : \mathbb{R}_+^S \to \mathbb{R}_+^S$ by the formulas:

$$\mathcal{L}h(x) = G(x) - \sum_{i=1}^{m} a_i(x) h(\xi_i(x))^{p(i)}, \qquad h \in X^S, x \in S,$$

$$\Psi\gamma(x) = \sum_{i=1}^{m} p(i) r^{p(i)-1} \|a_i(x)\| \gamma(\xi_i(x)), \qquad \gamma \in \mathbb{R}_+^S, x \in S.$$

Now we are in a position to present [16, Theorem 2].

Theorem 21 *Suppose that* $\delta \in \mathbb{R}_+^S$, $\gamma \in \mathcal{B}_r$,

$$\Big\|\gamma(x) + \sum_{j=1}^{m} a_j(x)\gamma(\xi_j(x))^{p(j)} - G(x)\Big\| \le \delta(x), \qquad x \in S,$$

$$\|G(x)\| \le r - \sum_{i=1}^{m} \|a_i(x)\| r^{p(i)}, \qquad x \in S,$$

$$\chi(x) := \sum_{n=0}^{\infty} \Psi^n \delta(x) < \infty, \qquad x \in S.$$

Then there is a unique solution $f \in \mathcal{B}_r$ *of* Eq. (4) *with*

$$\|f(x) - \gamma(x)\| \le \chi(x), \qquad x \in S;$$

in particular

$$f(x) = \lim_{n\to\infty} \mathcal{L}^n \gamma(x), \qquad x \in S.$$

If in Theorem 21 we take $S = \{t_0\}$, then it is easily seen that we obtain the following.

Corollary 1 *Suppose that* $\xi_0, \xi_1, \ldots, \xi_m \in X$, $z_0 \in X$, $\|z_0\| \le r$,

$$\|\xi_0\| \le r - \sum_{i=1}^{m} \|\xi_i\| r^i, \qquad \lambda_0 := \sum_{j=1}^{m} j r^{j-1} \|\xi_j\| < 1.$$

Then there is a unique $z \in X$ *such that* $\|z\| \le r$,

$$z = \sum_{j=0}^{m} \xi_j z^j, \qquad \|z - z_0\| \le \frac{1}{1-\lambda_0} \Big\| z_0 - \sum_{j=0}^{m} \xi_j z_0^j \Big\|.$$

In particular

$$z = \lim_{n\to\infty} L^n(z_0),$$

with

$$L(w) = \sum_{i=0}^{m} \xi_i w^i, \qquad w \in X.$$

References

1. Agarwal, R.P., Xu, B., Zhang, W.: Stability of functional equations in single variable. J. Math. Anal. Appl. **288**, 852–869 (2003)
2. Aoki, T.: On the stability of the linear transformation in Banach spaces. J. Math. Soc. Jpn. **2**, 64–66 (1950)
3. Baak, C., Boo, D.-H., Rassias, Th. M.: Generalized additive mappings in Banach modules and isomorphisms between C^*–algebras. J. Math. Anal. Appl. **314**, 150–161 (2006)
4. Badora, R., Brzdęk, J.: A note on a fixed point theorem and the Hyers–Ulam stability. J. Differ Eq. Appl. **18**, 1115–1119 (2012)
5. Baker, J.A.: The stability of certain functional equations. Proc. Am. Math. Soc. **112**, 729–732 (1991)
6. Bidkham, M., Soleiman Mezerji H.A., Eshaghi Gordji, M.: Hyers-Ulam stability of polynomial equations. Abstr. Appl. Anal. **2010**, Article ID 754120, 7 p. (2010)
7. Brillouët-Belluot, N., Brzdęk, J., Ciepliński, K.: On some recent developments in Ulam's type stability. Abstr. Appl. Anal. **2012**, Article ID 716936 (2012)

8. Brydak, D.: On the stability of the functional equation $\varphi[f(x)] = g(x)\varphi(x) + F(x)$. Proc. Am. Math. Soc. **26**, 455–460 (1970)
9. Brydak, D.: Iterative stability of the Böttcher equation. In: Rassias, Th.M., Tabor, J. (eds.) Stability of Mappings of Hyers-Ulam Type, pp. 15–18. Hadronic Press, Palm Harbor (1994)
10. Brzdęk, J.: A note on stability of additive mappings. In: Rassias, Th.M., Tabor, J. (eds.) Stability of Mappings of Hyers-Ulam Type, pp. 19–22. Hadronic Press , Palm Harbor (1994)
11. Brzdęk, J.: Hyperstability of the Cauchy equation on restricted domains. Acta Math. Hung. **41**, 58–67 (2013)
12. Brzdęk, J., Ciepliński, K.: A fixed point approach to the stability of functional equations in non-Archimedean metric spaces. Nonlinear Anal. **74**, 6861–6867 (2011)
13. Brzdęk, J., Ciepliński, K.: A fixed point theorem and the Hyers-Ulam stability in non-Archimedean spaces. J. Math. Anal. Appl. **400**, 68–75 (2013)
14. Brzdęk, J., Jung, S.-M.: A note on stability of a linear functional equation of second order connected with the Fibonacci numbers and Lucas sequences. J. Ineq. Appl. **2010**, Article ID 793947, 10 p. (2010)
15. Brzdęk, J., Jung, S.-M.: A note on stability of an operator linear equation of the second order. Abstr. Appl. Anal. **2011**, Article ID 602713, 15 p. (2011)
16. Brzdęk, J., Stević, S.: A note on stability of polynomial equations. Aequ. Math. **85**, 519–527 (2013)
17. Brzdęk, J., Popa, D., Xu, B.: Note on the nonstability of the linear recurrence. Abh. Math. Sem. Univ. Hambg. **76**, 183–189 (2006)
18. Brzdęk, J., Popa, D., Xu, B.: The Hyers-Ulam stability of nonlinear recurrences. J. Math. Anal. Appl. **335**, 443–449 (2007)
19. Brzdęk, J., Popa, D., Xu, B.: The Hyers-Ulam stability of linear equations of higher orders. Acta Math. Hung. **120**, 1–8 (2008)
20. Brzdęk, J., Popa, D., Xu, B.: Remarks on stability of the linear recurrence of higher order. Appl. Math. Lett. **23**, 1459–1463 (2010)
21. Brzdęk, J., Popa, D., Xu, B.: On nonstability of the linear recurrence of order one. J. Math. Anal. Appl. **367**, 146–153 (2010)
22. Brzdęk, J., Chudziak, J., Páles, Zs.: A fixed point approach to stability of functional equations. Nonlinear Anal. **74**, 6728–6732 (2011)
23. Brzdęk, J., Popa, D., Xu, B.: Note on nonstability of the linear functional equation of higher order. Comp. Math. Appl. **62**, 2648–2657 (2011)
24. Brzdęk, J., Popa, D., Xu, B.: On approximate solutions of the linear functional equation of higher order. J. Math. Anal. Appl. **373**, 680–689 (2011)
25. Brzdęk, J., Popa, D., Xu, B.: A note on stability of the linear functional equation of higher order and fixed points of an operator. Fixed Point Theory **13**, 347–356 (2012)
26. Brzdęk, J., Popa, D., Xu, B.: Remarks on stability of the linear functional equation in single variable. In: Pardalos, P., Srivastava, H.M., Georgiev, P. (eds.) Nonlinear Analysis: Stability, Approximation, and Inequalities, Springer Optimization and Its Applications, vol. 68, pp. 91–119. Springer, New York (2012)
27. Cădariu, L., Radu, V.: Fixed point methods for the generalized stability of functional equations in a single variable. Fixed Point Theory Appl. **2008**, Article ID 749392, 15 p. (2008)
28. Cădariu, L., Găvruţa, L., Găvruţa, P.: Fixed points and generalized Hyers-Ulam stability. Abstr. Appl. Anal. **2012**, Article ID 712743, 10 p. (2012)
29. Choczewski, B.: Stability of some iterative functional equations. In: General inequalities, 4 (Oberwolfach, 1983), pp. 249–255, Internat. Schriftenreihe Numer. Math., **71**. Birkhäuser, Basel (1984)
30. Choczewski, B., Turdza, E., Węgrzyk, R.: On the stability of a linear functional equation. Wyż. Szkoła Ped. Krakow. Rocznik Nauk.-Dydakt. Prace Mat. **9**, 15–21 (1979)
31. Czerni, M.: Stability of normal regions for linear homogeneous functional equations. Aequ. Math. **36**, 176–187 (1988)

32. Czerni, M.: On some relation between the Shanholt and the Hyers-Ulam types of stability of the nonlinear functional equation. In: Rassias, Th.M., Tabor, J. (eds.) Stability of Mappings of Hyers-Ulam Type, pp. 59–65, Hadronic Press, Palm Harbor (1994)
33. Czerni, M.: Stability of normal regions for nonlinear functional equation of iterative type. In: Rassias, Th.M., Tabor, J. (eds.) Stability of Mappings of Hyers-Ulam Type, pp. 67–79, Hadronic Press, Palm Harbor (1994)
34. Czerni, M.: Further results on stability of normal regions for linear homogeneous functional equations. Aequ. Math. **49**, 1–11 (1995)
35. Czerwik, S.: Functional Equations and Inequalities in Several Variables. World Scientific, London (2002)
36. Forti, G.L.: Hyers-Ulam stability of functional equations in several variables. Aequ. Math. **50**, 143–190 (1995)
37. Forti, G.L.: Comments on the core of the direct method for proving Hyers-Ulam stability of functional equations. J. Math. Anal. Appl. **295**, 127–133 (2004)
38. Gajda, Z.: On stability of additive mappings. Int. J. Math. Math. Sci. **14**, 431–434 (1991)
39. Găvruţa, P.: A generalization of the Hyers-Ulam-Rassias stability of approximately additive mappings. J. Math. Anal. Appl. **184**, 431–436 (1994)
40. Hyers, D.H.: On the stability of the linear functional equation. Proc. Nat. Acad. Sci. U S A **27**, 222–224 (1941)
41. Hyers, D.H., Isac, G., Rassias, Th.M.: Stability of Functional Equations in Several Variables. Birkhäuser, Boston (1998)
42. Jung, S.-M.: Hyers-Ulam-Rassias Stability of Functional Equations in Mathematical Analysis. Hadronic Press, Palm Harbor (2001)
43. Jung, S.-M.: Hyers-Ulam-Rassias Stability of Functional Equations in Nonlinear Analysis. Springer, New York (2011)
44. Jung, S.-M.: Hyers-Ulam stability of zeros of polynomials. Appl. Math. Lett. **24**, 1322–1325 (2011)
45. Jung, S-M., Popa, D., Rassias, M.Th: On the stability of the linear functional equation in a single variable on complete metric groups. J. Glob. Optim. (2014). doi:10.1007/s10898-013-0083-9
46. Kuczma, M.: Functional Equations in a Single Variable. PWN—Polish Scientific Publishers, Warszawa (1968)
47. Kuczma, M., Choczewski, B., Ger, R.: Iterative Functional Equations. Encyclopedia of Mathematics and its Applications. Cambridge University Press, Cambridge (1990)
48. Lee, Y.-H., Jung, S.-M., Rassias, M.Th: On an n–dimensional mixed type additive and quadratic functional equation. Appl. Math. Comput. (to appear)
49. Li, Y., Hua, L.: Hyers-Ulam stability of a polynomial equation. Banach J. Math. Anal. **3**, 86–90 (2009)
50. Milovanovic, G.V., Mitrinovic, D.S., Rassias, Th.M.: Topics in Polynomials: Extremal Problems, Inequalities, Zeros. World Scientific, Singapore (1994)
51. Moszner, Z.: Sur les définitions différentes de la stabilité des équations fonctionnelles. Aequ. Math. **68**, 260–274 (2004)
52. Moszner, Z.: On the stability of functional equations. Aequ. Math. **77**, 33–88 (2009)
53. Moszner, Z.: On stability of some functional equations and topology of their target spaces. Ann. Univ. Paedagog. Crac. Stud. Math. **11**, 69–94 (2012)
54. Pólya, Gy., Szegö, G.: Aufgaben und Lehrsätze aus der Analysis, vol. I.Springer, Berlin (1925)
55. Piszczek, M.: On selections of set-valued maps satisfying some inclusions in a single variable. Math. Slovaca (to appear)
56. Popa, D.: Hyers-Ulam-Rassias stability of the general linear equation. Nonlinear Funct. Anal. Appl. **7**, 581–588 (2002)
57. Popa, D.: Hyers-Ulam-Rassias stability of a linear recurrence. J. Math. Anal. Appl. **309**, 591–597 (2005)
58. Popa, D.: Hyers-Ulam stability of the linear recurrence with constant coefficients. Adv. Differ. Equ. **2005**, 101–107 (2005)

59. Rassias, Th.M.: On the stability of the linear mapping in Banach spaces. Proc. Am. Math. Soc. **72**, 297–300 (1978)
60. Rassias, Th.M.: On a modified Hyers–Ulam sequence. J. Math. Anal. Appl. **158**, 106–113 (1991)
61. Rassias, Th.M., Srivastava, H. M., Yanushauskas, A. (eds.): Topics in Polynomials of One and Several Variables and their Applications. World Scientific, Singapore (1993)
62. Rassias, Th.M., Tabor, J. (eds.): Stability of Mappings of Hyers–Ulam Type. Hadronic Press , Palm Harbor (1994)
63. Trif, T.: On the stability of a general gamma-type functional equation. Publ. Math. Debr. **60**, 47–61 (2002)
64. Trif, T.: Hyers-Ulam-Rassias stability of a linear functional equation with constant coefficients. Nonlinear Funct. Anal. Appl. **11**(5), 881–889 (2006)
65. Turdza, E.: On the stability of the functional equation of the first order. Ann. Polon. Math. **24**, 35–38 (1970/1971)
66. Turdza, E.: On the stability of the functional equation $\phi[f(x)] = g(x)\phi(x) + F(x)$. Proc. Am. Math. Soc. **30**, 484–486 (1971)
67. Turdza, E.: Some remarks on the stability of the non-linear functional equation of the first order. Collection of articles dedicated to Stanisław Gołąb on his 70th birthday, II. Demonstr Math. **6**, 883–891 (1973/1974)
68. Turdza, E.: Set stability for a functional equation of iterative type. Demonstr Math. **15**, 443–448 (1982)
69. Turdza, E.: The stability of an iterative linear equation. In: General inequalities, 4 (Oberwolfach, 1983), pp. 277–285, Internat. Schriftenreihe Numer. Math. **71**. Birkhäuser, Basel (1984)

Selections of Set-valued Maps Satisfying Some Inclusions and the Hyers–Ulam Stability

Janusz Brzdęk and Magdalena Piszczek

2010 Mathematics Subject Classification: 39B05, 39B82, 54C60, 54C65

Abstract We present a survey of several results on selections of some set-valued functions satisfying some inclusions and also on stability of those inclusions. Moreover, we show their consequences concerning stability of the corresponding functional equations.

Keywords Stability of functional equation · Set-valued map · Inclusion · Selection

1 Introduction

At present we know that the study of existence of selections of the set-valued maps, satisfying some inclusions, in many cases is connected to the stability problems of functional equations (see, e.g., [8, 26, 27, 29]). Let us remind the result on the stability of functional equation published in 1941 by D. H. Hyers in [6].

Let X be a linear normed space, Y a Banach space, and $\epsilon > 0$. Then, for every function $f : X \to Y$ satisfying the inequality

$$\|f(x+y) - f(x) - f(y)\| \le \epsilon, \qquad x, y \in X, \tag{1}$$

there exists a unique additive function $g : X \to Y$ such that

$$\|f(x) - g(x)\| \le \epsilon, \qquad x \in X. \tag{2}$$

J. Brzdęk (✉) · M. Piszczek
Department of Mathematics, Pedagogical University,
Podchorążych 2, 30-084 Kraków, Poland
e-mail: jbrzdek@up.krakow.pl

M. Piszczek
e-mail: magdap@up.krakow.pl

T. M. Rassias (ed.), *Handbook of Functional Equations*,
Springer Optimization and Its Applications 96, DOI 10.1007/978-1-4939-1286-5_4

For further information and references concerning that subject we refer to [1, 3, 5, 7, 10, 11, 15, 28].

W. Smajdor [29] and Z. Gajda, R. Ger [8] observed that inequality (2) can be written in the form

$$f(x+y) - f(x) - f(y) \in B(0,\epsilon), \qquad x, y \in X,$$

where $B(0,\epsilon)$ is the closed ball centered at 0 and of radius ϵ. Hence we have

$$f(x+y) + B(0,\epsilon) \subset f(x) + B(0,\epsilon) + f(y) + B(0,\epsilon), \qquad x, y \in X,$$

and the set-valued function

$$F(x) := f(x) + B(0,\epsilon), \qquad x \in X,$$

is subadditive, i.e.

$$F(x+y) \subset F(x) + F(y), \qquad x, y \in X;$$

moreover, the function g from inequality (2) satisfies

$$g(x) \in F(x), \qquad x \in X,$$

which means that F has the additive selection g.

There arises a natural question under what conditions a subadditive set-valed function admits an additive selection. An answer provides the result of Z. Gajda and R. Ger in [8] given below ($\delta(D)$ denotes the diameter of a nonempty set D).

Theorem 1 *Let $(S,+)$ be a commutative semigroup with zero, X a real Banach space and $F : S \to 2^X$ a set-valued map with nonempty, convex, and closed values such that*

$$F(x+y) \subset F(x) + F(y), \qquad x, y \in S,$$

and

$$\sup_{x \in S} \delta(F(x)) < \infty.$$

Then F admits a unique additive selection.

Some other results on the existence of the additive selections of subadditive, superadditive, or additive set-valued functions can be found in [16, 30–33].

2 Linear Inclusions

In this section X is a real vector space and Y is a real Banach space. We denote by $n(Y)$ the family of all nonempty subsets of Y and by $ccl(Y)$ the family of all nonempty closed and convex subsets of Y. The number

$$\delta(A) = \sup_{x,y \in A} \|x - y\|$$

is said to be the diameter of nonempty $A \subset Y$. For $A, B \subset Y$ and $\alpha, \beta \in \mathbb{R}$ (the set of reals) we write

$$A + B := \{a + b : a \in A, b \in B\}$$

and

$$\alpha A := \{\alpha x : x \in A\};$$

it is well known that

$$\alpha(A + B) = \alpha A + \alpha B$$

and

$$(\alpha + \beta)A \subset \alpha A + \beta A.$$

If $A \subset Y$ is convex and $\alpha\beta > 0$, then we have

$$(\alpha + \beta)A = \alpha A + \beta A.$$

A nonempty set $K \subset Y$ is said to be a convex cone if

$$K + K \subset K$$

and

$$tK \subset K, \qquad t > 0.$$

Any function $f : X \to Y$ such that

$$f(x) \in F(x), \qquad x \in X,$$

is said to be a selection of the multifunction $F : X \to n(Y)$.

Some generalization of Theorem 1 can be found in [20], where (α, β)-subadditive set-valued map was considered, i.e., the set valued function satisfying

$$F(\alpha x + \beta y) \subset \alpha F(x) + \beta F(y), \qquad x, y \in K.$$

It has been proved there that an (α, β)-subadditive set-valued map with closed, convex, and equibounded values in a Banach space has exactly one additive selection if α, β are positive reals and $\alpha+\beta \neq 1$. For $\alpha+\beta < 1$ a stronger result is true; namely, F is single valued and additive. The above results were extended by K. Nikodem and D. Popa [18, 22] to the case of the following general linear inclusions:

$$F(ax + by + k) \subset pF(x) + qF(y) + C, \qquad x, y \in K, \tag{3}$$

$$pF(x) + qF(y) \subset F(ax + by + k) + C, \qquad x, y \in K, \tag{4}$$

where a, b, p, q are positive reals, $K \subset X$ is a convex cone with zero, $F : K \to n(Y)$, $k \in K$, and $C \in n(Y)$. Namely, they have proved the following two theorems.

Theorem 2 *Suppose that $a+b \neq 1$, $p+q \neq 1$, and $F : K \to ccl(Y)$ satisfies the general linear inclusion*

$$F(ax+by+k) \subset pF(x)+qF(y), \qquad x,y \in K,$$

and

$$\sup_{x\in K} \delta(F(x)) < \infty. \tag{5}$$

Then,

(i) *in the case $p+q > 1$, there exists a unique selection $f : K \to Y$ of F that satisfies the general linear equation*

$$f(ax+by+k) = pf(x)+qf(y), \qquad x,y \in K; \tag{6}$$

(ii) *in the case $p+q < 1$, F is single valued.*

Making a suitable substitutions, we easily deduce from the above theorem the following corollary.

Corollary 1 *Suppose that $a+b \neq 1$, $p+q > 1$, $C \subset Y$ is nonempty, compact, and convex and $F : K \to ccl(Y)$ satisfies* (5) *and the general linear inclusion* (3).

Then there exists a unique single valued mapping $f : K \to Y$ satisfying Eq. (6) *and such that*

$$f(x) \in F(x) + \frac{1}{p+q-1}C, \qquad x \in K.$$

The next theorem is complementary to the above one.

Theorem 3 *Suppose that $p+q \neq 1$ and $F : K \to ccl(Y)$ satisfies the general linear inclusion*

$$pF(x)+qF(y) \subset F(ax+by), \qquad x,y \in K, \tag{7}$$

and

$$\sup_{x\in L_z} \delta(F(x)) < \infty, \qquad z \in K,$$

where

$$L_z = \{tz : t \geq 0\}.$$

Then,

(i) *in the case $p+q < 1$, there exists a unique selection $f : K \to Y$ of F satisfying the general linear equation*

$$pf(x)+qf(y) = f(ax+by), \qquad x,y \in K;$$

(ii) *in the case $p+q > 1$, F is single-valued.*

It can be easily shown that Theorem 3 yields the following.

Corollary 2 *Let $a+b \neq 1$, $p+q < 1$, $C \subset Y$ be nonempty, compact, and convex, and*

$$x_0 := \frac{k}{1-a-b}.$$

Suppose that $F : K + x_0 \to ccl(Y)$ satisfies the general linear inclusion (4) *for $x, y \in K + x_0$ and*

$$\sup_{x \in L_z + x_0} \delta(F(x)) < \infty, \qquad z \in K.$$

Then there exists a unique single valued mapping $f : K + x_0 \to Y$ satisfying Eq. (6) *for $x, y \in K + x_0$ and such that*

$$f(x) \in F(x) + \frac{1}{1-p-q} C, \qquad x \in K + x_0.$$

Now, we recall some results concerning the linear inclusions when $p + q = 1$. The special cases are the following two Jensen inclusions

$$F\left(\frac{x+y}{2}\right) \subset \frac{F(x)+F(y)}{2}$$

and

$$\frac{F(x)+F(y)}{2} \subset F\left(\frac{x+y}{2}\right).$$

First we show some examples. Namely, the multifunction $F : \mathbb{R} \to ccl(\mathbb{R})$ given by

$$F(x) = [x-1, x+1], \qquad x \in \mathbb{R},$$

satisfies the Jensen equation

$$F\left(\frac{x+y}{2}\right) = \frac{F(x)+F(y)}{2}, \qquad x, y \in \mathbb{R},$$

and each function $f : \mathbb{R} \to \mathbb{R}$,

$$f(x) = x + b, \qquad x \in \mathbb{R},$$

where $b \in [-1, 1]$ is fixed, is a selection of F and satisfies the Jensen functional equation.

Observe also that, in the case $p + q = 1$, a constant function $F : K \to ccl(Y)$, $F(x) = M$ for $x \in K$, where $K \subset X$ is a cone and $M \in ccl(Y)$ is fixed, satisfies the equation

$$F(ax + by) = pF(x) + qF(y), \qquad x, y \in K,$$

and each constant function $f : K \to Y$, $f(x) = m$ for $x \in K$, where $m \in M$ is fixed, satisfies

$$f(ax + by) = pf(x) + qf(y), \qquad x, y \in K.$$

The subsequent results, concerning this case, have been obtained by K. Nikodem [17] and by A. Smajdor and W. Smajdor in [34] (as before, $K \subset X$ is a convex cone containing zero).

Theorem 4 *Let $\alpha \in (0, 1)$, $a, b > 0$, C be a nonempty, compact, and convex subset of Y containing zero. Suppose that $F : K \to ccl(Y)$ satisfies*

$$(1 - \alpha)F(x) + \alpha F(y) \subset F(px + qy) + C, \qquad x, y \in K,$$

and

$$\sup_{x \in K} \delta(F(x)) < \infty.$$

Then there exists a function $f : K \to Y$ satisfying

$$(1 - \alpha)f(x) + \alpha f(y) = f(px + qy), \qquad x, y \in K,$$

and such that

$$f(x) \in F(x) + \frac{1}{\alpha}C, \qquad x \in K.$$

Recently D. Inoan and D. Popa in [9] generalized the above theorem onto the case of inclusion

$$(1 - \alpha)F(x) + \alpha F(y) \subset F(x \star y) + C, \qquad x, y \in G, \tag{8}$$

where $(G, \star)$ is a groupoid with an operation that is bisymmetric, i.e.,

$$(x_1 \star y_1) \star (x_2 \star y_2) = (x_1 \star x_2) \star (y_1 \star y_2), \qquad x_1, x_2, y_1, y_2 \in G,$$

and fulfills the property:
there exists an idempotent element $a \in G$ (i.e. $a \star a = a$) such that for every $x \in G$ there exists a unique $t_a(x) \in G$ with $t_a(x) \star a = x$.

They have proved the following (we write $t_a^{n+1}(x) := t_a(t_a^n(x))$ for $x \in G$ and each positive integer n).

Theorem 5 *Let $p \in (0,1)$ and $F : G \to n(Y)$ satisfy inclusion* (8) *and*

$$\sup_{n \in \mathbb{N}} \delta(F(t_a^n(x))) < \infty, \qquad x \in G.$$

Then there exists a function $f : G \to Y$ with the following properties:

$$f(x) \in \mathrm{cl}F(x) + \frac{1}{p}C, \qquad x \in G,$$

$$(1-p)f(x) + pf(y) = f(x \star y), \qquad x, y \in G.$$

To present the further generalizations of those results, we need to remind the notion of the square symmetric operation. Let $(G, \star)$ be a groupoid (i.e., G is a nonempty set endowed with a binary operation $\star : G^2 \to G$). We say that $\star$ is square symmetric provided

$$(x \star y) \star (x \star y) = (x \star x) \star (y \star y), \qquad x, y \in G.$$

D. Popa in [21, 23] have proved that a set-valued map $F : X \to n(Y)$ satisfying one of the following two functional inclusions

$$F(x \star y) \subset F(x) \diamond F(y), \qquad x, y \in X,$$

$$F(x) \diamond F(y) \subset F(x \star y), \qquad x, y \in X,$$

in appropriate conditions admits a unique selection $f : X \to Y$ satisfying the functional equation

$$f(x) \diamond f(y) = f(x \star y),$$

where $(X, \star)$, $(Y, \diamond)$ are square-symmetric groupoids.

Those results extend the previous ones, because it is easy to check that if $K \subset X$ is a convex cone, $k \in T$ and a, b are fixed positive reals, then $\star : T^2 \to T$ defined by

$$x \star y := ax + by + k, \qquad x, y \in T,$$

is square symmetric. Actually, even more general property is valid: the operation $* : T^2 \to T$, given by

$$x * y := \alpha(x) + \beta(y) + \gamma_0, \qquad x, y \in T,$$

is square symmetric, where $\alpha, \beta : T \to T$ are fixed additive mappings with

$$\alpha \circ \beta = \beta \circ \alpha$$

and γ_0 is a fixed element of T.

3 Inclusions in a Single Variable

Now, we present some results corresponding to inclusions in a single variable and applications to the inclusions in several variables.

In this section, K stands for a nonempty set and (Y, d) denotes a metric space, unless explicitly stated otherwise. For $F : K \to n(Y)$ we denote by $\mathrm{cl}F$ the multifunction defined by

$$(\mathrm{cl}F)(x) = \mathrm{cl}F(x), \qquad x \in K.$$

Given $\alpha : K \to K$ we write $\alpha^0(x) = x$ for $x \in K$ and

$$\alpha^{n+1} = \alpha^n \circ \alpha, \qquad n \in \mathbb{N}_0 := \mathbb{N} \cup \{0\}$$

($\mathbb{N}$ is the set of positive integers). The following result has been obtained in [24].

Theorem 6 *Let $F : K \to n(Y)$, $\Psi : Y \to Y$, $\alpha : K \to K$, $\lambda \in (0, +\infty)$,*

$$d(\Psi(x), \Psi(y)) \le \lambda d(x, y), \qquad x, y \in Y,$$

and

$$\lim_{n\to\infty} \lambda^n \delta(F(\alpha^n(x))) = 0, \qquad x \in K.$$

1) *If Y is complete and*

$$\Psi(F(\alpha(x))) \subset F(x), \qquad x \in K,$$

then, for each $x \in K$, the limit

$$\lim_{n\to\infty} \mathrm{cl}\Psi^n \circ F \circ \alpha^n(x) =: f(x)$$

exists and f is a unique selection of the multifunction $\mathrm{cl}F$ such that

$$\Psi \circ f \circ \alpha = f.$$

2) *If*

$$F(x) \subset \Psi(F(\alpha(x))), \qquad x \in K,$$

then F is a single-valued function and

$$\Psi \circ F \circ \alpha = F.$$

Obviously, if Ψ is a contraction (i.e., $\lambda < 1$) and

$$\sup_{x\in K} \delta(F(x)) < \infty,$$

then it is easily seen that

$$\lim_{n\to\infty} \lambda^n \delta(F(\alpha^n(x))) = 0$$

and consequently the assertions of Theorem 6 are satisfied.

It has been shown in [24] that from Theorem 6 we can derive results on the selections of the set-valued functions satisfying inclusions in several variables, especially the general linear inclusions. Indeed, it is enough to take

$$\Psi(x) = \frac{1}{p+q}x, \qquad \alpha(x) = (a+b)x, \qquad x \in K,$$

or

$$\Psi(x) = (p+q)x, \qquad \alpha(x) = \frac{1}{a+b}x, \qquad x \in K,$$

to obtain the results on selections for the inclusions

$$F(ax+by) \subset pF(x) + qF(y), \qquad x, y \in K,$$

and

$$pF(x) + qF(y) \subset F(ax+by), \qquad x, y \in K,$$

respectively. Analogously, we can also obtain results for the quadratic inclusions:

$$F(x+y) + F(x-y) \subset 2F(x) + 2F(y)$$

and

$$2F(x) + 2F(y) \subset F(x+y) + F(x-y),$$

the cubic inclusions:

$$F(2x+y) + F(2x-y) \subset 2F(x+y) + 2F(x-y) + 12F(x)$$

and

$$2F(x+y) + 2F(x-y) + 12F(x) \subset F(2x+y) + F(2x-y),$$

and the quartic inclusions:

$$F(2x+y) + F(2x-y) + 6F(y) \subset 4F(x+y) + 4F(x-y) + 24F(x), \tag{9}$$

$$4F(x+y) + 4F(x-y) + 24F(x) \subset F(2x+y) + F(2x-y) + 6F(y) \tag{10}$$

(some of them have been investigated in [19]), or the following one in three variables

$$F(x+y+z) \subset 2F\left(\frac{x+y}{2}\right) + F(z),$$

considered in [14].

From Theorem 6 we can deduce the same conclusions as in [14, 19] (cf. also, e.g., [13]), but under weaker assumptions. As an example we present below such a result for the quartic inclusions, with a proof.

Corollary 3 *Let Y be a real Banach space, $(K,+)$ be a commutative group, $F : K \to ccl(Y)$ and*

$$\sup_{x\in K} \delta(F(x)) < \infty.$$

(i) *If* (9) *holds for all $x, y \in K$, then there exists a unique selection $f : K \to Y$ of the multifunction F such that*

$$f(2x+y)+f(2x-y)+6f(y)=4f(x+y)+4f(x-y)+24f(x), \quad x,y\in K.$$

(ii) *If* (10) *holds for all $x, y \in K$, then F is single-valued.*

Proof (i) Setting $x = y = 0$ in (9) we have

$$8F(0) \subset 32F(0).$$

and, by the Rådström cancellation lemma, we get $0 \in F(0)$. Next setting $y = 0$ in (9) and using the last condition we obtain

$$2F(2x) \subset 2F(2x) + 6F(0) \subset 32F(x), \qquad x \in K,$$

whence we derive the inclusion

$$\frac{F(2x)}{16} \subset F(x), \qquad x \in K.$$

Next, by Theorem 6, with

$$\Psi(x) = \frac{1}{16}x, \qquad \alpha(x) = 2x, \qquad x \in K,$$

for each $x \in K$ there exists the limit

$$\lim_{n\to\infty} \Psi^n(F(\alpha^n(x))) = \lim_{n\to\infty} \frac{F(2^n x)}{16^n} = f(x);$$

moreover,

$$f(x) \in F(x), \qquad x \in K.$$

Since, for every $x, y \in K, n \in \mathbb{N}$,

$$\begin{aligned} &\frac{F(2^n(2x+y))}{16^n} + \frac{F(2^n(2x-y))}{16^n} + 6\frac{F(2^n y)}{16^n} \\ &\subset 4\frac{F(2^n(x+y))}{16^n} + 4\frac{F(2^n(x-y))}{16^n} + 24\frac{F(2^n x)}{16^n}, \end{aligned}$$

letting $n \to \infty$ we also get

$$f(2x+y)+f(2x-y)+6f(y)=4f(x+y)+4f(x-y)+24f(x), \qquad x, y \in K.$$

Also the uniqueness of f can be easily deduced from Theorem 6.

(ii) Setting $x = y = 0$ in (10) and using the Rådström cancellation lemma we get

$$F(0) = \{0\}.$$

Thus and by (10) (with $y = 0$) we have

$$32F(x) \subset 2F(2x) + 6F(0) = 2F(2x), \qquad x \in K,$$

and consequently

$$F(x) \subset \frac{F(2x)}{16}, \qquad x \in K.$$

So, using Theorem 6 with Ψ and α defined as in the previous case, we deduce that F must be single-valued. □

Some generalization of Theorem 6 can be found in [25]; they are given below.

Theorem 7 *Let* $F : K \to n(Y)$, $k \in \mathbb{N}$, $\alpha_1, \ldots, \alpha_k : K \to K$, $\lambda_1, \ldots, \lambda_k : K \to [0, \infty)$, $\Psi : K \times Y^k \to Y$,

$$d(\Psi(x, w_1, \ldots, w_k), \Psi(x, z_1, \ldots, z_k)) \le \sum_{i=1}^{k} \lambda_i(x) d(w_i, z_i)$$

for $x \in K$, $w_1, \ldots, w_k, z_1, \ldots, z_k \in Y$ *and*

$$\liminf_{n\to\infty} \sum_{i_1=1}^{k} \lambda_{i_1}(x) \sum_{i_2=1}^{k} (\lambda_{i_2} \circ \alpha_{i_1})(x) \ldots \sum_{i_n=1}^{k} (\lambda_{i_n} \circ \alpha_{i_{n-1}} \circ \ldots \circ \alpha_{i_1})(x)$$
$$\times \delta(F((\alpha_{i_n} \circ \ldots \circ \alpha_{i_1})(x))) = 0, \qquad x \in K.$$

(a) *If* Y *is complete and*

$$\Psi(x, F(\alpha_1(x)), \ldots, F(\alpha_k(x))) \subset F(x), \qquad x \in K,$$

then there exists a unique selection $f : K \to Y$ *of the multifunction* $\mathrm{cl}F$ *such that*

$$\Psi(x, f(\alpha_1(x)), \ldots, f(\alpha_k(x))) = f(x), \qquad x \in K.$$

(b) *If*

$$F(x) \subset \Psi(x, F(\alpha_1(x)), \ldots, F(\alpha_k(x))), \qquad x \in K,$$

then F *is a single-valued function and*

$$\Psi(x, F(\alpha_1(x)), \ldots, F(\alpha_k(x))) = F(x), \qquad x \in K.$$

From this theorem we can easily deduce similar results for the following two gamma-type inclusions in single variable

$$\phi(x)F(a(x)) \subset F(x), \qquad x \in K,$$

and

$$F(x) \subset \phi(x)F(a(x)), \qquad x \in K,$$

where $F : K \to n(Y)$, $a : K \to K$, $\phi : K \to \mathbb{R}$ (for some recent stability results connected with those inclusions see [12]); or for the subsequent two inclusions

$$\lambda_1 F(\alpha_1(x)) + \cdots + \lambda_k F(\alpha_k(x)) \subset F(x), \qquad x \in K,$$

and

$$F(x) \subset \lambda_1 F(\alpha_1(x)) + \cdots + \lambda_k F(\alpha_k(x)), \qquad x \in K,$$

where $\Psi : K \times Y^k \to Y$, $\alpha_1, \ldots, \alpha_k : K \to K$, $\lambda_1, \ldots, \lambda_k \in \mathbb{R}_+$ (nonegative reals), and $\lambda_1 + \cdots + \lambda_k \in (0, 1)$.

A different generalization of Theorem 6 have been suggested in [25], with the right side of inclusions as a sum of two set-valued functions. But in this situation we do not obtain existence of the selection but of a suitable single valued function close to F. Namely, we have the following two theorems.

Theorem 8 *Assume that* Y *is complete,* $F, G : K \to n(Y)$, $0 \in G(x)$ *for all* $x \in K$, $\Psi : Y \to Y$, $\alpha : K \to K$, $\lambda \in (0, 1)$,

$$d(\Psi(x), \Psi(y)) \leq \lambda d(x, y), \qquad x, y \in Y,$$

$$M := \sup_{x \in K} \delta(F(x) + G(x)) < \infty$$

and

$$\Psi(F(\alpha(x))) \subset F(x) + G(x), \qquad x \in K. \tag{11}$$

Then there exists a unique function $f : K \to Y$ *such that*

$$\Psi \circ f \circ \alpha = f$$

and

$$\sup_{y \in F(x)} d(f(x), y) \leq \frac{1}{1 - \lambda} M, \qquad x \in K.$$

Theorem 9 *Assume that* Y *is complete,* $F, G : K \to n(Y)$, $0 \in G(x)$ *for all* $x \in K$, $k \in \mathbb{N}$, $\Psi : K \times Y^k \to Y$, $\alpha_1, \ldots, \alpha_k : K \to K$, $\lambda_1, \ldots, \lambda_k : K \to [0, \infty)$,

$$d(\Psi(x, w_1, \ldots, w_k), \Psi(x, z_1, \ldots, z_k)) \leq \sum_{i=1}^{k} \lambda_i(x) d(w_i, z_i)$$

for $x \in K$, $w_1, \ldots, w_k, z_1, \ldots, z_k \in Y$,

$$\begin{aligned}k(x) &:= \delta(F(x)+G(x)) \\ &+ \sum_{l=1}^{\infty}\sum_{i_1=1}^{k}\lambda_{i_1}(x)\sum_{i_2=1}^{k}(\lambda_{i_2}\circ\alpha_{i_1})(x)\ldots\sum_{i_l=1}^{k}(\lambda_{i_l}\circ\alpha_{i_{l-1}}\circ\ldots\circ\alpha_{i_1})(x) \\ &\times \delta(F((\alpha_{i_l}\circ\ldots\circ\alpha_{i_1})(x))+G((\alpha_{i_l}\circ\ldots\circ\alpha_{i_1})(x))) < \infty\end{aligned}$$

for $x \in K$ and

$$\Psi(x, F(\alpha_1(x)), \ldots, F(\alpha_k(x))) \subset F(x)+G(x), \qquad x \in K.$$

Then there exists a unique function $f : K \to Y$ such that

$$\Psi(x, f(\alpha_1(x)), \ldots, f(\alpha_k(x))) = f(x), \qquad x \in K,$$

and

$$\sup_{y \in F(x)} d(f(x), y) \le k(x), \qquad x \in K.$$

A special case of inclusion (11), without the assumption $0 \in G(x)$, has been investigated in [4]. In what follows X is a Banach space over a field $\mathbb{K} \in \{\mathbb{R}, \mathbb{C}\}$, $a : K \to \mathbb{K}$, $b : K \to [0, \infty)$, $\phi : K \to K$, $\psi : K \to X$ are given functions and $B \in n(X)$ is a fixed balanced and convex set with $\delta(B) < \infty$. Moreover, we write

$$a_{-1}(x) := 1, \qquad a_n(x) := \prod_{j=0}^{n} a(\phi^j(x)),$$

$$c_n(x) := b(\phi^n(x))a_{n-1}(x),$$

and

$$s_{-1}(x) := 0, \qquad s_n(x) := -\sum_{k=0}^{n} a_{k-1}(x)\psi(\phi^k(x))$$

for every $n \in \mathbb{N}_0$, $x \in K$.

Theorem 10 *Assume that $F : K \to n(X)$ is a set-valued map and the following three conditions hold:*

$$a(x)F(\phi(x)) \subset F(x) + \psi(x) + b(x)B, \qquad x \in K,$$

$$\liminf_{n\to\infty} \delta(F(\phi^{n+1}(x)))|a_n(x)| = 0, \qquad x \in K,$$

$$\omega(x) := \sum_{n=0}^{\infty} |c_n(x)| < \infty, \qquad x \in K. \tag{12}$$

Let

$$\Phi_n(x) := \mathrm{cl}\left(a_{n-1}(x)F(\phi^n(x)) + s_{n-1}(x) + \left(\sum_{k=n}^{\infty}|c_k(x)|\right)B\right)$$

for $x \in K$, $n \in \mathbb{N}_0$. *Then, for each* $x \in K$, *the sequence* $(\Phi_n(x))_{n\in\mathbb{N}_0}$ *is decreasing (i.e.,* $\Phi_{n+1}(x) \subset \Phi_n(x)$*), the set*

$$\widehat{\Phi}(x) := \bigcap_{n=0}^{\infty} \Phi_n(x)$$

has exactly one point and the function $f : K \to X$ *given by* $f(x) \in \widehat{\Phi}(x)$ *is the unique solution of the equation*

$$a(x)f(\phi(x)) = f(x) + \psi(x), \qquad x \in K, \tag{13}$$

with

$$f(x) \in \Phi_0(x) = cl(F(x) + \omega(x)B), \qquad x \in K.$$

4 Applications

In this section we present a few applications of the results, presented in the previous sections, to the stability of some functional equations.

Let V be nonempty, compact, and convex subset of a real Banach space Y, $0 \in V$, and $a, b, p, q \in \mathbb{R}$.

Corollary 4 *Let* K *be a convex cone in a real vector space and* $c \in K$. *Suppose that* $a + b \neq 1$, $p + q > 1$, *and* $f : K \to Y$ *satisfies*

$$f(ax + by + c) - pf(x) - qf(y) \in V, \qquad x, y \in K.$$

Then there exists a unique function $h : K \to Y$ *such that*

$$h(ax + by + c) = ph(x) + qh(y), \qquad x, y \in K,$$

and

$$h(x) - f(x) \in \frac{1}{p+q-1}V, \qquad x \in K.$$

Proof Let

$$F(x) := f(x) + \frac{1}{p+q-1}V, \qquad x \in K.$$

Then

$$F(ax + by + c) = f(ax + by + c) + \frac{1}{p+q-1}V$$

$$\begin{aligned}&\subset pf(x)+qf(y)+\frac{p+q}{p+q-1}V\\&=p\left(f(x)+\frac{1}{p+q-1}V\right)+q\left(f(y)+\frac{1}{p+q-1}V\right)\\&=pF(x)+qF(y), \qquad x,y\in K.\end{aligned}$$

By Theorem 2 there exists a unique function $h : K \to Y$ with

$$h(x)\in f(x)+\frac{1}{p+q-1}V, \qquad x\in K,$$

and such that

$$h(ax+by+c)=ph(x)+qh(y), \qquad x,y\in K.$$

□

Corollary 5 *Let $(K,+)$ be a commutative group and $f : K \to Y$ satisfies*

$$f(2x+y)+f(2x-y)+6f(y)-4f(x+y)-4f(x-y)-24f(x)\in V$$

for every $x, y \in K$. Then there exists a unique function $h : K \to Y$ such that

$$h(2x+y)+h(2x-y)+6h(y)=4h(x+y)+4h(x-y)+24h(x), \qquad x,y\in K,$$

$$h(x)-f(x)\in\frac{1}{24}V, \qquad x\in K.$$

Proof Let $F(x) := f(x)+\frac{1}{24}V$ for $x \in K$. Then

$$\begin{aligned}&F(2x+y)+F(2x-y)+6F(y)\\&=f(2x+y)+f(2x-y)+6f(y)+\frac{8}{24}V\\&\subset 4f(x+y)+4f(x-y)+24f(x)+\frac{8}{24}V+V\\&=4\left(f(x+y)+\frac{1}{24}V\right)+4\left(f(x-y)+\frac{1}{24}V\right)+24\left(f(x)+\frac{1}{24}V\right)\\&=4F(x+y)+4F(x-y)+24F(x), \qquad x,y\in K.\end{aligned}$$

Now, according to Corollary 3 there exists a unique function $h : K \to X$ such that $h(2x+y)+h(2x-y)+6h(y) = 4h(x+y)+4h(x-y)+24h(x)$ for $x, y \in K$ and

$$h(x)\in f(x)+\frac{1}{24}V, \qquad x\in K.$$

□

In similar way we can obtain the stability results for some other equations. In particular, from Theorem 7 with

$$F(x) = f(x) + \frac{1}{1-(\lambda_1+\cdots+\lambda_k)}V, \qquad x \in K,$$

and $\lambda_1 + \cdots + \lambda_k \in (0,1)$, we can derive analogous as in Corollary 5 results for functions f satisfying the condition

$$\lambda_1 f(\alpha_1(x)) + \cdots + \lambda_k f(\alpha_k(x)) - f(x) \in V, \qquad x \in K.$$

The following corollary follows from Theorem 10 (see [4]).

Corollary 6 *Let* (12) *be valid and* $g : K \to X$ *satisfy*

$$a(x)g(\phi(x)) - g(x) - \psi(x) \in b(x)B, \qquad x \in K.$$

Then there exists a unique solution $f : K \to X$ *of* Eq. (13) *with*

$$f(x) - g(x) \in \omega(x)\mathrm{cl}B, \qquad x \in K.$$

Moreover, for each $x \in K$,

$$f(x) = \lim_{n\to\infty} [a_{n-1}(x)g(\phi^n(x)) + s_{n-1}(x)].$$

Finally, let us recall the result in [2].

Theorem 11 *Let* $(S,+)$ *be a left amenable semigroup and let* X *be a Hausdorff locally convex linear space. Let* $F : S \to n(X)$ *be set-valued function such that* $F(s)$ *is convex and weakly compact for all* $s \in S$. *Then* F *admits an additive selection* $a : S \to X$ *if and only if there exists* $f : S \to X$ *such that*

$$f(s+t) - f(t) \in F(s), \qquad s,t \in S.$$

As a consequence of it we obtain the following corollaries.

Corollary 7 *Let* $(S,+)$ *be a left amenable semigroup and let* X *be a reflexive Banach space. In addition, let* $\rho : S \to [0,\infty)$ *and* $g : S \to X$ *be arbitrary functions. Then there exists an additive function* $a : S \to X$ *such that*

$$\|a(s) - g(s)\| \le \rho(s), \qquad s \in S,$$

if and only if there exists a function $f : S \to X$ *such that*

$$\|f(s+t) - f(t) - g(s)\| \le \rho(s), \qquad s,t \in S.$$

Corollary 8 *Let* $(S,+)$ *be a left amenable semigroup,* X *be a reflexive Banach space, and let* $\rho : S \to [0,\infty)$ *be an arbitrary function. Assume that a function* $f : S \to X$ *satisfies*

$$\|f(s+t) - f(t) - f(s)\| \le \rho(s), \qquad s,t \in S.$$

Then there exists an additive function $a : S \to X$ *such that*

$$\|a(s) - f(s)\| \le \rho(s), \qquad s,t \in S.$$

References

1. Baak, C., Boo, D.-H., Rassias, Th. M.: Generalized additive mappings in Banach modules and isomorphisms between C^*–algebras. J. Math. Anal. Appl. **314**, 150–161 (2006)
2. Badora, R., Ger, R., Páles, Z.: Additive selections and the stability of the Cauchy functional equation. ANZIAM J. **44**, 323–337 (2003)
3. Brillouët-Belluot, N., Brzdęk, J., Ciepliński, K.: On some recent developments in Ulam's type stability. Abstr. Appl. Anal. **2012**, Article ID 716936 (2012)
4. Brzdęk, J., Popa, D., Xu, B.: Selections of set-valued maps satisfying a linear inclusions in single variable. Nonlinear Anal. **74**, 324–330 (2011)
5. Czerwik, S.: Functional Equations and Inequalities in Several Variables. World Scientific, London (2002)
6. Hyers, D.H.: On the stability of the linear functional equation. Proc. Natl. Acad. Sci. U S A **27**, 222–224 (1941)
7. Hyers, D.H., Isac, G., Rassias, Th.M.: Stability of Functional Equations in Several Variables. Birkhäuser, Boston (1998)
8. Gajda, Z., Ger, R.: Subadditive multifunctions and Hyers-Ulam stability. In: General Inequalities, 5 (Oberwolfach, 1986), pp. 281–291, Internat. Schriftenreihe Numer. Math. **80**. Birkhäuser, Basel (1987)
9. Inoan, D., Popa, D.: On selections of general convex set-valued maps. Aequ. Math. (2013). doi:10.1007/s00010-013-0219-5
10. Jung, S.-M.: Hyers-Ulam-Rassias Stability of Functional Equations in Mathematical Analysis. Hadronic Press, Palm Harbor (2001)
11. Jung, S.-M.: Hyers-Ulam-Rassias Stability of Functional Equations in Nonlinear Analysis. Springer, New York (2011)
12. Jung, S.-M., Popa, D., Rassias, M.Th.: On the stability of the linear functional equation in a single variable on complete metric groups. J. Glob. Optim. (2014). doi:10.1007/s10898-013-0083-9
13. Lee, Y.-H., Jung, S.-M., Rassias, M.Th.: On an n–dimensional mixed type additive and quadratic functional equation. Appl. Math. Comput. (to appear)
14. Lu, G., Park, C.: Hyers-Ulastability of additive set-valued functional equations. Appl. Math. Lett. **24**, 1312–1316 (2011)
15. Moszner, Z.: On stability of some functional equations and topology of their target spaces. Ann. Univ. Paedagog. Crac. Stud. Math. **11**, 69–94 (2012)
16. Nikodem, K.: Additive selections of additive set-valued functions. Univ. u Novum Sadu Zb. Rad. Prirod.-Mat. Fak. Ser. Mat. **18**, 143–148 (1988)
17. Nikodem, K.: Characterization of midconvex set-valued functions. Acta Univ. Carolin. Math. Phys. **30**, 125–129 (1989)
18. Nikodem, K., Popa, D.: On selections of general linear inclusions. Publ. Math. Debr. **75**, 239–249 (2009)
19. Park, C., O'Regan, D., Saadati, R.: Stability of some set-valued functional equations. Appl. Math. Lett. **24**, 1910–1914 (2011)
20. Popa, D.: Additive selections of (α, β)-subadditive set valued maps. Glas. Math. **36**, 11–16 (2001)
21. Popa, D.: Functional inclusions on square-symetric grupoid and Hyers-Ulam stability. Math. Inequal. Appl. **7**, 419–428 (2004)
22. Popa, D.: A stability result for a general linear inclusion. Nonlinear Funct. Anal. Appl. **3**, 405–414 (2004)
23. Popa, D.: A property of a functional incusion connected with Hyers-Ulam stability. J. Math. Inequal. **4**, 591–598 (2009)
24. Piszczek, M.: On selections of set-valued inclusions in a single variable with applications to several variables. Results Math. **64**, 1–12 (2013)
25. Piszczek, M.: On selections of set-valued maps satisfying some inclusions in a single variable. Math. Slov. (to appear)

26. Rassias, Th.M.: Stability and set-valued functions. In: Cazacu C.A., Lehto O.E., Rassias Th.M. (eds.) Analysis and Topology, pp. 585–614. World Scientific, River Edge (1998)
27. Rassias, Th.M.: Stability and set-valued functions. In: Andreian Cazacu, C., Lehto, O.E., Rassias, Th.M. (eds.) Analysis and Topology, pp. 585–614. World Scientific, River Edge (1998)
28. Rassias, Th.M., Tabor, J. (eds.): Stability of Mappings of Hyers - Ulam Type. Hadronic Press, Palm Harbor (1994)
29. Smajdor, W.: Superadditive set-valued functions. Glas. Mat. **21**, 343–348 (1986)
30. Smajdor, W.: Subadditive and subquadratic set-valued functions. Prace Naukowe Uniwersytetu Śląskiego w Katowicach [Scientific Publications of the University of Silesia], **889**, 75 pp., Uniwersytet Śląski, Katowice (1987)
31. Smajdor, W.: Superadditive set-valued functions and Banach-Steinhauss theorem. Radovi Mat. **3**, 203–214 (1987)
32. Smajdor, A.: Additive selections of superadditive set-valued functions. Aequ. Math. **39**, 121–128 (1990)
33. Smajdor, A.: Additive selections of a composition of additive set-valued functions. In: Iteration theory (Batschuns, 1992), pp. 251–254. Word Scientific, River Edge (1996)
34. Smajdor, A., Smajdor, W.: Affine selections of convex set-valued functions. Aequ. Math. **51**, 12–20 (1996)

Generalized Ulam–Hyers Stability Results: A Fixed Point Approach

Liviu Cădariu

Abstract We show that a recent fixed point result in (Cădariu et al., Abstr. Appl. Anal., 2012) can be used to prove some generalized Ulam–Hyers stability theorems for additive Cauchy functional equation as well as for the monomial functional equation in $\beta-$normed spaces.

Keywords Generalized · Ulam–Hyers stability · Fixed point approach · Monomial functional equation · Cauchy functional equation · β-norm

1 Preliminaries

Starting from a question of S. M. Ulam concerning the stability of group homomorphisms, D. H. Hyers gave a purely constructive solution in the case of Cauchy functional equation in Banach spaces (see, e.g., [26, 27]). The result of Hyers was generalized by T. Aoki [1] for approximately additive mappings and by Th. M. Rassias [37] for approximately linear mappings. G. L. Forti [19] extended a part of Th. M. Rassias' result for a general class of functional equations. P. Găvruţa [23] obtained a generalization of Th. M. Rassias' theorem, by replacing the Cauchy differences by a control mapping satisfying a simple condition of convergence. These papers had a great influence in the development of what is now known as *generalized Hyers–Ulam–Rassias stability of the functional equations*. For a comprehensive presentation of this field we refer the reader to the papers [3, 20–22, 24, 28, 30–32, 35, 38, 39] and to the books [17, 18, 27, 29].

Almost all proofs in this topic used the direct method (of Hyers): the exact solution of the functional equation is explicitly constructed as a limit of a sequence, starting from the given approximate solution f. On the other hand, J. A. Baker [2] used in 1991 the Banach fixed point theorem to give Hyers–Ulam stability results for a nonlinear functional equation. In 2003, V. Radu [36] proposed a new method, successively developed in [8–10,12] to obtain the existence of the exact solutions and

L. Cădariu (✉)
Department of Mathematics, Politehnica University of Timişoara,
Piaţa Victoriei no. 2, 300006 Timişoara, Romania
e-mail: liviu.cadariu-brailoiu@upt.ro; lcadariu@yahoo.com

T. M. Rassias (ed.), *Handbook of Functional Equations*,
Springer Optimization and Its Applications 96, DOI 10.1007/978-1-4939-1286-5_5

the error estimations, based on the fixed point alternative. After that, D. Miheţ [33] applied the Luxemburg–Jung fixed point theorem in generalized metric spaces while L. Găvruţa [25] used the Matkowski's fixed point theorem to obtain some general results concerning the Hyers–Ulam stability of several types of functional equations in a single variable. L. Cădariu and V. Radu used two fixed point alternatives together with the error estimations for generalized contractions of type Bianchini–Grandolfi and Matkowski for proving stability of Cauchy functional equation [13] and of the monomial functional equation [14], in $\beta-$normed spaces (see, e.g., [16] for more details).

In the last 3 years several papers based on the idea to construct some fixed points theorems in very general conditions for operators with suitable properties have been published. Afterwards, these theorems have been used to obtain properties of generalized Ulam–Hyers stability for several classes of functional equations. In this context, J. Brzdęk et al. proved in [7] a fixed point theorem for (not necessarily) linear operators and they used it for proving Hyers–Ulam stability results for a class of functional equations in a single variable. A fixed point result of the same type was proved by J. Brzdęk and K. Ciepliński [6], in complete non-Archimedean metric spaces as well as in complete metric spaces. These results were extended in [15] by L. Cădariu et al. Moreover, they gave an affirmative answer to the open problem of J. Brzdęk and K. Ciepliński [6], concerning the uniqueness of the fixed point of some operator $\mathcal{T}$, defined in the following lines.

In this chapter, we will show that some classical results of generalized Ulam–Hyers stability can be obtained directly from these fixed point theorems.

We now recall some necessary notions and results, used in the sequel. Let S be a vector space over the real or the complex field $\mathbb{K}$ and $\beta \in (0, 1]$.

Definition 1 A mapping $||\cdot||_\beta : S \to \mathbb{R}_+$ is called a *$\beta-$norm* iff it has the following properties:

$$n^I_\beta : ||x||_\beta = 0 \Longleftrightarrow x = 0;$$

$$n^{II}_\beta : ||\lambda \cdot x||_\beta = |\lambda|^\beta \cdot ||x||_\beta, \qquad \text{for all } x \in S, \lambda \in \mathbb{K};$$

$$n^{III}_\beta : ||x + y||_\beta \le ||x||_\beta + ||y||_\beta, \quad \text{for all } x, y \in S.$$

We also recall some fixed point results that will be used to prove the generalized Ulam–Hyers stability theorems for the additive Cauchy functional equation as well as for the monomial functional equation.

We consider a nonempty set X, a complete metric space (Y, d), and the mappings $\Lambda : \mathbb{R}_+^X \to \mathbb{R}_+^X$ and $\mathcal{T} : Y^X \to Y^X$. We remind that Y^X denotes the space of all mappings from X into Y.

Definition 2 We say that $\mathcal{T}$ is $\Lambda-$ contractive if for $u, v : X \to Y$ and $\delta \in \mathbb{R}_+^X$ with

$$d(u(t), v(t)) \le \delta(t), \ \forall t \in X,$$

it follows

$$d((\mathcal{T}u)(t), (\mathcal{T}v)(t)) \le (\Lambda\delta)(t), \ \forall t \in X.$$

In the following, we assume that Λ satisfies the condition:
For every sequence $(\delta_n)_{n\in\mathbb{N}}$ of elements of $\mathbb{R}_+^X$ and every $t \in X$,

$$\lim_{n\to\infty} \delta_n(t) = 0 \Rightarrow \lim_{n\to\infty} (\Lambda\delta_n)(t) = 0. \tag{C_1}$$

Also, we suppose that $\varepsilon \in \mathbb{R}_+^X$ is a given function such that

$$\varepsilon^*(t) := \sum_{k=0}^{\infty} \left(\Lambda^k \varepsilon\right)(t) < \infty, \; t \in X. \tag{C_2}$$

Now we are in position to recall one of the main results in [15]:

Theorem 1 ([15], Theorem 2.1) *We suppose that the operator $\mathcal{T}$ is Λ−contractive and the conditions (C_1) and (C_2) hold. We consider a mapping $f \in Y^X$ such that*

$$d((\mathcal{T} f)(t), f(t)) \le \varepsilon(t), \; \forall t \in X. \tag{1}$$

Then, for every $t \in X$, the limit

$$g(t) := \lim_{n\to\infty} (\mathcal{T}^n f)(t), \tag{2}$$

exists and the mapping g is the unique fixed point of $\mathcal{T}$ with the property

$$d((\mathcal{T}^m f)(t), g(t)) \le \sum_{k=m}^{\infty} \left(\Lambda^k \varepsilon\right)(t), \; t \in X, \; m \in \mathbb{N} = \{0, 1, 2, \ldots\}. \tag{3}$$

Moreover, if we have

$$\lim_{n\to\infty} (\Lambda^n \varepsilon^*)(t) = 0, \forall t \in X, \tag{C_3}$$

then g is the unique fixed point of $\mathcal{T}$ with the property

$$d(f(t), g(t)) \le \varepsilon^*(t), \forall t \in X. \tag{4}$$

As a direct consequence of the Theorem 1 we obtained the following result:

Proposition 1 ([15], Corollary 2.3) *Let G be a nonempty set, (X, d) be a complete metric space, $\mathbb{R}_+$ be the set of all nonnegative real numbers, and $\Lambda : \mathbb{R}_+^G \to \mathbb{R}_+^G$ be a non-decreasing operator satisfying the hypothesis (C_1). If $\mathcal{T} : X^G \to X^G$ is an operator satisfying the inequality*

$$d((\mathcal{T}\xi)(x), (\mathcal{T}\mu)(x)) \le \Lambda(d(\xi(x), \mu(x))), \xi, \mu \in X^G, x \in G, \tag{5}$$

and the functions $\varepsilon : G \to \mathbb{R}_+$ and $g : G \to X$ are such that

$$d((\mathcal{T} g)(x), g(x)) \le \varepsilon(x), \; x \in G, \tag{6}$$

and

$$\varepsilon^*(x) := \sum_{k=0}^{\infty} \left(\Lambda^k \varepsilon\right)(x) < \infty,\ x \in G, \tag{C_2'}$$

then, for every $x \in G$, the limit

$$A(x) := \lim_{n\to\infty} (\mathcal{T}^n g)(x)$$

exists and the function $A \in X^G$, defined in this way, is a fixed point of $\mathcal{T}$, with

$$d(g(x), A(x)) \leq \varepsilon^*(x),\ x \in G.$$

Moreover, if the condition

$$\lim_{n\to\infty} (\Lambda^n \varepsilon^*)(x) = 0, \forall x \in G, \tag{C_3'}$$

holds, then A is the unique fixed point of $\mathcal{T}$ with the property

$$d(g(x), A(x)) \leq \varepsilon^*(x),\ x \in G.$$

2 Results

In this section we show that generalized Ulam–Hyers stability properties for two well-known functional equations can be obtained directly from the fixed point result in Proposition 1. The first outcome refers to the generalized Ulam–Hyers stability for additive Cauchy functional equation in $\beta-$normed spaces. This result was proved in [23], by using the direct method, for functions defined on the Abelian groups into Banach spaces. In [9] a variant of this theorem was obtained, for functions with values in $\beta-$ normed spaces, by using the fixed point alternative (see also [13]).

We denote by $(G, +)$ an Abelian group, by $(X, ||\cdot||_\beta)$ a complete $\beta-$normed space and by $\varphi : G \times G \to [0, \infty)$ a mapping such that

$$\Phi(x) := \sum_{k=0}^{\infty} \frac{\varphi(2^k x, 2^k x)}{2^{\beta k}} < \infty, \forall x \in G \tag{7}$$

and

$$\lim_{n\to\infty} \frac{\varphi(2^n x, 2^n y)}{2^{\beta n}} = 0, \forall x, y \in G. \tag{8}$$

Theorem 2 *Let $f : G \to X$, such that*

$$||f(x+y) - f(x) - f(y)||_\beta \leq \varphi(x, y), \forall x, y \in G. \tag{9}$$

Then there exists a unique mapping $A : G \to X$, which satisfies the additive Cauchy functional equation and

$$||f(x) - A(x)||_\beta \le \frac{1}{2^\beta}\Phi(x), \forall x \in G. \tag{10}$$

Proof We apply the Proposition 1 taking the mapping

$$\Lambda : \mathbb{R}_+^G \to \mathbb{R}_+^G, (\Lambda\delta)(x) := \frac{\delta(2x)}{2^\beta}, \ (\delta : G \to \mathbb{R}_+),$$

and the operator

$$\mathcal{T} : X^G \to X^G, (\mathcal{T}\psi)(x) := \frac{\psi(2x)}{2}, \ (\psi : G \to X).$$

From the definition of Λ, the relation (C_1) is obvious and condition (5) from Proposition 1 holds with equality.

If we take

$$\varepsilon(x) := \frac{\varphi(x,x)}{2^\beta}, \forall x \in G,$$

the relation (7) implies that the series

$$\varepsilon^*(x) = \sum_{k=0}^{\infty} \left(\Lambda^k \varepsilon\right)(x) = \sum_{k=0}^{\infty} \frac{\varepsilon(2^k x)}{2^{k\beta}} = \frac{1}{2^\beta}\sum_{k=0}^{\infty} \frac{\varphi(2^k x, 2^k x)}{2^{k\beta}} = \frac{\Phi(x)}{2^\beta}, \forall x \in G$$

is convergent, so $(C_2^{'})$ is verified.

Also, we have that

$$\left\|\frac{f(2x)}{2} - f(x)\right\|_\beta \le \frac{1}{2^\beta}\varphi(x,x), \forall x \in G,$$

and f satisfied the hypotheses of Proposition 1. This means that

$$d_\beta\left((\mathcal{T}f)(x), f(x)\right) \le \varepsilon(x), \ \forall x \in G,$$

with $d_\beta(u,v) := ||u - v||_\beta, u, v \in X^G$.

On the other hand,

$$(\Lambda^n \varepsilon^*)(x) = \frac{\Phi(2^n x)}{2^{(n+1)\beta}} = \frac{1}{2^\beta}\sum_{k=0}^{\infty} \frac{\varphi(2^{n+k}x, 2^{n+k}x)}{2^{(n+k)\beta}} = \frac{1}{2^\beta}\sum_{p=n}^{\infty} \frac{\varphi(2^p x, 2^p x)}{2^{\beta p}}, \forall x \in G.$$

Taking on the limit in the above relation as $n \to \infty$, we obtain that $(C_3^{'})$ is verified. Applying Proposition 1, it results that the limit

$$\lim_{n\to\infty} (\mathcal{T}^n f)(x) = \lim_{n\to\infty} \frac{f(2^n x)}{2^n}$$

exists for every $x \in G$. Moreover, the mapping $A : G \to X$,

$$A(x) := \lim_{n\to\infty} (\mathcal{T}^n f(x)), \forall x \in G$$

is the unique fixed point of $\mathcal{T}$, with

$$d_\beta(f(x), A(x)) \leq \varepsilon^*(x), \forall x \in G,$$

which implies that

$$||f(x) - A(x)||_\beta \leq \frac{1}{2^\beta}\Phi(x), \forall x \in G.$$

As in [23], to prove that the function A is a solution of the Cauchy functional equation, we use (8) and the definition of A. □

The last part of the chapter is devoted to the study of generalized Ulam–Hyers stability of the monomial functional equation. We now recall some necessary notions used in the sequel.

Let X and Y be vector spaces and consider the difference operators defined, for every $y \in X$ and any mapping $f : X \to Y$, by

$$\Delta_y^1 f(x) := f(x + y) - f(x), \ \forall x \in X,$$

and, inductively, $\Delta_y^{n+1} = \Delta_y^1 \circ \Delta_y^n$, for all natural number $n \geq 1$.

A mapping $f : X \to Y$ is called a *monomial function of degree N* if it is a solution of the monomial functional equation

$$\Delta_y^N f(x) - (N!)f(y) = 0, \ \forall\, x, y \in X. \tag{11}$$

The monomial equation of degree 1 is exactly the Cauchy equation, for $N = 2$ the monomial equation is equivalent to the well-known quadratic functional equation

$$f(x + y) + f(x - y) = 2\, f(x) + 2\, f(y), \forall x, y \in X, \tag{12}$$

for $N = 3$ the monomial equation is of cubic type, etc. In the sequel, the degree N will be fixed. Recall also the following noteworthy formula for the difference operator:

$$\Delta_y^N f(x) = \sum_{j=0}^{N} (-1)^{N+j} \binom{N}{N-j} f(x + jy). \tag{13}$$

In [10] we proved by the direct method, some properties of generalized Ulam–Hyers stability of the monomial functional equation (see also [11]). In what follows we give a very simple proof of the above mentioned result, by using also the fixed point result in Proposition 1.

Let us consider an Abelian group G, a complete β-normed space X, and a controlling mapping $\varphi : G \times G \to [0, \infty)$ such that

$$\Phi_i(x) := \sum_{k=0}^{\infty} \frac{\varphi(2^k i x, 2^k x)}{(2^{N\beta})^k} < \infty, \forall x \in G, \text{for } i = 0, 1, \ldots, N, \tag{14}$$

and

$$\lim_{m\to\infty} \frac{\varphi(2^m x, 2^m y)}{(2^{N\beta})^m} = 0, \ \forall x, y \in G. \tag{15}$$

Theorem 3 ([10], Theorem 2.1) *For every mapping $f : G \to X$ which satisfies the control condition*

$$\|\Delta_y^N f(x) - (N!) f(y)\|_\beta \leq \varphi(x, y) \ \forall x, y \in G, \tag{16}$$

there exists a unique monomial function $A : G \to X$ of degree N such that, for all $x \in G$,

$$\|f(x) - A(x)\|_\beta \leq \frac{1}{2^{N\beta} \cdot (N!)^\beta} \left(\Phi_0(2x) + \sum_{i=0}^{N} \binom{N}{N-i} \cdot \Phi_i(x) \right). \tag{17}$$

For proving the above theorem, we also need the following result:

Proposition 2 ([10], Lemma 2.2) *Let G be an Abelian group, X a β-normed space, and $\varphi : G \times G \to [0, \infty)$ a given mapping. If the function $f : G \to X$ satisfies* (16) *then, for all $x \in G$,*

$$\left\| \frac{f(2x)}{2^N} - f(x) \right\|_\beta \leq \frac{\varphi(0, 2x) + \sum_{i=0}^{N} \binom{N}{N-i} \cdot \varphi(ix, x)}{2^{N\beta} \cdot (N!)^\beta}. \tag{18}$$

Proof of Theorem 3. We apply Proposition 1 taking the mapping

$$\Lambda : \mathbb{R}_+^G \to \mathbb{R}_+^G, (\Lambda\delta)(x) := \frac{\delta(2x)}{2^{N\beta}}, \ (\delta : G \to \mathbb{R}_+),$$

and the operator

$$\mathcal{T} : X^G \to X^G, (\mathcal{T}\psi)(x) := \frac{\psi(2x)}{2^N}, \ (\psi : G \to X).$$

From the definition of Λ, the relation (C_1) is obvious and condition (5) from Proposition 1 holds with equality. From Proposition 2, we have that

$$\left\| \frac{f(2x)}{2^N} - f(x) \right\|_\beta \leq \frac{1}{2^{N\beta} \cdot (N!)^\beta} \left(\varphi(0, 2x) + \sum_{i=0}^{N} \binom{N}{N-i} \cdot \varphi(ix, x) \right) := \varepsilon(x),$$

for all $x \in G$, hence the relation

$$d_\beta((\mathcal{T}f)(x), f(x)) \leq \varepsilon(x),\ x \in G,$$

holds. On the other hand,

$$\varepsilon^*(x) = \sum_{k=0}^{\infty} \left(\Lambda^k \varepsilon\right)(x) =$$

$$= \sum_{k=0}^{\infty} \left(\frac{1}{2^{N\beta} \cdot (N!)^\beta} \left(\frac{\varphi(0, 2^{k+1}x)}{2^{N\beta k}} + \sum_{i=0}^{N} \binom{N}{N-i} \cdot \frac{\varphi\left(2^k i x, 2^k x\right)}{2^{N\beta k}} \right) \right) =$$

$$= \frac{1}{2^{N\beta} \cdot (N!)^\beta} \left(\Phi_0(2x) + \sum_{i=0}^{N} \binom{N}{N-i} \cdot \Phi_i(x) \right), \forall x \in G,$$

which is convergent from (14). So we have that $(C_2^{'})$ is verified. Moreover,

$$(\Lambda^m \varepsilon^*)(x) = \frac{1}{2^{N\beta} \cdot (N!)^\beta} \left(\frac{\Phi_0(2^{m+1}x)}{2^{mN\beta}} + \sum_{i=0}^{N} \binom{N}{N-i} \cdot \frac{\Phi_i\left(2^m x\right)}{2^{mN\beta}} \right) =$$

$$= \frac{1}{2^{N\beta} \cdot (N!)^\beta} \sum_{k=0}^{\infty} \left(\frac{\varphi(0, 2^{m+k+1}x)}{2^{N\beta(m+k)}} + \sum_{i=0}^{N} \binom{N}{N-i} \cdot \frac{\varphi\left(2^{m+k} i x, 2^{m+k} x\right)}{2^{N\beta(m+k)}} \right) =$$

$$= \frac{1}{2^{N\beta} \cdot (N!)^\beta} \sum_{p=m}^{\infty} \left(\frac{\varphi(0, 2^{p+1}x)}{2^{N\beta p}} + \sum_{i=0}^{N} \binom{N}{N-i} \cdot \frac{\varphi\left(2^p i x, 2^p x\right)}{2^{N\beta p}} \right), \forall x \in G.$$

By letting $m \to \infty$ in the above relation, we obtain

$$\lim_{m \to \infty} (\Lambda^m \varepsilon^*)(x) = 0, \forall x \in G,$$

so $(C_3^{'})$ is verified. By using Proposition 1, it results that the limit

$$\lim_{m \to \infty} (\mathcal{T}^m g)(x) = \lim_{m \to \infty} \frac{f(2^m x)}{2^{mN}}$$

exists for every $x \in G$. Moreover, the mapping $A : G \to X$,

$$A(x) := \lim_{m \to \infty} (\mathcal{T}^m f(x)) = \lim_{m \to \infty} \frac{f(2^m x)}{2^{mN}}, \forall x \in G$$

is the unique fixed point of $\mathcal{T}$, with

$$d_\beta(f(x), A(x)) \leq \varepsilon^*(x), \forall x \in G,$$

which implies that the estimation relation (17) holds.

To prove that the mapping A is a monomial function of degree N, we use the same method as in [10]. In fact, we replace x by $2^m x$ and y by $2^m y$ in (16), then divide the obtained relation by 2^{mN} and it results

$$\left\|\frac{\Delta_{2^m y}^N f(2^m x)}{2^{mN}} - (N!)\frac{f(2^m y))}{2^{mN}}\right\|_\beta \leq \frac{\varphi(2^m x, 2^m y)}{2^{mN\beta}} \quad \forall x, y \in G.$$

On the other hand, we denote

$$A_m(x) := \frac{f(2^m x)}{2^{mN}}, \forall x \in G.$$

By (13) we have that

$$\frac{\Delta_{2^m y}^N f(2^m x)}{2^{mN}} = \sum_{k=0}^{N} (-1)^{N-k} \binom{N}{k} \frac{f(2^m x + k2^m y)}{2^{mN}} =$$
$$= \sum_{k=0}^{N} (-1)^{N-k} \binom{N}{k} A_m(x + ky) = \Delta_y^N A_m(x),$$

for all $x, y \in G$. And we get

$$\left\|\Delta_y^N A_m(x) - (N!) \cdot A_m(y)\right\|_\beta \leq \frac{\varphi(2^m x, 2^m y)}{2^{mN\beta}}, \forall x, y \in G.$$

By letting $m \to \infty$ and having in mind (15), we obtain

$$\Delta_y^N A(x) - (N!) \cdot A(y) = 0, \forall x, y \in G.$$

Remark 1 It is worth noting that the generalized Ulam–Hyers stability properties for a large class of functional equations (Cauchy and Jensen, quadratic, cubic, quartic, quintic, etc.) can be obtained directly from Proposition 1, for suitable operator $\mathcal{T}$ and mapping Λ. Moreover, several results of generalized stability for functional equations in a single variable can be obtained by the same method.

Remark 2 In some recent papers there are proved properties of hyperstability for Cauchy functional equation on restricted domain [5] as well as for linear functional equations [34], by using a fixed point result in [7]. Moreover, the same fixed point theorem is the main tool for proving a stability result of Th. M. Rassias–Aoki's type for p-Drygas functional equation in [4]. Following the ideas of these papers, our future goal is to identify several classes of functional equations whose generalized stability properties can be obtained directly by suitable fixed point theorems. A first step in this regard is the present chapter.

References

1. Aoki, T.: On the stability of the linear transformation in Banach spaces. J. Math. Soc. Jpn. **2**, 64–66 (1950)
2. Baker, J.A..: The stability of certain functional equations. Proc. AMS **112**(3), 729–732 (1991)
3. Bourgin, D.G..: Classes of transformations and bordering transformations. Bull. Am. Math. Soc. **57**, 223–237 (1951)
4. Brzdęk, J.: Stability of the equation of the p-Wright affine functions. Aequ. Math. **85**(3), 497–503 (2013)
5. Brzdęk, J.: Hyperstability of the Cauchy equation on restricted domains. Acta Math. Hung. **141**, 58–67 (2013)
6. Brzdęk, J., Ciepliński, K.: A fixed point approach to the stability of functional equations in non-Archimedean metric spaces. Nonlinear Anal. TMA **74**, 6861–6867 (2011)
7. Brzdęk, J., Chudziak, J., Páles, Z.: A fixed point approach to stability of functional equations. Nonlinear Anal. TMA **74**, 6728–6732 (2011)
8. Cădariu, L., Radu, V.: Fixed points and the stability of Jensen's functional equation. J. Inequal. Pure Appl. Math. **4**(1), Art. 4 (2003)
9. Cădariu, L., Radu, V.: On the stability of the Cauchy functional equation: a fixed points approach. In: Sousa Ramos, J., Gronau, D., Mira, C., Reich, L., Sharkovsky, A.N. (eds.) Iteration Theory (ECIT 02), Grazer Math. Bericht. **346**, 43–52 (2004)
10. Cădariu, L., Radu, V.: Stability properties for monomial functional equations. Anal. Univ. Vest Timiş. Seria Mat.-Inform. **43**(1), 23–38 (2005)
11. Cădariu, L., Radu, V.: Remarks on the stability of monomial functional equations. Fixed Point Theory **8**(2), 201–218 (2007)
12. Cădariu, L., Radu, V.: Fixed point methods for the generalized stability of functional equations in a single variable. Fixed Point Theory Appl. 2008, Article ID 749392, 15 p. (2008)
13. Cădariu, L., Radu, V.: A general fixed point method for the stability of Cauchy functional equation. In Rassias, Th.M., Brzdek, J. (eds.) Functional Equations in Mathematical Analysis. Springer Optimization and Its Applications, vol. 52, pp. 19–32. Springer, New York (2011)
14. Cădariu, L., Radu, V.: A general fixed point method for the stability of the monomial functional equation. Carpath. J. Math. **28**(1), 25–36 (2012)
15. Cădariu, L., Găvruţa, L., Găvruţa, P.: Fixed points and generalized Hyers-Ulam stability. Abstr. Appl. Anal. 2012, Article ID 712743, 10 p. (2012)
16. Ciepliński, K.: Applications of Fixed Point Theorems to the Hyers-Ulam Stability of Functional Equations—A Survey. Ann. Funct. Anal. **3**(1), 151–164 (2012)
17. Cho, Y.J., Rassias, Th.M., Saadati, R.: Stability of Functional Equations in Random Normed Spaces. Springer Optimization and Its Applications, vol. 86. Springer, New York (2013)
18. Czerwik, S.: Functional Equations and Inequalities in Several Variables. World Scientific, New Jersey (2002)
19. Forti, G.L.: An existence and stability theorem for a class of functional equations. Stochastica **4**, 23–30 (1980)
20. Forti, G.L.: Hyers-Ulam stability of functional equations in several variables. Aequ. Math. **50**, 143–190 (1995)
21. Forti, G.L.: Comments on the core of the direct method for proving Hyers-Ulam stability of functional equations. J. Math. Anal. Appl. **295**(1), 127–133 (2004)
22. Gajda, Z.: On stability of additive mappings. Internat. J. Math. Math. Sci. **14**, 431–434 (1991)
23. Găvruţa, P.: A generalization of the Hyers-Ulam-Rassias stability of approximately additive mappings. J. Math. Anal. Appl. **184**, 431–436 (1994)
24. Găvruţa, P.: On a problem of G. Isac and Th. M. Rassias concerning the stability of mappings. J. Math. Anal. Appl. **261**, 543–553 (2001)
25. Găvruţa, L.: Matkowski contractions and Hyers-Ulam stability. Bul. Şt. Univ. Politeh. Timişoara. Seria Mat.-Fiz. **53**(67) no. 2, 32–35 (2008)

26. Hyers, D.H..: On the stability of the linear functional equation. Proc. Natl. Acad. Sci. U S A **27**, 222–224 (1941)
27. Hyers, D.H., Isac, G., Rassias, Th.M.: Stability of Functional Equations in Several Variables. Birkhâuser, Basel (1998)
28. Jung, S.-M.: On the Hyers-Ulam-Rassias stability of approximately additive mappings. J. Math. Anal. Appl. **204**, 221–226 (1996)
29. Jung, S.-M.: Hyers-Ulam-Rassias Stability of Functional Equations in Nonlinear Analysis. Springer Optimization and Its Applications, vol. 48. Springer, New York (2011)
30. Jung, S.-M., Rassias, M.Th.: A linear functional equation of third order associated to the Fibonacci numbers Abstr. Appl. Anal. 2014, Article ID 137468, 7 p. (2014)
31. Jung, S.-M., Popa, D., Rassias, M.Th. : On the stability of the linear functional equation in a single variable on complete metric groups. J. Glob. Optim. **59**(1), 165–171 (2014)
32. Lee, Y.H., Jung, S.-M., Rassias, M.Th. : On an n-dimensional mixed type additive and quadratic functional equation. Appl. Math. Comput. **228**(1), 13–16 (2014)
33. Miheţ, M.: The Hyers-Ulam stability for two functional equations in a single variable. Banach J. Math. Anal. Appl. **2**(1), 48–52 (2008)
34. Piszczek, M.: Remark on hyperstability of the general linear equation. Aequ. Math. **88**(1-2), 163–168 (2014)
35. Páles, Z.: Hyers-Ulam stability of the Cauchy functional equation on square-symmetric grupoids. Publ. Math. Debr. **58**(4), 651–666 (2001)
36. Radu, V.: The fixed point alternative and the stability of functional equations. Fixed Point Theory **4**(1), 91–96 (2003)
37. Rassias, Th.M.: On the stability of the linear mapping in Banach spaces. Proc. Am. Math. Soc. **72**, 297–300 (1978)
38. Rassias, Th.M.: On the stability of functional equations and a problem of Ulam. Acta Appl. Math. **62**, 23–130 (2000)
39. Rassias, Th.M.: On the stability of functional equations in Banach spaces. J. Math. Anal. Appl. **251**, 264–284 (2000)

On a Weak Version of Hyers–Ulam Stability Theorem in Restricted Domains

Jaeyoung Chung and Jeongwook Chang

Abstract In this chapter we consider a weak version of the Hyers–Ulam stability problem for the Pexider equation, Cauchy equation satisfied in restricted domains in a group when the target space of the functions is a 2-divisible commutative group. As the main result we find an approximate sequence for the unknown function satisfying the Pexider functional inequality, the limit of which is the approximate function in the Hyers–Ulam stability theorem.

Keywords Hyers–Ulam stability · Functional equations · Restricted domains · Pexider equation · 2-divisible commutative group

1 Introduction

The Hyers–Ulam stability problems of functional equations were originated by S. M. Ulam in 1940 when he proposed the following question [36]:

Let f be a mapping from a group G_1 to a metric group G_2 with metric $d(\cdot,\cdot)$ such that

$$d(f(xy),\ f(x)f(y)) \le \varepsilon.$$

Then does there exist a group homomorphism h and $\delta_\epsilon > 0$ such that

$$d(f(x),\ h(x)) \le \delta_\epsilon$$

for all $x \in G_1$?

One of the first assertions to be obtained is the following result, essentially due to D. H. Hyers [20], that gives an answer for the question of Ulam.

J. Chung (✉)
Department of Mathematics, Kunsan National University,
573-701 Kunsan, Republic of Korea
e-mail: jychung@kunsan.ac.kr

J. Chang
Department of Mathematics Education, Dankook University,
448-701 Yongin, Republic of Korea
e-mail: jchang@dankook.ac.kr

T. M. Rassias (ed.), *Handbook of Functional Equations,*
Springer Optimization and Its Applications 96, DOI 10.1007/978-1-4939-1286-5_6

Theorem 1 *Suppose that S is a commutative semigroup, B is a Banach space, $\epsilon \geq 0$, and $f : S \to B$ satisfies the inequality*

$$\|f(x+y) - f(x) - f(y)\| \leq \epsilon \tag{1}$$

for all $x, y \in S$. Then there exists a unique function $A : S \to B$ satisfying

$$A(x+y) = A(x) + A(y) \tag{2}$$

and

$$\|f(x) - A(x)\| \leq \epsilon \tag{3}$$

for all $x \in S$.

In 1950, this result was generalized by T. Aoki [4] and D.G. Bourgin [9, 8]. In 1978 T.M. Rassias generalized the Hyers' result to new approximately linear mappings [?]. Since then the stability problems have been investigated in various directions for many other functional equations. Among the results, the stability problem in a restricted domain was investigated by F. Skof, who proved the stability problem of the inequality (1) in a restricted domain [35]. Several papers have been published on the Hyers–Ulam stability in restricted domains for a large variety of functional equations including the Jensen functional equation [24], quadratic type functional equations [23], mixed type functional equations [30], and Jensen type functional equations [31]. The results can be summarized as follows: Let X and B be a real normed space and a real Banach space, respectively. For fixed $d \geq 0$, if $f : X \to B$ satisfies the functional inequalities (such as that of Cauchy, quadratic, Jensen, and Jensen type, etc.) for all $x, y \in X$ with $\|x\| + \|y\| \geq d$, then the inequalities hold for all $x, y \in X$.

In [14, 15], generalizing the restricted domains such as $\|x\|+\|y\| \geq d$ in a normed space to some abstract domains in a group, we consider the stability problem of Pexider equation and Jensen-type equations in the restricted domains. In the present paper, we consider a weak version of Hyers–Ulam stability of the Pexider equation when the target space of the functions in given functional inequalities are not a normed space but a 2-divisible commutative group. Note that the existence of the approximate additive function A in Theorem 1 is due to the completeness of the target space B. For example, if Y is a noncomplete normed space and $f : S \to Y$ satisfies (1), then we can only find a Cauchy sequence $a_n : S \to Y$ such that

$$|a_n(x+y) - a_n(x) - a_n(y)| \leq 2^{-n}\epsilon \tag{4}$$

for all $x, y \in S$, $n = 1, 2, 3, \ldots$, and

$$|f(x) - a_n(x)| \leq \epsilon \tag{5}$$

for all $x \in S$ and $n = 1, 2, 3, \ldots$. Throughout this paper, we denote a commutative group by G and a 2-divisible commutative group by H respectively, $0 \in V \subset H$ and $W \subset G \times G$. Also, we denote a Banach space and a real normed space by B

and Y, respectively, and $f, g, h : G \to H$(or Y, B). In Sect. 2 of this chapter, we consider the behavior of $f : G \to H$ satisfying

$$f(x + y) - f(x) - f(y) \in V \tag{6}$$

for all $x, y \in G$. As a result we prove that there exists a Cauchy-type sequence $a_n : G \to H$ (which is a Cauchy sequence when $H = Y$) such that

$$f(x) - a_n(x) \in 2^{-n}(V + 2V + \ldots + 2^{n-1}V) \tag{7}$$

for all $x \in G$. In Sect. 3, we consider

$$f(x + y) - g(x) - h(y) \in V \tag{8}$$

for all $(x, y) \in W \subset G \times G$. As the main result we prove that under some assumptions on W, if f, g, h satisfy (8) then there exist approximate Cauchy-type sequences a_n, b_n, and c_n for f, g, and h respectively. From our result we obtain the Hyers–Ulam stability theorem for Pexider equation when $f, g, h : G \to B$.

2 A Weak Stability of Pexider Equation

For subsets V, V_1, V_2 of H, $v \in V$, and $n \in \mathbb{N}$, we define

$$nv = \underbrace{v + \cdots + v}_{n-\text{times}}, \quad nV = \{nv : v \in V\}, \quad 2^{-n}V = \{h \in H : 2^n h \in V\},$$

and

$$V_1 + V_2 = \{v_1 + v_2 : v_1 \in V_1, v_2 \in V_2\}.$$

We call $a_n : G \to H$ a V-Cauchy sequence if

$$a_{m+n}(x) - a_m(x) \in 2^{-m-n}(V + 2V + \ldots + 2^{n-1}V)$$

for all $m, n = 1, 2, 3, \ldots$, and $x \in G$.

First we consider the weak version of the Hyers–Ulam stability theorem for the Cauchy equation.

Theorem 2 *Suppose that $f : G \to H$ satisfies*

$$f(x + y) - f(x) - f(y) \in V \tag{9}$$

for all $x, y \in G$. Then there exists a V-Cauchy sequence $a_n : G \to H$ satisfying

$$a_n(x + y) - a_n(x) - a_n(y) \in 2^{-n}V, \tag{10}$$

and

$$a_n(x) - f(x) \in 2^{-n}(V + 2V + \ldots + 2^{n-1}V) \tag{11}$$

for all $x, y \in G$ and $n \in \mathbb{N}$.

Proof Note that since H is 2-divisible, for each $n \in \mathbb{N}$ and $x \in G$ we can choose an $a_n(x)$ such that

$$2^n a_n(x) = f(2^n x). \tag{12}$$

Replacing y by x in (9) and using induction argument we have

$$2^{n-1} f(2x) - 2^n f(x) \in 2^{n-1} V$$
$$2^{n-2} f(4x) - 2^{n-1} f(2x) \in 2^{n-2} V$$
$$\cdots\cdots\cdots\cdots\cdots\cdots$$
$$2f(2^{n-1}x) - 4f(2^{n-2}x) \in 2V$$
$$f(2^n x) - 2f(2^{n-1}x) \in V$$

for all $x \in G$. Thus it follows that

$$f(2^n x) - 2^n f(x) \in V + 2V + \ldots + 2^{n-1} V \tag{13}$$

for all $x \in G$. Now it follows from (12) and (13) that

$$a_n(x) - f(x) \in 2^{-n}(V + 2V + \ldots + 2^{n-1} V) \tag{14}$$

for all $x \in G$. Replacing x by $2^m x$ in (13) and using (12) we have

$$a_{m+n}(x) - a_m(x) \in 2^{-m-n}(V + 2V + \ldots + 2^{n-1} V) \tag{15}$$

for all $x \in G$, which implies that a_n is V-Cauchy. Replacing x by $2^n x$ and y by $2^n y$ in (9) and using (12) we have

$$a_n(x + y) - a_n(x) - a_n(y) \in 2^{-n} V \tag{16}$$

for all $n \in \mathbb{N}$ and $x \in G$. This completes the proof.

Let $\langle Y, \|\cdot\|\rangle$ be a normed space and $V = \{x \in Y : \|x\| \le \epsilon\}$. Then we have

$$2^{-n}(V + 2V + \ldots + 2^{n-1} V) \subset \{x \in Y : \|x\| \le \epsilon\}$$

for all $n \in \mathbb{N}$, and

$$2^{-m-n}(V + 2V + \ldots + 2^{n-1} V) \subset \{x \in Y : \|x\| \le 2^{-m}\epsilon\}$$

for all $m, n \in \mathbb{N}$. Thus in this case, every V-Cauchy sequence is a Cauchy sequence. Now as a direct consequence of Theorem 2 we have the following.

Corollary 1 *Let $\epsilon > 0$. Suppose that $f : G \to Y$ satisfies*

$$\|f(x + y) - f(x) - f(y)\| \le \epsilon \tag{17}$$

for all $x, y \in G$. Then there exists a Cauchy sequence $a_n : G \to Y$ satisfying

$$\|a_n(x+y) - a_n(x) - a_n(y)\| \le 2^{-n}\epsilon \tag{18}$$

for all $n \in \mathbb{N}$ and $x, y \in G$, and

$$\|a_n(x) - f(x)\| \le \epsilon \tag{19}$$

for all $x \in G$.

In particular, if $f : G \to B$, then there exists $A : G \to B$ such that

$$\lim_{n\to\infty} a_n(x) = A(x).$$

Letting $n \to \infty$ in (18) we have

$$A(x+y) - A(x) - A(y) = 0 \tag{20}$$

for all $x, y \in G$. We call a function $A : G \to B$ satisfying (20) an *additive function*. Thus as a direct consequence of Corollary 1 we have the well known Hyers–Ulam stability theorem.

Corollary 2 *Let $\epsilon > 0$. Suppose that $f : G \to B$ satisfies*

$$\|f(x+y) - f(x) - f(y)\| \le \epsilon \tag{21}$$

for all $x, y \in G$. Then there exists an additive function $A : G \to B$ such that

$$\|f(x) - A(x)\| \le \epsilon \tag{22}$$

for all $x \in G$.

Throughout this chapter we denote

$$V^* = \{v_1 + v_2 - v_3 - v_4 : v_j \in V,\ j = 1,2,3,4\}.$$

Theorem 3 *Suppose that $f, g, h : G \to H$ satisfy*

$$f(x+y) - g(x) - h(y) \in V \tag{23}$$

for all $x, y \in G$. Then there exist V^*-Cauchy sequences $a_n, b_n, c_n : G \to H$ satisfying

$$a_n(x+y) - a_n(x) - a_n(y) \in 2^{-n}V^* \tag{24}$$

$$b_n(x+y) - b_n(x) - b_n(y) \in 2^{-n}V^* \tag{25}$$

$$c_n(x+y) - c_n(x) - c_n(y) \in 2^{-n}V^* \tag{26}$$

for all $n \in \mathbb{N}$ and $x, y \in G$, and

$$a_n(x) - f(x) + f(0) \in V_n^*, \tag{27}$$

$$b_n(x) - g(x) + g(0) \in V_n^*, \tag{28}$$

$$c_n(x) - h(x) + h(0) \in V_n^*, \tag{29}$$

and

$$a_n(x+y) - b_n(x) - c_n(y) \in V_n^{**} \tag{30}$$

for all $n \in \mathbb{N}$ *and* $x, y \in G$, *where*

$$V_n^* = 2^{-n}(V^* + 2V^* + \ldots + 2^{n-1}V^*),$$

$$V_n^{**} = V - V + V_n^* - V_n^* - V_n^*.$$

Proof Let $D(x, y) = f(x + y) - g(x) - h(y)$. Then we have

$$f(x+y) - f(x) - f(y) + f(0) = D(x, y) + D(0, 0) - D(x, 0) - D(y, 0) \in V^* \tag{31}$$

$$g(x+y) - g(x) - g(y) + g(0) = D(x, y) + D(y, 0) - D(x+y, 0) - D(0, y) \in V^* \tag{32}$$

$$h(x+y) - h(x) - h(y) + h(0) = D(x, y) + D(0, x) - D(0, x+y) - D(x, 0) \in V^* \tag{33}$$

for all $x, y \in G$. Thus, in view of (31), (32), and (33), using Theorem 2 for $f(x) - f(0)$, $g(x) - g(0)$, $h(x) - h(0)$, we obtain (24)–(29). Now, putting $x = y = 0$ in (23), we have

$$f(0) - g(0) - h(0) \in V. \tag{34}$$

Then, by (23), (27), (28), (29), and (34) we get (30).

This completes the proof.

In particular, let $V = \{x \in Y : \|x\| \le \epsilon\}$. Then we have

$$V_n^* \subset \{x \in Y : \|x\| \le 4\epsilon\}, \quad V_n^{**} \subset \{x \in Y : \|x\| \le 14\epsilon\}$$

for all $n \in \mathbb{N}$. Thus as a direct consequence of Theorem 3 we have the following.

Corollary 3 *Let* $\epsilon > 0$. *Suppose that* $f, g, h : G \to Y$ *satisfy*

$$\|f(x+y) - g(x) - h(y)\| \le \epsilon \tag{35}$$

for all $x, y \in G$. *Then there exist Cauchy sequences* $a_n, b_n, c_n : G \to Y$ *satisfying*

$$\|a_n(x+y) - a_n(x) - a_n(y)\| \le 2^{-n+2}\epsilon \tag{36}$$

$$\|b_n(x+y) - b_n(x) - b_n(y)\| \le 2^{-n+2}\epsilon \tag{37}$$

$$\|c_n(x+y) - c_n(x) - c_n(y)\| \le 2^{-n+2}\epsilon \tag{38}$$

for all $n \in \mathbb{N}$ and $x, y \in G$, and

$$\|f(x) - a_n(x) - f(0)\| \le 4\epsilon, \tag{39}$$

$$\|g(x) - b_n(x) - g(0)\| \le 4\epsilon, \tag{40}$$

$$\|h(x) - c_n(x) - h(0)\| \le 4\epsilon \tag{41}$$

and

$$\|a_n(x+y) - b_n(x) - c_n(y)\| \le 14\epsilon \tag{42}$$

for all $n \in \mathbb{N}$ and $x, y \in G$.

Corollary 4 *Let $\epsilon > 0$. Suppose that $f, g, h : G \to B$ satisfy*

$$\|f(x+y) - g(x) - h(y)\| \le \epsilon \tag{43}$$

for all $x, y \in G$. Then there exists an additive function $A : G \to B$ such that

$$\|f(x) - A(x) - f(0)\| \le 4\epsilon,$$
$$\|g(x) - A(x) - g(0)\| \le 4\epsilon,$$
$$\|h(x) - A(x) - h(0)\| \le 4\epsilon$$

for all $x \in G$.

Proof Let $A_1(x) = \lim_{n\to\infty} a_n(x)$, $A_2(x) = \lim_{n\to\infty} b_n(x)$, $A_3(x) = \lim_{n\to\infty} c_n(x)$. Then it follows from (36)–(38) that for each $j = 1, 2, 3$, A_j is an additive function. Letting $n \to \infty$ in (39)–(41) we have

$$\|f(x) - A_1(x) - f(0)\| \le 4\epsilon,$$
$$\|g(x) - A_2(x) - g(0)\| \le 4\epsilon,$$
$$\|h(x) - A_3(x) - h(0)\| \le 4\epsilon$$

for all $x \in G$. Finally, letting $n \to \infty$ in (42) we have

$$\|A_1(x+y) - A_2(x) - A_3(y)\| \le 14\epsilon \tag{44}$$

for all $x, y \in G$. Putting $y = 0$ and $x = 0$ in (44) separately, we have

$$\|A_1(x) - A_2(x)\| \le 14\epsilon$$
$$\|A_1(y) - A_3(y)\| \le 14\epsilon$$

for all $x, y \in G$, which implies that $A_1 = A_2$ and $A_1 = A_3$. This completes the proof.

3 Weak Stability of Pexider Equation in Restricted Domains

It is a frequent situation to consider a functional equation satisfied in a restricted domain or satisfied under a restricted condition [3, 5–7, 10–12, 15, 18, 28, 32–35]. In this section we consider the weak version of the Hyers–Ulam stability theorem in some restricted domains in G. We use the following usual notations. Let $G \times G = \{(a_1, a_2) : a_1, a_2 \in G\}$ be the product group. For a subset K of $G \times G$ and $a \in G \times G$, we define $a + K = \{a + k : k \in K\}$. For given $x, y \in G$ we denote the sets of points of the forms (not necessarily distinct) in $G \times G$ by $P_{x,y}$, $Q_{x,y}$, and$R_{x,y}$, respectively as,

$$\begin{aligned} P_{x,y} &= \{(0,0), (x,0), (0,y), (x,y)\}, \\ Q_{x,y} &= \{(y,0), (0,y), (x,y), (x+y,0)\}, \\ R_{x,y} &= \{(x,0), (0,x), (x,y), (0,x+y)\}, \end{aligned}$$

where 0 is the identity element of G. The set $P_{x,y}$ can be viewed as the vertices of a rectangle in $G \times G$, and $Q_{x,y}$ and $R_{x,y}$ can be viewed as the vertices of parallelograms in $G \times G$.

Definition 1 Let $W \subset G \times G$. We introduce the following conditions $(C1)$, $(C2)$, and $(C3)$ on W: For any $x, y \in G$, there exist $z_1, z_2, z_3 \in G$ such that

$$\begin{aligned} &(C1) \quad (-z_1, z_1) + P_{x,y} \subset W, \\ &(C2) \quad (0, z_2) + Q_{x,y} \subset W, \\ &(C3) \quad (z_3, 0) + R_{x,y} \subset W, \end{aligned}$$

respectively.

Example 1 Let G be a real normed space. For $\alpha, \beta, d \in \mathbb{R}$, let

$$U = \{(x,y) \in G \times G : \alpha\|x\| + \beta\|y\| \geq d\}, \tag{45}$$

$$V = \{(x,y) \in G \times G : \|\alpha x + \beta y\| \geq d\}. \tag{46}$$

Then U satisfies $(C1)$ if $\alpha + \beta > 0$, $(C2)$ if $\beta > 0$ and $(C3)$ if $\alpha > 0$, and V satisfies $(C1)$ if $\alpha \neq \beta$, $(C2)$ if $\beta \neq 0$ and $(C3)$ if $\alpha \neq 0$.

Example 2 Let G be a real inner product space. For $d \geq 0$, $x_0, y_0 \in G$

$$U = \{(x,y) \in G \times G : \langle x_0, x\rangle + \langle y_0, y\rangle \geq d\}. \tag{47}$$

Then U satisfies $(C1)$, if $x_0 \neq y_0$, $(C2)$ if $y_0 \neq 0$ and $(C3)$ if $x_0 \neq 0$.

Example 3 Let G be the group of nonsingular square matrices with the operation of matrix multiplication. For $\alpha, \beta \in \mathbb{R}$, $\delta, d \geq 0$, let

$$U = \{(P_1, P_2) \in G \times G : |\det P_1|^{\alpha} |\det P_2|^{\beta} \le \delta\}, \tag{48}$$

$$U = \{(P_1, P_2) \in G \times G : |\det P_1|^{\alpha} |\det P_2|^{\beta} \ge d\}. \tag{49}$$

Then U satisfies $(C1)$ if $\alpha \neq \beta$, $(C2)$ if $\beta \neq 0$, and $(C3)$ if $\alpha \neq 0$.

In the following one can see that if $P_{x,y}$, $Q_{x,y}$, and $R_{x,y}$ are replaced by arbitrary subsets of four points (not necessarily distinct) in $G \times G$, respectively, the conditions become stronger, that is, there are subsets U_j, $j = 1, 2, 3$, which satisfy the conditions $(C1)$, $(C2)$, and $(C3)$, respectively, but U_j, $j = 1, 2, 3$, fail to fulfill the following conditions (2.6), (2.7), and (2.8), respectively: For any subset $\{p_1, p_2, p_3, p_4\}$ of points (not necessarily distinct) in $G \times G$, there exists a $z \in G$ such that

$$(e, z)\{p_1, p_2, p_3, p_4\}(z^{-1}, e) \subset U_1, \tag{50}$$

$$\{p_1, p_2, p_3, p_4\}(e, z) \subset U_2, \tag{51}$$

$$(z, e)\{p_1, p_2, p_3, p_4\} \subset U_3, \tag{52}$$

respectively.

Now we give examples of U_1, U_2, U_3 which satisfy $(C1), (C2)$, and $(C3)$, respectively, but not (50), (51), and (52), respectively.

Example 4 Let $G = \mathbb{Z}$ be the group of integers. Enumerating

$$\mathbb{Z} \times \mathbb{Z} = \{(a_1, b_1), (a_2, b_2), \ldots, (a_n, b_n), \ldots\}$$

such that

$$|a_1| + |b_1| \le |a_2| + |b_2| \le \cdots \le |a_n| + |b_n| \le \cdots,$$

and let $P_n = \{(0, 0), (a_n, 0), (0, b_n), (a_n, b_n)\}$, $n = 1, 2, \ldots$. Then it is easy to see that $U_1 = \bigcup_{n=1}^{\infty} (P_n + (-2^n, 2^n))$ satisfies the condition $(C1)$. Now let $P = \{(x_1, y_1), (x_2, y_2)\} \subset \mathbb{Z} \times \mathbb{Z}$ with $x_2 > x_1$, $y_2 > y_1$, $(x_1 + y_1)(x_2 + y_2) > 0$. Then $P + (-z, z)$ is not contained in U_1 for all $z \in \mathbb{Z}$. Indeed, let $(a, b) \in P_n + (-2^n, 2^n), (c, d) \in P_{n+1} + (-2^{n+1}, 2^{n+1})$. Then we have $a > c$, $b < d$ for all $n = 1, 2, \ldots$. Thus it follows from $x_2 > x_1$, $y_2 > y_1$ that if $P + (-z, z) \subset U_1$, then $P + (-z, z) \subset P_n + (-2^n, 2^n)$ for some $n \in \mathbb{N}$, which implies that the line segment joining the points of $P + (-z, z)$ intersects the line $y = -x$ in $\mathbb{R}^2$, contradicting to the condition $(x_1 + y_1)(x_2 + y_2) > 0$. Similarly, let $Q_n = \{(b_n, 0), (0, b_n), (a_n, b_n), (a_n + b_n, 0)\}$ and $R_n = \{(a_n, 0), (0, a_n), (a_n, b_n), (0, a_n + b_n)\}$, $n = 1, 2, \ldots$. Then it is easy to see that $U_2 = \bigcup_{n=1}^{\infty} (Q_n + (0, 2^n))$ satisfies the condition $(C2)$ but not (2.7) and $U_3 = \bigcup_{n=1}^{\infty} (R_n + (2^n, 0))$ satisfies the condition $(C3)$ but not (52).

As in Sect. 2, we denote

$$V^* = \{v_1 + v_2 - v_3 - v_4 : v_j \in V,\ j = 1, 2, 3, 4\}.$$

Theorem 4 *Let W satisfy the condition $(C1)$. Suppose that $f, g, h : G \to H$ satisfy*

$$f(x + y) - g(x) - h(y) \in V \tag{53}$$

for all $(x, y) \in W$. *Then there exists a* V^**-Cauchy sequence* $a_n : G \to H$ *satisfying*

$$a_n(x + y) - a_n(x) - a_n(y) \in 2^{-n} V^* \tag{54}$$

for all $n \in \mathbb{N}$ *and* $x, y \in G$ *and*

$$a_n(x) - f(x) + f(0) \in 2^{-n}(V^* + 2V^* + \ldots + 2^{n-1} V^*) \tag{55}$$

for all $x \in G$.

Proof For given $x, y \in G$, choose $z \in G$ such that $(-z, z) + P_{x,y} \subset W$. Then we have

$$\begin{aligned} f(x + y) - g(x - z) - h(z + y) &\in V, \\ - f(x) + g(x - z) + h(z) &\in -V, \\ - f(y) + g(-z) + h(z + y) &\in -V, \\ + f(0) - g(-z) - h(z) &\in V. \end{aligned}$$

Thus it follows that

$$f(x + y) - f(x) - f(y) + f(0) \in V + (-V) + (-V) + V = V^* \tag{56}$$

for all $x, y \in G$.

Now by Theorem 2, there exists a V^*-Cauchy sequence $a_n : G \to H$ satisfying (54) and (55). This completes the proof.

In particular, let $V = \{x \in Y : \|x\| \le \epsilon\}$. Then we have

$$V^* \subset \{x \in Y : \|x\| \le 4\epsilon\}, \quad 2^{-n}(V^* + 2V^* + \ldots + 2^{n-1} V^*) \subset \{x \in Y : \|x\| \le 4\epsilon\}$$

for all $n \in \mathbb{N}$, and

$$2^{-m-n}(V^* + 2V^* + \ldots + 2^{n-1} V^*) \subset \{x \in Y : \|x\| \le 2^{-m+2}\epsilon\}$$

for all $m, n \in \mathbb{N}$. Thus in this case, every V^*-Cauchy sequence is a Cauchy sequence. Now as a direct consequence of Theorem 4 we have the following.

Corollary 5 *Let* W *satisfy the condition* $(C1)$ *and* $\epsilon \ge 0$. *Suppose that* $f, g, h : G \to Y$ *satisfy*

$$\|f(x + y) - g(x) - h(y)\| \le \epsilon \tag{57}$$

for all $(x, y) \in W$. *Then there exists a Cauchy sequence* $a_n : G \to Y$ *satisfying*

$$\|a_n(x + y) - a_n(x) - a_n(y)\| \le 2^{-n+2}\epsilon \tag{58}$$

for all $n \in \mathbb{N}$ *and* $x, y \in G$, *and*

$$\|a_n(x) - f(x) + f(0)\| \le 4\epsilon \tag{59}$$

for all $x \in G$.

As a direct consequence of Corollary 5 we have the following.

Corollary 6 *Let* W *satisfy the condition* $(C1)$ *and* $\epsilon \geq 0$. *Suppose that* $f, g, h : G \to B$ *satisfy*

$$\|f(x+y) - g(x) - h(y)\| \leq \epsilon \tag{60}$$

for all $(x, y) \in W$. *Then there exists an additive function* $A_1 : G \to B$ *and*

$$\|f(x) - A_1(x) - f(0)\| \leq 4\epsilon \tag{61}$$

for all $x \in G$.

Theorem 5 *Let* W *satisfy the condition* $(C2)$. *Suppose that* $f, g, h : G \to H$ *satisfy*

$$f(x+y) - g(x) - h(y) \in V \tag{62}$$

for all $(x, y) \in W$. *Then there exists a* V^**-Cauchy sequence* $b_n : G \to H$ *satisfying*

$$b_n(x+y) - b_n(x) - b_n(y) \in 2^{-n} V^* \tag{63}$$

for all $n \in \mathbb{N}$ *and* $x, y \in G$, *and*

$$b_n(x) - g(x) + g(0) \in 2^{-n}(V^* + 2V^* + \ldots + 2^{n-1}V^*) \tag{64}$$

for all $x \in G$.

Proof For given $x, y \in G$, choose $z \in G$ such that $(0, z) + Q_{x,y} \subset W$. Then we have

$$\begin{aligned} -f(x+y+z) + g(x+y) + h(z) &\in -V, \\ f(x+y+z) - g(x) - h(y+z) &\in V, \\ f(y+z) - g(y) - h(z) &\in V, \\ -f(y+z) + g(0) + h(y+z) &\in -V. \end{aligned}$$

Thus it follows that

$$g(x+y) - g(x) - g(y) + g(0) \in -V + V + V - V = V^* \tag{65}$$

for all $x, y \in G$. Now by Theorem 2, there exists a sequence $b_n : G \to H$ satisfying (63) and (64). This completes the proof.

In particular, if $f, g, h : G \to Y$ we have the following.

Corollary 7 *Let* W *satisfy the condition* $(C2)$ *and* $\epsilon \geq 0$. *Suppose that* $f, g, h : G \to Y$ *satisfy*

$$\|f(x+y) - g(x) - h(y)\| \leq \epsilon \tag{66}$$

for all $(x, y) \in W$. *Then there exists a Cauchy sequence* $b_n : G \to Y$ *satisfying*

$$\|b_n(x+y) - b_n(x) - b_n(y)\| \leq 2^{-n+2}\epsilon \tag{67}$$

for all $n \in \mathbb{N}$ *and* $x, y \in G$, *and*

$$\|b_n(x) - g(x) + g(0)\| \leq 4\epsilon \tag{68}$$

for all $x \in G$.

In particular, if $f, g, h : G \to B$ we have the following.

Corollary 8 *Let* W *satisfy the condition* $(C2)$ *and* $\epsilon \geq 0$. *Suppose that* $f, g, h : G \to B$ *satisfy*

$$\|f(x+y) - g(x) - h(y)\| \leq \epsilon \tag{69}$$

for all $(x, y) \in W$. *Then there exists a unique additive function* $A_2 : G \to B$ *such that*

$$\|g(x) - A_2(x) - g(0)\| \leq 4\epsilon \tag{70}$$

for all $x \in G$.

Theorem 6 *Let* W *satisfy the condition* $(C3)$. *Suppose that* $f, g, h : G \to H$ *satisfy*

$$f(x+y) - g(x) - h(y) \in V \tag{71}$$

for all $(x, y) \in W$. *Then there exists a* V^*-*Cauchy sequence* $c_n : G \to H$ *satisfying*

$$c_n(x+y) - c_n(x) - c_n(y) \in 2^{-n} V^* \tag{72}$$

for all $n \in \mathbb{N}$ *and* $x, y \in G$ *and*

$$c_n(x) - h(x) + h(0) \in 2^{-n}(V^* + 2V^* + \ldots + 2^{n-1} V^*) \tag{73}$$

for all $x \in G$.

Proof For given $x, y \in G$, choose $z \in G$ such that $(0, z) + Q_{x,y} \subset W$. Then we have

$$\begin{aligned} -f(z+x+y) + g(z) + h(x+y) &\in -V, \\ f(z+x+y) - g(z+x) - h(y) &\in V, \\ f(z+x) - g(z) - h(x) &\in V, \\ -f(z+x) + g(z+x) + h(0) &\in -V. \end{aligned}$$

Thus it follows that

$$h(x+y) - h(x) - h(y) + h(0) \in -V + V + V - V = V^* \tag{74}$$

for all $x, y \in G$. Now by Theorem 2, there exists a sequence $c_n : G \to H$ satisfying (72) and (73). This completes the proof.

In particular, if $f, g, h : G \to Y$ we have the following.

Corollary 9 *Let W satisfy the condition $(C3)$ and $\epsilon \geq 0$. Suppose that $f, g, h : G \to Y$ satisfy*

$$\|f(x+y) - g(x) - h(y)\| \leq \epsilon \tag{75}$$

for all $(x, y) \in W$. Then there exists a Cauchy sequence $c_n : G \to Y$ satisfying

$$\|c_n(x+y) - c_n(x) - c_n(y)\| \leq 2^{-n+2}\epsilon \tag{76}$$

for all $n \in \mathbb{N}$ and $x, y \in G$, and

$$\|c_n(x) - h(x) + h(0)\| \leq 4\epsilon \tag{77}$$

for all $x \in G$.

In particular, if $f, g, h : G \to B$ we have the following.

Corollary 10 *Let W satisfy the condition $(C3)$ and $\epsilon \geq 0$. Suppose that $f, g, h : G \to B$ satisfy*

$$\|f(x+y) - g(x) - h(y)\| \leq \epsilon \tag{78}$$

for all $(x, y) \in W$. Then there exists a unique additive function $A_3 : G \to B$ such that

$$\|h(x) - A_3(x) - h(0)\| \leq 4\epsilon \tag{79}$$

for all $x \in G$.

Theorem 7 *Let W satisfy all the conditions $(C1)$, $(C2)$, and $(C3)$. Suppose that $f, g, h : G \to H$ satisfy*

$$f(x+y) - g(x) - h(y) \in V \tag{80}$$

for all $(x, y) \in W$. Then there exist V^*-Cauchy sequences $a_n, b_n, c_n : G \to H$ satisfying

$$a_n(x+y) - a_n(x) - a_n(y) \in 2^{-n}V^* \tag{81}$$

$$b_n(x+y) - b_n(x) - b_n(y) \in 2^{-n}V^* \tag{82}$$

$$c_n(x+y) - c_n(x) - c_n(y) \in 2^{-n}V^* \tag{83}$$

for all $n \in \mathbb{N}$ and $x, y \in G$, and

$$a_n(x) - f(x) + f(0) \in V_n^*, \quad (84)$$

$$b_n(x) - g(x) + g(0) \in V_n^*, \quad (85)$$

$$c_n(x) - h(x) + h(0) \in V_n^* \quad (86)$$

for all $n \in \mathbb{N}$ and $x \in G$, and

$$a_n(x + y) - b_n(x) - c_n(y) \in V_n^{**}. \quad (87)$$

for all $n \in \mathbb{N}$ and $x, y \in G$, where

$$V_n^* = 2^{-n}(V^* + 2V^* + \ldots + 2^{n-1}V^*),$$

$$V_n^{**} = V + V + V + V + V - V - V - V - V - V + V_n^* - V_n^* - V_n^*.$$

Proof From Theorems 4, 5, and 6, it remains to show (87). By the condition $(C1)$, for given $x, y \in G$, choose $z \in G$ such that $(-z, z), (x - z, z + y) \in W$. Then from (80) we have

$$f(x + y) - g(x - z) - h(z + y) \in V, \quad (88)$$

$$-f(0) + g(-z) + h(z) \in -V. \quad (89)$$

Also, by (65) and (74) we have

$$g(x - z) - g(x) - g(-z) + g(0) \in V + V - V - V, \quad (90)$$

$$h(z + y) - h(z) - h(y) + h(0) \in V + V - V - V. \quad (91)$$

for all $x, y, z \in G$. From (88)–(91), we have

$$f(x+y)-g(x)-h(y)-f(0)+g(0)+h(0) \in V+V+V+V+V-V-V-V-V-V \quad (92)$$

for all $x, y \in G$. Using (84), (85), (86), and (92) we have

$$a_n(x+y)-b_n(x)-c_n(y) \in V+V+V+V+V-V-V-V-V-V+V_n^*-V_n^*-V_n^*. \quad (93)$$

This completes the proof.

In particular, let $V = \{x \in Y : \|x\| \leq \epsilon\}$. Then we have

$$V_n^* \subset \{x \in Y : \|x\| \leq 4\epsilon\}, \quad V_n^{**} \subset \{x \in Y : \|x\| \leq 22\epsilon\}$$

for all $n \in \mathbb{N}$. Thus as a direct consequence of Theorem 7 we have the following.

Corollary 11 *Let W satisfy the conditions $(C1)$, $(C2)$, and $(C3)$ and $\epsilon \geq 0$. Suppose that $f, g, h : G \to Y$ satisfy*

$$\|f(x + y) - g(x) - h(y)\| \leq \epsilon \quad (94)$$

for all $(x, y) \in W$. *Then there exist Cauchy sequences* $a_n, b_n, c_n : G \to Y$ *satisfying*

$$\|a_n(x+y) - a_n(x) - a_n(y)\| \le 2^{-n+2}\epsilon \tag{95}$$

$$\|b_n(x+y) - b_n(x) - b_n(y)\| \le 2^{-n+2}\epsilon \tag{96}$$

$$\|c_n(x+y) - c_n(x) - c_n(y)\| \le 2^{-n+2}\epsilon \tag{97}$$

for all $n \in \mathbb{N}$ *and* $x, y \in G$,

$$\|f(x) - a_n(x) - f(0)\| \le 4\epsilon, \tag{98}$$

$$\|g(x) - b_n(x) - g(0)\| \le 4\epsilon, \tag{99}$$

$$\|h(x) - c_n(x) - h(0)\| \le 4\epsilon \tag{100}$$

for all $n \in \mathbb{N}$ *and* $x \in G$, *and*

$$\|a_n(x+y) - b_n(x) - c_n(y)\| \le 22\epsilon \tag{101}$$

for all $n \in \mathbb{N}$ *and* $x, y \in G$.

Corollary 12 *Let* W *satisfy the conditions* $(C1)$, $(C2)$, *and* $(C3)$ *and* $\epsilon \ge 0$. *Suppose that* $f, g, h : G \to B$ *satisfy*

$$\|f(x+y) - g(x) - h(y)\| \le \epsilon \tag{102}$$

for all $(x, y) \in W$. *Then there exists an additive function* $A : G \to B$ *such that*

$$\|f(x) - A(x) - f(0)\| \le 4\epsilon,$$

$$\|g(x) - A(x) - g(0)\| \le 4\epsilon,$$

$$\|h(x) - A(x) - h(0)\| \le 4\epsilon$$

for all $x \in G$.

Proof Let $A_1(x) = \lim_{n\to\infty} a_n(x)$, $A_2(x) = \lim_{n\to\infty} b_n(x)$, $A_3(x) = \lim_{n\to\infty} c_n(x)$. Then it follows from (95)–(97) that for each $j = 1, 2, 3$, A_j is additive. Letting $n \to \infty$ in (98)–(100) we have

$$\|f(x) - A_1(x) - f(0)\| \le 4\epsilon,$$

$$\|g(x) - A_2(x) - g(0)\| \le 4\epsilon,$$

$$\|h(x) - A_3(x) - h(0)\| \le 4\epsilon$$

for all $x \in G$. Finally letting $n \to \infty$ in (101) we have

$$\|A_1(x+y) - A_2(x) - A_3(y)\| \le 22\epsilon \tag{103}$$

for all $x, y \in G$. Putting $y = 0$ and $x = 0$ in (103) separately, we have

$$\|A_1(x) - A_2(x)\| \le 22\epsilon$$

$$\|A_1(y) - A_3(y)\| \le 22\epsilon$$

for all $x, y \in G$, which implies that $A_1 = A_2$ and $A_1 = A_3$. This completes the proof.

In particular, if G is a normed vector space we have the following.

Corollary 13 *Let $d > 0$. Suppose that $f, g, h : G \to B$ satisfy*

$$\|f(x+y) - g(x) - h(y)\| \le \epsilon \tag{104}$$

for all $\|x\| + \|y\| \ge d$. Then there exists an additive function $A : G \to B$ such that

$$\|f(x) - A(x) - f(0)\| \le 4\epsilon,$$
$$\|g(x) - A(x) - g(0)\| \le 4\epsilon,$$
$$\|h(x) - A(x) - h(0)\| \le 4\epsilon$$

for all $x \in G$.

Finally we give another interesting example of the set $W \subset \mathbb{R}^n \times \mathbb{R}^n$ with finite Lebesgue measure satisfying all the conditions $(C1)$.

Lemma 1 *Let $D := \{(x_1, y_1), (x_2, y_2), (x_3, y_3), \ldots\}$ be a countable dense subset of $\mathbb{R}^2$. For each $j = 1, 2, 3, \ldots$, we denote by*

$$R_j = \{(x, y) \in \mathbb{R}^2 : |x - x_j| < 1,\ |y - y_j| < 2^{-j}\epsilon\}$$

the rectangle in $\mathbb{R}^2$ with center (x_j, y_j) and let $W = \bigcup_{j=1}^{\infty} R_j$. It is easy to see that the Lebesgue measure $m(W)$ of U satisfies $m(W) \le \epsilon$. Now for $d > 0$, let

$$W_d = W \cap \{(x, y) \in \mathbb{R}^2 : |x| + |y| > d\}.$$

Then W_d satisfies $(C1)$.

Proof For given $x, y \in \mathbb{R}$ we choose a $p \in \mathbb{R}$ such that

$$|p| \ge d + |x| + |y| + 1. \tag{105}$$

We first choose $(x_{i_1}, y_{i_1}) \in K$ such that

$$|-p - x_{i_1}| + |p - y_{i_1}| < \frac{1}{4}, \tag{106}$$

and then we choose $(x_{i_2}, y_{i_2}) \in K$, $(x_{i_3}, y_{i_3}) \in K$ and $(x_{i_4}, y_{i_4}) \in K$ with $1 < i_1 < i_2 < i_3 < i_4$, step by step, satisfying

$$|x - y_{i_1} - x_{i_2}| + |y_{i_1} - y_{i_2}| < 2^{-i_1-1}, \tag{107}$$

$$|x - y_{i_2} - x_{i_3}| + |y + y_{i_2} - y_{i_3}| < 2^{-i_2-1}, \tag{108}$$

$$|y - y_{i_3} - x_{i_4}| + |y_{i_3} - y_{i_4}| < 2^{-i_3-1}. \tag{109}$$

Let

$$
\begin{aligned}
z_1 &= y_{i_1} - p,\\
z_2 &= y_{i_2} - y_{i_1},\\
z_3 &= y_{i_3} - y_{i_2} - y,\\
z_4 &= y_{i_4} - y_{i_3},
\end{aligned}
$$

and

$$z = z_1 + z_2 + z_3 + z_4.$$

Then from (106)–(109) we have

$$|z_1| < \frac{1}{4},\ |z_2| < 2^{-i_1-1},\ |z_3| < 2^{-i_2-1},\ |z_4| < 2^{-i_3-1},\ |z| < \frac{1}{2}. \tag{110}$$

Thus from (105), (106), and (110) we have

$$
\begin{aligned}
|-p-z| + |p+z| &\geq 2(|p| - |z|) \geq 2(|p| - \frac{1}{2}) \qquad (111)\\
&> 2d \geq d,\\
|-p-z-x_{i_1}| &\leq |-p-x_{i_1}| + |z| \qquad (112)\\
&< \frac{1}{4} + \frac{1}{2} < 1,
\end{aligned}
$$

and

$$|p+z-y_{i_1}| = |z_2+z_3+z_4| < 2^{-i_1-1} + 2^{-i_2-1} + 2^{-i_3-1} < 2^{-i_1}. \tag{113}$$

The inequalities (111), (112), and (113) imply

$$(-p-z, p+z) \in W_d. \tag{114}$$

Also from the inequalities

$$
\begin{aligned}
|x-p-z| + |p+z| &\geq 2(|p| - |x| - |z|) > 2(|p| - |x| - \frac{1}{2}) > d,\\
|x-p-z-x_{i_2}| &\leq |x - y_{i_1} - x_{i_2}| + |z_2| + |z_3| + |z_4|\\
&< \frac{1}{8} + \frac{1}{8} + \frac{1}{16} + \frac{1}{32} < 1,
\end{aligned}
$$

and

$$|p+z-y_{i_2}| = |z_3+z_4| < 2^{-i_2-1} + 2^{-i_3-1} < 2^{-i_2},$$

we have

$$(x-p-z, p+z) \in W_d. \tag{115}$$

Similarly, using the inequalities

$$|x - p - z - x_{i_3}| \le |x - y_{i_2} - x_{i_3}| + |z_3| + |z_4| < 1,$$
$$|y + p + z - y_{i_3}| = |z_4| < 2^{-i_3},$$
$$|- p - z - x_{i_4}| \le |y - y_{i_3} - x_{i_4}| + |z_4| < 1,$$
$$|y + p + z - y_{i_4}| = 0,$$

we have

$$(x - p - z, y + p + z), (- p - z, y + p + z) \in W_d. \tag{116}$$

Let $\{(x_1, y_1), (x_2, y_2), (x_3, y_3), \dots\}$ be defined as above. For each $j = 1, 2, 3, \dots,$ let

$$S_j = \{(x, y) : x, y \in \mathbb{R} : |x + y - x_j - y_j| < 1, |x - y - x_j + y_j| < 2^{-j}\epsilon\}$$

and let $V = \bigcup_{j=1}^{\infty} S_j$. Then V satisfies $m(V) \le \epsilon$. For fixed $d > 0$, let

$$V_d = V \cap \{(x, y) \in \mathbb{R}^2 : |x| + |y| > d\}.$$

Using the similar method as in the proof of Lemma 1 we can show that V_d satisfies the conditions $(C1), (C2)$, and $(C3)$.

As a direct consequence of Lemma 1 we have the following.

Theorem 8 *Let $d > 0$. Suppose that $f : \mathbb{R} \to \mathbb{R}$ satisfies*

$$|f(x + y) - f(x) - f(y)| \le \epsilon \tag{117}$$

for all $(x, y) \in W_d$. Then there exists a unique additive function $A : \mathbb{R} \to \mathbb{R}$ such that

$$|f(x) - A(x)| \le 3\epsilon \tag{118}$$

for all $x \in \mathbb{R}$.

Proof It follows from (115) and (116) that for given $x, y \in \mathbb{R}$ there exist $p, z \in \mathbb{R}$ satisfying

$$\begin{aligned} |f(x + y) - f(x) - f(y)| &\le |- f(x) + f(x - p - z) + f(p + z)| \\ &\quad + |f(x + y) - f(x - p - z) - f(y + p + z)| \\ &\quad + |- f(y) + f(- p - z) + f(y + p + z)| \\ &\le 3\epsilon. \end{aligned}$$

Using Theorem A we get the result.

As a consequence of Theorem 8 we obtain an asymptotic behavior of

$$C_d(f) := \sup_{(x,y)\in W_d} |f(x + y) - f(x) - f(y)| \to 0 \tag{119}$$

as $d \to \infty$.

Theorem 9 *Suppose that* $f : \mathbb{R} \to \mathbb{R}$ *satisfies the condition*

$$C_d(f) \to 0 \tag{120}$$

as $d \to \infty$. *Then* f *is an additive function.*

Proof By the condition (120), for each $j \in \mathbb{N}$, there exists $d_j > 0$ such that

$$|f(x+y) - f(x) - f(y)| \le \frac{1}{j}$$

for all $(x, y) \in W_{d_j}$. By Theorem 8, there exists a unique additive function $A_j : \mathbb{R} \to \mathbb{R}$ such that

$$|f(x) - A_j(x)| \le \frac{3}{j} \tag{121}$$

for all $x \in \mathbb{R}$. From (121), using the triangle inequality we have

$$|A_j(x) - A_k(x)| \le \frac{3}{j} + \frac{3}{k} \le 6 \tag{122}$$

for all $x \in \mathbb{R}$ and all positive integers j, k. Now, the inequality (122) implies $A_j = A_k$. Indeed, for all $x \in \mathbb{R}$ and all rational numbers $r > 0$ we have

$$|A_j(x) - A_k(x)| = \frac{1}{r}|A_j(rx) - A_k(rx)| \le \frac{6}{r}. \tag{123}$$

Letting $r \to \infty$ in (123) we have $A_j = A_k$. Thus, letting $j \to \infty$ in (121) we get the result.

Acknowledgements This work was supported by Basic Science Research Program through the National Research Foundation of Korea (NRF) funded by the Ministry of Education, Science and Technology (MEST) (no. 2012008507).

References

1. Aczél, J., Dhombres, J.: Functional Equations in Several Variables. Cambridge University Press, New York-Sydney (1989)
2. Aczél, J., Chung J.K.: Integrable solutions of functional equations of a general type. Studia Sci. Math. Hung. **17**, 51–67 (1982)
3. Alsina, C., Garcia-Roig, J.L.: On a conditional Cauchy equation on rhombuses. In: Rassias, J.M. (ed.), Functional Analysis, Approximation Theory and Numerical Analysis. World Scientific, Singapore (1994)
4. Aoki, T.: On the stability of the linear transformation in Banach spaces. J. Math. Soc. Jpn **2**, 64–66 (1950)

5. Bahyrycz, A., Brzdęk, J.: On solutions of the d'Alembert equation on a restricted domain. Aequat. Math. **85**, 169–183 (2013)
6. Batko, B.: Stability of an alternative functional equation. J. Math. Anal. Appl. **339**, 303–311 (2008)
7. Batko, B.: On approximation of approximate solutions of Dhombres' equation. J. Math. Anal. Appl. **340**, 424–432 (2008)
8. Bourgin, D.G.: Multiplicative transformations. Proc. Nat. Acad. Sci. U S A **36**, 564–570 (1950)
9. Bourgin, D.G.: Class of transformations and bordering transformations. Bull. Amer. Math. Soc. **57**, 223–237 (1951)
10. Brzdęk, J.: On the quotient stability of a family of functional equations, Nonlin. Anal. TMA **71**, 4396–4404 (2009)
11. Brzdęk, J.: On a method of proving the Hyers-Ulam stability of functional equations on restricted domains. Aust. J. Math. Anal. Appl. **6**, 1–10 (2009)
12. Brzdęk, J., Sikorska, J.: A conditional exponential functional equation and its stability. Nonlin. Anal. TMA **72**, 2929–2934 (2010)
13. Chung, J.: Stability of functional equations on restricted domains in a group and their asymptotic behaviors. Comput. Math. Appl. **60**, 2653–2665 (2010)
14. Chung, J.: Stability of conditional Cauchy functional equations. Aequat. Math. **83**, 313–320 (2012)
15. Chung, J.: Stability of Jensen-type functional equation on restricted domains in a group and their asymptotic behaviors. J. Appl. Math. **2012**, 12, Article ID 691981, (2012)
16. Czerwik, S.: Stability of Functional Equations of Hyers-Ulam-Rassias Type. Hadronic Press, Palm Harbor (2003).
17. Fochi, M.: An alternative functional equation on restricted domain. Aequat. Math. **70**, 2010–212 (2005)
18. Ger, R., Sikorska, J.: On the Cauchy equation on spheres. Ann. Math. Sil. **11**, 89–99 (1997)
19. Gordji, M.E., Rassias, T. M.: Ternary homomorphisms between unital ternary C^*-algebras. Proc. Roman. Acad. Series A **12**(3), 189–196 (2011)
20. Hyers, D.H.: On the stability of the linear functional equations. Proc. Nat. Acad. Sci. U S A **27**, 222–224 (1941)
21. Hyers, D.H., Rassias T.M.: Approximate homomorphisms. Aequat. Math. **44**, 125–153 (1992)
22. Hyers, D.H., Isac, G., Rassias T.M.: Stability of Functional Equations in Several Variables. Birkhauser, Boston (1998)
23. Jung, S.M.: On the Hyers–Ulam stability of functional equations that have the quadratic property. J. Math. Anal. Appl. **222**, 126–137 (1998)
24. Jung, S.M.: Hyers–Ulam stability of Jensen's equation and its application. Proc. Amer. Math. Soc. **126**, 3137–3143 (1998)
25. Jung, S.-M.: Hyers–Ulam–Rassias Stability of Functional Equations in Nonlinear Analysis. Springer, New York (2011)
26. Jung, S.-M., Rassias, T. M.: Ulam's problem for approximate homomorphisms in connection with Bernoulli's differential equation. Appl. Math. Comput. **187**(1), 223–227 (2007)
27. Jung, S.-M., Rassias T.M.: Approximation of analytic functions by Chebyshev functions. Abst. Appl. Anal. **2011**, 10 p, Article ID 432961, (2011)
28. Kuczma, M.: Functional equations on restricted domains. Aequat. Math. **18**, 1–34 (1978)
29. Park, C., W.-G. Park, Lee, J.R., Rassias, T.M.: Stability of the Jensen type functional equation in Banach algebras: A fixed point approach. Korean J. Math. **19**(2), 149–161 (2011)
30. Rassias, T.M.: On the stability of linear mapping in Banach spaces. Proc. Amer. Math. Soc. **72**, 297–300 (1978)
31. Rassias, T.M.: On the stability of functional equations in Banach spaces. J. Math. Anal. Appl. **251**, 264–284 (2000)
32. Rassias, J.M., Rassias, M.J.: On the Ulam stability of Jensen and Jensen type mappings on restricted domains. J. Math. Anal. Appl. **281**, 516–524 (2003)

33. Rassias, M.J., Rassias, J.M. On the Ulam stability for Euler-Lagrange type quadratic functional equations. Austral. J. Math. Anal. Appl. **2**, 1–10 (2005)
34. Sikorska, J.: On two conditional Pexider functinal equations and their stabilities. Nonlin. Anal. TMA **70**, 2673–2684 (2009)
35. Skof, F.: Sull'approssimazione delle applicazioni localmente $\delta-$additive. Atii Accad. Sci. Torino Cl. Sci. Fis. Mat. Natur. **117**, 377–389 (1983)
36. Ulam, S.M.: A Collection of Mathematical Problems. Interscience, New York (1960)

On the Stability of Drygas Functional Equation on Amenable Semigroups

Elhoucien Elqorachi, Youssef Manar and Themistocles M. Rassias

Abstract In this chapter, we will prove the Hyers–Ulam stability of Drygas functional equation

$$f(xy) + f(x\sigma(y)) = 2f(x) + f(y) + f(\sigma(y)), \quad x, y \in G,$$

where G is an amenable semigroup, σ is an involution of G and $f : G \to E$ is approximatively central (i.e., $|f(xy) - f(yx)| \leq \delta$).

Keywords Hyers–Ulam stability · Drygas functional equation · Amenable semigroups · Invariant means · Semigroups · Abelian groups

1 Introduction

The stability problem of functional equations originated from a question of Ulam [50] in 1940, concerning the stability of group homomorphisms

> Given a group G_1 and a metric group G_2 with a metric $d(\cdot,\cdot)$. Given $\epsilon > 0$, does there exist a $\delta > 0$ such that if $f : G_1 \to G_2$ satisfies $d(f(xy), f(x)f(y)) < \delta$ for all $x, y \in G_1$, then a homomorphism $g : G_1 \to G_2$ exist with $d(f(x), g(x)) < \epsilon$ for all $x \in G_1$?

In other words, under what condition does there exists a homomorphism near an approximate homomorphism? The concept of stability for functional equation arises when we replace the functional equation by an inequality which outs as a perturbation of the equation. The case of approximately additive mappings was solved by Hyers [23] under the assumption that G_1 and G_2 are banach spaces. In 1950, Aoki [1]

E. Elqorachi · Y. Manar (✉)
Department of Mathematics, Faculty of Sciences, University Ibn Zohr, Agadir, Morocco
e-mail: elqorachi@hotmail.com

Y. Manar
e-mail: manaryoussef1984@gmail.com

T. M. Rassias
Department of Mathematics, National Technical University of Athens,
Zografou Campus, 15780 Athens, Greece
e-mail: trassias@math.ntua.gr

T. M. Rassias (ed.), *Handbook of Functional Equations*,
Springer Optimization and Its Applications 96, DOI 10.1007/978-1-4939-1286-5_7

provided a generalization of the Hyers' Theorem for additive mappings. In 1978, Rassias [41] succeeded in extending the result of Hyers for linear mappings by allowing the Cauchy difference to be unbounded. Since then, several mathematicians have been attracted to the results of Hyers and Rassias and stimulated to investigate the stability problems of functional equations. The interested reader can get a rapid overview on the subject by consulting the books of Czerwik [10], Hyers et al. [24] and Jung [27], Kannappan [29] or the survey papers of Forti [17], Ger [21], Hyers and Rassias [25] and Székelyhidi [49]. We refer also to the references: [3, 4, 7–10, 20, 21, 27, 32–46]. Let G be a semigroup and E a Banach space. A mapping $f : G \to E$ will be called a solution of the generalized Drygas functional equation if it satisfies

$$f(xy) + f(x\sigma(y)) = 2f(x) + f(y) + f(\sigma(y)), \quad x, y \in G, \tag{1}$$

when $\sigma : G \to G$ is an involution of G, i.e., $\sigma(xy) = \sigma(y)\sigma(x)$ and $\sigma \circ \sigma(x) = x$ for all $x, y \in G$. The functional Eq. (1) is a generalization of the classical Drygas functional equation

$$f(xy) + f(xy^{-1}) = 2f(x) + f(y) + f(y^{-1}), \quad x, y \in G \tag{2}$$

introduced in [11]. The functional Eq. (2) has been studied by Szabo [47], Ebanks et al. [12], and Faĭziev and Sahoo [13]. The solutions of Eq. (1) in abelian group are obtained by Stetkær in [46]. A stability result for the Drygas functional Eq. (2) was derived by Jung and Sahoo [28] when G is a linear space, while later on Yang [51] proved the stability when G is an abelian group. Recently, Faĭziev and Sahoo [14, 15] proved the stability of Eq. (2) under an additional condition implied by the centrality. Li, Kim and Chung [30] obtained the stability of Eq. (2) in the space of generalized functions. Bouikhalene et al. [5] obtained a stability theorem for the functional Eq. (1) an abelian groups. In the same years, in [4], they proved a stability theorem for Eq. (1) in non-abelian group, when σ is an automorphism of G, i.e., $\sigma \circ \sigma(x) = x$ and $\sigma(xy) = \sigma(x)\sigma(y)$ for all $x, y \in G$. Székelyhidi [48] extended the Hyers' result to amenable semigroups. He replaced the original proof given by Hyers by a new one based on the use of invariant means. The connections between stability and amenability of groups (or semigroups) and sufficient condition for amenability in term of stability is proved by Forti in [16]. In [18], Forti and Sikorska obtained the stability of the Drygas functional Eq. (2) in amenable group G, under the assumption that f is approximatively central (i.e., $|f(xy) - f(yx)| \leq \delta$ for some $\delta \geq 0$ and for all $x, y \in G$). In the present chapter, we study the stability of the generalized Drygas functional Eq. (1) when the domain G is an amenable semigroup and the function f is approximatively central. The result of this chapter can be compared with the ones of Forti and Sikorska [18] because we formulate them in the same way by using some ideas from [18]. In contrast to [18], we work with a general involution σ on the semigroup G. We recall that a linear functional m on the space $B(G, \mathbf{C})$. The space of all bounded functions on G is called a left (right) invariant mean if and only if

$$\inf_{x\in G} f(x) \leq m(f) \leq \sup_{x\in G} f(x); \quad m({}_af) = m(f); \quad (m(f_a) = m(f))$$

for all $f \in B(G, \mathbf{C})$ and $a \in G$, where ${}_af$ and f_a are the left and right translates of f defined by ${}_af(x) = f(ax)$, $f_a(x) = f(xa)$, $x \in G$. A semigroup G which admits a left (right) invariant mean on $B(G, \mathbf{C})$ will be termed left (right) amenable. If on the space $B(G, \mathbf{C})$, there exists a real linear functional which is simultaneously a left and right invariant mean then we say that G is two-sided amenable or just amenable. We refer to [22] for the definition and properties of invariant means. Throughout this chapter, f^o and f^e denote the odd and even parts of f, respectively, i.e., $f^o(x) = \dfrac{f(x) - f(\sigma(x))}{2}$, $f^e(x) = \dfrac{f(x) + f(\sigma(x))}{2}$ for all $x \in G$.

2 Hyers–Ulam Stability of the Drygas Functional Equation in Amenable Semigroups

In the following lemma, we establish a connection between solutions and approximate solutions of Drygas functional Eq. (1).

Lemma 1 *Let G be a semigroup and E be a Banach space. Let $f : G \to E$ be a function for which there exists a solution g of the Drygas functional Eq. (1) such that $\|f(x) - g(x)\| \leq \delta$ for all $x \in G$ and for some $\delta \geq 0$. Then*

$$g(x) = \lim_{n\to+\infty} 2^{-2n}\left\{f^e(x^{2^n}) + \frac{1}{2}\sum_{k=1}^{n} 2^{k-1}\left[f^e((x^{2^{n-k}}\sigma(x)^{2^{n-k}})^{2^{k-1}}) + f^e((\sigma(x)^{2^{n-k}}x^{2^{n-k}})^{2^{k-1}})\right]\right\}$$
$$+ 2^{-n}\left\{f^o(x^{2^n}) + \frac{1}{2}\sum_{k=1}^{n}\left[f^e((x^{2^{k-1}}\sigma(x)^{2^{k-1}})^{2^{n-k}}) - f^e((\sigma(x)^{2^{k-1}}x^{2^{k-1}})^{2^{n-k}})\right]\right\} \quad (3)$$

where f^e, f^o are the even and odd part of f.

Proof By using Drygas functional Eq. (1), we get

$$[g^e(xy) + g^e(x\sigma(y)) - 2g^e(x) - 2g^e(y)] + [g^o(xy) + g^o(x\sigma(y)) - 2g^o(x)] = 0 \quad (4)$$

Setting $x = y$ in (4) and using $g^o(x\sigma(x)) = g^o(\sigma(x)x) = 0$, we obtain

$$[g^e(x^2) + g^e(x\sigma(x)) - 4g^e(x)] + [g^o(x^2) - 2g^o(x)] = 0. \quad (5)$$

Changing x into $\sigma(x)$ in (5), we get

$$[g^e(x^2) + g^e(\sigma(x)x) - 4g^e(x)] + [-\, g^o(x^2) + 2g^o(x)] = 0. \quad (6)$$

By adding and subtracting (5) to (6), we obtain respectively,

$$g^e(x^2) + \frac{1}{2}[g^e(x\sigma(x)) + g^e(\sigma(x)x)] = 4g^e(x) \quad (7)$$

and

$$g^o(x^2) + \frac{1}{2}[g^e(x\sigma(x)) - g^e(\sigma(x)x)] = 2g^o(x). \quad (8)$$

From (7), we get

$$g^e(x^2) + g^e(x\sigma(x)) - 4g^e(x) = \frac{1}{2}[g^e(x\sigma(x)) - g^e(\sigma(x)x)] \quad (9)$$
$$= \frac{1}{2}[g(x\sigma(x)) - g(\sigma(x)x)] = c(x),$$

where $c(x) = \frac{1}{2}[g(x\sigma(x)) - g(\sigma(x)x)]$. First, we use induction on n to prove the following relation

$$g^e(x^{2^n}) = 4^n g^e(x) - \sum_{k=1}^{n} 4^{k-1}\left[g^e(x^{2^{n-k}}\sigma(x)^{2^{n-k}}) - c(x^{2^{n-k}})\right] \quad (10)$$

for all $x \in G$. Equation (9) proves that the assertion (10) is true for $n = 1$. Now, assume that (10) holds for n. By using (9) and (10), we obtain

$$g^e(x^{2^{n+1}}) = 4g^e(x^{2^n}) - g^e(x^{2^n}\sigma(x)^{2^n}) + c(x^{2^n})$$
$$= 4\left[4^n g^e(x) - \sum_{k=1}^{n} 4^{k-1}\left[g^e(x^{2^{n-k}}\sigma(x)^{2^{n-k}}) - c(x^{2^{n-k}})\right]\right]$$
$$- g^e(x^{2^n}\sigma(x)^{2^n}) + c(x^{2^n})$$
$$= 4^{n+1} g^e(x) - \sum_{k=1}^{n+1} 4^{k-1}\left[g^e(x^{2^{n+1-k}}\sigma(x)^{2^{n+1-k}}) - c(x^{2^{n+1-k}})\right],$$

which proves the validity of (10) for $n + 1$. If we replace x by $x\sigma(x)$ in (9), we get

$$4g^e(x\sigma(x)) = 2g^e((x\sigma(x))^2), \quad x \in G. \quad (11)$$

By applying the inductive assumption, we obtain

$$4^n g^e(x\sigma(x)) = 2^n g^e((x\sigma(x))^{2^n}) \quad (12)$$

for some positive integer n. It follows from (11) that (12) is true for $n = 1$. The inductive step must now be demonstrated to hold true for $n + 1$, that is,

$$4^{n+1} g^e(x\sigma(x)) = 4[2^n g^e((x\sigma(x))^{2^n})]$$
$$= 2^n \times 2g^e((x\sigma(x))^{2^{n+1}}) = 2^{n+1} g^e((x\sigma(x))^{2^{n+1}})$$

This proves the validity of the relation (12). By the hypothesis, $f = g + b$, where b is a bounded function. So, we have

$$f^e = g^e + b^e \ \ and \ \ f^o = g^o + b^o. \quad (13)$$

The first relation in (13) gives

$$4^{-n}f^e(x^{2^n}) = 4^{-n}g^e(x^{2^n}) + 4^{-n}b^e(x^{2^n}).$$

By using (10) and (12), we get

$$\begin{aligned}4^{-n}f^e(x^{2^n}) &= 4^{-n}\left[4^n g^e(x) - \sum_{k=1}^{n} 4^{k-1}\left[g^e(x^{2^{n-k}}\sigma(x)^{2^{n-k}}) - c(x^{2^{n-k}})\right]\right] + 4^{-n}b^e(x^{2^n})\\ &= g^e(x) - 4^{-n}\sum_{k=1}^{n} 2^{k-1}g^e((x^{2^{n-k}}\sigma(x)^{2^{n-k}})^{2^{k-1}}) + 4^{-n}\sum_{k=1}^{n} 4^{k-1}c(x^{2^{n-k}})\\ &\quad + 4^{-n}b^e(x^{2^n}).\end{aligned}$$

Therefore, we have

$$\begin{aligned}&4^{-n}\left[f^e(x^{2^n}) + \sum_{k=1}^{n} 2^{k-1}f^e((x^{2^{n-k}}\sigma(x)^{2^{n-k}})^{2^{k-1}})\right] \qquad (14)\\ &= g^e(x) + 4^{-n}\sum_{k=1}^{n} 2^{k-1}\left[f^e((x^{2^{n-k}}\sigma(x)^{2^{n-k}})^{2^{k-1}}) - g^e((x^{2^{n-k}}\sigma(x)^{2^{n-k}})^{2^{k-1}})\right]\\ &\quad + 4^{-n}\sum_{k=1}^{n} 4^{k-1}c(x^{2^{n-k}}) + 4^{-n}b^e(x^{2^n}).\end{aligned}$$

Since the function c is odd ($c(\sigma(x)) = -c(x)$), so by substituting x by $\sigma(x)$ in (14), and adding the new result to (14), we obtain

$$\begin{aligned}&4^{-n}\left[f^e(x^{2^n}) + \frac{1}{2}\sum_{k=1}^{n} 2^{k-1}\left[f^e((x^{2^{n-k}}\sigma(x)^{2^{n-k}})^{2^{k-1}}) + f^e((\sigma(x)^{2^{n-k}}x^{2^{n-k}})^{2^{k-1}})\right]\right]\\ &= g^e(x) + \frac{1}{2}4^{-n}\sum_{k=1}^{n} 2^{k-1}\left[f^e((x^{2^{n-k}}\sigma(x)^{2^{n-k}})^{2^{k-1}}) - g^e((x^{2^{n-k}}\sigma(x)^{2^{n-k}})^{2^{k-1}})\right]\\ &\quad + \frac{1}{2}4^{-n}\sum_{k=1}^{n} 2^{k-1}\left[f^e((\sigma(x)^{2^{n-k}}x^{2^{n-k}})^{2^{k-1}}) - g^e((\sigma(x)^{2^{n-k}}x^{2^{n-k}})^{2^{k-1}})\right] + 4^{-n}b^e(x^{2^n}).\end{aligned}$$

Therefore, we get

$$\begin{aligned}&\|g^e(x) - 4^{-n}\left[f^e(x^{2^n}) + \frac{1}{2}\sum_{k=1}^{n} 2^{k-1}\left[f^e((x^{2^{n-k}}\sigma(x)^{2^{n-k}})^{2^{k-1}}) + f^e((\sigma(x)^{2^{n-k}}x^{2^{n-k}})^{2^{k-1}})\right]\right]\|\\ &\leq \frac{1}{2}4^{-n}\sum_{k=1}^{n} 2^{k-1}\|f^e((x^{2^{n-k}}\sigma(x)^{2^{n-k}})^{2^{k-1}}) - g^e((x^{2^{n-k}}\sigma(x)^{2^{n-k}})^{2^{k-1}})\| \qquad (15)\\ &\quad + \frac{1}{2}4^{-n}\sum_{k=1}^{n} 2^{k-1}\|f^e((\sigma(x)^{2^{n-k}}x^{2^{n-k}})^{2^{k-1}}) - g^e((\sigma(x)^{2^{n-k}}x^{2^{n-k}})^{2^{k-1}})\| + 4^{-n}\|b^e(x^{2^n})\|\end{aligned}$$

$$\leq 4^{-n}\delta\sum_{k=1}^{n}2^{k-1}+4^{-n}\|b^e(x^{2^n})\|$$
$$\leq \frac{2^n-1}{4^n}\delta+4^{-n}\|b^e(x^{2^n})\|.$$

So, by letting $n\to+\infty$ in the formula (15), we obtain

$$g^e(x)=\lim_{n\to+\infty}4^{-n}\left\{f^e(x^{2^n})+\frac{1}{2}\sum_{k=1}^{n}2^{k-1}\left[f^e((x^{2^{n-k}}\sigma(x)^{2^{n-k}})^{2^{k-1}})+f^e((\sigma(x)^{2^{n-k}}x^{2^{n-k}})^{2^{k-1}})\right]\right\}.$$

Consider now the odd parts. First, we check by induction that

$$g^o(x^{2^n})=2^ng^o(x)-\sum_{k=1}^{n}2^{n-k}c(x^{2^{k-1}}) \tag{16}$$

for all $x\in G$. From (8), it follows that (16) is true for $n=1$. Now, assume that (16) holds for n. Using (8) (by replacing x by x^{2^n}), Eq. (16) and the definition of the function c, we get

$$\begin{aligned}g^o(x^{2^{n+1}})&=2g^o(x^{2^n})-c(x^{2^n})\\&=2\left[2^ng^o(x)-\sum_{k=1}^{n}2^{n-k}c(x^{2^{k-1}})\right]-c(x^{2^n})\\&=2^{n+1}g^o(x)-\sum_{k=1}^{n+1}2^{n+1-k}c(x^{2^{k-1}}),\end{aligned}$$

which proves the validity of (16). By using (12), (16), and the definition of the function c, we get

$$\begin{aligned}2^{-n}f^o(x^{2^n})&=2^{-n}g^o(x^{2^n})+2^{-n}b^o(x^{2^n})\\&=2^{-n}\left[2^ng^o(x)-\sum_{k=1}^{n}2^{n-k}c(x^{2^{k-1}})\right]+2^{-n}b^o(x^{2^n})\\&=g^o(x)-2^{-n}\left[\frac{1}{2}\sum_{k=1}^{n}g^e((x^{2^{k-1}}\sigma(x)^{2^{k-1}})^{2^{n-k}})-g^e((\sigma(x)^{2^{k-1}}x^{2^{k-1}})^{2^{n-k}})\right]\\&\quad+2^{-n}b^o(x^{2^n}).\end{aligned}$$

Then, we obtain

$$\|g^o(x)-2^{-n}\left\{f^o(x^{2^n})+\frac{1}{2}\sum_{k=1}^{n}f^e((x^{2^{k-1}}\sigma(x)^{2^{k-1}})^{2^{n-k}})-f^e((\sigma(x)^{2^{k-1}}x^{2^{k-1}})^{2^{n-k}})\right\}\|$$
$$\leq 2^{-n}\frac{1}{2}\sum_{k=1}^{n}\|f^e((x^{2^{k-1}}\sigma(x)^{2^{k-1}})^{2^{n-k}})-g^e((x^{2^{k-1}}\sigma(x)^{2^{k-1}})^{2^{n-k}})\| \tag{17}$$

$$+\|f^e((\sigma(x)^{2^{k-1}}x^{2^{k-1}})^{2^{n-k}}) - g^e((\sigma(x)^{2^{k-1}}x^{2^{k-1}})^{2^{n-k}})\| + 2^{-n}\|b^o(x^{2^n})\|$$

$$\leq 2^{-n}[\frac{1}{2}n\delta + \frac{1}{2}n\delta] + 2^{-n}\|b^o(x^{2^n})\| = 2^{-n}n\delta + 2^{-n}\|b^o(x^{2^n})\|.$$

If we let $n \to +\infty$ in (17), we get

$$g^o(x) = \lim_{n\to+\infty} 2^{-n}\left\{f^o(x^{2^n}) + \frac{1}{2}\sum_{k=1}^{n}\left[f^e((x^{2^{k-1}}\sigma(x)^{2^{k-1}})^{2^{n-k}}) - f^e((\sigma(x)^{2^{k-1}}x^{2^{k-1}})^{2^{n-k}})\right]\right\}.$$

Since $g(x) = g^e(x) + g^o(x)$, the rest of the proof follows.

Remark 1 Let G be a group, $\sigma = -I$ and let E be a Banach space. Let $f : G \to E$ be a function for which there exists a solution g of Drygas functional Eq. (2) such that $\|f(x) - g(x)\| \leq \delta$ for all $x \in G$ and for some $\delta \geq 0$. Then, we obtain the result proved in [18]:

$$g(x) = \lim_{n\to+\infty} 2^{-2n}\{f^e(x^{2^n})\} + 2^{-n}\{f^o(x^{2^n})\}$$

Remark 2 In Lemma 1, if a function g is a solution of Drygas functional equation

$$g(xy) + g(\sigma(x)y) = 2g(y) + g(x) + g(\sigma(x)), \quad x, y \in G \tag{18}$$

and $\|f(x) - g(x)\| \leq \delta$ for all $x \in G$ and for some $\delta \geq 0$. Then

$$g(x)$$

$$= \lim_{n\to+\infty} 2^{-2n}\left\{f^e(x^{2^n}) + \frac{1}{2}\sum_{k=1}^{n}2^{k-1}\left[f^e((x^{2^{n-k}}\sigma(x)^{2^{n-k}})^{2^{k-1}}) + f^e((\sigma(x)^{2^{n-k}}x^{2^{n-k}})^{2^{k-1}})\right]\right\}$$

$$+ 2^{-n}\left\{f^o(x^{2^n}) - \frac{1}{2}\sum_{k=1}^{n}\left[f^e((x^{2^{k-1}}\sigma(x)^{2^{k-1}})^{2^{n-k}}) - f^e((\sigma(x)^{2^{k-1}}x^{2^{k-1}})^{2^{n-k}})\right]\right\}.$$

Remark 3 if $f : G \longrightarrow \mathbf{C}$ is a solution of Drygas functional Eq. (1) with f is central, i.e., $f(xy) = f(yx)$ then the even and odd parts of f satisfies the quadratic functional equation

$$f^e(xy) + f^e(x\sigma(y)) = 2f^e(x) + 2f^e(y), \quad x, y \in G \tag{19}$$

respectively, the Jensen functional equation

$$f^o(xy) + f^o(x\sigma(y)) = 2f^o(x), \quad x, y \in G. \tag{20}$$

Later, we use the following result.

Lemma 2 *Let G be a semigroup and E be a Banach space. Suppose that $f : G \to E$ be an even function, (i.e., $f(\sigma(x)) = f(x)$) satisfying the inequality*

$$\|f(xy) + f(x\sigma(y)) - 2f(x) - 2f(y)\| \leq \delta \tag{21}$$

for all $x \in G$ *and for some* $\delta \geq 0$. *Then, for every* $x \in G$, *the limit*

$$q(x) = \lim_{n \to +\infty} 2^{-2n} \{ f(x^{2^n}) + (2^n - 1) f(x^{2^{n-1}} \sigma(x)^{2^{n-1}}) \} \tag{22}$$

exists. Moreover, the mapping q *satisfies the inequality*

$$\| f(x) - q(x) \| \leq \frac{11}{6} \delta \tag{23}$$

for all $x \in G$.

Proof In the proof, we use some ideas from Bouikhalene et al. [4]. By interchanging x by y in (21), we obtain with $f(x\sigma(y)) = f(y\sigma(x))$ that

$$\| f(xy) - f(yx) \| \leq 2\delta. \tag{24}$$

Now, from (21), we get

$$\| 2f(x^{2^{n-1}} \sigma(x)^{2^{n-1}} \sigma(x)^{2^{n-1}} x^{2^{n-1}}) - 2f(x^{2^{n-1}} \sigma(x)^{2^{n-1}}) - 2f(\sigma(x)^{2^{n-1}} x^{2^{n-1}}) \| \leq \delta. \tag{25}$$

Since

$$\| f(x^{2^{n-1}} \sigma(x)^{2^{n-1}}) - f(\sigma(x)^{2^{n-1}} x^{2^{n-1}}) \| \leq 2\delta \tag{26}$$

and

$$\| f(x^{2^{n-1}} \sigma(x)^{2^n} x^{2^{n-1}}) - f(x^{2^n} \sigma(x)^{2^n}) \| \leq 2\delta. \tag{27}$$

Then

$$\| 2f(x^{2^{n-1}} \sigma(x)^{2^n} x^{2^{n-1}}) - 4f(x^{2^{n-1}} \sigma(x)^{2^{n-1}}) \| \leq 5\delta \tag{28}$$

and

$$\begin{aligned} & \| 2f(x^{2^n} \sigma(x)^{2^n}) - 4f(x^{2^{n-1}} \sigma(x)^{2^{n-1}}) \| \\ & \quad \leq \| 2f(x^{2^{n-1}} \sigma(x)^{2^n} x^{2^{n-1}}) - 2f(x^{2^n} \sigma(x)^{2^n}) \| \\ & \qquad + \| 2f(x^{2^{n-1}} \sigma(x)^{2^n} x^{2^{n-1}}) - 4f(x^{2^{n-1}} \sigma(x)^{2^{n-1}}) \| \\ & \quad \leq 9\delta. \end{aligned} \tag{29}$$

First, by using induction on n, we prove the following inequality

$$\| f(x) - \frac{1}{2^{2n}} \{ f(x^{2^n}) + (2^n - 1) f(x^{2^{n-1}} \sigma(x)^{2^{n-1}}) \} \| \leq \frac{11}{6} \delta + (\frac{8}{3} \frac{1}{2^{2n}} - \frac{9}{2} \frac{1}{2^n}) \delta. \tag{30}$$

By letting $x = y$ in (21), we get

$$\| f(x^2) + f(x\sigma(x)) - 4f(x) \| \leq \delta. \tag{31}$$

So

$$\|f(x) - \frac{1}{2^2}\{f(x^2) + (2-1)f(x\sigma(x))\}\| \le \frac{\delta}{2^2}. \tag{32}$$

This proves (30) for $n = 1$. The inductive step must now be demonstrated to hold true for the integer $n+1$ that is,

$$\|f(x) - \frac{1}{2^{2(n+1)}}\{f(x^{2^{n+1}}) + (2^{n+1}-1)f(x^{2^n}\sigma(x)^{2^n})\}\|$$

$$\le \frac{1}{2^{2(n+1)}}\|f(x^{2^{n+1}}) + f(x^{2^n}\sigma(x)^{2^n}) - 4f(x^{2^n})\|$$

$$+ \frac{1}{2^{2(n+1)}}\|2(2^n-1)f(x^{2^n}\sigma(x)^{2^n}) - 4(2^n-1)f(x^{2^{n-1}}\sigma(x)^{2^{n-1}})\|$$

$$+ \frac{1}{2^{2(n+1)}}\|4f(x^{2^n}) + 4(2^n-1)f(x^{2^{n-1}}\sigma(x)^{2^{n-1}}) - 2^{2(n+1)}f(x)\|$$

$$\le \frac{\delta}{2^{2(n+1)}} + \frac{(2^n-1)}{2^{2(n+1)}}9\delta + \|f(x) - \frac{1}{2^{2n}}\{f(x^{2^n}) + (2^n-1)f(x^{2^{n-1}}\sigma(x)^{2^{n-1}})\}\|$$

$$\le \frac{9\delta}{2^2}\frac{1}{2^n} - \frac{2\delta}{2^{2n}} + \frac{11}{6}\delta + (\frac{8}{3}\frac{1}{2^{2n}} - \frac{9}{2}\frac{1}{2^n})\delta = \frac{11}{6}\delta + (\frac{8}{3}\frac{1}{2^{2(n+1)}} - \frac{9}{2}\frac{1}{2^{n+1}})\delta.$$

This completes the proof of the induction assumption (30). Now, Let us define

$$q_n(x) = \frac{1}{2^{2n}}\left\{f(x^{2^n}) + (2^n-1)f(x^{2^{n-1}}\sigma(x)^{2^{n-1}})\right\} \tag{33}$$

for any positive integer n, and $x \in G$. By using (21), (29), and (33), we get

$$\|q_{n+1}(x) - q_n(x)\|$$

$$= \frac{1}{2^{2(n+1)}}\|f(x^{2^{n+1}}) + (2^{n+1}-1)f(x^{2^n}\sigma(x)^{2^n}) - 4f(x^{2^n}) - 4(2^n-1)f(x^{2^{n-1}}\sigma(x)^{2^{n-1}})\|$$

$$\le \frac{1}{2^{2(n+1)}}\|f(x^{2^{n+1}}) + f(x^{2^n}\sigma(x)^{2^n}) - 4f(x^{2^n})\|$$

$$+ \frac{1}{2^{2(n+1)}}\|2(2^n-1)f(x^{2^n}\sigma(x)^{2^n}) - 4(2^n-1)f(x^{2^{n-1}}\sigma(x)^{2^{n-1}})\|$$

$$\le \frac{\delta}{2^{2(n+1)}} + 9\delta\frac{(2^n-1)}{2^{2(n+1)}} \le \frac{5}{2^{n+3}}\delta.$$

It follows that $\{q_n(x)\}_n$ is a Cauchy sequence for every $x \in G$. Since E is complete, we can define $q(x) = \lim_{n\to+\infty} q_n(x)$ for any $x \in G$. In view of (30), one can verify that q satisfies the inequality (23). This completes the proof.

In the following theorem, we prove a partial stability result of the quadratic and Jensen functional equations on amenable semigroups.

Theorem 1 *Let G be an amenable semigroup and E be a Banach space. Assume that $f : G \to E$ be a function satisfying the following inequalities:*

$$\|f^e(xy) + f^e(x\sigma(y)) - 2f^e(x) - 2f^e(y)\| \le \delta, \tag{34}$$

$$\|f^{o}(xy)+f^{o}(x\sigma(y))-2f^{o}(x)\|\leq\gamma \tag{35}$$

for all $x, y \in G$ *and for some* $\delta, \gamma \geq 0$. *Then, there exists a unique solution* $g = Q+D$ *of Drygas functional Eq. (1), with* Q *solution of the quadratic functional equation*

$$Q(xy)+Q(x\sigma(y))=2Q(x)+2Q(y)$$

and D *solution of Jensen functional equation*

$$D(xy)+D(x\sigma(y))=2D(x)$$

such that

$$\|f(x)-g(x))\|\leq\frac{1}{2}(\delta+\gamma) \tag{36}$$

for all $x \in G$.

Proof We follows the ideas and the computations used in [4] and [51]. By using (34), we get

$$\|f^{e}((xy)^{n})-f^{e}((yx)^{n})\|=\|f^{e}([(xyxy\ldots xy)x]y)-f^{e}(y[(xyxy\ldots xy)x])\|\leq 2\delta. \tag{37}$$

From (34), (37), and the triangle inequality, we deduce that

$$\begin{aligned}
&\|f^{e}((xy)^{2^n}(\sigma(xy)^{2^n}))-f^{e}((yx)^{2^n}(\sigma(yx)^{2^n}))\| \qquad (38)\\
&\leq\|f^{e}((xy)^{2^n}(\sigma(xy)^{2^n}))+f^{e}((xy)^{2^n}(xy)^{2^n})-4f^{e}((xy)^{2^n})\|\\
&\quad+\|f^{e}((yx)^{2^n}(\sigma(yx)^{2^n}))+f^{e}((yx)^{2^n}(yx)^{2^n})-4f^{e}((yx)^{2^n})\|\\
&\quad+\|f^{e}((xy)^{2^n}(xy)^{2^n})-f^{e}((yx)^{2^n}(yx)^{2^n})\|\\
&\quad+4\|f^{e}((xy)^{2^n})-f^{e}((yx)^{2^n})\|\\
&\leq\delta+\delta+2\delta+8\delta=12\delta.
\end{aligned}$$

From Lemma 2, for every $x \in G$, the limit

$$q(x)=\lim_{n\to+\infty}2^{-2n}\{f^{e}(x^{2^n})+(2^n-1)f^{e}(x^{2^{n-1}}\sigma(x)^{2^{n-1}})\} \tag{39}$$

exists and

$$\|f^{e}(x)-q(x)\|\leq\frac{11}{6}\delta. \tag{40}$$

Furthermore, in view of (37) and (2.36), we have

$$\|q(xy)-q(yx)\|$$

$$= \lim_{n\to+\infty} 2^{-2n} \| f^e((xy)^{2^n}) + (2^n - 1) f^e((xy)^{2^{n-1}} \sigma(xy)^{2^{n-1}})$$
$$- f^e((yx)^{2^n}) - (2^n - 1) f^e((yx)^{2^{n-1}} \sigma(yx)^{2^{n-1}}) \|$$
$$\leq \lim_{n\to+\infty} 2^{-2n} \| f^e((xy)^{2^n}) - f^e((yx)^{2^n}) \|$$
$$+ \frac{(2^n - 1)}{2^{2n}} \| f^e((xy)^{2^{n-1}} \sigma(xy)^{2^{n-1}}) - f^e((yx)^{2^{n-1}} \sigma(yx)^{2^{n-1}}) \|$$
$$\leq \lim_{n\to+\infty} 2^{-2n+1}\delta + 12\frac{(2^n - 1)}{2^{2n}}\delta = 0.$$

Then, the mapping q satisfies the relation

$$q(xy) = q(yx), \quad x, y \in G. \tag{41}$$

So, by (40) and (34), the function q satisfies the inequality

$$\|q(xy) + q(x\sigma(y)) - 2q(x) - 2q(y)\| \leq 12\delta \tag{42}$$

for all $x, y \in G$. Consequently, for any fixed $y \in G$, the function $x \longmapsto q(xy) + q(x\sigma(y)) - 2q(x)$ is bounded. Since G is amenable then there exists an invariant mean m on the space of bounded function on G and we have

$$m\{_{zy}q +_{\sigma(z)y} q - 2_y q\} = m\{_y(_z q +_{\sigma(z)} q - 2q)\} = m\{_z q +_{\sigma(z)} q - 2q\},$$
$$m\{q_{yz} + q_{y\sigma(z)} - 2q_y\} = m\{(q_z + q_{\sigma(z)} - 2q)_y\} = m\{q_z + q_{\sigma(z)} - 2q\},$$

when $q_y(z) = q(zy)$, $z \in G$. Define

$$\Psi(y) = m\{q_y + q_{\sigma(y)} - 2q\} \tag{43}$$

for all $y \in G$. By using (43) and (41), we obtain

$$\begin{aligned}\Psi(zy) + \Psi(\sigma(z)y) &= m\{q_{zy} + q_{\sigma(y)\sigma(z)} - 2q\} + m\{q_{\sigma(z)y} + q_{\sigma(y)z} - 2q\}\\
&= m\{_{zy}q +_{\sigma(z)y} q - 2_y q\} + m\{q_{\sigma(y)\sigma(z)} + q_{\sigma(y)z} - 2q_{\sigma(y)}\}\\
&\quad + m\{2q_y + 2q_{\sigma(y)} - 4q\}\\
&= m\{_z q +_{\sigma(z)} q - 2q\} + m\{q_{\sigma(z)} + q_z - 2q\}\\
&\quad + 2m\{q_y + q_{\sigma(y)} - 2q\}\\
&= 2\Psi(z) + 2\Psi(y).\end{aligned}$$

So, $Q(y) = \Psi(y)/2$ satisfies the functional equation

$$Q(xy) + Q(\sigma(x)y) = 2Q(x) + 2Q(y) \tag{44}$$

for all $x, y \in G$ and the following inequality

$$\|Q(y) - q(y)\| = \frac{1}{2}\|m\{q_y + q_{\sigma(y)} - 2q - 2q(y)\}\| \tag{45}$$

$$\leq \sup_{x\in G} \frac{1}{2} \|q(xy) + q(x\sigma(y)) - 2q(x) - 2q(y)\|$$
$$\leq 6\delta.$$

If we let $y = e$ in (44), we get Q is even and by a simple computation, we deduce that $Q(xy) = Q(yx)$ for all $x, y \in G$. Consequently, there exists a mapping Q which satisfies the quadratic functional equation

$$Q(xy) + Q(x\sigma(y)) = 2Q(x) + 2Q(y), \quad x, y \in G \tag{46}$$

and the inequality $\|f^e(y) - Q(y)\| \leq \frac{47}{6}\delta$. Assume now that there exists another mapping $H : G \to E$ solution of (46) satisfying $\|f^e(x) - H(x)\| \leq \frac{47}{6}\delta$ for all $x \in G$. First, by mathematical induction, we show that

$$Q(x) = 2^{-2n} \left\{ Q(x^{2^n}) + \sum_{k=1}^{n} 2^{k-1} Q((x^{2^{n-k}} \sigma(x)^{2^{n-k}})^{2^{k-1}}) \right\} \tag{47}$$

for each $n \in \mathbf{N}$. By letting $x = y$ in (46) we obtain

$$Q(x^2) + Q(x\sigma(x)) = 2^2 Q(x). \tag{48}$$

This proves (47) for $n = 1$. Substituting x by $x\sigma(x)$ in (48), we get

$$2Q((x\sigma(x))^2) = 4Q(x\sigma(x)). \tag{49}$$

By using (48) and (49), we obtain (47) for $n = 2$, that is:

$$\begin{aligned} Q(x^{2^2}) + Q(x^2\sigma(x)^2) + 2Q((x\sigma(x))^2) &= 2^2[Q(x^2) + Q(x\sigma(x))] \\ &= 2^4 Q(x). \end{aligned}$$

Suppose that (47) is true for n. Hence, by using (48) and (49), we have

$$\begin{aligned} &Q(x^{2^{n+1}}) + \sum_{k=1}^{n+1} 2^{k-1} Q((x^{2^{n+1-k}} \sigma(x)^{2^{n+1-k}})^{2^{k-1}}) \\ &= Q(x^{2^{n+1}}) + Q(x^{2^n} \sigma(x)^{2^n}) + \sum_{k=2}^{n+1} 2^{k-1} Q((x^{2^{n+1-k}} \sigma(x)^{2^{n+1-k}})^{2^{k-1}}) \\ &= 2^2 Q(x^{2^n}) + \sum_{k=2}^{n+1} 2^{k-1} Q((x^{2^{n+1-k}} \sigma(x)^{2^{n+1-k}})^{2^{k-1}}) \\ &= 2^2 \left[Q(x^{2^n}) + \sum_{k=1}^{n} 2^{k-1} Q((x^{2^{n-k}} \sigma(x)^{2^{n-k}})^{2^{k-1}}) \right] \\ &= 2^{2(n+1)} Q(x). \end{aligned}$$

From (47), we obtain

$$\begin{aligned}
&\|Q(x) - H(x)\| \\
&= 2^{-2n} \| Q(x^{2^n}) - H(x^{2^n}) + \sum_{k=1}^{n} 2^{k-1} \left[Q((x^{2^{n-k}} \sigma(x)^{2^{n-k}})^{2^{k-1}}) - H((x^{2^{n-k}} \sigma(x)^{2^{n-k}})^{2^{k-1}}) \right] \| \\
&\leq 2^{-2n} \| Q(x^{2^n}) - f^e(x^{2^n}) \| + 2^{-2n} \| H(x^{2^n}) - f^e(x^{2^n}) \| \\
&\quad + 2^{-2n} \sum_{k=1}^{n} 2^{k-1} \| Q((x^{2^{n-k}} \sigma(x)^{2^{n-k}})^{2^{k-1}}) - f^e((x^{2^{n-k}} \sigma(x)^{2^{n-k}})^{2^{k-1}}) \| \\
&\quad + 2^{-2n} \sum_{k=1}^{n} 2^{k-1} \| H((x^{2^{n-k}} \sigma(x)^{2^{n-k}})^{2^{k-1}}) - f^e((x^{2^{n-k}} \sigma(x)^{2^{n-k}})^{2^{k-1}}) \| \\
&\leq 2^{-2n} \frac{94}{6} \delta + 2^{-n} \frac{94}{6} \delta.
\end{aligned}$$

By letting $n \to +\infty$, we get $Q = H$. This proves the uniqueness of the mapping Q. Now, we define by induction the sequence function $f_0^e(x) = f^e(x)$ and

$$f_n^e(x) = 2^{-n} \left\{ f^e(x^{2^n}) + \sum_{k=1}^{n} 2^{k-1} f^e((x^{2^{n-k}} \sigma(x)^{2^{n-k}})^{2^{k-1}}) \right\}$$

for all $n \geq 1$. By a direct computation, we can easily verify that $f_n^e(x) = \frac{1}{2}[f_{n-1}^e(x^2) + f_{n-1}^e(x\sigma(x))]$ for $n \geq 1$. By letting $x = y$ in (34) we get

$$\| \frac{1}{2}[f^e(x^2) + f^e(x\sigma(x))] - 2f^e(x) \| \leq \frac{\delta}{2}, \tag{50}$$

so

$$\| f_1^e(x) - 2f_0^e(x) \| \leq \frac{\delta}{2} \tag{51}$$

for all $x \in G$. In the following, we prove by induction the inequalities

$$\| f_n^e(x) - 2f_{n-1}^e(x) \| \leq \frac{\delta}{2}, \tag{52}$$

$$\| f_n^e(x) - 2^n f^e(x) \| \leq \frac{(2^n - 1)}{2} \delta \tag{53}$$

for all $n \in \mathbf{N}$ and $x \in G$. From (51), we get (52) for $n = 1$. The inductive step must now be demonstrated to hold true for the integer $n + 1$, that is

$$\| f_{n+1}^e(x) - 2f_n^e(x) \| = \frac{1}{2} \| f_n^e(x^2) + f_n^e(x\sigma(x)) - 2f_{n-1}^e(x^2) - 2f_{n-1}^e(x\sigma(x)) \|$$

$$\leq \frac{1}{2}\|f_n^e(x^2) - 2f_{n-1}^e(x^2)\| + \frac{1}{2}\|f_n^e(x\sigma(x)) - 2f_{n-1}^e(x\sigma(x))\|$$
$$\leq \frac{1}{2}[\frac{\delta}{2} + \frac{\delta}{2}] = \frac{\delta}{2}.$$

This proves that (52) is true for all $n \geq 1$. Now, by using the inequality

$$\|f_n^e(x) - 2^n f^e(x)\| \leq \|f_n^e(x) - 2f_{n-1}^e(x)\| + 2\|f_{n-1}^e(x) - 2^{n-1} f^e(x)\| \quad (54)$$

we check that (53) holds true for any $n \in \mathbf{N}$. Let us define

$$g_n(x) = \frac{f_n^e(x)}{2^n} = 2^{-2n}\left\{f^e(x^{2^n}) + \sum_{k=1}^{n} 2^{k-1} f^e((x^{2^{n-k}}\sigma(x)^{2^{n-k}})^{2^{k-1}})\right\}$$

for any positive integer n and $x \in G$. By using (47) and inequality $\|f^e(x) - Q(x)\| \leq \frac{47}{6}\delta$, we can prove that

$$Q(x) = \lim_{n\to+\infty} g_n(x).$$

From (53), one can verify that Q satisfies $\|f^e(x) - Q(x)\| \leq \frac{\delta}{2}$ for all $x \in G$. Consider now (35), we have

$$\|f^o(yx) - f^o(x\sigma(y)) - 2f^o(y)\| = \|f^o(yx) + f^o(y\sigma(x)) - 2f^o(y)\| \leq \gamma. \quad (55)$$

Whence, for every $y \in G$, the function $x \to f^o(yx) - f^o(x\sigma(y))$ is bounded. By using (35) and (55), we obtain

$$\begin{aligned} &\|f^o(yx) + f^o(\sigma(y)x) - 2f^o(x)\| \quad (56)\\ &\leq \|f^o(yx) - f^o(x\sigma(y)) - 2f^o(y)\| + \|f^o(\sigma(y)x) + f^o(\sigma(y)\sigma(x)) - 2f^o(\sigma(y))\|\\ &\quad + \|f^o(xy) + f^o(x\sigma(y)) - 2f^o(x)\|\\ &\leq 3\gamma. \end{aligned}$$

The function $x \to f^o(yx) + f^o(\sigma(y)x) - 2f^o(x)$ is bounded. So, we have

$$m\{_{zy}f^o +_{\sigma(z)y} f^o - 2_y f^o\} = m\{_y(_z f^o +_{\sigma(z)} f^o - 2f^o)\} = m\{_z f^o +_{\sigma(z)} f^o - 2f^o\},$$

$$m\{f^o_{\sigma(y)\sigma(z)} + f^o_{\sigma(y)z} - 2f^o_{\sigma(y)}\} = m\{(f^o_{\sigma(z)} + f^o_z - 2f^o)_{\sigma(y)}\} = m\{f^o_{\sigma(z)} + f^o_z - 2f^o\}.$$

Define

$$\varphi(y) = m\{_y f^o - f^o_{\sigma(y)}\},\ y \in G. \quad (57)$$

Therefore, we have

$$\begin{aligned}\varphi(zy) + \varphi(\sigma(z)y) &= m\{_{zy}f^o - f^o_{\sigma(y)\sigma(z)}\} + m\{_{\sigma(z)y}f^o - f^o_{\sigma(y)z}\}\\ &= m\{_{zy}f^o +_{\sigma(z)y} f^o - 2(_y f^o)\} - m\{f^o_{\sigma(y)\sigma(z)} + f^o_{\sigma(y)z} - 2f^o_{\sigma(y)}\}\end{aligned}$$

$$
\begin{aligned}
&+ 2m\{{}_y f^o - f^o_{\sigma(y)}\} \\
&= m\{{}_z f^o + {}_{\sigma(z)} f^o - 2f^o\} - m\{f^o_{\sigma(z)} + f^o_z - 2f^o\} \\
&+ 2m\{{}_y f^o - f^o_{\sigma(y)}\} \\
&= m\{{}_z f^o - f^o_{\sigma(z)}\} + m\{{}_{\sigma(z)} f^o - f^o_z\} + 2m\{{}_y f^o - f^o_{\sigma(y)}\} \\
&= \varphi(z) + \varphi(\sigma(z)) + 2\varphi(y).
\end{aligned}
$$

Thus, φ is a solution of Drygas functional Eq. (18). The function $D(y) = \frac{1}{2}\varphi(y)$ is a solution of Drygas functional Eq. (18). Moreover, we have

$$
\begin{aligned}
\|D(y) - f^o(y)\| &= \frac{1}{2}\|\varphi(y) - 2f^o(y)\| = \frac{1}{2}\|m\{{}_y f^o - f^o_{\sigma(y)} - 2f^o(y)\}\| \qquad (58)\\
&\leq \frac{1}{2}\sup_{x\in G}\|f^o(yx) - f^o(x\sigma(y)) - 2f^o(y)\}\| \\
&\leq \frac{1}{2}\gamma.
\end{aligned}
$$

So, D is a solution of (18) such that (58). Then, by Remark 2

$$
\begin{aligned}
D(x) = &\frac{1}{2}\lim_{n\to+\infty} 2^{-2n}\Big\{f^o(x^{2^n}) + f^o(\sigma(x^{2^n})) \\
&+ \frac{1}{2}\sum_{k=1}^{n} 2^{k-1}\Big[f^o((x^{2^{n-k}}\sigma(x)^{2^{n-k}})^{2^{k-1}}) + f^o((\sigma(x)^{2^{n-k}}x^{2^{n-k}})^{2^{k-1}})\Big] \\
&+ \frac{1}{2}\sum_{k=1}^{n} 2^{k-1}\Big[f^o(\sigma((x^{2^{n-k}}\sigma(x)^{2^{n-k}})^{2^{k-1}})) + f^o(\sigma((\sigma(x)^{2^{n-k}}x^{2^{n-k}})^{2^{k-1}}))\Big]\Big\} \\
&+ 2^{-n}\Big\{f^o(x^{2^n}) - f^o(\sigma(x^{2^n})) \\
&- \frac{1}{2}\sum_{k=1}^{n}\Big[f^o((x^{2^{k-1}}\sigma(x)^{2^{k-1}})^{2^{n-k}}) + f^o((\sigma(x)^{2^{k-1}}x^{2^{k-1}})^{2^{n-k}})\Big] \\
&- \frac{1}{2}\sum_{k=1}^{n}\Big[f^o(\sigma((x^{2^{k-1}}\sigma(x)^{2^{k-1}})^{2^{n-k}})) + f^o(\sigma((\sigma(x)^{2^{k-1}}x^{2^{k-1}})^{2^{n-k}}))\Big]\Big\} \\
= &\lim_{n\to+\infty} 2^{-n} f^o(x^{2^n}).
\end{aligned}
$$

Then, D is odd. Thus, D satisfies the following equation

$$
D(xy) + D(\sigma(x)y) = 2D(y), \quad x, y \in G, \qquad (59)
$$

which implies that

$$
-D(\sigma(y)\sigma(x)) - D(\sigma(y)x) = -2D(\sigma(y)), \qquad (60)
$$

so D is a solution of Jensen functional equation

$$D(xy) + D(x\sigma(y)) = 2D(x), \quad x, y \in G. \tag{61}$$

The uniqueness of the function D follows as usual. Finally, we conclude that $g = Q + D$ is the unique solution of Drygas functional Eq. (1) such that (36).

Now, we are able to prove the main result of the present chapter.

Theorem 2 *Let G be an amenable semigroup and E a Banach space. Suppose that $f : G \to E$ be a function satisfying the following inequalities*

$$\|f(xy) + f(x\sigma(y)) - 2f(x) - f(y) - f(\sigma(y))\| \le \delta \tag{62}$$

and

$$\|f(xy) - f(yx))\| \le \mu \tag{63}$$

for all $x, y \in G$ and for some $\delta, \mu \ge 0$. Then, there exists a unique solution $g = Q+D$ of Drygas functional Eq. (1), with Q solution of the quadratic functional Eq. (46) and D solution of Jensen functional Eq. (61), such that

$$\|f(x) - g(x)\| \le \delta + \mu \tag{64}$$

for all $x \in G$.

Proof From (62) and (63) we have

$$\begin{aligned}
&\|f^e(xy) + f^e(x\sigma(y)) - 2f^e(x) - 2f^e(y)\| \\
&= \frac{1}{2}\|f(xy) + f(\sigma(y)\sigma(x)) + f(x\sigma(y)) + f(y\sigma(x)) - 2f(x) - 2f(\sigma(x)) - 2f(y) - 2f(\sigma(y))\| \\
&\le \frac{1}{2}\|f(xy) + f(x\sigma(y)) - 2f(x) - f(y) - f(\sigma(y))\| \\
&\quad + \frac{1}{2}\|f(\sigma(x)y) + f(\sigma(x)\sigma(y)) - 2f(\sigma(x)) - f(y) - f(\sigma(y))\| \\
&\quad + \frac{1}{2}\|f(\sigma(y)\sigma(x)) - f(\sigma(x)\sigma(y))\| + \frac{1}{2}\|f(y\sigma(x)) - f(\sigma(x)y)\| \\
&\le \delta + \mu
\end{aligned}$$

and analogous approximation we have

$$\|f^o(xy) + f^o(x\sigma(y)) - 2f^o(x)\| \le \delta + \mu \tag{65}$$

for all $x, y \in G$. Hence, by Theorem 1 we get our main result.

Corollary 1 [18] *Let G be an amenable semigroup and E a Banach space. Suppose that $f : G \to E$ be a function satisfying the following inequalities*

$$\|f(xy) + f(xy^{-1}) - 2f(x) - f(y) - f(y^{-1})\| \le \delta \tag{66}$$

and

$$\|f(xy) - f(yx))\| \leq \mu \tag{67}$$

for all $x, y \in G$ *and for some* $\delta, \mu \geq 0$. *Then there exists a unique solution* $g = Q + D$ *of Drygas functional Eq. (2), with* Q *quadratic and* D *Jensen such that*

$$\|f(x) - g(x)\| \leq \delta + \mu \tag{68}$$

for all $x \in G$.

Corollary 2 [48] *Let* G *be an amenable semigroup and* E *a Banach space. Suppose that* $f : G \to E$ *be a function satisfying the following inequality*

$$\|f(xy) - f(x) - f(y)\| \leq \delta \tag{69}$$

for all $x, y \in G$ *and for some* $\delta \geq 0$. *Then, there exists a unique additive mapping* $a : G \to E$ *such that*

$$\|f(x) - a(x)\| \leq \frac{\delta}{2} \tag{70}$$

for all $x \in G$.

In fact, from the proof of Theorem 2, we see that instead of the condition $\|f(xy) - f(yx))\| \leq \mu$, we may assume a weaker one and which follows from the approximate centrality, namely

$$\|f(xy) + f(x\sigma(y)) - f(yx) - f(\sigma(y)x)\| \leq \gamma, \quad x, y \in G.$$

Corollary 3 *Let* G *be an amenable semigroup and* E *a Banach space. Suppose that* $f : G \to E$ *be a function satisfying the following inequalities*

$$\|f(xy) + f(x\sigma(y)) - 2f(x) - f(y) - f(\sigma(y))\| \leq \delta \tag{71}$$

and

$$\|f(xy) + f(x\sigma(y)) - f(yx) - f(\sigma(y)x)\| \leq \gamma \tag{72}$$

for all $x, y \in G$ *and for some* $\delta, \gamma \geq 0$. *Then, there exists a unique solution* $g = Q + D$ *of Drygas functional* Eq. (2), *with* Q *solution of the quadratic functional* Eq. (46) *and* D *solution of the Jensen functional* Eq. (61) *such that*

$$\|f(x) - g(x)\| \leq \delta + \frac{\gamma}{2} \tag{73}$$

for all $x \in G$.

In [52], D. Yang presented some rich ideas on the stability of Jensen's functional equation

$$f(xy) + f(xy^{-1}) = 2f(x), \quad x, y \in G \tag{74}$$

on amenable groups. However, the proof of his result is incorrect. The function ψ defined by Eq. (11) in [52] satisfies Drygas functional equation, the deduction that the odd parts $\psi_\circ$ of the function ψ satisfies Jensen functional Eq. (74) is not true. In the following, we correct the error that occurs in the proof of [[52], Proposition 2].

Corollary 4 *Let G be an amenable semigroup with neutral element e. Let $f: G \longrightarrow \mathbf{C}$ be a function satisfying the following inequality:*

$$|f(xy) + f(xy^{-1}) - 2f(x)| \leq \delta \tag{75}$$

for all $x, y \in G$ and for some nonnegative δ. Then, there exists a unique solution g of Jensen Eq. (74) *such that*

$$|f(x) - g(x) - f(e)| \leq 3\delta \tag{76}$$

for all $x \in G$.

Proof In the proof, we use some ideas from Yang [52] and Forti and Sikorska [18]. Setting $x = e$ in (75), we have

$$|f^e(y) - f(e)| \leq \frac{\delta}{2} \tag{77}$$

for all $y \in G$. The inequalities (75), (77), and the triangle inequality gives

$$\begin{aligned} &|f(xy) + f(yx) - 2f(x) - 2f(y) + 2f(e)| \\ &\quad \leq |f(xy) + f(xy^{-1}) - 2f(x)| + |f(yx) + f(yx^{-1}) - 2f(y)| \\ &\quad + |\, 2f(e) - f(xy^{-1}) - f(yx^{-1})| \\ &\quad \leq 3\delta. \end{aligned} \tag{78}$$

Hence, from (75), (77), and (78), we get

$$\begin{aligned} &|f(yx) + f(y^{-1}x) - 2f(x)| \\ &\quad \leq |f(yx) + f(xy) - 2f(y) - 2f(x) + 2f(e)| \\ &\quad + |f(y^{-1}x) + f(xy^{-1}) - 2f(y^{-1}) - 2f(x) + 2f(e)| \\ &\quad + |-f(xy) - f(xy^{-1}) + 2f(x)| + |2f(y) + 2f(y^{-1}) - 4f(e)| \\ &\quad \leq 9\delta. \end{aligned} \tag{79}$$

Now, from (75) and (79), we obtain

$$\begin{aligned} &|f(yx) - f(x^{-1}y^{-1}) + f(yx^{-1}) - f(xy^{-1}) - 2(f(y) - f(y^{-1}))| \\ &\quad \leq |f(yx) + f(yx^{-1}) - 2f(y)| + |f(xy^{-1}) + f(x^{-1}y^{-1}) - 2f(y^{-1})| \\ &\quad \leq 10\delta. \end{aligned} \tag{80}$$

Consequently, we get

$$|f^o(yx) + f^o(yx^{-1}) - 2f^o(y)| \leq 5\delta \tag{81}$$

for all $x, y \in G$. Now, by using the proof of Theorem 2.6, we get the rest of the proof.

References

1. Aoki, T.: On the stability of the linear transformation in Banach spaces. J. Math. Soc. Jpn. **2**, 64–66 (1950)
2. Akkouchi, M.: Stability of certain functional equations via a fixed point of Ćirić. Filomat. **25**, 121–127 (2011)
3. Baker, J.A.: The stability of certain functional equations. Proc. Am. Math. Soc. **112**, 729–732 (1991)
4. Bouikhalene, B., Elqorachi, E., Redouani, A.: Hyers-Ulam stability on the generalized quadratic functional equation in amenable semigroups. Jipam **8**(2) Article 56, 18 (2007).
5. Bouikhalene, B., Elqorachi, E., Rassias, Th.M.: On the Hyers-Ulam stability of approximately Pexider mappings. Math. Inequal. Appl. **11**, 805–818 (2008)
6. Brzdęk, J.: On a method of proving the Hyers-Ulam stability of functional equations on restricted domains. Aust. J. Math. Anal. Appl. **6**(1), Article 4, 1–10 (2009)
7. Că̆dariu, L., Radu, V.: Fixed points and the stability of Jensen's functional equation. J. Inequal. Pure Appl. Math. **4**(1), Article 4 (2003).
8. Cholewa, P.W.: Remarks on the stability of functional equations. Aequ. Math. **27**, 76–86 (1984)
9. Czerwik, S.: On the stability of the quadratic mapping in normed spaces. Abh. Math. Sem. Univ. Hambg. **62**, 59–64 (1992)
10. Czerwik, S.: Functional Equations and Inequalities in Several Variables. World Scientific, London, (2002)
11. Drygas, H.: Quasi-inner products and their applications. In: Gupta, A.K. (ed.) Advances in Multivariate Statistical Analysis, pp. 13–30. D. Reidel Publishing Co., Dordrecht, (1987)
12. Ebanks, B.R., Kannappan, Pl., Sahoo, P.K.: A common generalization of functional equations characterizing normed and quasi-inner product spaces. Canad. Math. Bull. **35**(3), 321–327 (1992)
13. Faǐziev, V.A., Sahoo, P.K.: On Drygas functional equation on groups. Int. J. Appl. Math. Stat. **7**, 59–69 (2007)
14. Faǐziev, V.A., Sahoo, P.K.: Stability of Drygas functional equation on T (3, R). Int. J. Appl. Math. Stat. **7**, 70–81 (2007)
15. Faǐziev, V.A., Sahoo, P.K.: On the stability of Drygas functional equation on groups. Banach J. Math. **1**, 1–18 (2007)
16. Forti, G.L.: The stability of Homomorphisms and Amenability, with applications to functional equations. Abhandlungen aus dem Mathematischen Seminar der Universität Hamburg. **57**, 215–226 (1987)
17. Forti, G.L.: Hyers-Ulam stability of functional equations in several variables. Aequ. Math. **50**, 143–190 (1995)
18. Forti, G.L., Sikorska, J.: Variations on the Drygas equation and its stability. Nonlinear Anal. **74**, 343–350 (2011)
19. Gajda, Z.: On stability of additive mappings. Internat. J. Math. Math. Sci. **14**, 431–434 (1991)
20. Gă̆vruta, P.: A generalization of the Hyers-Ulam-Rassias stability of approximately additive mappings. J. Math. Anal. Appl. **184**, 431–436 (1994)
21. Ger, R.: A survey of recent results on stability of functional equations. Proceedings of the 4th International Conference on Functional Equations and Inequalities Pedagogical University of Cracow, 5–36 (1994)
22. Greenleaf, F.P.: Invariant Means on Topological Groups and there applications. Van Nostrand, New York, (1969)
23. Hyers, D.H.: On the stability of the linear functional equation. Proc. Natl. Acad. Sci. U. S. A. **27**, 222–224 (1941)
24. Hyers, D.H., Isac, G.I., Rassias, Th.M.: Stability of Functional Equations in Several Variables. Birkhäuser, Basel, 1998
25. Hyers, D.H., Rassias, Th.M.: Approximate homomorphisms. Aequ. Math. **44**, 125–153 (1992)

26. Hyers, D.H., Isac, G., Rassias, Th.M.: On the asymptoticity aspect of Hyers-Ulam stability of mappings. Proc. Am. Math. Soc. **126**, 425–430 (1998)
27. Jung, S.-M.: Hyers-Ulam-Rassias Stability of Functional Equations in Mathematical Analysis. Hadronic Press, Palm Harbor, (2003)
28. Jung, S.-M., Sahoo, P.K.: Stability of a functional equation of Drygas. Aequ. Math. **64** (3), 263–273
29. Kannappan, Pl.: Functional Equations and Inequalities with Applications. Springer, New York, (2009)
30. Li, L.S., Kim, D., Chung, J.: Stability of functional equations of Drygas in the space of Schwartz distributions. J. Math. Anal. Appl. **320**(1), 163–173 (2006)
31. Moslehian, M.S.: The Jensen functional equation in non-Archimedean normed spaces. J. Funct. Spaces Appl. **7**, 13–24 (2009)
32. Moslehian, M.S., Najati, A.: Application of a fixed point theorem to a functional inequality. Fixed Point Theory, **10**, 141–149 (2009)
33. Moslehian, M.S., Sadeghi, Gh.: Stability of linear mappings in quasi-Banach modules. Math. Inequal. Appl. **11**, 549–557 (2008)
34. Moslehian, M.S., Rassias, Th.M.: Stability of functional equations on non-Archimedean spaces. Applicable Anal. Discrete Math. **1**, 325–334 (2007)
35. Najati, A., Moghimi, M.B.: Stability of a functional equation deriving from quadratic and additive functions in quasi-Banach spaces. J. Math. Anal. Appl. **337**, 399–415 (2008)
36. Najati, A.: On the stability of a quartic functional equation. J. Math. Anal. Appl. **340**, 569–574 (2008)
37. Park, C.: On the stability of the linear mapping in Banach modules. J. Math. Anal. Appl. **275**, 711–720 (2002)
38. Park, C.: Hyers-Ulam-Rassias stability of homomorphisms in quasi-Banach algebras. Bull. Sci. Math. **132**, 87–96 (2008)
39. Pourpasha, M.M., Rassias, J.M., Saadati, R., Vaezpour, S.M.: A fixed point approach to the stability of Pexider quadratic functional equation with involution. J. Ineq. Appl. (2010). Article ID 839639, doi:10.1155/2010/839639.
40. Rassias, J.M.: On approximation of approximately linear mappings by linear mappings. J. Funct. Anal. **46**, 126–130 (1982)
41. Rassias, Th.M.: On the stability of linear mapping in Banach spaces. Proc. Am. Math. Soc. **72**, 297–300 (1978)
42. Rassias, Th.M.: The problem of S. M. Ulam for approximately multiplicative mappings. J. Math. Anal. Appl. **246**, 352–378 (2000)
43. Rassias, Th.M.: On the stability of the functional equations and a problem of Ulam. Acta Appl. Math. **62**, 23–130 (2000)
44. Rassias, Th.M., Šemrl, P.: On the behavior of mappings which do not satisfy Hyers-Ulam stability. Proc. Am. Math. Soc. **114**, 989–993 (1992)
45. Rassias, Th.M., Tabor, J.: Stability of Mappings of Hyers-Ulam Type. Hardronic Press, Palm Harbor, (1994)
46. Stetkær, H.: Functional equations on abelian groups with involution II. Aequ. Math. **55**, 227–240 (1998)
47. Szabo, Gy.: some functional equations related to quadratic functions. Glasnik Math. **38**, 107–118 (1983)
48. Székelyhidi, L.: Note on a stability theorem. Can. Math. Bull. **25**, 500–501 (1982)
49. Székelyhidi, L.: Ulam's problem, Hyers's solution-and to where they led. In: Rassias, Th.M., Tabor, Jo. (eds.), Functional Equations and Inequalities, Math. Appl. **518**, Kluwer, Dordrecht, 259–285 (2000)
50. Ulam, S.M.: A Collection of Mathematical Problems. Interscience Publishers, New York, 1961. Problems in Modern Mathematics, Wiley, New York, (1964)
51. Yang, D.L.: Remarks on the stability of Drygas' equation and the Pexider-quadratic equation. Aequ. Math. **68**, 108–116 (2004)
52. Yang, D.L.: The stability of Jensen's functional equation on locally compact groups. Result. Math. **46**, 381–388 (2004)

Stability of Quadratic and Drygas Functional Equations, with an Application for Solving an Alternative Quadratic Equation

Gian Luigi Forti

Abstract The aim of this survey is to present stability results obtained in the last years (roughly after 1995) for the quadratic equation and its various generalizations, and the Drygas equation. The number of papers on this subject is very high, hence, the author of the present chapter made a (quite arbitrary) choice of some of them to be shown in detail. The last section is devoted to an application of stability for solving an alternative form of the quadratic equation.

Keywords Ulam–Hyers stability · Quadratic equation · Drygas equation · Alternative quadratic equation

1 Introduction

The well-known characterization of inner product spaces among normed spaces is given by the so-called parallelogram law

$$\|x + y\|^2 + \|x - y\|^2 = 2\|x\|^2 + 2\|y\|^2$$

and this leads naturally to the following functional equation

$$f(x + y) + f(x - y) = 2f(x) + 2f(y). \tag{1}$$

In the case of $f : \mathbf{R} \to \mathbf{R}$, the regular solutions of the equation above have the form $f(x) = \lambda x^2$ and from this fact Eq. (1) has been named *quadratic equation* and its solutions *quadratic functions* (see [2]).

The similar functional equation

$$g(x + y) + g(x - y) = 2g(x) + g(y) + g(-y) \tag{2}$$

was introduced in 1987 by H. Drygas in [28], where the author was looking for characterizations of quasi inner product spaces, which in turn lead to solutions of

G. L. Forti (✉)
Dipartimento di Matematica, Università degli Studi di Milano, Milano, Italy
e-mail: gianluigi.forti@unimi.it

T. M. Rassias (ed.), *Handbook of Functional Equations*,
Springer Optimization and Its Applications 96, DOI 10.1007/978-1-4939-1286-5_8

some problems in statistics and mathematical programming. Equation (2) is now known in the literature as *Drygas equation*.

Both functional equations have been the subject of investigations in various settings: linear spaces, commutative and non–commutative groups, etc. Moreover, several authors dealt with the Ulam–Hyers stability of them.

The aim of this survey is to present stability results obtained after 1995 for the two previous equations and some of their generalizations. For the period before 1995, refer to [26, 42, 45, 86, 87, 91]. Moreover, various results are in the more recent books [27, 57, 61, 88, 89, 91, 98]. The number of papers on this subject is very high, hence, it is not realistic to try to describe the content of all of them. The author of the present chapter made a (quite arbitrary) choice of some of them to be shown in detail and decided not to treat the case of restricted domains.

The last section is devoted to an application of stability for solving an alternative form of the quadratic equation.

Throughout this chapter $\mathbf{N}, \mathbf{R}, \mathbf{C}$ denote the natural, real and complex numbers, respectively.

2 Stability of the Quadratic Equation

The natural setting for studying stability of Eq. (1) is that of functions mapping a group $\mathcal{G}$ into a Banach space $\mathcal{B}$. When $\mathcal{G}$ is not necessarily Abelian, we write Eq. (1) in the multiplicative form

$$f(xy) + f(xy^{-1}) = 2f(x) + 2f(y) \tag{3}$$

otherwise we use the additive form.

Following [41] and [100], we shall use the following definition.

Definition 1 Let $\mathcal{G}$ be a group and $\mathcal{B}$ a Banach space. We say that *the couple* $(\mathcal{G}, \mathcal{B})$ *has the property of the stability of the quadratic functional equation* (write $(\mathcal{G}, \mathcal{B})$ is QS for short) if for every function $f : \mathcal{G} \to B$ such that

$$\|f(xy) + f(xy^{-1}) - 2f(x) - 2f(y)\| \leq \delta \text{ for all } x, y \in \mathcal{G} \text{ and for some } \delta \geq 0 \tag{4}$$

there exists a quadratic function $q : \mathcal{G} \to \mathcal{B}$ and a constant $\varepsilon \geq 0$ depending only on δ such that

$$\|f(x) - q(x)\| \leq \varepsilon \text{ for all } x \in \mathcal{G}. \tag{5}$$

The first author treating the problem above was F. Skof, who in [96] proved the following.

Theorem 1 *Let $\mathcal{X}$ be a normed vector space and $\mathcal{B}$ a Banach space. If $f : \mathcal{X} \to \mathcal{B}$ fulfils* (4)*, then for every $x \in \mathcal{X}$ the limit*

$$q(x) = \lim_{n\to\infty} \frac{f(2^n x)}{2^{2n}}$$

exists and q *is the unique quadratic function satisfying* (5) *with* $\varepsilon = \delta/2$.

P. W. Cholewa in [18] proved that the previous theorem remains true if we substitute the vector space $\mathcal{X}$ with an Abelian group $\mathcal{G}$.

Having in mind what has been done for the additive equation, we face the problem of eliminating or weakening the request of commutativity of $\mathcal{G}$. The first step in order to simplify the study of that problem is provided by the following:

Theorem 2 [100] *Suppose that the couple* $(\mathcal{G}, \mathbf{C})$ *(or* $(\mathcal{G}, \mathbf{R})$*) is QS. Then, for every complex (real) Banach space* $\mathcal{B}$*, the couple* $(\mathcal{G}, \mathcal{B})$ *is QS. Moreover, if the* $(\mathcal{G}, \mathcal{B})$ *is QS and* f, q, δ *and* ε *are as in Definition 1, then* q *is unique and* $\varepsilon = \delta/2$.

Following the ideas of L. Székelyhidi [97] for the Cauchy equation, D. Yang was able to prove the following.

Theorem 3 [100] *Let* $\mathcal{G}$ *be an amenable group. Then,* $(\mathcal{G}, \mathbf{C})$ *is QS.*

Moreover, this paper of D. Yang contains a counterexample, suggested by that presented in [41] for the Cauchy equation, proving that on the free group generated by two elements the quadratic equation is not stable.

Also V. A. Faĭziev and P. K. Sahoo attacked in [39] the problem of reducing the requirement of commutativity (without quoting the former result of D. Yang). In order to state the result, we need to introduce some classes of groups.

Definition 2 Given an integer n, a group $\mathcal{G}$ is said to be n-Abelian if for every $x, y \in \mathcal{G}$, we have

$$(xy)^n = x^n y^n.$$

By $\mathcal{K}_n$, we denote the class of groups such that for every x, y the relation

$$(xy)^n = x^n y^n = y^n x^n$$

is satisfied. Note that the 2-Abelian groups are commutative.

Another group considered in [39] is $T(2, K)$, that is the group of matrices of the form

$$\begin{bmatrix} y & t \\ 0 & x \end{bmatrix}$$

where $x, y, t \in K$ and K is a commutative field.

Then, the stability of the quadratic equation is proved in [39] when the domain is either a n-Abelian group or the group $T(2, K)$.

Along the lines traced by T. Aoki [3] for the additive mappings and Th. M. Rassias for the linear ones [83], many authors have considered the stability of the quadratic equation in the more general case where the constant bound is substituted by a control function.

Probably the first author to prove stability in this setting was St. Czerwik, who in [25] proved the following.

Theorem 4 *Let $\mathcal{X}$ be a normed space and $\mathcal{B}$ a Banach space and let $f : \mathcal{X} \to \mathcal{B}$ be a function satisfying the inequality*

$$\|f(x+y)+f(x-y)-2f(x)-2f(y)\| \le \Lambda(x,y). \tag{6}$$

with either

(i) $\Lambda(x,y) = \eta + \theta(\|x\|^s + \|y\|^s), \quad s<2 \quad x,y \in \mathcal{X}\setminus\{0\},$

or

(ii) $\Lambda(x,y) = \theta(\|x\|^s + \|y\|^s), \quad s>2 \quad x,y \in \mathcal{X},$

for some $\eta, \theta \ge 0$.

Then, there exists exactly one quadratic function q such that

$$\|f(x)-q(x)\| \le \frac{1}{3}(\eta + \|f(0)\|) + \frac{2\theta}{4-2^s}\|x\|^s, \quad x \in \mathcal{X}\setminus\{0\}$$

in case (i) *or*

$$\|f(x)-q(x)\| \le \frac{2\theta}{2^s-4}\|x\|^s, \quad x \in \mathcal{X}$$

in case (ii). *Moreover, if the function $t \mapsto f(tx)$, $t \in \mathbf{R}$, is continuous for each $x \in \mathcal{X}$, then q satisfies the equation*

$$q(tx) = t^2 q(x), \quad x \in \mathcal{X}, \quad t \in \mathbf{R}.$$

In [9], as a particular case of a stability theorem for a wider class of functional equations, the following result has been obtained.

Theorem 5 *Let $\mathcal{G}$ be an Abelian group, $\mathcal{B}$ a Banach space and let $f : \mathcal{G} \to \mathcal{B}$ be a function with $f(0) = 0$ and fulfilling* (6). *Assume that one of the series*

$$\sum_{i=1}^{+\infty} 2^{-2i}\Lambda(2^{i-1}x, 2^{i-1}x) \quad or \quad \sum_{i=1}^{+\infty} 2^{2(i-1)}\Lambda(2^{-i}x, 2^{-i}x)$$

converges for every x and call $\Gamma(x)$ its sum. If, for every x, y

$$2^{2i}\Lambda(2^{i-1}x, 2^{i-1}y) \to 0 \quad or \quad 2^{2(i-1)}\Lambda(2^{-i}x, 2^{-i}y) \to 0,$$

respectively, as $i \to \infty$, then there exists a unique quadratic function q such that

$$\|f(x)-q(x)\| \le \Gamma(x), \quad x \in \mathcal{X}.$$

Some authors studied the stability of the quadratic equation in different contexts concerning both the domain and the range of the functions involved. C.-G. Park in [77] considered the case of Banach modules; in [113], the domain is a Banach module over a C^*-algebra; M. S. Moslehian, K. Nikodem and D. Popa in [75] used multi-normed spaces.

A. K. Mirmostafaee and M. S. Moslehian in [73] investigated the stability when domain and range are fuzzy normed space.

Definition 3 A function $N : \mathcal{X} \times \mathbf{R} \to [0, 1]$, $\mathcal{X}$ real linear space, is said to be a *fuzzy norm* on $\mathcal{X}$ if for all $x, y \in \mathcal{X}$ and all $s, t \in \mathbf{R}$, the following conditions hold:

$$\begin{cases} N(x, c) = 0 \text{ for } c \leq 0; \\ x = 0 \text{ if and only if } N(x, c) = 1 \text{ for all } c > 0; \\ N(cx, t) = N\big(x, \frac{t}{|c|}\big) \text{ if } c \neq 0; \\ N(x + y, s + t) = \min[N(x, s), N(y, t)]; \\ N(x, \cdot) \text{ is non-decreasing and } \lim_{t \to \infty} N(x, t) = 1. \end{cases}$$

The pair $(\mathcal{X}, N)$ is called a *fuzzy normed linear space*; a complete fuzzy normed linear space is called *fuzzy Banach*.

The result in this setting, presented in [73], says:

Theorem 6 *Let $(\mathcal{X}, N)$ be a fuzzy normed linear space and $(\mathcal{B}, N')$ a fuzzy Banach space. Let $p > 1/2$ and assume that $f : \mathcal{X} \to \mathcal{B}$ satisfies the inequality*

$$N'(f(x + y) + f(x - y) - 2f(x) - 2f(y), t + s) \geq \min[N(x, t^p), N(y, s^p)]$$

for all $x, y \in \mathcal{X}$ and $s, t \in [0, \infty)$. Then, there exists a unique quadratic function q such that

$$N'(q(x) - f(x), t) \geq N\left(x, \left(\frac{2^{2-1/p} - 1}{4}\right)^p t^p\right).$$

A similar result is valid for $p < 1/2$.

Other stability theorems in the same framework of fuzzy spaces are in [29, 33, 48, 62]. For results in non-Archimedean spaces see [8, 63, 95].

As for other functional equations, also the quadratic one has been investigated in the frame of distributions. A stability result in this setting has been published by J.-Y. Chung in [19]. Clearly the first, and crucial, step consists of transforming a functional inequality in an inequality meaningful for distributions.

Let A, B, P_1 and P_2 be the functions

$$A(x, y) = x + y, \ B(x, y) = x - y, \ P_1(x, y) = x, \ P_2(x, y) = y, \ x, y \in \mathbf{R}^n.$$

Then, the inequality (6) can be naturally transformed as

$$\|u \circ A + u \circ B - 2u \circ P_1 - 2u \circ P_2\| \leq \Lambda(x, y) \tag{7}$$

where $u \circ A, u \circ B, u \circ P_1, u \circ P_2$ are the pullback of u by A, B, P_1, P_2, respectively, and $\|v\| \leq \Lambda(x, y)$ means that $|\langle v, \phi \rangle| \leq \|\Lambda\phi\|_{L_1}$ for all test functions ϕ.

The main result reads as follows:

Theorem 7 [19] *Let $u \in \mathcal{S}'$ satisfies the inequality* (7). *Then, there exists a unique quadratic function*

$$q(x) = \sum_{1 \leq j \leq k \leq n} a_{jk} x_j x_k$$

such that

$$\|u(x) - q(x)\| \le \frac{1}{|4-2^p|}\Lambda(x,x),\ 0 < p < 2 \text{ or } p > 4;$$
$$\|u(x) - q(x)\| \le \tfrac{1}{2}\Lambda(0,0),\ p = 0.$$

(S' is the space of Schwartz tempered distributions.)

Other results of this kind can be found in [20–22, 24, 69, 70].

The term $f(x - y)$ in the quadratic equation or, more precisely, the $-y$ in the argument, can be interpreted in a more abstract way as an involution σ applied to y. Some results in this direction are given in [10, 11, 31].

The first natural generalization of the quadratic equation consists of the so-called *Pexideration* of the functional equation, that is, while preserving the structure of the equation, in one or more terms appear different functions.

S.-M. Jung in [56] proves the following stability theorem for a Pexider-type equation:

Theorem 8 *Assume that $f_1, f_2, f_3, f_4 : \mathcal{X} \to \mathcal{B}$ satisfy the inequality*

$$\|f_1(x+y) + f_2(x-y) - f_3(x) - f_4(y)\| \le \phi(x,y)$$

where ϕ is symmetric, $\phi(x,-y) = \phi(x,y)$ and there exists an integer $s \ge 2$ such that either

$$\sum_{i=0}^{\infty} \frac{1}{s^i}\phi(s^i x, s^i y) < \infty,$$

or

$$\sum_{i=0}^{\infty} s^{2i}\phi\left(\frac{x}{s^i}, \frac{y}{s^i}\right) < \infty.$$

Then, there exist a quadratic function q and additive functions a_1, a_2 such that

$$\|f_i(x) - q(x) - a_1(x) - a_2(x) - f_i(0)\| \le \Psi_i(x),\ i = 1,2,3$$
$$\|f_4(x) - 2q(x) - 2a_1(x) - f_4(0)\| \le \Psi_4(x),$$

for some functions $\Psi_i, i = 1,2,3,4$ depending only on ϕ. In the special case $\phi(x,y) = \varepsilon$, the four functions Ψ_i are the constants $\frac{137}{3}\varepsilon, \frac{125}{3}\varepsilon, \frac{136}{3}\varepsilon$ and $\frac{124}{3}\varepsilon$, respectively.

Quite curiously, this very same result, in the case of constant bound, constitutes the only content of the paper [58], published later than [56] and by the same author.

Other results, very similar among them, are in [50–53].

A different approach to the stability is in the paper [10] where the following system of equations is studied:

$$\begin{cases} f(xy) + f(x\sigma(y)) = 2f(x) + 2f(y) \\ f(xy) + g(x\sigma(y)) = f(x) + g(y) \end{cases}$$

Here, the domain of the functions f and g is an amenable semigroup $\mathcal{G}$ and σ is an automorphisms on $\mathcal{G}$ such that $\sigma\sigma = I$. Some of the results therein contained concern the Pexiderized quadratic equation

$$f_1(xy) + f_2(x\sigma(y)) = f_3(x) + f_4(y).$$

Note that Z. Kominek in [68] has been able to investigate the stability of the quadratic equation in semigroups by modifying its form as follows:

$$f(x + 2y) + f(x) = 2f(x + y) + 2f(y).$$

In the last ten years, several other functional equations called "quadratic" (or Euler–Lagrange: see [84–90]) have been investigated in order to determine the general solution and to prove stability results in the sense of Ulam–Hyers in various different situations. The name "quadratic" has been given since, in the case of real functions of real variable, the quadratic polynomials are among their solutions.

Showing all these results would be excessively long. In the following, we present a choice of these equations and related stability results.

J. M. Rassias in [85] proved the following:

Theorem 9 *Let $f : \mathcal{X} \to \mathcal{B}$, where $\mathcal{X}$ is a linear normed space and $\mathcal{B}$ a real Banach space. For reals a_i and positive reals m_i, $i = 1, 2$, define*

$$m_0 = \frac{m_1 m_2 + 1}{m_1 + m_2}, \qquad m = \frac{m_1 a_1^2 + m_2 a_2^2}{m_0}$$

and

$$\overline{f}(x) = \tfrac{m_0}{m}\left[\tfrac{1}{m_1} f\left(\tfrac{m_1}{m_0} a_1 x\right) + \tfrac{1}{m_2} f\left(\tfrac{m_2}{m_0} a_2 x\right)\right],$$
$$\overline{\overline{f}}(x) = \tfrac{1}{m m_0}\left[m_1 f(a_1 x) + \tfrac{1}{m_2} f(m_2 a_2 x)\right].$$

Assume that $m \neq 1$ and $\|\overline{\overline{f}}(x) - \overline{f}(x)\| \leq c'$ and

$$\|m_1 m_2 f(a_1 x_1 + a_2 x_2) + f(m_2 a_2 x_1 - m_1 a_1 x_2) - m m_0 (m_2 f(x_1) + m_1 f(x_2))\| \leq c,$$

for some positive constants c and c'. Then, there exists a unique quadratic mapping q satisfying the functional equation

$$m_1 m_2 q(a_1 x_1 + a_2 x_2) + q(m_2 a_2 x_1 - m_1 a_1 x_2) = m m_0 (m_2 q(x_1) + m_1 q(x_2)),$$

and such that $\|f(x) - q(x)\| \leq c_1$, for some c_1 depending only on c and c'.

The same author treated analogous functional equations in several other papers.

S.-M. Jung in [54] considered the functional equation

$$f(x + y + z) + f(x) + f(y) + f(z) = f(x + y) + f(y + z) + f(z + x),$$

where f maps a real linear space $\mathcal{X}$ into a Banach space $\mathcal{B}$. If $\mathcal{X} = \mathcal{B} = \mathbf{R}$, the function $f(x) = x^2$ is a solution, hence, again we have a "quadratic" equation.

The stability result proved therein is the following:

Theorem 10 *Assume that f satisfies the system of inequalities*

$$\begin{cases} \|f(x+y+z)+f(x)+f(y)+f(z)-f(x+y)-f(y+z)-f(z+x)\| \le \varepsilon \\ \|f(x)-f(-x)\| \le \theta \end{cases}$$

for some $\varepsilon, \theta \ge 0$. Then, there exists a unique quadratic mapping q which satisfies the previous equation and the inequality

$$\|f(x)-q(x)\| \le 3\varepsilon.$$

If, moreover, f is measurable or $f(tx)$ is continuous in t for each fixed $x \in \mathcal{X}$, then $q(tx) = t^2 q(x)$ for all $x \in \mathcal{X}$ and $t \in \mathbf{R}$.

The same equation has been studied in [14].

Again S.-M. Jung in [55] investigates the functional equation

$$f(x-y-z)+f(x)+f(y)+f(z)-f(x-y)-f(y+z)-f(z-x)=0, \qquad f : \mathcal{X} \to \mathcal{B},$$

and, after showing that it is equivalent to the quadratic Eq. (1), proves the following:

Theorem 11 *Assume that f satisfies the inequality*

$$\|f(x-y-z)+f(x)+f(y)+f(z)-f(x-y)-f(y+z)-f(z-x)\| \le \varepsilon(\|x\|^p+\|y\|^p+\|z\|^p)$$

for some $\varepsilon \ge 0$, some $p > 0$, $p \ne 2$ and for all $x, y, z \in \mathcal{X}$. Then, there exists a unique quadratic mapping q which satisfies the inequality

$$\|f(x)-q(x)\| \le \frac{8\varepsilon}{|2^p-4|}\|x\|^p.$$

If, moreover, f is measurable or $f(tx)$ is continuous in t for each fixed $x \in \mathcal{X}$, then $q(tx) = t^2 q(x)$ for all $x \in \mathcal{X}$ and $t \in \mathbf{R}$.

A similar result is true for $p < 0$ if the following inequality is added to the hypotheses:

$$\|f(x)-f(-x)\| \le \delta.$$

In this case, we have the different bound

$$\|f(x)-q(x)\| \le \frac{2}{3}(\delta+\|f(0)\|)+\frac{4\varepsilon}{|2^p-4|}\|x\|^p.$$

J.-H. Bae and H.-M. Kim in [4] studied the stability of equation

$$f(x+y+z)+f(x-y)+f(y-z)+f(x-z)=3[f(x)+f(y)+f(z)].$$

I.-S. Chang and H.-M. Kim in [15] considered the functional equations

$$f(2x+y)+f(2x-y)=f(x+y)+f(x-y)+6f(x)$$
$$f(2x+y)+f(x+2y)=4f(x+y)+f(x)+f(y),$$

I.-S. Chang, E. H. Lee and H.-M. Kim in [17] considered the equation

$$f(x+y+z+w)+2f(x)+2f(y)+2f(z)+2f(w)=$$
$$f(x+y)+f(y+z)+f(z+x)+f(x+w)+f(y+w)+f(z+w).$$

and H.-M. Kim in [64] considered the following one

$$f(x+y+z)+f(x-y)+f(x-z)=f(x-y-z)+f(x+y)+f(x+z).$$

In [49], K.-W. Jun and H.-K. Kim solved the functional equation

$$\sum_{i=1}^{n} f\left(\sum_{j\neq i} x_j-(n-1)x_i\right)+nf\left(\sum_{i=1}^{n} x_i\right)=n^2\sum_{i=1}^{n} f(x_i),$$

proving that its solutions are exactly the quadratic functions. Then, this stability theorem is given:

Theorem 12 *Let $f:\mathcal{X}\to\mathcal{B}$ be a function such that*

$$\left\|\sum_{i=1}^{n} f\left(\sum_{j\neq i} x_j-(n-1)x_i\right)+nf\left(\sum_{i=1}^{n} x_i\right)-n^2\sum_{i=1}^{n} f(x_i)\right\|\le\phi(x_1,\cdots,x_n)$$

and assume that either

$$\Phi(x_1,\cdots,x_n):=\sum_{i=0}^{\infty}\frac{1}{n^{2i}}\phi(n^i x_1,\cdots,n^i x_n)<\infty$$

or

$$\Phi(x_1,\cdots,x_n):=\sum_{i=0}^{\infty} n^{2i}\phi(\frac{x_1}{n^i},\cdots,\frac{x_n}{n^i})<\infty$$

for all $x_i\in\mathcal{X}$, $i=1,\cdots,n$.

Then, there exists a unique quadratic mapping q such that

$$\left\|f(x)-\frac{f(0)}{n^2-1}-q(x)\right\|\le\frac{1}{n^3}\Phi(x,\cdots,x)$$

for all $x\in\mathcal{X}$.

An analogous result is obtained when A is a unital Banach $*$-algebra and $f: M_1\to M_2$, where M_1, M_2 are Banach left A-modules.

In this line of generalizations, various so-called *mixed* equations, i.e., equations whose regular solutions, in the real-to-real case, are polynomials containing linear, quadratic, cubic, quartic, etc. monomials, have been investigated.

An example of these kind of equations and related results is in [1], where we can find the following:

Theorem 13 *Let $\mathcal{X}$ be a quasi-Banach space with quasi-norm $\|\cdot\|_{\mathcal{X}}$ and $\mathcal{Y}$ a p-Banach space with p-norm $\|\cdot\|_{\mathcal{Y}}$. Suppose that a mapping $f:\mathcal{X}\to\mathcal{Y}$, with $f(0)=0$, satisfies the inequality*

$$\begin{aligned}&\|f(nx+y)+f(nx-y)-n^2 f(x+y)-n^2 f(x-y)\\&-2f(nx)+2n^2 f(x)+2(n^2-1)f(y)\|_{\mathcal{Y}}\le\phi(x,y).\end{aligned}$$

where $\phi:\mathcal{X}\times\mathcal{X}\to[0,+\infty)$ satisfies the following conditions:

$$\lim_{m\to\infty}4^m\phi\left(\frac{x}{2^m},\frac{y}{2^m}\right)=0=\frac{1}{16^m}\phi(2^m x,2^m y)$$

for all x, y and

$$\begin{aligned}&\textstyle\sum_{i=1}^{\infty}4^{pi}\phi^p\left(\frac{x}{2^i},\frac{y}{2^i}\right)<\infty,\\&\textstyle\sum_{i=0}^{\infty}\frac{1}{16^{pi}}\phi^p(2^i x,2^i y)<\infty\end{aligned}$$

for all x and all

$$y\in\{x,2x,3x,nx,(n-1)x,(n+1)x,(n-2)x,(n+2)x,(n-3)x\}.$$

Then, there exists a unique quadratic mapping q and a unique quartic mapping t such that

$$\|f(x)-q(x)-t(x)\|_{\mathcal{Y}}\le\Phi(x),$$

where Φ is a function explicitly computed from ϕ.

(For the definition and properties of quasi-normed and quasi-Banach spaces see [7] and [93].)

Other stability problems have been investigated in the framework of random normed spaces. See the papers [6, 32–36, 47, 81, 94] for the definition of random normed space and stability results.

We feel obliged to remark that most papers dealing with generalizations of the quadratic equation and/or less usual domains and ranges do not give any motivation for the choice of the equations to be studied and the setting where they are investigated.

This is the point where few words about the methods are necessary. Most of the results previously stated have been obtained either by using the so-called *direct method* (see, for instance, [43]) or by using the translation invariant means in the case of amenable groups.

In [16], L. Cădariu and V. Radu introduced the use of fixed point theorems for the stability of some functional equations. While in the opinion of the author of the present paper this new method is essentially a change in language with respect to the direct one, several mathematicians started to use it, sometimes proving again already known results. Here is a (certainly non-exhaustive) list of papers dealing with that method for the quadratic equation and its generalizations: [5, 12–16, 34, 46, 60, 66, 67, 71, 72, 74, 76, 78–82, 92].

Another method which can be applied for the investigation of stability make use of the so-called *shadowing property* introduced in [99]. A stability theorem concerning a functional equation of quadratic type proved by using this method can be found in [65].

To finish this section, we prove a result which is the analogous of Theorem 4 proved in [41] for the Cauchy functional equation and which will be used in the last section of this chapter. If

$$Qf(x,y) := f(x+y) + f(x-y) - 2f(x) - 2f(y)$$

is bounded then, due to any of the previous stability theorems, we have the decomposition $f(x) = q(x) + k(x)$, with q quadratic and k bounded. Our aim is to provide information on the range of the bounded function k.

Theorem 14 *Let $f : \mathcal{G} \to \mathcal{B}$, where $\mathcal{G}$ is an Abelian group and $\mathcal{B}$ a Banach space and let M be a bounded subset of $\mathcal{B}$. If $Qf(x,y) \in M$, then $f(x) = q(x) + k(x)$, where q is quadratic and the range of k is contained in $\frac{1}{2}\overline{C(-M)}$, where $\overline{C(-M)}$ is the closure of the convex hull of $-M$.*

Proof By any of the stability results, we have the decomposition

$$f(x) = q(x) + k(x)$$

with q quadratic and k bounded. Since $q(0) = 0$, we have $f(0) = k(0) = -\frac{1}{2}m_0$, for certain $m_0 \in M$. Fix $x \in \mathcal{G}$ and consider the value $k(x) =: u$. We have

$$k(2x) - 4k(x) = Qf(x,x) - f(0), \text{ hence, } k(2x) = 4u + \frac{1}{2}m_0 + m_1 \text{ for some } m_1 \in M.$$

By induction, we obtain

$$k(sx) = s^2 u + \frac{s-1}{2} m_0 + \sum_{i=1}^{s-1} (s-i) m_i,$$

for some $m_i \in M$. By dividing by s^2, we have

$$\frac{k(sx)}{s^2} = u + \frac{1}{2} \sum_{i=1}^{s-1} \frac{2(s-i)}{s^2} m_i + \frac{1}{2s} m_0 - \frac{1}{2s^2} m_0.$$

Clearly, $\sum_{i=1}^{s-1} \frac{2(s-i)}{s^2} m_i + \frac{1}{2s} m_0 \in C(M)$; taking the limit as $s \to \infty$ and remembering that k is bounded, we get

$$u + \frac{1}{2}\mu = 0, \text{ where } \mu = \lim_{s\to\infty} \sum_{i=1}^{s-1} \frac{2(s-i)}{s^2} m_i + \frac{1}{2s} m_0,$$

thus, $u \in \frac{1}{2}\overline{C(-M)}$. □

Theorem 15 *In the hypotheses of Theorem 14, the range of k is contained in the set* $K = \left\{-\sum_{i=1}^{\infty} \frac{m_i}{4^i} - \frac{m_0}{6} : m_i \in M,\ m_0 = -2k(0)\right\}$.

Proof By Theorem 14, the range of k is contained in $\frac{1}{2}\overline{C(-M)}$ and we have $k(0) = -\frac{m_0}{2}$ for some $m_0 \in M$. From

$$Qk(x,x) = k(2x) + k(0) - 4k(x) = k(2x) - 4k(x) - \frac{m_0}{2} \in M,$$

we obtain

$$k(2x) = 4k(x) + \frac{m_0}{2} + m_1 \in \frac{1}{2}\overline{C(-M)}, \text{ for some } m_1 \in M,$$

hence

$$k(x) \in \left[\frac{1}{8}\overline{C(-M)} - \frac{m_0}{8} - \frac{m_1}{4}\right] \cap \left[\frac{1}{2}\overline{C(-M)}\right].$$

It is easy to see that

$$\frac{1}{8}\overline{C(-M)} - \frac{m_0}{8} - \frac{m_1}{4} \subset \frac{1}{2}\overline{C(-M)},$$

thus

$$k(x) \in \bigcup_{m_1 \in M} \left[\frac{1}{8}\overline{C(-M)} - \frac{m_0}{8} - \frac{m_1}{4}\right].$$

We claim that

$$k(x) \in \bigcup_{m_1, m_2, \cdots, m_n \in M} \left[\frac{1}{2\cdot 4^n}\overline{C(-M)} - \frac{m_0}{2}\sum_{i=1}^{n}\frac{1}{4^i} - \sum_{i=1}^{n}\frac{m_{n+1-i}}{4^i}\right] =: K_n.$$

The proof is by induction. Consider $n+1$ and

$$4k(x) + \frac{m_0}{2} + m_{n+1} \in \frac{1}{2\cdot 4^n}\overline{C(-M)} - \frac{m_0}{2}\sum_{i=1}^{n}\frac{1}{4^i} - \sum_{i=1}^{n}\frac{m_{n+1-i}}{4^i}$$

for some $m_1, m_2, \cdots, m_{n+1} \in M$. Hence,

$$k(x) \in \frac{1}{2\cdot 4^{n+1}}\overline{C(-M)} - \frac{m_0}{2}\sum_{i=1}^{n+1}\frac{1}{4^i} - \sum_{i=1}^{n}\frac{m_{n+1-i}}{4^i} - \frac{m_{n+1}}{4}$$

and

$$k(x) \in \bigcup_{m_1, m_2, \cdots, m_{n+1} \in M} \left[\frac{1}{2 \cdot 4^{n+1}} \overline{C(-M)} - \frac{m_0}{2} \sum_{i=1}^{n+1} \frac{1}{4^i} - \sum_{i=1}^{n+1} \frac{m_{n+2-i}}{4^i} \right] = K_{n+1}.$$

It is not difficult to prove that $K_{n+1} \subset K_n$, then

$$k(x) \in \bigcap_{n=1}^{\infty} K_n =: K$$

and

$$K = \left\{ -\sum_{i=1}^{\infty} \frac{m_i}{4^1} - \frac{m_0}{6} : m_i \in M,\ m_0 = -2k(0) \right\}.$$

□

3 Stability of the Drygas Equation

In this section, we intend to present some of the stability results concerning the Drygas equation

$$g(x+y) + g(x-y) = 2g(x) + g(y) + g(-y).$$

As for the quadratic equation, if the domain is a group $\mathcal{G}$ non-necessarily commutative, we shall use the multiplicative notation

$$g(xy) + g(xy^{-1}) = 2g(x) + g(y) + g(y^{-1}). \tag{8}$$

The range is always at least a commutative field $\mathcal{F}$ with characteristic different from 2.

Before going to stability, we cite the important result obtained by B. R. Ebanks, P. L. Kannappan and P. K. Sahoo ([30]) which gives the structure of the solutions of Eq. (8).

Theorem 16 *Let* $g : \mathcal{G} \to \mathcal{F}$ *be a solution of* Eq. (8), *satisfying the additional condition* $g(zyx) = g(zxy)$ *for all* $x, y, z \in \mathcal{G}$. *Then,* g *has the following form*

$$g(x) = a(x) + H(x, x),$$

where $a : \mathcal{G} \to \mathcal{F}$ *is a homomorphism and* $H : \mathcal{G} \times \mathcal{G} \to \mathcal{F}$ *is biadditive and symmetric (hence,* $H(x, x)$ *is quadratic).*

The condition $g(zyx) = g(zxy)$ is known in the literature as the *Kannappan condition* and constitutes a weak substitute of the commutativity of the domain.

Thus, if $\mathcal{G}$ is Abelian, Theorem 16 says that any solution of the Drygas equation is the sum of an additive function and a quadratic function.

Further results in this direction have been obtained in [37] by V. A. Faĭziev and P. K. Sahoo in the case of some special non-commutative groups for the system

$$\begin{cases} g(xy) + g(xy^{-1}) - 2g(x) - g(y) - g(y^{-1}) = 0 \\ g(yx) + g(y^{-1}x) - 2g(x) - g(y) - g(y^{-1}) = 0 \end{cases} \tag{9}$$

where g is real valued. A glance at the system above shows that a sort of weak commutativity is introduced by the couple of equations.

The first stability result concerning Drygas equation has been proved by S.-M. Jung and P. K. Sahoo in [59], when the relevant domain is a real vector space:

Theorem 17 *Let $\mathcal{X}$ be a real vector space and $\mathcal{B}$ a Banach space. If $g : \mathcal{X} \to \mathcal{B}$ satisfies the inequality*

$$\|g(x + y) + g(x - y) - 2g(x) - g(y) - g(-y)\| \leq \varepsilon \tag{10}$$

for some $\varepsilon \geq 0$ and all $x, y \in \mathcal{X}$, then there exist a unique additive function $a : \mathcal{X} \to \mathcal{B}$ and a unique quadratic function $q : \mathcal{X} \to \mathcal{B}$ such that

$$\|g(x) - q(x) - a(x)\| \leq \frac{25}{3}\varepsilon$$

for all $x \in \mathcal{X}$. In other words, there exists a unique solution d of Drygas equation such that

$$\|g(x) - d(x)\| \leq \frac{25}{3}\varepsilon.$$

D. Yang in [101] as a corollary of a more general stability result for a functional equation involving several unknown functions, obtained the following improvement of Theorem 17:

Theorem 18 *Let $\mathcal{G}$ be a group and $\mathcal{B}$ a Banach space. If $g : \mathcal{G} \to \mathcal{B}$ satisfies the inequality*

$$\|g(xy) + g(xy^{-1}) - 2g(x) - g(y) - g(y^{-1})\| \leq \varepsilon$$

for some $\varepsilon \geq 0$ and all $x, y \in \mathcal{G}$ and $g(zyx) = g(zxy)$ for all $x, y, z \in \mathcal{G}$, then there exists a unique solution d of Drygas equation such that

$$\|g(x) - d(x)\| \leq \frac{3}{2}\varepsilon.$$

for all $x \in \mathcal{G}$.

V. A. Faĭziev and P. K. Sahoo in [38] attacked the problem of weakening the requirement of commutativity of the domain or that that the function involved satisfies the Kannappan condition. In analogy with what has been done for the equation, they considered the following system of functional inequalities

$$\begin{cases} |g(xy) + g(xy^{-1}) - 2g(x) - g(y) - g(y^{-1})| \le \varepsilon \\ |g(yx) + g(y^{-1}x) - 2g(x) - g(y) - g(y^{-1})| \le \varepsilon \end{cases} \tag{11}$$

for some non-negative ε and for real valued g.

As for the quadratic equation they used n-Abelian groups (see Definition 2 in the previous section) or the Heisemberg group $UT(3, K)$, that is the group of the matrices of the form

$$\begin{bmatrix} 1 & y & t \\ 0 & 1 & x \\ 0 & 0 & 1 \end{bmatrix}$$

where $x, y, t \in K$ and K is a commutative field.

The stability result can be formulated as follows:

Theorem 19 [38] *Let $\mathcal{G} \in \mathcal{K}_n$ or $\mathcal{G} = UT(3, K)$, then the system* (11) *is stable, that is the function g is the sum of a solution of the system and a bounded function.*

Theorem 3 shows that the quadratic equation is stable if the domain is an amenable group. It is natural to ask if a similar result is true also for the Drygas equation. The investigations in this direction have been conducted by J. Sikorska and the present author in [44]. We present here two results. The first is given by the following.

Theorem 20 *Let $\mathcal{G}$ be an amenable group and $\mathcal{B}$ a Banach space. Assume that $g : \mathcal{G} \to \mathcal{B}$ is a function satisfying the following inequalities, where g^e and g^o denote its even and odd part, respectively:*

$$\begin{aligned} &\|g^e(xy) + g^e(xy^{-1}) - 2g^e(x) - g^e(y) - g^e(y^{-1})\| \\ &\quad = \|g^e(xy) + g^e(xy^{-1}) - 2g^e(x) - 2g^e(y)\| \le \varepsilon, \end{aligned}$$

and

$$\begin{aligned} &\|g^o(xy) + g^o(xy^{-1}) - 2g^o(x) - g^o(y) - g^o(y^{-1})\| \\ &\quad = \|g^o(xy) + g^o(xy^{-1}) - 2g^o(x)\| \le \mu, \end{aligned}$$

for some non-negative ε and μ. Then, there exists a unique solution d of the Drygas equation such that

$$\|g(x) - d(x)\| \le \frac{1}{2}(\varepsilon + \mu), \quad x \in \mathcal{G}.$$

The proof of Theorem 20 is inspired by those contained in [101] and [102].

Instead of considering the inequalities concerning the even and odd part of the function g, the next theorem starts with the boundedness of the Drygas difference of the function g and the condition of approximate centrality of it.

Theorem 21 *Let $\mathcal{G}$ be an amenable group and $\mathcal{B}$ a Banach space. Assume that $g : \mathcal{G} \to \mathcal{B}$ is a function satisfying the following inequalities:*

$$\|g(xy) + g(xy^{-1}) - 2g(x) - g(y) - g(y^{-1})\| \leq \varepsilon,$$
$$\|g(xy) - g(yx)\| \leq \delta.$$

for all $x, y \in \mathcal{G}$ and some non-negative ε and δ. Then, there exists a unique solution d of the Drygas equation such that

$$\|g(x) - d(x)\| \leq \varepsilon + \delta, \quad x \in \mathcal{G}.$$

As a consequence of the previous results, we obtain the following:

Theorem 22 *Let $\mathcal{G}$ be an amenable group and $\mathcal{B}$ a Banach space. Assume that $g : \mathcal{G} \to \mathcal{B}$ is a function satisfying the Drygas equation and such that*

$$\|g(xy) + g(xy^{-1}) - g(yx) - g(y^{-1}x)\| \leq \gamma, \quad x, y \in \mathcal{G}$$

for some non-negative γ. Then g is of the form $g = q + a$ with q quadratic and a additive.

Moreover, as for the additive and quadratic equation, it is also proved that the Drygas equation is not stable on the free group generated by two elements.

As a last result concerning stability, we present that obtained by J.-Y. Chung, L. Li and D. Kim ([23]) in the frame of Schwartz distributions $\mathcal{D}'$. As for the quadratic equation (see the previous section) let A, B, P_1 and P_2 be the functions

$$A(x, y) = x + y, \ B(x, y) = x - y, \ P_1(x, y) = x, \ P_2(x, y) = y, \ x, y \in \mathbf{R}^n.$$

Then, the inequality (10) can be naturally transformed as

$$\|u \circ A + u \circ B - 2u \circ P_1 - u \circ P_2 - u \circ (-P_2)\| \leq \varepsilon \tag{12}$$

where $u \circ A, u \circ B, u \circ P_1, u \circ P_2, u \circ (-P_2)$ are the pullback of u by A, B, P_1, P_2, $-P_2$, respectively, and $\|u\| \leq \varepsilon$ means that $|\langle u, \phi \rangle| \leq \varepsilon \|\phi\|_{L_1}$ for all test functions ϕ.

The main result is stated in the following.

Theorem 23 *Let $u \in \mathcal{D}'$ satisfy the inequality* (12). *Then, there exist a unique $a \in \mathbf{C}^n$ and a unique quadratic form*

$$q(x) = \sum_{1 \leq j \leq k \leq n} a_{jk} x_j x_k$$

such that $u = a \cdot x + q(x) + h(x)$, where h is a bounded measurable function satisfying $\|h\|_{L^\infty} \leq \frac{3}{2}\varepsilon$.

We finish this section devoted to the Drygas equation with two theorems analog to Theorems 14 and 15.

Indeed, we have

Theorem 24 *Let $f : \mathcal{G} \to \mathcal{B}$, where $\mathcal{G}$ is an Abelian group and $\mathcal{B}$ a Banach space and let M be a bounded subset of $\mathcal{B}$. If $g(x+y)+g(x-y)-2g(x)-g(y)-g(-y) \in M$, then $g(x) = d(x) + r(x)$, where d is a solution of Drygas equation and $(r(x)+r(-x))/2$ is contained in $\frac{1}{2}\overline{C(-M)}$, where $\overline{C(-M)}$ is the closure of the convex hull of $-M$.*

Proof By the stability we have the decomposition $g(x) = d(x)+r(x)$ with d solution of Drygas equation and r bounded. Since $d(0) = 0$, we have $g(0) = r(0) = -\frac{1}{2}m_0$, for certain $m_0 \in M$. Fix $x \in \mathcal{G}$ and consider the values $r(x) =: u$ and $r(-x) =: v$. We have

$$r(2x) + r(0) - 3r(x) - r(-x) \in M,$$

hence

$$r(2x) = 3u + v + \frac{1}{2}m_0 + m_1 \text{ for some } m_1 \in M.$$

By induction, we obtain

$$r(sx) = \frac{s(s+1)}{2}u + \frac{s(s-1)}{2}v + \frac{s-1}{2}m_0 + \sum_{i=1}^{s-1}(s-i)m_i,$$

for some $m_i \in M$. By dividing by s^2 and taking the limit as $s \to \infty$, we have, as in the proof of Theorem 14,

$$\frac{u+v}{2} \in \frac{1}{2}\overline{C(-M)}.$$

□

Theorem 25 *In the hypotheses of Theorem 24, we have that $(r(x)+r(-x))/2$ is contained in the set*

$$R = \left\{ -\frac{1}{2}\sum_{i=1}^{\infty} \frac{m_i + t_i}{4^1} - \frac{m_0}{6} : m_i, t_i \in M,\ m_0 = -2r(0) \right\}.$$

Proof By adding the two relations

$$r(2x) + r(0) - 3r(x) - r(-x) \in M \text{ and } r(-2x) + r(0) - 3r(-x) - r(x) \in M$$

we obtain

$$\frac{r(2x) + r(-2x)}{2} = 2[r(x) + r(-x)] - r(0) + \frac{m_1 + t_1}{2}$$

for some $m_1, t_1 \in M$.

From now on, we proceed as in the proof of Theorem 15. □

4 Alternative Quadratic Equation

The author of the present chapter became aware of the existence of Hyers' theorem about stability of the additive equation while working, in 1978, on a problem proposed by Marek Kuczma concerning the alternative Cauchy equation. The stability result has been the main tool for solving that problem (see [40]).

In order to conclude the present paper devoted to the quadratic equation, we intend to investigate the analogous alternative equation. Namely, we intend to find the solutions of the following alternative equation:

$$Qf(x,y) = f(x+y) + f(x-y) - 2f(x) - 2f(y) \in \{0,1\} \tag{13}$$

We assume that $f : \mathcal{G} \to \mathbf{R}$, where $\mathcal{G}$ is an Abelian group, hence, we use the additive notation.

Due to the stability results, we transform the previous problem into the following

$$Qk(x,y) = k(x+y) + k(x-y) - 2k(x) - 2k(y) \in \{0,1\}, \tag{14}$$

where the function k is bounded and, by Theorem 14, has its range in the interval $[-1/2, 0]$. By setting $x = y = 0$, we have $k(0) \in \{-\frac{1}{2}, 0\}$.

By setting $p(x) := -k(x) - \frac{1}{2}$, we see that $k(0) = -\frac{1}{2}$ implies $p(0) = 0$ and $Qp(x,y) \in \{0,1\}$. Thus, we can consider only the case $k(0) = 0$ and investigate the problem

$$\begin{cases} k(x+y) + k(x-y) - 2k(x) - 2k(y) \in \{0,1\} \\ k : \mathcal{G} \to \left[-\frac{1}{2}, 0\right],\ k(0) = 0. \end{cases} \tag{15}$$

Theorem 15 applied to this situation gives that the range of k is contained in the set $K = \{-\sum_{n=1}^{\infty} \frac{\alpha_n}{4^n} : \alpha_n \in \{0,1\}\}$.

Writing the set K in the form

$$K = \left\{ -\frac{1}{3} \sum_{n=1}^{\infty} \frac{3\alpha_n}{4^n} : \alpha_n \in \{0,1\} \right\}$$

we see that it is obtained by a procedure similar to that of the construction of the ternary Cantor set. In this case, we take the unit interval, divide it in four equal parts, say $[0, 1/4], [1/4, 1/2], [1/2, 3/4]$ and $[3/4, 1]$ and eliminate the open central interval $(1/4, 3/4)$. Proceeding in this way and multiplying the resulting set by $-\frac{1}{3}$, we obtain K.

It should be noted that the numbers in K have a unique representation in the form $-\sum_{n=1}^{\infty} \frac{\alpha_n}{4^n}$ with $\alpha_n \in \{0,1\}$.

Consider the set $Z_k = \{x \in \mathcal{G} : k(x) = 0\}$ and put $x, y \in Z_k$ in Eq. (14): we have

$$k(x+y) + k(x-y) \in \{0,1\}$$

and this forces $k(x+y) = k(x-y) = 0$, i.e., Z_k is a subgroup of $\mathcal{G}$.

Take now $x \notin Z_k$ and let $k(x) = -\sum_{n=1}^{\infty} \frac{\alpha_n}{4^n}$ for some sequence $\{\alpha_n\} \in \{0,1\}^{\mathbf{N}}$. Then

$$k(2x) - 4k(x) \in \{0,1\} \Leftrightarrow k(2x) \in \left\{ -\sum_{n=1}^{\infty} \frac{\alpha_n}{4^{n-1}}, 1 - \sum_{n=1}^{\infty} \frac{\alpha_n}{4^{n-1}} \right\}.$$

If $\alpha_1 = 0$, then $\sum_{n=1}^{\infty} \frac{\alpha_n}{4^{n-1}} = \sum_{n=2}^{\infty} \frac{\alpha_n}{4^{n-1}} \leq \frac{1}{3}$, hence, $1 - \sum_{n=1}^{\infty} \frac{\alpha_n}{4^{n-1}} \geq \frac{2}{3}$. Thus,

$$k(2x) = -\sum_{n=1}^{\infty} \frac{\alpha_n}{4^{n-1}} = -\sum_{n=2}^{\infty} \frac{\alpha_n}{4^{n-1}}.$$

If $\alpha_1 = 1$, then $\sum_{n=1}^{\infty} \frac{\alpha_n}{4^{n-1}} = 1 + \sum_{n=2}^{\infty} \frac{\alpha_n}{4^{n-1}} \geq 1$, hence, $1 - \sum_{n=1}^{\infty} \frac{\alpha_n}{4^{n-1}} \leq 0$. Thus,

$$k(2x) = 1 - \sum_{n=1}^{\infty} \frac{\alpha_n}{4^{n-1}} = -\sum_{n=2}^{\infty} \frac{\alpha_n}{4^{n-1}}.$$

If we identify $k(x)$ with the sequence $\{\alpha_n\}_{n=1}^{\infty}$, then $k(2x)$ is identified by $\{\alpha_{n+1}\}_{n=1}^{\infty}$ and we always have $k(2x) \in K$.

Now we compute $k(3x)$. From Eq. (14) with $2x$ instead of x and x instead of y, we obtain

$$k(3x) + k(x) - 2k(2x) - 2k(x) = k(3x) - 2k(2x) - k(x) \in \{0,1\}$$

whence

$$k(3x) \in \left\{ -\sum_{n=1}^{\infty} \frac{\alpha_n + 2\alpha_{n+1}}{4^n}, 1 - \sum_{n=1}^{\infty} \frac{\alpha_n + 2\alpha_{n+1}}{4^n} \right\}.$$

If

$$k(3x) = -\sum_{n=1}^{\infty} \frac{\alpha_n + 2\alpha_{n+1}}{4^n},$$

then for having $k(3x) \in K$, by Theorem 15, we must have

$$\sum_{n=1}^{\infty} \frac{\alpha_n + 2\alpha_{n+1}}{4^n} = \sum_{n=1}^{\infty} \frac{a_n}{4^n}$$

for some sequence $\{a_n\}$ with $a_n = 0, 1$. We prove that this is possible if and only if $\alpha_n + 2\alpha_{n+1} \in \{0,1\}$. If not, let n_0 be the first index such that $\alpha_n + 2\alpha_{n+1} \neq a_n$; we have two possibilities: either $a_{n_0} < \alpha_{n_0} + 2\alpha_{n_0+1}$ or $1 = a_{n_0} > \alpha_{n_0} + 2\alpha_{n_0+1} = 0$.

In the first case, we have

$$\sum_{n=n_0}^{\infty} \frac{a_n}{4^n} \leq \frac{a_{n_0}}{4^{n_0}} + \frac{1}{3 \cdot 4^{n_0}} = \frac{a_{n_0} + 1/3}{4^{n_0}} < \frac{a_{n_0} + 1}{4^{n_0}} \leq \frac{\alpha_{n_0} + 2\alpha_{n_0+1}}{4^{n_0}} \leq \sum_{n=n_0}^{\infty} \frac{\alpha_n + 2\alpha_{n+1}}{4^n},$$

a contradiction.

In the second case, we have

$$\sum_{n=n_0}^{\infty} \frac{\alpha_n + 2\alpha_{n+1}}{4^n} \le \frac{1}{4^{n_0}} \le \frac{1}{4^{n_0}} + \sum_{n=n_0+1}^{\infty} \frac{a_n}{4^n} = \sum_{n=n_0}^{\infty} \frac{a_n}{4^n}.$$

This forces the equality and $a_n = 0$, $\alpha_n + 2\alpha_{n+1} = 3$ for all $n > n_0$, i.e., $\alpha_n = 1$ for all $n > n_0$. Hence, $0 = \alpha_{n_0} + 2\alpha_{n_0+1} \ge 2$: a contradiction.

Thus, $\alpha_n + 2\alpha_{n+1} \in \{0, 1\}$ and this is possible if and only if $\alpha_{n+1} = 0$ for every $n \ge 1$. Thus, either $\alpha_n = 0$ for every $n \ge 0$, i.e., $x \in Z_k$, impossible, or $\alpha_1 = 1$ and $\alpha_n = 0$ for every $n \ge 2$. This means that $k(x) = -\frac{1}{4}$. In this case, $k(2x) = 0$, i.e., $2x \in Z_k$ and $k(3x) = -\frac{1}{4}$.

The other possibility is

$$k(3x) = 1 - \sum_{n=1}^{\infty} \frac{\alpha_n + 2\alpha_{n+1}}{4^n} = \sum_{n=1}^{\infty} \frac{3}{4^n} - \sum_{n=1}^{\infty} \frac{\alpha_n + 2\alpha_{n+1}}{4^n} = -\sum_{n=1}^{\infty} \frac{\alpha_n + 2\alpha_{n+1} - 3}{4^n}.$$

The condition $k(3x) \in K$ implies $\alpha_n + 2\alpha_{n+1} - 3 = 0$, i.e., $\alpha_n + 2\alpha_{n+1} = 3$ for every $n \ge 1$, hence, $\alpha_n = 1$ for every $n \ge 1$. This means that $k(x) = -\frac{1}{3}$. In this case, $k(2x) = -\frac{1}{3}$ and $k(3x) = 0$, i.e., $3x \in Z_k$.

Let now $x, y \notin Z_k$, with $k(x) = -\frac{1}{4}$ and $k(y) = -\frac{1}{3}$. Then

$$k(x+y) + k(x-y) + \frac{1}{2} + \frac{2}{3} \in \{0, 1\} \Leftrightarrow k(x+y) + k(x-y) \in \left\{-\frac{7}{6}, -\frac{1}{6}\right\}.$$

Since $k(x+y), k(x-y) \in \{-\frac{1}{3}, -\frac{1}{4}, 0\}$, we cannot obtain the values $-\frac{7}{6}$ and $-\frac{1}{6}$. Thus, we have proved the following

Theorem 26 *If k is a solution of problem* (14), *then the range of k is contained in the set $\{-\frac{1}{3}, 0\}$ or $\{-\frac{1}{4}, 0\}$. Thus, problem* (14) *has a non-identically zero solution if and only if either $\mathcal{G}$ has a subgroup Z such that $\mathcal{G}/Z$ is cyclic of order* 3 *and the non-trivial solution takes the values* 0 *on Z and $-\frac{1}{3}$ outside Z or $\mathcal{G}$ has a subgroup Z such that $\mathcal{G}/Z$ is cyclic of order* 2 *and the non-trivial solution takes the values* 0 *on Z and $-\frac{1}{4}$ outside Z.*

References

1. Abbaszadeh, S., Eshaghi Gordji, M., Park, C.-G.: On the stability of a generalized quadratic and quartic type functional equation in quasi-Banach spaces. J. Inequal. Appl. (2009). doi:10.1155/2009/153084
2. Aczél, J., Dhombres, J.: Functional Equations in Several Variables. Encyclopedia of Mathematics and its Applications 31. Cambridge University Press, Cambridge (1989)
3. Aoki, T.: On the stability of the linear transformation in Banach spaces. J. Math. Soc. Jpn. **2**, 64–66 (1950)
4. Bae, J.-H., Kim, H.-M.: On the generalized Hyers-Ulam stability of a quadratic mapping. Far East J. Math. Sci. **3**, 599–608 (2001)

5. Bae, J.-H., Park, W.-G.: A fixed-point approach to the stability of a functional equation on quadratic forms. J. Inequal. Appl. (2011). doi:10.1186/1029-242X-2011-82
6. Baktash, E., Cho, Y.J., Jalili, M., Saadati, R., Vaezpour, S.M.: On the stability of cubic mappings and quadratic mappings in random normed spaces. J. Inequal. Appl. (2008). doi:10.1155/2008/902187
7. Benyamini, Y., Lindenstrauss, J.: Geometric Nonlinear Functional Analysis, vol. 1, 48. American Mathematical Society, Providence (2000)
8. Bidkham, M., Eshaghi Gordji, M., Savadkouhi, M.B.: Stability of a mixed type additive and quadratic functional equation in non-Archimedean spaces. J. Comput. Anal. Appl. **12**, 454–462 (2010)
9. Borelli, C., Forti, G.L.: On a general Hyers–Ulam stability result. Internat. J. Math. Math. Sci. **18**, 229–236 (1995)
10. Bouikhalene, B., Elqorachi, E., Redouani, A.: Hyers-Ulam stability of the generalized quadratic functional equation in amenable semigroups. J. Inequal. Pure Appl. Math. **8**, Art. 47 (2007)
11. Bouikhalene, B., Elqorachi, E., Rassias, Th.M.: On the generalized Hyers-Ulam stability of the quadratic functional equation with a general involution. Nonlinear Funct. Anal. Appl. **12**, 247–262 (2007)
12. Brzdęk, J., Ciepliński, K.: A fixed point approach to the stability of functional equations in non-Archimedean metric spaces. Nonlinear Anal. **74**, 6861–6867 (2011)
13. Brzdęk, J., Chudziak, J., Páles, Z.: A fixed point approach to stability of functional equations. Nonlinear Anal. **74**, 6728–6732 (2011)
14. Chang, I.-S., Kim, H.-M.: Hyers-Ulam-Rassias stability of a quadratic functional equation. Kyungpook Math. J. **42**, 71–86 (2002)
15. Chang, I.-S., Kim, H.-M.: On the Hyers-Ulam stability of quadratic functional equations. J. Inequal. Pure Appl. Math. **3**, Art. 33 (2002)
16. Cădariu, L., Radu, V.: Fixed points and the stability of quadratic functional equations. An. Univ. Timişoara Ser. Mat.-Inform. **41**, 25–48 (2003)
17. Chang, I.-S., Lee, E.H., Kim, H.-M.: On Hyers-Ulam-Rassias stability of a quadratic functional equation. Math. Inequal. Appl. **6**, 87–95 (2003)
18. Cholewa, P.W.: Remarks on the stability of functional equations. Aequ. Math. **27**, 76–86 (1984)
19. Chung, J.-Y.: Stability of approximately quadratic Schwartz distributions. Nonlinear Anal. **67**, 175–186 (2007)
20. Chung, J.-Y.: Stability of a generalized quadratic functional equation in Schwartz distributions. Acta Math. Sin. (Engl. Ser.) **25**, 1459–1468 (2009)
21. Chung, J.-Y.: On an L^{∞}–version of a Pexiderized quadratic functional inequality. Honam Math. J. **33**, 73–84 (2011)
22. Chung, S.-Y., Lee, Y.-S.: Stability of cubic functional equation in the spaces of generalized functions. J. Inequal. Appl. (2007). doi:10.1155/2007/79893
23. Chung, J.-Y., Li, L., Kim, D. : Stability of functional equations of Drygas in the space of Schwartz distributions. J. Math. Anal. Appl. **320**, 163–173 (2006)
24. Chung, J.-Y., Kim, D., Rassias, J.M.: Hyers-Ulam stability on a generalized quadratic functional equation in distributions and hyperfunctions. J. Math. Phys. **50**, 1089–7658 (2009)
25. Czerwik, S.: On the stability of the quadratic mapping in normed spaces. Abh. Math. Sem. Univ. Hamburg. **62**, 59–64 (1992)
26. Czerwik, S.: On the stability of the quadratic functional equation and related topics. In: Recent Progress in Inequalities (Nis, 1996). Mathematics and Its Applications, vol. 430, pp. 449–455. Kluwer, Dordrecht (1998)
27. Czerwik, St.: Functional Equations and Inequalities in Several Variables. World Scientific, River Edge (2002)
28. Drygas, H.: Quasi–inner products and their applications. In: Gupta, A.K. (ed.) Advances in Multivariate Statistical Analysis, pp. 13–30. Reidel, Dordrecht (1987)

29. Ebadian, A., Eshaghi Gordji, M., Ghobadipour, N., Savadkouhi, M.B.: Stability of a mixed type cubic and quartic functional equation in non-Archimedean ℓ–fuzzy normed spaces. Thai J. Math. **9**, 249–265 (2011)
30. Ebanks, B.R., Kannappan, P.L., Sahoo, P.K.: A common generalization of functional equations characterizing normed and quasi–inner–product spaces. Can. Math. Bull. **35**, 321–327 (1992)
31. Elqorachi, E., Rassias, Th. M.: On the generalized Hyers-Ulam stability of the quadratic functional equation with a general involution. Nonlinear Funct. Anal. Appl. **12**, 247–262 (2007)
32. Eshaghi Gordji, M., Rassias, J.M., Savadkouhi, M.B.: Approximation of the quadratic and cubic functional equations in RN-spaces. Eur. J. Pure Appl. Math. **2**, 494–507 (2009)
33. Eshaghi Gordji, M., Kamyar, M., Khodaei, H., Park, C.-G., Shin, D.Y. : Fuzzy stability of generalized mixed type cubic, quadratic, and additive functional equation. J. Inequal. Appl. (2011). doi:10.1186/1029-242X-2011-95
34. Eshaghi Gordji, M., Khodaei, H., Rassias, J.M.: Fixed point methods for the stability of general quadratic functional equation. Fixed Point Theory **12**, 71–82 (2011)
35. Eshaghi Gordji, M., Savadkouhi, M.B. : Stability of a mixed type additive, quadratic and cubic functional equation in random normed spaces. Filomat **25**, 43–54 (2011)
36. Eshaghi Gordji, M. Savadkouhi, M.B., Vaezpour, S.M.: Stability of an additive-quadratic functional equation in Menger probabilistic normed spaces. J. Appl. Funct. Anal. **7**, 138–147 (2012)
37. Faĭziev, V.A., Sahoo, P.K.: On Drygas functional equation on groups. Int. J. Appl. Math. Stat. **7**, 59–69 (2007)
38. Faĭziev, V.A., Sahoo, P.K.: On the stability of Drygas functional equation on groups. Banach J. Math. Anal. **1**, 1–18 (2007)
39. Faĭziev, V.A., Sahoo, P.K.: On the stability of the quadratic equation on groups. Bull. Belg. Math. Soc. Simon Stevin **15**, 135–151 (2008)
40. Forti, G.L.: On an alternative functional equation related to the Cauchy equation. Aequ. Math. **24**, 195–206 (1982)
41. Forti, G.L.: The stability of homomorphisms and amenability, with applications to functional equations. Abh. Math. Sem. Univ. Hamburg. **57**, 215–226 (1987)
42. Forti, G.L.: Hyers–Ulam stability of functional equations in several variables. Aequ. Math. **50**, 143–190 (1995)
43. Forti, G.L.: Comments on the core of the direct method for proving Hyers–Ulam stability of functional equations. J. Math. Anal. Appl. **295**, 127–133 (2004)
44. Forti, G.L., Sikorska, J.: Variations on the Drygas equation and its stability. Nonlinear Anal. **74**, 343–350 (2011)
45. Hyers, D.H., Isac, G., Rassias, Th.M.: Stability of functional equations in several variables. Progress in Nonlinear Differential Equations and their Applications, 34. Birkhäuser, Boston (1998)
46. Jang, S.-Y., Kenary, H.A., Park, C.-G., Rezaei, H.: Stability of a generalized quadratic functional equation in various spaces: a fixed point alternative approach. Adv. Differ. Equ. (2011). doi:10.1186/1687-1847-2011-62
47. Jang, S.Y., Lee, J.R., Park, C.-G., Shin, D.Y.: On the stability of an AQCQ-functional equation in random normed spaces. Abstr. Appl. Anal. (2011). doi:10.1186/1029-242X-2011-34
48. Jin, S.S., Lee, Y.-H.: Fuzzy stability of a mixed type functional equation. J. Inequal. Appl. (2011). doi:10.1186/1029-242X-2011-70
49. Jun, K.-W., Kim, H.-M.: Ulam stability problem for generalized A-quadratic mappings. J. Math. Anal. Appl. **305**, 466–476 (2005)
50. Jun, K.-W., Lee, Y.-H.: On the Hyers-Ulam-Rassias stability of a generalized quadratic equation. Bull. Korean Math. Soc. **38**, 261–272 (2001)

51. Jun, K.-W., Lee, Y.-H.: On the Hyers-Ulam-Rassias stability of a Pexiderized quadratic equation. II. In: Rassias T. M. (ed.) Functional Equations, Inequalities and Applications, pp. 39–65. Kluwer, Dordrecht (2003)
52. Jun, K.-W., Lee, Y.-H.: A generalization of the Hyers-Ulam-Rassias stability of the Pexiderized quadratic equations. J. Math. Anal. Appl. **297**, 70–86 (2004)
53. Jun, K.-W., Lee, Y.-H.: A generalization of the Hyers-Ulam-Rassias stability of the Pexiderized quadratic equations. II. Kyungpook Math. J. **47**, 91–103 (2007)
54. Jung, S.-M.: On the Hyers-Ulam stability of the functional equations that have the quadratic property. J. Math. Anal. Appl. **222**, 126–137 (1998)
55. Jung, S.-M.: On the Hyers-Ulam stability of a quadratic functional equation. J. Math. Anal. Appl. **232**, 384–393 (1999)
56. Jung, S.-M.: Stability of the quadratic equation of Pexider type. Abh. Math. Sem. Univ. Hambg. **70**, 175–190 (2000)
57. Jung, S.-M.: Hyers-Ulam-Rassias Stability of Functional Equations in Nonlinear Analysis. Springer Optimization and Its Applications, vol. 48. Springer, New York (2011)
58. Jung, S.-M., Sahoo, P.K.: Hyers–Ulam stability of the quadratic equation of Pexider type. J. Korea Math. Soc. **38**, 645–656 (2001)
59. Jung, S.-M., Sahoo, P.K.: Stability of a functional equation of Drygas. Aequ. Math. **64**, 263–273 (2002)
60. Jung, S.-M., Kim, T.-S., Lee, K.-S.: A fixed point approach to the stability of quadratic functional equation. Bull. Korean Math. Soc. **43**, 531–541 (2006)
61. Kannappan, Pl.: Functional Equations and Inequalities with Applications. Springer Monographs in Mathematics. Springer, New York (2009)
62. Kenary, H.A., Lee, J.R., Rezaei, H., Shin, D.Y., Talebzadeh, S.: Fuzzy Hyers-Ulam approximation of a mixed AQ mapping. J. Inequal. Appl. (2012). doi:10.1186/1029-242X-2012-64
63. Kenary, H.A., Rassias, Th.M., Rezaei, H., Talebzadeh, S., Park, W.-G.: Non-Archimedean Hyers-Ulam stability of an additive-quadratic mapping. Discrete Dyn. Nat. Soc. (2012). doi:10.1155/2012/824257
64. Kim, H.-M.: Hyers-Ulam stability of a general quadratic functional equation. Publ. Inst. Math. (Beograd) (N.S.) **73**(87), 129–137 (2003)
65. Koh, H., Ku, S.-H., Lee, S.-H.: Investigation of the stability via shadowing property. J. Inequal. Appl. (2009). doi:10.1155/2009/156167
66. Kim, H.-M., Lee, J., Son, E.: Stability of quadratic functional equations via the fixed point and direct method. J. Inequal. Appl. (2010). doi:10.1155/2010/635720
67. Kim, J.-H., Lee, J.R., Park, C.-G.: A fixed point approach to the stability of an additive-quadratic-cubic-quartic functional equation. Fixed Point Theory Appl. (2010). doi:10.1155/2010/185780
68. Kominek, Z.: Stability of a quadratic functional equation on semigroups. Publ. Math. Debr. **75**, 173–178 (2009)
69. Lee,Y.-S.: Stability of a quadratic functional equation in the spaces of generalized functions. J. Inequal. Appl. (2008). doi:10.1155/2008/210615
70. Lee,Y.-S.: Stability of quadratic functional equations in tempered distributions. J. Inequal. Appl. (2012). doi:10.1186/1029-242X-2012-177
71. Lee,Y.-H., Jung, S.-M.: A fixed point approach to the stability of an n-dimensional mixed-type additive and quadratic functional equation. Abstr. Appl. Anal. (2012). doi:10.1155/2012/482936
72. Mirmostafaee, A.K.: A fixed point approach to the stability of quadratic equations in quasi normed spaces. Kyungpook Math. J. **49**, 691–700 (2009)
73. Mirmostafaee, A.K., Moslehian, M.S.: Fuzzy almost quadratic functions. Results Math. **52**, 161–177 (2008)
74. Moghimi, M.B., Najati, A., Park, C.-G.: A fixed point approach to the stability of a quadratic functional equation in C^*–algebras. Adv. Differ. Equ. (2009). doi:10.1155/2009/256165

75. Moslehian, M.S., Nikodem, K., Popa, D.: Asymptotic aspect of the quadratic functional equation in multi-normed spaces. J. Math. Anal. Appl. **355**, 717–724 (2009)
76. Najati, A., Park, C.-G.: Fixed points and stability of a generalized quadratic functional equation. J. Inequal. Appl. (2009). doi:10.1155/2009/193035
77. Park, C.-G.: On the stability of the quadratic mapping in Banach modules. J. Math. Anal. Appl. **276**, 135–144 (2002)
78. Park, C.-G.: A fixed point approach to the fuzzy stability of an additive-quadratic-cubic functional equation. Fixed Point Theory Appl. (2009). doi:10.1155/2009/918785
79. Park, C.-G., Kim, J.-H.: The stability of a quadratic functional equation with the fixed point alternative. Abstr. Appl. Anal. (2009). doi:10.1155/2009/907167
80. Park, C.-G., Rassias, Th.M.: Fixed points and generalized Hyers-Ulam stability of quadratic functional equations. J. Math. Inequal. **1**, 515–528 (2007)
81. Park, C.-G., Saadati, R., Vahidi, J.: A functional equation related to inner product spaces in non-Archimedean L–random normed spaces. J. Inequal. Appl. (2012). doi:10.1186/1029-242X-2012-168
82. Pourpasha, M.M., Rassias, J.M., Saadati, R., Vaezpour, S.M.: A fixed point approach to the stability of Pexider quadratic functional equation with involution. J. Inequal. Appl. (2010). doi:10.1155/2010/839639
83. Rassias, Th.M.: On the stability of the linear mapping in Banach spaces. Proc. Am. Math. Soc. **72**, 297–300 (1978)
84. Rassias, J.M.: On the stability of the Euler-Lagrange functional equation. C. R. Acad. Bulgare Sci. **45**, 17–20 (1992)
85. Rassias, J.M.: Solution of the Ulam stability problem for Euler-Lagrange quadratic mappings. J. Math. Anal. Appl. **220**, 613–639 (1998)
86. Rassias, Th.M.: On the stability of the quadratic functional equation. Studia Univ. Babes-Bolyai Math. **45**, 77–114 (2000)
87. Rassias, Th.M. (ed.): Functional Equations, Inequalities and Applications. Kluwer, Dordrecht (2003)
88. Rassias, Th.M. (ed.): Stability of functional equations and applications. Tamsui Oxf. J. Math. Sci. **24** (2008) (Aletheia University, Taipei (2008))
89. Rassias, Th.M., Brzdęk, J. (eds.): Functional Equations in Mathematical Analysis. Springer, New York (2012)
90. Rassias, J.M., Rassias, M.J.: Refined Ulam stability for Euler-Lagrange type mappings in Hilbert spaces. Int. J. Appl. Math. Stat. **7**, 126–132 (2007)
91. Rassias, Th.M., Tabor, J. (eds.): Stability of Mappings of Hyers-Ulam Type. Hadronic Press Collection of Original Articles. Hadronic Press, Palm Harbor (1994)
92. Rassias, J.M., Rassias, M.J., Xu, T.Z., Xu, W.X.: A fixed point approach to the stability of quintic and sextic functional equations in quasi–normed spaces. J. Inequal. Appl. (2010). doi:10.1155/2010/423231
93. Rolewicz, S.: Metric Linear Spaces, 2nd edn. PWN Polish Scientific, Warsaw (1984)
94. Saadati, R., Vaezpour, S.M., Zohdi, M.M.: Nonlinear L-random stability of an ACQ functional equation. J. Inequal. Appl. (2011). doi:10.1155/2011/194394
95. Saadati, R., Sadeghi, Z., Vaezpour, S.M.: On the stability of Pexider functional equation in non-archimedean spaces. J. Inequal. Appl. (2011). doi:10.1186/1029-242X-2011-17
96. Skof, F.: Proprietà locali e approssimazione di operatori. In: Geometry of Banach Spaces and Related Topics (Milan, 1983). Rend. Sem. Mat. Fis. Milano, vol. 53, 113–129 (1986)
97. Székelyhidi, L.: Note on a stability theorem. Canad. Math. Bull. **25**, 500–501 (1982)
98. Székelyhidi, L.: Ulam's problem, Hyers's solution-and to where they led. In: Rassias T. M. (ed.) Functional Equations and Inequalities. Mathematics and Its Applications, vol. 518, pp. 259–285. Kluwer, Dordrecht (2000)
99. Tabor, J., Tabor, J.: General stability of functional equations of linear type. J. Math. Anal. Appl. **328**, 192–200 (2007)

100. Yang, D.: The stability of the quadratic functional equation on amenable groups. J. Math. Anal. Appl. **291**, 666–672 (2004)
101. Yang, D.: Remarks on the stability of Drygas' equation and the Pexider–quadratic equation. Aequ. Math. **68**, 108–116 (2004)
102. Yang, D.: The stability of Jensen's equation on amenable locally compact groups. Result. Math. **46**, 381–388 (2004)

A Functional Equation Having Monomials and Its Stability

M. E. Gordji, H. Khodaei and Themistocles M. Rassias

Abstract We use some results about the Fréchet functional equation to consider the following functional equation:

$$f\left(\left(\sum_{i=1}^{m} a_i x_i^p\right)^{\frac{1}{p}}\right) = \sum_{i=1}^{m} a_i f(x_i).$$

We also apply a fixed point method and homogeneous functions of degree α to investigate some stability results for this functional equation in β-Banach spaces.

Keywords Frechet functional equation · Homogeneous functions · Dynamical systems · Hyers–Ulam–Rassias stability · Fixed point theorem

1 Introduction

In 1909, M. Fréchet [10] studied an important generalization of Cauchy's equation, which characterizes the polynomials among the continuous functions. This functional equation has the form

$$\Delta_{h_1,\dots,h_n} f(x) = 0, \tag{1}$$

where $\Delta_h f(x) = f(x+h) - f(x)$, $\Delta_{h_1,\dots,h_n} f(x) = \Delta_{h_1} \cdots \Delta_{h_n} f(x)$ and $n = 2, 3, \cdots$. In particular, if $h_1 = \cdots = h_n = h$, then (1) can be rewritten succinctly as

$$\Delta_h^n f(x) = 0. \tag{2}$$

M. E. Gordji (✉)
Department of Mathematics, Semnan University, P.O. Box 35195-363, Semnan, Iran
e-mail: meshaghi@semnan.ac.ir; madjid.eshaghi@gmail.com

H. Khodaei
Department of Mathematics, Malayer University, P.O. Box 65719-95863, Malayer, Iran
e-mail: hkhodaei.math@yahoo.com

T. M. Rassias
Department of Mathematics, National Technical University of Athens,
Zografou Campus, 15780 Athens, Greece
e-mail: trassias@math.ntua.gr

T. M. Rassias (ed.), *Handbook of Functional Equations*,
Springer Optimization and Its Applications 96, DOI 10.1007/978-1-4939-1286-5_9

From the result of Fréchet, it follows that a continuous function $f : \mathbb{R} \to \mathbb{R}$ satisfies (2) for all $x, h \in \mathbb{R}$ if and only if f is a polynomial function of degree at most $n - 1$; for details, see Theorem 1 of [2] and also Theorem 13.5 of [20]. In explicit form, the functional Eq. (2) can be written as

$$\Delta_y^n f(x) = \sum_{j=0}^{n} (-1)^{n-j} \binom{n}{j} f(x + jy) = 0.$$

The problem of stability of functional equations is one of the main topics in the theory of functional equations and is connected with perturbation theory and the notions of shadowing in dynamical systems and controlled chaos (see [13, 26, 32]). Zhou [34] used a stability property of the functional equation

$$f(x + y) + f(x - y) = 2f(x) \tag{3}$$

to prove a conjecture of Z. Ditzian about the relationship between the smoothness of a mapping and the degree of its approximation by the associated Bernstein polynomials. The starting point of the stability theory of functional equations was a problem formulated in the celebrated book by Pólya and Szegő [29], and the problem of Ulam concerning the stability of group homomorphisms [33]. Recall that an equation is called stable in the Hyers–Ulam–Rassias sense if for any solution of the perturbed equation, called an approximate solution, there exists a solution of the equation close to it. For definitions, approaches, and results on Hyers–Ulam–Rassias stability, we refer the reader to, e.g., [4, 5, 7, 9, 11, 12, 14, 16, 17, 25, 28, 30].

We use some results about the Fréchet functional equation to consider Cauchy–Jensen, quadratic, cubic, quartic, and quintic functional equations, and then we deal with the following monomial functional equation

$$f\left(\left(\sum_{i=1}^{m} a_i x_i^p\right)^{\frac{1}{p}} \right) = \sum_{i=1}^{m} a_i f(x_i). \tag{4}$$

Finally, we apply a fixed point theorem to investigate the stability by using contractively subhomogeneous and expansively superhomogeneous functions of degree α for the Eq. (4) in β-Banach spaces.

2 Preliminaries

Throughout this paper $\mathbb{N}$, $\mathbb{Z}$, $\mathbb{Q}^+$, $\mathbb{Q}$, $\mathbb{R}^+$, $\mathbb{R}$, and $\mathbb{C}$ stand, as usual, for the set of positive integers, integers, positive rationals, rationals, positive reals, reals, and complex numbers, respectively. In the rest of this paper, unless otherwise explicitly stated, we will assume that $\mathbb{K}$ denote either $\mathbb{Q}$ or $\mathbb{R}$, X and Y are linear spaces over $\mathbb{K}$, $m \in \mathbb{N}\backslash\{1\}$, $p \in \{1, \ldots, 5\}$, $a_1, \ldots, a_m$ are fixed nonzero reals when $p \in \{1, 3, 5\}$ and are fixed positive reals when $p \in \{2, 4\}$, α is a fixed nonzero real number, $L \in (0, 1)$ and $\beta \in (0, 1]$.

We now recall the definition and some necessary notions of *multi-additive mappings*, used in the sequel. A mapping $A_n : X^n \to Y$ is called *n-additive* if it is additive (satisfies Cauchy's functional equation) in each variable, that is

$$A_n(x_1, \dots, x_{i-1}, x_i + x_i', x_{i+1}, \dots, x_n) = A_n(x_1, \dots, x_n) + A_n(x_1, \dots, x_{i-1}, x_i', x_{i+1}, \dots, x_n)$$

for all $i \in \{1, \dots, n\}$, $x_1, \dots, x_{i-1}, x_i, x_i', x_{i+1}, \dots, x_n \in X$. Some basic facts on such mappings can be found for instance in [21], where their application to the representation of polynomial functions is also presented (see also [23, 24]). A mapping A_n is called *symmetric* if

$$A_n(x_1, \dots, x_n) = A_n(x_{i_1}, \dots, x_{i_n})$$

for any permutation $\{i_1, \dots, i_n\}$ of $\{1, \dots, n\}$. If $A_n(x_1, \dots, x_n)$ is an n-additive symmetric mapping, then $A^n : X^n \to Y$ denotes the diagonal $A_n(x, \dots, x)$, that is, $A^n(x) := A_n(x, \dots, x)$ for $x \in X$; note that $A^n(rx) = r^n A^n(x)$ whenever $x \in X$ and $r \in \mathbb{Q}$. Such a mapping $A^n(x)$ is called a monomial mapping of degree n (under the assumption that $A^n(x) \neq 0$). Furthermore, the mapping obtained by the substitution $x_1 = \cdots = x_l = x$, $x_{l+1} = \cdots = x_n = y$ in $A_n(x_1, \dots, x_n)$ is denoted by $A^{l,n-l}(x, y)$.

Lemma 1 (see Czerwik [7], p. 74). *Let $A_n : X^n \to Y$ be a n-additive symmetric mapping and $A^n : X^n \to Y$ be the diagonal of A_n. Then, for every $k \geq n$ and for every $x, y_1, \dots, y_k \in X$, we have*

$$\Delta_{y_1,\dots,y_k} A^n(x) = \begin{cases} n! A_n(y_1, \dots, y_n) & \text{if } k = n; \\ 0 & \text{if } k > n. \end{cases}$$

The following theorem was proved by Mazur and Orlicz [23, 24], and in greater generality by Djoković [8]; also see Baker [3].

Theorem 1 *If $n \in \mathbb{N}$ and $f : X \to Y$, then the following assertions are equivalent:*

(1) $\Delta_y^{n+1} f(x) = 0$ *for all* $x, y \in X$;
(2) $\Delta_{y_1,\dots,y_{n+1}} f(x) = 0$ *for all* $x, y_1, \dots, y_{n+1} \in X$;
(3) f *is a generalized polynomial of degree at most n, that is,*

$$f(x) = A^0(x) + A^1(x) + \cdots + A^n(x)$$

for all $x \in X$, where $A^0(x) = A^0$ is an arbitrary element of Y and A^i, $i = 1, \dots, n$ is the diagonal of an i-additive symmetric mapping $A_i : X^i \to Y$.

The following fixed point theorem will play a crucial role in proving our stability results.

Theorem 2 (Banach's Contraction Principle). *Let (X, d) be a complete metric space and consider a mapping $\Lambda : X \to X$ as a strictly contractive mapping, that is*

$$d(\Lambda x, \Lambda y) \leq L d(x, y), \qquad \forall x, y \in X,$$

for some (Lipschitz constant) $L \in [0, 1)$. Then:

(i) The mapping Λ has a unique fixed point $x^* = \Lambda(x^*)$;
(ii) The fixed point x^* is globally attractive, that is

$$\lim_{n\to\infty} \Lambda^n x = x^*$$

for any starting point $x \in X$;

(iii) One has the following estimation inequalities

$$d(\Lambda^n x, x^*) \leq L^n d(x, x^*),$$

$$d(\Lambda^n x, x^*) \leq \frac{1}{1-L} d(\Lambda^n x, \Lambda^{n+1} x),$$

$$d(x, x^*) \leq \frac{1}{1-L} d\,(x, \Lambda x)$$

for all $x \in X$ and nonnegative integers n.

Definition 1 A mapping $\phi : X \to Y$ is called,

- A *homogeneous* mapping of degree α if $\phi(cx) = c^\alpha \phi(x)$ (for the case of $\alpha = 1$, the corresponding mapping is simply said to be homogeneous),
- A *contractively subhomogeneous* mapping of degree α if there exists a constant L such that

$$\phi(cx) \leq c^\alpha L \phi(x),$$

- An *expansively superhomogeneous* mapping of degree α if there exists a constant L such that

$$\phi(cx) \geq \frac{c^\alpha}{L} \phi(x)$$

for any $x \in X$ and any positive reals c.

Remark 1 It follows by the contractively subhomogeneous of degree α (:$\ell = 1$) and expansively superhomogeneous of degree α (:$\ell = -1$) conditions of ϕ that

$$\phi(c^{\ell k} x) \leq \left(c^{\ell\alpha} L\right)^k \phi(x), \qquad k \in \mathbb{N}.$$

Definition 2 We fix a real number β and let X be a real or complex linear space. A β-norm on X is a function $x \mapsto \|x\|_\beta$ from X to $[0, \infty)$ which satisfies

(βN_1) $\|x\|_\beta = 0$ if and only if $x = 0$;
(βN_2) $\|\lambda x\|_\beta = |\lambda|^\beta . \|x\|$ for all scalars λ and all $x \in X$;
(βN_3) $\|x + y\|_\beta \leq \|x\|_\beta + \|y\|_\beta$ for all $x, y \in X$.

The pair $(X, \| . \|_\beta)$ is then said to be a β-normed space, which is called a β-Banach space if it is complete. In special case, when $\beta = 1$, $\left(X, \| . \|_\beta\right)$ turns into a normed linear space.

Example 1 Let L_β be the space of all measurable functions $f(t)$ on $I = [a, b]$ with $\int_a^b |f(t)|^\beta dt < \infty$ (we identify functions which are equal almost everywhere). For all $f \in L_\beta$, let the function $\|f\|_\beta$ be defined by $\|f\|_\beta = \left(\int_a^b |f(t)|^\beta dt\right)^{\frac{1}{\beta}}$. Hence, $\|f\|_\beta$ is a quasi-norm on a topological linear space [19] and the βth power of the quasi-norm $\|f\|_\beta$ is a β-norm on L_β.

3 Multi-additive and Monomial Mappings

We start with the following lemma which plays a crucial role in this section.

Lemma 2 *Let $f : X \to Y$ be a mapping. Then:*

(i) *f satisfies the functional equation (3) if and only if f is of the form $f(x) = A^0 + A^1(x)$;*

(ii) *f satisfies the functional equation*

$$f(x+y) + f(x-y) = 2f(x) + 2f(y) \tag{5}$$

if and only if f is of the form $f(x) = A^2(x)$ (see also [1]);

(iii) *f satisfies the functional equation*

$$f(2x+y) + f(2x-y) = 2f(x+y) + 2f(x-y) + 12f(x) \tag{6}$$

if and only if f is of the form $f(x) = A^3(x)$ (compare with [15, 31]);

(iv) *f satisfies the functional equation*

$$f(2x+y) + f(2x-y) + 6f(y) = 4f(x+y) + 4f(x-y) + 24f(x) \tag{7}$$

if and only if f is of the form $f(x) = A^4(x)$ (compare with [6, 22, 27]);

(v) *f satisfies the functional equation*

$$\begin{aligned} f(3x+y) + f(2x-y) + 10f(x+y) &= 5f(2x+y) + 5f(x-y) \\ &+ 120f(x) + 10f(y) \end{aligned} \tag{8}$$

if and only if f is of the form $f(x) = A^5(x)$;
for all $x, y \in X$, where A^0 is an arbitrary element of Y and A^i, $i = 1, \ldots, 5$ is the diagonal of an i-additive symmetric mapping $A_i : X^i \to Y$.

Proof (i) Assume that f is a solution of the Eq. (3). Letting $x = x + y$ in (3), we get

$$f(x+2y) - 2f(x+y) + f(x) = 0,$$

that is, $\Delta_y^2 f(x) = 0$. Consequently, by Theorem 1, f is a generalized polynomial of degree at most 1; that is, $f(x) = A^0 + A^1(x)$ for all $x \in X$.

Conversely, if f has the form $f(x) = A^0 + A^1(x)$, where A^1 is additive, then we have $\Delta_y^2 f(x) = 0$ by Theorem 1. Letting $x = x - y$ in $\Delta_y^2 f(x) = 0$, we get the Eq. (3).

(ii) Assume that f is a solution of the Eq. (5). Letting $x = y = 0$ and $x = 0$ separately in (5), we get $f(0) = 0$ and $f(-y) = f(y)$ for all $y \in X$. Letting $y = x + y$ and $y = 2x + y$ separately in (5) and subtracting the latter from the former, we obtain

$$f(3x + y) - 3f(2x + y) + 3f(x + y) - f(y) = 0,$$

that is, $\Delta_x^3 f(y) = 0$. Due to Theorem 1, f is a generalized polynomial of degree at most 2; that is, $f(x) = A^0 + A^1(x) + A^2(x)$, where $A^i(x)$ is the diagonal of an i-additive symmetric mapping $A_i : X^i \to Y$ for $i = 1, 2$ and A^0 is an arbitrary constant. From $f(0) = 0$, we have $A^0 = 0$. Since f is an even mapping, $A^1(x)$ must vanish. Thus, $f(x) = A^2(x)$ for all $x \in X$.

Conversely, assume that there exists a 2-additive symmetric mapping $A_2 : X^2 \to Y$ such that $f(x) = A^2(x)$ for all $x \in X$. By Lemma 1, we obtain $\Delta_x^2 A^2(y) = 2!A^2(x)$, that is,

$$A^2(2x + y) - 2A^2(x + y) + A^2(y) = 2A^2(x).$$

Replacing y by $y - x$ in the last equation, we obtain

$$A^2(x + y) - 2A^2(y) + A^2(y - x) = 2A^2(x).$$

On account of the additivity of $A_2(x_1, x_2)$, we have $A^2(rx) = r^2A^2(x)$ for all $r \in \mathbb{Q}$, so

$$A^2(x + y) + A^2(x - y) - 2A^2(x) - 2A^2(y) = 0.$$

By the assumption, we arrive at the functional Eq. (5).

(iii) Assume that f is a solution of the Eq. (6). Letting $x = y = 0$, $x = 0$ and $y = 0$ separately in (6), we get $f(0) = 0$, $f(-x) = -f(x)$ and $f(2x) = 8f(x)$ for all $x \in X$. Interchanging x and y in (6) and using oddness of f, we obtain

$$f(x + 2y) - f(x - 2y) - 2f(x + y) + 2f(x - y) - 12f(y) = 0. \tag{9}$$

Replacing x by $2x$ in (9) and using $f(2x) = 8f(x)$, we obtain

$$f(2x + y) - f(2x - y) - 4f(x + y) + 4f(x - y) + 6f(y) = 0. \tag{10}$$

Letting $y = x + y$ and $y = 2x + y$ separately in (6), then the two resulting equations yield

$$\begin{aligned} &f(4x + y) - 4f(3x + y) + 4f(2x + y) + 2f(x + y) - 2f(x - y) \\ &\quad + 12f(x) - 5f(y) = 0. \end{aligned} \tag{11}$$

It follows from (6), (10), and (11) that

$$f(4x+y)-4f(3x+y)+6f(2x+y)-4f(x+y)+f(y)=0,$$

that is, $\Delta_x^4 f(y)=0$. Due to Theorem 1, f is a generalized polynomial of degree at most 3; that is, $f(x)=A^0+A^1(x)+A^2(x)+A^3(x)$, where $A^i(x)$ is the diagonal of an i-additive symmetric mapping $A_i : X^i \to Y$ for $i=1,2,3$ and A^0 is an arbitrary constant. From $f(0)=0$, we have $A^0=0$. Since f is an odd mapping, $A^2(x)$ must vanish. Thus, $f(x)=A^1(x)+A^3(x)$ for all $x \in X$. By $f(2x)=8f(x)$ and $A^m(rx)=r^m A^m(x)$ whenever $x \in X$ and $r \in \mathbb{Q}$, we get $A^1(x)=0$. Therefore, $f(x)=A^3(x)$ for all $x \in X$.

Conversely, assume that there exists a 3-additive symmetric mapping $A_3 : X^3 \to Y$ such that $f(x)=A^3(x)$ for all $x \in X$. By Lemma 1, we obtain $\Delta_x^3 A^3(y)=3!A^3(x)$, that is,

$$A^3(3x+y)-3A^3(2x+y)+3A^3(x+y)-A^3(y)=6A^3(x). \tag{12}$$

Setting $y=y-x$ in (12), we get

$$A^3(2x+y)-3A^3(x+y)+3A^3(y)-A^3(y-x)=6A^3(x). \tag{13}$$

Setting $y=-y$ in (13), we get

$$A^3(2x-y)-3A^3(x-y)+3A^3(-y)-A^3(-y-x)=6A^3(x). \tag{14}$$

Adding (13) to (14) and using oddness of A^3, we get

$$A^3(2x+y)+A^3(2x-y)-2A^3(x+y)-2A^3(x-y)-12A^3(x)=0.$$

By the assumption, we arrive at the functional Eq. (6).

(iv) Assume that f is a solution of the Eq. (7). Letting $x=y=0$, $x=0$ and $y=0$ separately in (7), we get $f(0)=0$, $f(-x)=f(x)$ and $f(2x)=16f(x)$ for all $x \in X$. Letting $y=2x+y$ and $y=3x+y$ separately in (7) and subtracting the latter from the former, we obtain

$$f(5x+y)-5f(4x+y)+10f(3x+y)-10f(2x+y)+5f(x+y)-f(y)=0,$$

that is, $\Delta_x^5 f(y)=0$. Due to Theorem 1, f is a generalized polynomial of degree at most 4; that is, $f(x)=A^0+A^1(x)+A^2(x)+A^3(x)+A^4(x)$, where $A^i(x)$ is the diagonal of an i-additive symmetric mapping $A_i : X^i \to Y$ for $i=1,2,3,4$ and A^0 is an arbitrary constant. From $f(0)=0$, we have $A^0=0$. Since f is an even mapping, $A^1(x)$ and $A^3(x)$ must vanish. Thus, $f(x)=A^2(x)+A^4(x)$ for all $x \in X$. By $f(2x)=16f(x)$ and $A^m(rx)=r^m A^m(x)$ whenever $x \in X$ and $r \in \mathbb{Q}$, we get $A^2(x)=0$. Therefore, $f(x)=A^4(x)$ for all $x \in X$.

Conversely, assume that there exists a 4-additive symmetric mapping $A_4 : X^4 \to Y$ such that $f(x)=A^4(x)$ for all $x \in X$. By Lemma 1, we get $\Delta_x^4 A^4(y)=4!A^4(x)$, that is,

$$A^4(4x+y)-4A^4(3x+y)+6A^4(2x+y)-4A^4(x+y)+A^4(y)=24A^4(x). \tag{15}$$

Setting $y = y - 2x$ in (15) and using evenness of A^4, we get

$$A^4(2x+y) + A^4(2x-y) - 4A^4(x+y) - 4A^4(x-y) - 24A^4(x) + 6A^4(y) = 0.$$

By the assumption, we arrive at the functional Eq. (7).

(v) Assume that f is a solution of the Eq. (8). Letting $x = y = 0$, $x = 0$, $y = 0$, $y = x$, $y = 2x$ and $y = 3x$ separately in (8), we get $f(0) = 0$, $f(-x) = -f(x)$, $f(3x) = 4f(2x) + 115f(x)$ $(*_1)$, $f(4x) = 10f(2x) + 704f(x)$ $(*_2)$, $f(5x) = 20f(2x) + 2485f(x)$ and $f(6x) = 35f(2x) + 6656f(x)$ $(*_3)$ for all $x \in X$. From $(*_1)$, $(*_2)$, and $(*_3)$, we obtain $f(2x) = 32f(x)$ for all $x \in X$. Letting $y = 2x + y$ and $y = 3x + y$ separately in (8) and subtracting the latter from the former, we obtain

$$\begin{aligned} &f(6x+y) - 6f(5x+y) + 15f(4x+y) - 20f(3x+y) + 15f(2x+y) \\ &\quad - 6f(x+y) + f(y) = 0, \end{aligned}$$

that is, $\Delta_x^6 f(y) = 0$. Due to Theorem 1, f is a generalized polynomial of degree at most 5; that is, $f(x) = A^0 + A^1(x) + A^2(x) + A^3(x) + A^4(x) + A^5(x)$, where $A^i(x)$ is the diagonal of an i-additive symmetric mapping $A_i : X^i \to Y$ for $i = 1, 2, 3, 4, 5$ and A^0 is an arbitrary constant. From $f(0) = 0$, we have $A^0 = 0$. Since f is an odd mapping, $A^2(x)$ and $A^4(x)$ must vanish. Thus, $f(x) = A^1(x) + A^3(x) + A^5(x)$ for all $x \in X$. By $f(2x) = 32f(x)$ and $A^m(rx) = r^m A^m(x)$ whenever $x \in X$ and $r \in \mathbb{Q}$, we get $A^1(x) = A^3(x) = 0$. Therefore, $f(x) = A^5(x)$ for all $x \in X$.

Conversely, assume that there exists a 5-additive symmetric mapping $A_5 : X^5 \to Y$ such that $f(x) = A^5(x)$ for all $x \in X$. Due to Lemma 1, we obtain $\Delta_x^5 A^5(y) = 5!A^5(x)$, that is,

$$\begin{aligned} &A^5(5x+y) - 5A^5(4x+y) + 10A^5(3x+y) - 10A^5(2x+y) + 5A^5(x+y) \\ &\quad - A^5(y) = 120A^5(x). \end{aligned} \tag{16}$$

Setting $y = y - 2x$ in (16) and using oddness of A^5, we get

$$\begin{aligned} &A^5(3x+y) - 5A^5(2x+y) + A^5(2x-y) + 10A^5(x+y) - 5A^5(x-y) \\ &\quad - 120A^5(x) - 10A^5(y) = 0. \end{aligned}$$

By the assumption, we arrive at the functional Eq. (8).

Remark 2 (i) The functional Eq. (3) is called the Cauchy–Jensen functional equation, and every solution of the Eq. (3) is called a Cauchy–Jensen mapping.

(ii) The functional Eq. (5) is called the quadratic functional equation, and every solution of the Eq. (5) is called a quadratic mapping.

(iii) The functional Eq. (6) is called the cubic functional equation, and every solution of the Eq. (6) is called a cubic mapping.

(iv) The functional Eq. (7) is called the quartic functional equation, and every solution of the Eq. (7) is called a quartic mapping.

(v) The functional Eq. (8) is called the quintic functional equation, and every solution of the Eq. (8) is called a quintic mapping.

Lemma 3 *If a mapping $f : \mathbb{R} \to X$ satisfies the functional equation*

$$f\left((\sum_{i=1}^{m} x_i^p)^{\frac{1}{p}} \right) = \sum_{i=1}^{m} f(x_i), \tag{17}$$

then for each $p = 1, \ldots, 5$, f is Cauchy-additive, quadratic, cubic, quartic, and quintic, respectively.

Proof Letting $x_1 = \cdots = x_m = 0$ in (17), we get $f(0) = 0$. Setting $x_1 = x$, $x_2 = -x$ and $x_3 = \cdots = x_m = 0$ in (17), we get $0 = f(0) = f(x) + f(-x)$; that is, $f(-x) = -f(x)$ for all $x \in \mathbb{R}$ and $p \in \{1, 3, 5\}$. Putting $x_1 = -x_1$ in (17), we have

$$f\left((\sum_{i=1}^{m} x_i^p)^{\frac{1}{p}} \right) = f(-x_1) + \sum_{i=2}^{m} f(x_i) \tag{18}$$

for all $x_1, \ldots, x_m \in \mathbb{R}$ and $p \in \{2, 4\}$. If we compare (17) with (18), then we obtain that $f(-x_1) = f(x_1)$; that is, $f(-x) = f(x)$ for all $x \in \mathbb{R}$ and $p \in \{2, 4\}$. It is easy to see that $f(\sqrt[p]{k}x) = kf(x)$, and so $f(\frac{x}{\sqrt[p]{k}}) = \frac{1}{k}f(x)$ for $k \in \mathbb{Z}$ when $p \in \{1, 3, 5\}$ and $k \in \mathbb{N}$ when $p \in \{2, 4\}$. Hence, $f(\sqrt[p]{r}x) = rf(x)$ for $r \in \mathbb{Q}$ when $p \in \{1, 3, 5\}$ and $r \in \mathbb{Q}^+$ when $p \in \{2, 4\}$. Thus, for every $n \in \mathbb{Z}$, we have $f(r^{\frac{n}{p}}x) = r^n f(x)$ for $r \in \mathbb{Q} \setminus \{0\}$ when $p \in \{1, 3, 5\}$ and $r \in \mathbb{Q}^+$ when $p \in \{2, 4\}$.

For $p = 1$, Eq. (17) yields the generalized Cauchy-additive equation $f\left(\sum_{i=1}^{m} x_i\right) = \sum_{i=1}^{m} f(x_i)$. If $x_3 = \cdots = x_m = 0$ in the last equality, then f is Cauchy-additive.

Using the proof of Lemma 2.1 of [18], we conclude that if $p = 2$ and $x_3 = \cdots = x_m = 0$ in (17), then f satisfies (5); i.e., f is quadratic.

For $p = 3$, letting $x_1 = x$, $x_2 = y$ and $x_3 = \cdots = x_m = 0$ in (17) and using $f(0) = 0$, one gets

$$f\left(\sqrt[3]{x^3 + y^3}\right) = f(x) + f(y). \tag{19}$$

Replacing x by $x + y$ and y by $x - y$ in (19) and using $f(\sqrt[3]{2}x) = 2f(x)$, we obtain

$$f\left(\sqrt[3]{x^3 + 3xy^2}\right) = \frac{1}{2}[f(x + y) + f(x - y)], \tag{20}$$

which by replacing x by $2x$ yields

$$f\left(\sqrt[3]{4x^3 + 3xy^2}\right) = \frac{1}{4}[f(2x + y) + f(2x - y)]. \tag{21}$$

Setting $y = \left(x^3 + 3xy^2\right)^{\frac{1}{3}}$ in (19) and using (20), we get

$$f\left(\sqrt[3]{2x^3 + 3xy^2}\right) = \frac{1}{2}[f(x + y) + f(x - y)] + f(x). \tag{22}$$

Putting $y = \left(2x^3 + 3xy^2\right)^{\frac{1}{3}}$ in (19) and using (22), we get

$$f\left(\sqrt[3]{3x^3 + 3xy^2}\right) = \frac{1}{2}[f(x+y) + f(x-y)] + 2f(x). \tag{23}$$

Letting $y = \left(3x^3 + 3xy^2\right)^{\frac{1}{3}}$ in (19) and using (23), we get

$$f\left(\sqrt[3]{4x^3 + 3xy^2}\right) = \frac{1}{2}[f(x+y) + f(x-y)] + 3f(x). \tag{24}$$

From (21) and (24), we conclude that f satisfies (6). Hence, by Lemma 2, f is cubic.

For $p = 4$, letting $x_1 = x$, $x_2 = y$ and $x_3 = \cdots = x_m = 0$ in (17) and using $f(0) = 0$, one finds

$$f\left(\sqrt[4]{x^4 + y^4}\right) = f(x) + f(y). \tag{25}$$

Replacing x by $x + y$ and y by $x - y$ in (25), we obtain

$$f\left(\sqrt[4]{x^4 + 6x^2y^2 + y^4}\right) = \frac{1}{2}[f(x+y) + f(x-y)]. \tag{26}$$

Setting $y = \left(x^4 + 6x^2y^2 + y^4\right)^{\frac{1}{4}}$ in (25) and using (26), we obtain

$$f\left(\sqrt[4]{2x^4 + 6x^2y^2 + y^4}\right) = \frac{1}{2}[f(x+y) + f(x-y)] + f(x), \tag{27}$$

which by replacing x by $\sqrt{2}x$ and using $f\left(r^{\frac{n}{4}}x\right) = r^n f(x)$ for $r \in \mathbb{Q}^+$ becomes

$$f\left(\sqrt[4]{8x^4 + 12x^2y^2 + y^4}\right) = \frac{1}{2}\left[f\left(\sqrt{2}x + y\right) + f\left(\sqrt{2}x - y\right)\right] + 4f(x). \tag{28}$$

Interchange x and y in (27) and using evenness of f to get

$$f\left(\sqrt[4]{x^4 + 6x^2y^2 + 2y^4}\right) = \frac{1}{2}[f(x+y) + f(x-y)] + f(y),$$

which by putting $x = 2x$ gives

$$f\left(\sqrt[4]{8x^4 + 12x^2y^2 + y^4}\right) = \frac{1}{4}[f(2x+y) + f(2x-y) + 2f(y)]. \tag{29}$$

It follows from (28) and (29) that

$$f(2x+y) + f(2x-y) = 2f\left(\sqrt{2}x + y\right) + 2f\left(\sqrt{2}x - y\right) + 16f(x) - 2f(y), \tag{30}$$

which by letting $y = \sqrt{2}y$ yields

$$f\left(\sqrt{2}x + y\right) + f\left(\sqrt{2}x - y\right) = 2f(x+y) + 2f(x-y) + 4f(x) - 2f(y). \tag{31}$$

From (30) and (31), we conclude that f satisfies (7). Hence, by Lemma 2, f is quartic.

For $p = 5$, letting $x_1 = x$, $x_2 = y$ and $x_3 = \cdots = x_m = 0$ in (17) and using $f(0) = 0$, we see that

$$f\left(\sqrt[5]{x^5 + y^5}\right) = f(x) + f(y). \tag{32}$$

Replacing x by $x + y$ and y by 0 in (32), we obtain

$$f\left(\sqrt[5]{x^5 + 5x^4y + 10x^3y^2 + 10x^2y^3 + 5xy^4 + y^5}\right) = f(x + y), \tag{33}$$

which by replacing x by $3x$ and then using (32) and $f\left(\sqrt[5]{3}x\right) = 3f(x)$ becomes

$$f\left(\sqrt[5]{81x^5 + 135x^4y + 90x^3y^2 + 30x^2y^3 + 5xy^4}\right) = \frac{1}{3}[f(3x + y) - f(y)],$$

which, by using (32) again, gives

$$f\left(\sqrt[5]{27x^4y + 18x^3y^2 + 6x^2y^3 + xy^4}\right) = \frac{1}{15}[f(3x + y) - 243f(x) - f(y)]. \tag{34}$$

Replacing x by $x + y$ and y by $x - y$ in (32), we obtain

$$f\left(\sqrt[5]{x^5 + 10x^3y^2 + 5xy^4}\right) = \frac{1}{2}[f(x + y) + f(x - y)], \tag{35}$$

which by replacing x by $2x$ yields

$$f\left(\sqrt[5]{16x^5 + 40x^3y^2 + 5xy^4}\right) = \frac{1}{4}[f(2x + y) + f(2x - y)]. \tag{36}$$

Set $y = x + y$ in (35) and using oddness of f to get

$$f\left(\sqrt[5]{16x^5 + 40x^4y + 40x^3y^2 + 20x^2y^3 + 5xy^4}\right) = \frac{1}{2}[f(2x + y) - f(y)]. \tag{37}$$

From (32) and (37), we have

$$f\left(\sqrt[5]{8x^4y + 8x^3y^2 + 4x^2y^3 + xy^4}\right) = \frac{1}{10}[f(2x + y) - 32f(x) - f(y)]. \tag{38}$$

It follows from (32), (36), and (37) that

$$f\left(\sqrt[5]{2x^4y + x^2y^3}\right) = \frac{1}{80}[f(2x + y) - f(2x - y) - 2f(y)], \tag{39}$$

which by replacing y by $x + y$ and then using (32) becomes

$$f\left(\sqrt[5]{5x^4y + 3x^3y^2 + x^2y^3}\right) = \frac{1}{80}[f(3x + y) - 2f(x + y) - f(x - y)] - 3f(x). \tag{40}$$

By (32), (39), and (40), we obtain

$$f\left(\sqrt[5]{x^4y+x^3y^2}\right) = \frac{1}{240}[f(3x+y) - f(2x+y) + f(2x-y) - 2f(x+y) \\ - f(x-y) + 2f(y)] - f(x). \quad (41)$$

By (32), (38), and (41), we obtain

$$f\left(\sqrt[5]{4x^2y^3+xy^4}\right) = \frac{-1}{30}[f(3x+y) - 4f(2x+y) + f(2x-y) - 2f(x+y) \\ - f(x-y) - 144f(x) + 5f(y)]. \quad (42)$$

By (32), (34), and (42), we obtain

$$f\left(\sqrt[5]{27x^4y+18x^3y^2+2x^2y^3}\right) = \frac{1}{30}[3f(3x+y) - 4f(2x+y) + f(2x-y) \\ - 2f(x+y) - f(x-y) + 3f(y)] - 21f(x). \quad (43)$$

By (32), (41), and (43), we obtain

$$f\left(\sqrt[5]{9x^4y+2x^2y^3}\right) = \frac{1}{120}[3f(3x+y) - 7f(2x+y) - 5f(2x-y) \\ + 10f(x+y) + 5f(x-y) - 6f(y)] - 3f(x). \quad (44)$$

By (32), (39), and (44), we obtain

$$f\left(\sqrt[5]{x^4y}\right) = \frac{1}{600}[3f(3x+y) - 10f(2x+y) - 2f(2x-y) + 10f(x+y) \\ + 5f(x-y) - 360f(x)]. \quad (45)$$

By (32), (39), and (45), we obtain

$$f\left(\sqrt[5]{x^2y^3}\right) = \frac{-1}{1200}[12f(3x+y) - 55f(2x+y) + 7f(2x-y) + 40f(x+y) \\ + 20f(x-y) - 1440f(x) + 30f(y)]. \quad (46)$$

It follows from (32), (33), (35), (45), and (46) that f satisfies (8). Hence, by Lemma 2, f is quintic.

We now solve the functional equation (4), which is the generalized form of (17).

Theorem 3 *If a mapping $f : \mathbb{R} \to X$ satisfies the functional equation* (4) *and $\sum_{i=1}^{m} a_i \neq 1$, then for each $p = 1, \ldots, 5$, f is Cauchy-additive, quadratic, cubic, quartic, and quintic, respectively.*

Proof Substituting $x_i = 0 (1 \le i \le m)$ in (4) yields $f(0) = 0$ since $\sum_{i=1}^{m} a_i \ne 1$. If we put $x_1 = -x_1$ in (4), then we have

$$f\left(\left(\sum_{i=1}^{m} a_i x_i^p\right)^{\frac{1}{p}}\right) = a_1 f(-x_1) + \sum_{i=2}^{m} a_i f(x_i) \tag{47}$$

for all $x_1, \dots, x_m \in \mathbb{R}$ and $p \in \{2, 4\}$. If we compare (4) with (47), we obtain that $f(-x) = f(x)$ for all $x \in \mathbb{R}$ and $p \in \{2, 4\}$. Setting $x_i = 0 (1 \le i \ne j \le m)$ and $x_j = x$ in (4), we get $f\left(\sqrt[p]{a_j} x\right) = a_j f(x)$ for all $x \in \mathbb{R}$. Hence,

$$f\left(\sqrt[p]{\prod_{j=\ell}^{k} a_j}\, x\right) = \prod_{j=\ell}^{k} a_j f(x) \quad (1 \le \ell \le k \le m) \tag{48}$$

for all $x \in \mathbb{R}$. Letting $x_i = \prod_{j=1, j\ne i}^{m} a_j x_i (1 \le i \le m)$ in (4) and using (48), we obtain that f satisfies (17). Hence, by Lemma 3, for each $p = 1, \dots, 5$ f is Cauchy-additive, quadratic, cubic, quartic, and quintic, respectively.

Corollary 1 *If a mapping $f : \mathbb{R} \to X$ satisfies the functional* Eq. (4) *and $\sum_{i=1}^{m} a_i \ne 1$, then for each $p = 1, \dots, 5$, f is of the form $f(x) = A^p(x)$ for all $x \in \mathbb{R}$, where A^p is the diagonal of an p-additive symmetric mapping $A_p : \mathbb{R}^p \to X$.*

4 Fixed Points and Stability of Monomial Functional Equations

Let ϕ be a function from $\mathbb{R}^m$ to $\mathbb{R}^+$ and Y be a β-Banach space. A mapping $f : \mathbb{R} \to Y$ is called a ϕ-approximately monomial, if

$$\left\| f\left(\left(\sum_{i=1}^{m} a_i x_i^p\right)^{\frac{1}{p}}\right) - \sum_{i=1}^{m} a_i f(x_i) \right\|_\beta \le \phi(x_1, \dots, x_m) \tag{49}$$

for all $x_1, \dots, x_m \in \mathbb{R}$, where $\sum_{i=1}^{m} a_i \in \mathbb{R}^+ \backslash \{1\}$. Now, we apply the Banach fixed point theorem and our results in the previous section to get the following.

Theorem 4 *Suppose that $f : \mathbb{R} \to Y$ is a ϕ-approximately monomial mapping and that the function ϕ is contractively subhomogeneous of degree α with a constant L, where $\alpha \le p\beta$. Moreover, assume that f is even when p is even. Then, there exists a unique mapping $\mathcal{F}_p : \mathbb{R} \to Y$ satisfying* (4) *such that*

$$\left\| f(x) - \mathcal{F}_p(x) \right\|_\beta \le \frac{1}{\lambda^{p\beta}(1-L)} \Phi(x) \tag{50}$$

for all $x \in \mathbb{R}$, where $\lambda := \sum_{i=1}^{m} a_i$ and

$$\Phi(x) := \lambda^{(p-1)\beta} \sum_{\ell=0}^{p-1} \frac{1}{\lambda^{\ell\beta}} \phi\Big(\overbrace{\lambda^{\frac{\ell}{p}} x, \dots, \lambda^{\frac{\ell}{p}} x}^{n-times}\Big).$$

The mapping $\mathcal{F}_p$ is given by $\mathcal{F}_p(x) = \lim_{n\to\infty} \frac{1}{\lambda^{pn}} f(\lambda^n x)$ for all $x \in \mathbb{R}$, and $\mathcal{F}_p$ for each $p = 1, \ldots, 5$ is Cauchy-additive, quadratic, cubic, quartic, and quintic, respectively.

Proof Consider the set

$$\mathcal{W} := \left\{ g : \mathbb{R} \to Y, \quad \sup_{x\in\mathbb{R}} \frac{\|g(x) - f(x)\|_\beta}{\Phi(x)} < \infty \right\}$$

and introduce the following *metric* on $\mathcal{W}$:

$$d(g,h) = \sup_{x\in\mathbb{R}} \frac{\|g(x) - h(x)\|_\beta}{\Phi(x)}.$$

We assert that $(\mathcal{W}, d)$ is complete. Let $\{g_k\}$ be a Cauchy sequence in $(\mathcal{W}, d)$. Then, for any $\varepsilon > 0$, there exists a positive integer N_ε such that $d(g_j, g_k) \le \varepsilon$ for all $j, k \ge N_\varepsilon$. By the definition of d, for each $j, k \ge N_\varepsilon$,

$$\left\| g_j(x) - g_k(x) \right\|_\beta \le \varepsilon \Phi(x), \quad x \in \mathbb{R}. \tag{51}$$

So, for each $x \in \mathbb{R}$, $\{g_k(x)\}$ is a Cauchy sequence in Y. Since Y is complete, for each $x \in \mathbb{R}$, there exists $g(x) \in Y$ such that $g_k(x) \to g(x)$ as $k \to \infty$. Hence, by (51) for each $k \ge N_\varepsilon$,

$$\|g_k(x) - g(x)\|_\beta = \lim_{j\to\infty} \left\| g_k(x) - g_j(x) \right\|_\beta \le \varepsilon \Phi(x), \quad x \in \mathbb{R},$$

that is, $d(g_k, g) \le \varepsilon$ for each $k \ge N_\varepsilon$. Hence, $g_k \to g \in \mathcal{W}$ as $k \to \infty$. Thus, $(\mathcal{W}, d)$ is complete.

Now we consider the mapping $\mathcal{P} : \mathcal{W} \to \mathcal{W}$ defined by $(\mathcal{P}g)(x) = \frac{1}{\lambda^p} g(\lambda x)$ for all $g \in \mathcal{W}$ and all $x \in \mathbb{R}$. Let $g, h \in \mathcal{W}$ and let $C \in [0, \infty)$ be an arbitrary constant with $d(g,h) < C$. From the definition of d, we have

$$\frac{\|g(x) - h(x)\|_\beta}{\Phi(x)} \le C$$

for all $x \in \mathbb{R}$. By the assumption and the last inequality, we have

$$\frac{\|(\mathcal{P}g)(x) - (\mathcal{P}h)(x)\|_\beta}{\Phi(x)} = \frac{\|g(\lambda x) - h(\lambda x)\|_\beta}{\lambda^{p\beta}\Phi(x)} \le \frac{\lambda^{\alpha - p\beta} L \, \|g(\lambda x) - h(\lambda x)\|_\beta}{\Phi\,(\lambda x)} \le LC$$

for all $x \in \mathbb{R}$. So

$$d(\mathcal{P}g, \mathcal{P}h) \le L d(g,h)$$

for all $g, h \in \mathcal{W}$, which means that $\mathcal{P}$ is a strictly contractive self-mapping of $\mathcal{W}$ with the Lipschitz constant L.

Substituting $x_1, \ldots, x_m := x$ in the functional inequality (49), we obtain

$$\left\| f\left(\lambda^{\frac{1}{p}} x\right) - \lambda f(x) \right\|_\beta \le \phi(\overbrace{x, \ldots, x}^{n-times})$$

for all $x \in \mathbb{R}$. Thus,

$$\left\| f(\lambda x) - \lambda^p f(x) \right\|_\beta \le \Phi(x) \tag{52}$$

for all $x \in \mathbb{R}$. Hence,

$$\frac{\left\| f(x) - \frac{1}{\lambda^p} f(\lambda x) \right\|_\beta}{\Phi(x)} \le \frac{1}{\lambda^{p\beta}}$$

for all $x \in \mathbb{R}$. So $d(f, \mathcal{P}f) \le \dfrac{1}{\lambda^{p\beta}}$.

Due to Theorem 2, there exists a unique mapping $\mathcal{F}_p \in \mathcal{W}$ such that $\mathcal{F}_p(\lambda x) = \lambda^p \mathcal{F}_p(x)$ for all $x \in \mathbb{R}$, i.e., $\mathcal{F}_p$ is a unique fixed point of $\mathcal{P}$. Moreover,

$$\mathcal{F}_p(x) = \lim_{n\to\infty} (\mathcal{P}^n f)(x) = \lim_{n\to\infty} \frac{1}{\lambda^{pn}} f(\lambda^n x) \tag{53}$$

for all $x \in \mathbb{R}$, and

$$d(f, \mathcal{F}_p) \le \frac{1}{1-L} d(f, \mathcal{P}f) \le \frac{1}{\lambda^{p\beta}(1-L)},$$

i.e., inequality (50) holds true for all $x \in \mathbb{R}$.

In addition it is clear from (49) and (53) that the equality

$$\begin{aligned}
&\left\| \mathcal{F}_p\left(\left(\sum_{i=1}^m a_i x_i^p \right)^{\frac{1}{p}} \right) - \sum_{i=1}^m a_i \mathcal{F}_p(x_i) \right\|_\beta \\
&= \lim_{n\to\infty} \frac{1}{\lambda^{p\beta n}} \left\| f\left(\left(\sum_{i=1}^m \lambda^{pn} a_i x_i^p \right)^{\frac{1}{p}} \right) - \sum_{i=1}^m a_i f(\lambda^n x_i) \right\|_\beta \\
&\le \lim_{n\to\infty} \frac{1}{\lambda^{p\beta n}} \phi\left(\lambda^n x_1, \ldots, \lambda^n x_m \right) \\
&\le \lim_{n\to\infty} \left(\lambda^{\alpha - p\beta} L \right)^n \phi(x_1, \ldots, x_m) = 0
\end{aligned}$$

holds for all $x_1, \ldots, x_m \in \mathbb{R}$, and so by Theorem 3 the mapping $\mathcal{F}_p \in \mathcal{W}$ for each $p = 1, \ldots, 5$ is Cauchy-additive, quadratic, cubic, quartic, and quintic, respectively, as desired.

Theorem 5 *Suppose that $f : \mathbb{R} \to Y$ is a ϕ-approximately monomial mapping and that the function ϕ is expansively superhomogeneous of degree α with a constant L, where $\alpha \ge p\beta$. Moreover, assume that f is even when p is even. Then, there exists a unique mapping $\mathcal{F}_p : \mathbb{R} \to Y$ satisfying* (4) *such that*

$$\left\| f(x) - \mathcal{F}_p(x) \right\|_\beta \le \frac{L}{\lambda^\alpha (1-L)} \Phi(x) \tag{54}$$

for all $x \in \mathbb{R}$, where λ and $\Phi(x)$ are defined as in Theorem 4. The mapping $\mathcal{F}_p$ is given by $\mathcal{F}_p(x) = \lim_{n\to\infty} \lambda^{pn} f\left(\frac{x}{\lambda^n}\right)$ for all $x \in \mathbb{R}$, and F_p for each $p = 1, \ldots, 5$ is Cauchy-additive, quadratic, cubic, quartic, and quintic, respectively.

Proof We introduce the same definitions for $\mathcal{W}$ and d as in the proof of Theorem 4 such that $(\mathcal{W}, d)$ becomes a complete metric space. Let $\mathcal{P} : \mathcal{W} \to \mathcal{W}$ be the mapping defined by $(\mathcal{P}g)(x) = \lambda^p g\left(\frac{x}{\lambda}\right)$ for all $g \in \mathcal{W}$ and all $x \in \mathbb{R}$. One can show that $d(\mathcal{P}g, \mathcal{P}h) \leq Ld(g,h)$ for any $g, h \in \mathcal{W}$. Similar to the proof of Theorem 4, we obtain that f satisfies (52) for all $x \in \mathbb{R}$. Hence,

$$\frac{\left\| f(x) - \lambda^p f\left(\frac{x}{\lambda}\right) \right\|_\beta}{\Phi(x)} \leq \frac{L}{\lambda^\alpha}$$

for all $x \in \mathbb{R}$. So $d(f, \mathcal{P}f) \leq \frac{L}{\lambda^\alpha}$.

Due to Theorem 2, there exists a unique mapping $\mathcal{F} \in \mathcal{W}$ such that $\mathcal{F}_p(\lambda x) = \lambda^p \mathcal{F}_p(x)$ for all $x \in \mathbb{R}$, i.e., $\mathcal{F}_p$ is a unique fixed point of $\mathcal{P}$. Moreover,

$$\mathcal{F}_p(x) = \lim_{n\to\infty} (\mathcal{P}^n f)(x) = \lim_{n\to\infty} \lambda^{pn} f\left(\frac{x}{\lambda^n}\right)$$

for all $x \in \mathbb{R}$, and

$$d(f, \mathcal{F}_p) \leq \frac{1}{1-L} d(f, \mathcal{P}f) \leq \frac{L}{\lambda^\alpha(1-L)},$$

i.e., inequality (54) holds true for all $x \in \mathbb{R}$.

The remaining assertion goes through in a similar way to the corresponding part of Theorem 4.

References

1. Aczél, J., Dhombres, J.: Functional Equations Inseveral Variables. Cambridge University Press, New York (1989)
2. Almira, J.M., Lopez-Moreno, A.J.: On solutions of the Fréchet functional equation. J. Math. Anal. Appl. **332**, 119–133 (2007)
3. Baker, J.A.: A general functional equation and its stability. Proc. Amer. Math. Soc. **133**, 1657–1664 (2005)
4. Brzdęk, J., Popa, D., Xu, B.: The Hyers–Ulam stability of nonlinear recurrences. J. Math. Anal. Appl. **335**, 443–449 (2007)
5. Cădariu, L., Radu, V.: Fixed point methods for the generalized stability of functional equations in a single variable. Fixed Point Theory Appl. 2008, Art. ID 749392, (2008)
6. Chung, J.K., Sahoo, P.K.: On the general solution of a quartic functional equation. Bull. Korean Math. Soc. **40**, 565–576 (2003)
7. Czerwik, S.: Functional Equations and Inequalities in Several Variables. World Scientific Publishing Company, New Jersey (2002)
8. Djoković, D.Ž.: A representation theorem for $(X_1 - 1)(X_2 - 1) \cdots (X_n - 1)$ and its applications. Ann. Pol. Math. **22**, 189–198 (1969)
9. Forti, G.L.: Hyers–Ulam stability of functional equations in several variables. Aequ. Math. **50**, 143–190 (1995)
10. Fréchet, M.: Un definition fonctionnelle des polynômes. Nouv. Ann. de Math. **9**, 145–162 (1909)
11. Gordji, M.E., Khodaei, H.: Solution and stability of generalized mixed type cubic, quadratic and additive functional equation in quasi-Banach spaces. Nonlin. Anal. **71**, 5629–5643 (2009)

12. Gordji, M.E., Khodaei, H.: A fixed point technique for investigating the stability of (α, β, γ)-derivations on Lie C^*-algebras. Nonlin. Anal. **76**, 52–57 (2013)
13. Hayes, W., Jackson, K.R.: A survey of shadowing methods for numerical solutions of ordinary differential equations. Appl. Numer. Math. **53**, 299–321 (2005)
14. Hyers, D.H., Isac, G., Rassias, Th.M.: Stability of Functional Equations in Several Variables. Birkhäuser, Basel (1998)
15. Jun, K.W., Kim, H.M.: The generalized Hyers–Ulam–Rassias stability of a cubic functional equation. J. Math. Anal. Appl. **274**, 867–878 (2002)
16. Jung, S.-M.: Hyers–Ulam–Rassias Stability of Functional Equations in Nonlinear Analysis. Springer, New York (2011)
17. Khodaei, H., Rassias, Th.M: Approximately generalized additive functions in several variables.. Int. J. Nonlin. Anal. Appl. **1**, 22–41 (2010)
18. Khodaei, H., Gordji, M.E., Kim, S.S., Cho, Y.J.: Approximation of radical functional equations related to quadratic and quartic mappings. J. Math. Anal. Appl. **395**, 284–297 (2012)
19. Köthe, G.: Topological Vector Spaces I. Springer-Verlag, Berlin (1969)
20. Kuczma, M.: Functional Equations in a Single Variable. PWN-Polish Scientific Publishers, Warszawa (1968)
21. Kuczma, M.: An Introduction to the Theory of Functional Equations and Inequalities. Cauchy's Equation and Jensen's Inequality. Birkhäuser, Basel (2009)
22. Lee, S.H., Im, S.M., Hawng, I.S.: Quartic functional equation. J. Math. Anal. Appl. **307**, 387–394 (2005)
23. Mazur, S., Orlicz, W.: Grundlegende Eigenschaften der Polynomischen Operationen, Erst Mitteilung. Studia Math. **5**, 50–68 (1934)
24. Mazur, S., Orlicz, W.: Grundlegende Eigenschaften der Polynomischen Operationen, Zweite Mitteilung, ibidem. Studia Math. **5**, 179–189 (1934)
25. Moszner, Z.: On the stability of functional equations. Aequ. Math. **77**, 33–88 (2009)
26. Palmer, K.: Shadowing in Dynamical Systems. Theory and Applications. In: Mathematics and its Applications, vol. 501. Kluwer Academic Publishers, Dordrecht (2000)
27. Park, W.-G., Bae, J.-H.: On a bi-quadratic functional equation and its stability. Nonlin. Anal. **62**, 643–654 (2005)
28. Park, C., O'Regan, D., Saadati, R.: Stability of some set-valued functional equations. Appl. Math. Lett. **24**, 1910–1914 (2011)
29. Pólya, Gy., Szegő, G.: Aufgaben und Lehrsätze aus der Analysis. Springer, Berlin (1925)
30. Rassias, Th.M.: On the stability of the linear mapping in Banach spaces. Proc. Amer. Math. Soc. **72**, 297–300 (1978)
31. Sahoo, P.K.: On a functional equation characterizing polynomials of degree three. Bull. Inst. Math. Acad. Sin. **32**, 35–44 (2004)
32. Stević, S.: Bounded solutions of a class of difference equations in Banach spaces producing controlled chaos. Chaos Solitons Fractals. **35**, 238–245 (2008)
33. Ulam, S.M.: A Collection of Mathematical Problems. Interscience Publ., New York, (1960); reprinted as: Problems in Modern Mathematics. Wiley, New York (1964)
34. Zhou, D.-X.: On a conjecture of Z. Ditzian. J. Approx. Theory **69**, 167–172 (1992)

Some Functional Equations Related to the Characterizations of Information Measures and Their Stability

Eszter Gselmann and Gyula Maksa

Abstract The main purpose of this chapter is to investigate the stability problem of some functional equations that appear in the characterization problem of information measures.

Keywords Information measures · Information quantity · Probability distribution · Shannon entropy · Stability

1 Introduction and Preliminaries

Throughout this chapter, $\mathbb{N}, \mathbb{Z}, \mathbb{Q}, \mathbb{R}$, and $\mathbb{C}$ will stand for the set of the positive integers, the integers, the rational numbers, the reals and the set of the complex numbers, respectively. Furthermore, $\mathbb{R}_+$ and $\mathbb{R}_{++}$ will denote the set of the nonnegative and the positive real numbers, respectively.

In this section, firstly we summarize some notations and preliminaries that will be used subsequently. We begin with the introduction of the information measures. Here, their definition and some results concerning them will follow.

The second section of our chapter will be devoted to the topic of information functions. Here—among others—the general solution of the (parametric) fundamental equation of information will be described. Furthermore, some results concerning the so-called sum form information measures will also be listed.

Finally, in the last part of this chapter, we will investigate the stability problem for the functional equations that appeared in the second section. Here, some open problems will also be presented.

E. Gselmann (✉) · G. Maksa
Department of Analysis, Institute of Mathematics, University of Debrecen, Debrecen, P. O. Box: 12, 4010 Hungary
e-mail: gselmann@science.unideb.hu

G. Maksa
e-mail: maksa@science.unideb.hu

T. M. Rassias (ed.), *Handbook of Functional Equations*,
Springer Optimization and Its Applications 96, DOI 10.1007/978-1-4939-1286-5_10

1.1 Information Measures

The question "How information can be measured?," was first raised by Hartley in 1928. In his paper [37], Hartley considered only those systems of events, in which every event occurs with the same probability. After that, in 1948, the celebrated paper of Shannon [72] appeared where the information quantity contained in a complete (discrete) probability distribution was defined.

In what follows, based on the notions and the results of the monograph Aczél–Daróczy [7], a short introduction to information measures will follow.

Let $n \in \mathbb{N}$, $n \geq 2$ be arbitrarily fixed and define the sets

$$\Gamma_n^{\circ} = \left\{(p_1, \ldots, p_n) \in \mathbb{R}^n | p_i > 0, i = 1, \ldots, n, \sum_{i=1}^{n} p_i = 1\right\}$$

and

$$\Gamma_n = \left\{(p_1, \ldots, p_n) \in \mathbb{R}^n | p_i \geq 0, i = 1, \ldots, n, \sum_{i=1}^{n} p_i = 1\right\},$$

respectively. We say that the sequence of functions $(I_n)_{n=2}^{\infty}$ (or simply (I_n)) is an *information measure*, if either $I_n : \Gamma_n^{\circ} \to \mathbb{R}$ for all $n \geq 2$ or $I_n : \Gamma_n \to \mathbb{R}$ for all $n \geq 2$.

We have to mention that, in the literature, "information measures" depending on not only probabilities but on the events themselves (inset information measures) (see e.g., Aczél–Daróczy [5]) or depending on several probability distributions (see Ebanks–Sahoo–Sander [25]) are also investigated. Here, we do not involve these cases. On the other hand, originally the zero probabilities were allowed adopting the conventions

$$0 \log_2 (0) = \frac{0}{0+0} = 0^{\alpha} = 0 \qquad (\alpha \in \mathbb{R}). \tag{1}$$

We follow these conventions, and we denote Γ_n° or Γ_n by $\mathcal{G}_n$ provided that it does not matter that the zero probabilities are excluded or not.

Certainly, the most known information measures are the *Shannon entropy* (see Shannon [72]), i.e.,

$$H_n^1(p_1, \ldots, p_n) = -\sum_{i=1}^{n} p_i \log_2 (p_i), \quad ((p_1, \ldots, p_n) \in \mathcal{G}_n, n \geq 2)$$

and the so-called entropy of degree α, or the *Havrda–Charvát entropy* (see Aczél–Daróczy [7], Daróczy [15], Kullback [53], Tsallis [78]), i.e.,

$$H_n^{\alpha}(p_1, \ldots, p_n) = \begin{cases} \left(2^{1-\alpha} - 1\right)^{-1} \left(\sum_{i=1}^{n} p_i^{\alpha} - 1\right), & \text{if} \quad \alpha \neq 1 \\ H_n^1(p_1, \ldots, p_n), & \text{if} \quad \alpha = 1 \end{cases},$$

where $n \in \mathbb{N},\ n \geq 2,\ \alpha \in \mathbb{R}$ and $(p_1, \dots, p_n) \in \mathcal{G}_n$.

(H_n^1) was first introduced to the statistical thermodynamics by Boltzmann and Gipps, to the information theory by Shannon [72], while (H_n^α) (for $\alpha \neq 1$) was first investigated from cybernetic point of view by Havrda and Charvát [38], from information theoretical point of view by Daróczy [15], and was rediscovered by Tsallis [78] for the Physics community.

It is easy to see that, for arbitrarily fixed $n \geq 2$ and $(p_1, \dots, p_n) \in \mathcal{G}_n$,

$$\lim_{\alpha \to 1} H_n^\alpha (p_1, \dots, p_n) = H_n^1 (p_1, \dots, p_n),$$

which shows that the Shannon entropy can continuously be embedded to the family of entropies of degree α. As it is formulated in [7], the *characterization problem* for the information measure (H_n^α) is the following: What properties have to be imposed upon an information measure (I_n) in order that $(I_n) = (H_n^\alpha)$ be valid?

In what follows, we list the properties which seem to be reasonable for characterizing (H_n^α). It is not difficult to check that the information measure (H_n^α) has these properties.

An information measure (I_n) is called *symmetric* if

$$I_n (p_1, \dots, p_n) = I_n \left(p_{\sigma(1)}, \dots, p_{\sigma(n)}\right) \tag{2}$$

is satisfied for all $n \geq 2$, $(p_1, \dots, p_n) \in \mathcal{G}_n$ and for arbitrary permutation $\sigma : \{1, \dots, n\} \to \{1, \dots, n\}$. Further, we say that (I_n) is *3-semi-symmetric* if

$$I_3(p_1, p_2, p_3) = I_3(p_1, p_3, p_2) \tag{3}$$

holds for all $(p_1, p_2, p_3) \in \mathcal{G}_3$.

(I_n) is called *normalized* if

$$I_2 \left(\frac{1}{2}, \frac{1}{2}\right) = 1, \tag{4}$$

and it is called *α-recursive* if

$$\begin{aligned} &I_n (p_1, \dots, p_n) \\ &\quad = I_{n-1} (p_1 + p_2, p_3, \dots, p_n) + (p_1 + p_2)^\alpha I_2 \left(\frac{p_1}{p_1 + p_2}, \frac{p_2}{p_1 + p_2}\right) \end{aligned} \tag{5}$$

holds for all for all $n \geq 3$ and $(p_1, \dots, p_n) \in \mathcal{G}_n$. In case $\alpha = 1$, we say simply that (I_n) is *recursive*.

For a fixed $\alpha \in \mathbb{R}$ and $2 \leq n \in \mathbb{N}, 2 \leq m \in \mathbb{N}$, the information measure (I_n) is said to be *(α, n, m)- additive*, if

$$I_{nm} (P * Q) = I_n (P) + I_m (Q) + (2^{1-\alpha} - 1) I_n (P) I_m (Q) \tag{6}$$

holds for all $P = (p_1, \dots, p_n) \in \mathcal{G}_n,\ Q = (q_1, \dots, q_m) \in \mathcal{G}_m$ where $P * Q = (p_1q_1, \dots, p_1q_m, \dots, p_nq_1, \dots, p_nq_m) \in \mathcal{G}_{nm}$. Finally, we say that an information measure (I_n) has the *sum property*, if there exists a function $f : I \to \mathbb{R}$ such that

$$I_n(p_1, \ldots, p_n) = \sum_{i=1}^{n} f(p_i) \qquad ((p_1, \ldots, p_n) \in \mathcal{G}_n) \tag{7}$$

for all $2 \le n \in \mathbb{N}$. Here (and through the chapter) I denotes the closed unit interval $[0, 1]$ if $\mathcal{G}_n = \Gamma_n$ for all $2 \le n \in \mathbb{N}$ and the open unit interval $]0, 1[$ if $\mathcal{G}_n = \Gamma_n^\circ$ for all $2 \le n \in \mathbb{N}$. Such a function f satisfying (7) is called a *generating function* of (I_n).

1.2 The Characterization Problem and Functional Equations

The properties listed above are of algebraic nature. This is the reason why they lead to functional equations. In this section, we present how they imply the so-called parametric fundamental equation of information and the sum form functional equations. Following the ideas of Daróczy [14] (see also [7]), suppose first that the information measure (I_n) is (5) α-recursive and (3) 3-semi-symmetric, and define the function f on I by

$$f(x) = I_2(x, 1 - x)$$

and the set $D^\circ = \{(x, y) \mid x, y, x + y \in I\}$, if $I =]0, 1[$ and $D = \{(x, y) \mid x, y \in [0, 1[, \ x + y \in I\}$ if $I = [0, 1]$, respectively. Let now

$$(x, y) \in \begin{cases} D & \text{if } I = [0, 1] \\ D^\circ & \text{if } I =]0, 1[\end{cases}$$

and $n = 3$, $p_1 = 1 - x - y$, $p_2 = y$, $p_3 = x$ in (5). Then, we have that

$$\begin{aligned} I_3(1 - x - y, y, x) &= I_2(1 - x, x) + (1 - x)^\alpha I_2\left(1 - \frac{y}{1 - x}, \frac{y}{1 - x}\right) \\ &= f(x) + (1 - x)^\alpha f\left(\frac{y}{1 - x}\right) \end{aligned}$$

which, by (3), implies that

$$f(x) + (1 - x)^\alpha f\left(\frac{y}{1 - x}\right) = f(y) + (1 - y)^\alpha f\left(\frac{x}{1 - y}\right) \tag{8}$$

holds on D° and on D, respectively. Functional Eq. (8) is called the *parametric fundamental equation of information*, (in case $\alpha = 1$ simply the *fundamental equation of information*).

Furthermore, in case $\alpha = 1$ and the domain D, its solutions $f : [0,1] \to \mathbb{R}$ satisfying the additional requirements $f(0) = f(1)$, $f\left(\frac{1}{2}\right) = 1$ are the *information functions*.

The role of the α-recursivity is very important since, with the aid of this property, we can determine the entire information measure from its initial element I_2. On the other hand, this idea shows the importance of Eq. (8), as well.

The appearance of the sum form functional equations in the characterization problems of information measures is more evident. Indeed, the (6) (α, n, m)- additivity and the (7) sum property immediately imply the functional equation

$$\sum_{i=1}^{n}\sum_{j=1}^{m} f(p_i q_j) = \sum_{i=1}^{n} f(p_i) + \sum_{j=1}^{m} f(q_j) + (2^{1-\alpha} - 1)\sum_{i=1}^{n} f(p_i)\sum_{j=1}^{m} f(q_j) \quad (9)$$

for the generating function f.

As we shall see in the sections below, the solutions of (8) and (in many cases) also of (9) can be expressed by the solutions of some well-known and well-discussed functional equations. In what follows we remind the reader some basic facts from this part of the theory of functional equations.

1.3 Prerequisites from the Theory of Functional Equations

All the results of this subsection can be found in the monographs Aczél [3] and Kuczma [52].

Let $A \subset \mathbb{R}$ be an arbitrary nonempty set and

$$\mathcal{A} = \left\{(x, y) \in \mathbb{R}^2 \mid x, y, x + y \in A\right\}.$$

A function $a : I \to \mathbb{R}$ is called *additive on* A, if for all $(x, y) \in \mathcal{A}$

$$a(x + y) = a(x) + a(y). \quad (10)$$

If $A = \mathbb{R}$, then the function a will be called simply *additive*. It is well known that the solutions of the equation above, under some mild regularity condition, are of the form

$$a(x) = cx \qquad (x \in I),$$

with a certain real constant c. For example, it is true that those additive functions which are bounded above or below on a set of positive Lebesgue measure have the form

$$a(x) = cx \qquad (x \in \mathbb{R})$$

with some $c \in \mathbb{R}$. It is also known, however, that there are additive functions the graph of which is dense in the plain. A great number of basic functional equations can easily be reduced to (10). In the following, we list some of them.

Let

$$\mathcal{M} = \left\{(x, y) \in \mathbb{R}^2 \mid x, y, xy \in A\right\}.$$

A function $m : A \to \mathbb{R}$ is called *multiplicative on* A, if for all $(x, y) \in \mathcal{M}$

$$m(xy) = m(x)m(y).$$

If $A = \mathbb{R}_+$ or $A = \mathbb{R}_{++}$ then the function m is called simply *multiplicative*.

Furthermore, we say that the function $\ell : A \to \mathbb{R}$ is *logarithmic on* A if for any $(x, y) \in \mathcal{M}$,

$$\ell(xy) = \ell(x) + \ell(y)$$

The functional equation

$$\varphi(xy) = x\varphi(y) + y\varphi(x) \tag{11}$$

has an important role in the following and it can easily be reduced to the functional equation of logarithmic functions by introducing the function $\ell(x) = \frac{\varphi(x)}{x}$. Finally, we will use functions $d : \mathbb{R} \to \mathbb{R}$ that are both additive and they are solutions of functional Eq. (11), that is,

$$d(xy) = xd(y) + yd(x)$$

is also satisfied for all $x, y \in \mathbb{R}$. This kind of functions are called *real derivations*. Their complete description can be found in Kuczma [52] from which it turns out the somewhat surprising fact that *there are nonidentically zero real derivations*. Of course, if a real derivation bounded from one side on a set of positive Lebesgue measure then it must be identically zero, otherwise its graph is dense in the plain.

In the subsequent sections it will occur that the equations introduced above are fulfilled only on restricted domains. Most of these cases it can be proved that the functions in question are the restrictions of some functions which satisfy the above equations on its natural domains. The results of this type are the so-called *extension theorems*, and the first classical ones are due to Aczél–Erdős [8] and Daróczy–Losonczi [20]. As a typical and important example, we cite the following extension theorem (see [20]).

Theorem 1 *Assume that the function* $a_0 : [0, 1] \to \mathbb{R}$ *is additive on* $[0, 1]$. *Then there exists a uniquely determined function* $a : \mathbb{R} \to \mathbb{R}$ *which is additive on* $\mathbb{R}$ *such that*

$$a_0(x) = a(x)$$

holds for all $x \in [0, 1]$.

Since all the other functional equations mentioned above in this subsection can be reduced to (10), we can easily get extension theorems for them as consequences, and their regular (say bounded on a set of positive Lebesgue measure) solutions can also be obtained easily. In particular, the typical regular (say bounded from one side on a set of positive Lebesque measure) solutions $\varphi : [0, +\infty[\to \mathbb{R}$ of (11) are of the form $\varphi(x) = cx \log_2(x)$ for all $0 \leq x \in \mathbb{R}$ and for some $c \in \mathbb{R}$.

2 Results on the Fundamental Equation of Information and on the Sum Form Equations

2.1 Information Functions

The first characterization theorem concerning the Shannon entropy (the case $\alpha = 1$) considered on Γ_n is due to Shannon himself, see [72]. The second one, which is more abstract and mathematically well-based, can be found in Khinchin [47]. In 1956, Faddeev succeed to reduce the system of axioms used by the two previous authors, see [27]. Faddeev assumed only symmetry, the normalization property, recursivity and that the function $f : [0, 1] \to \mathbb{R}$ defined by

$$f(x) = I_2(x, 1-x) \qquad (x, y \in [0, 1])$$

is continuous. After that, the regularity assumption in the result of Faddeev was replaced by weaker and weaker assumptions. For example, together with the above three algebraic properties, Tverberg [79] assumed (Lebesgue) integrability, Lee [55] measurability, Daróczy [14] continuity at zero ('small for small probabilities'), and Diderrich [24] boundedness on a set of positive measure, and they showed that the above properties determine uniquely the Shannon entropy. We mention here the result of Kendall [46] and Borges [12] who suppose monotonicity on the interval $[0, 1/2[$ and proved the same.

Concerning the characterization of the Shannon entropy, a 1969 paper of Daróczy [14] meant a breakthrough. He recognized that this characterization problem is equivalent with finding information functions that are identical with the Shannon information function S defined by

$$S(x) = x \log_2(x) + (1-x)\log_2(1-x) \qquad (x \in I).$$

The other important contribution was to find the general form of information functions (see [7]) which is the following.

Theorem 2 *A function $f : [0, 1] \to \mathbb{R}$ is an information function if, and only if,*

$$f(x) = \varphi(x) + \varphi(1-x) \quad (x \in [0, 1]) \tag{12}$$

with some function $\varphi : [0, +\infty[\to \mathbb{R}$ satisfying the functional equation

$$\varphi(xy) = x\varphi(y) + y\varphi(x) \qquad (x, y \in [0, +\infty[) \tag{13}$$

and $\varphi\left(\frac{1}{2}\right) = \frac{1}{2}$.

The proof of this theorem is based on some results and ideas of purely algebraic nature in Jessen, Karpf, and Thorup [43] on the cocycle equation

$$F(x+y, z) + F(x, y) = F(x, y+z) + F(y, z)$$

that is satisfied, provided that

$$F(x, y) = (x+y)f\left(\frac{y}{x+y}\right) \qquad (x, y \in \mathbb{R}_+, x+y \in \mathbb{R}_{++})$$

where f is an information function. Supposing that

$$f(x)+(1-x)f\left(\frac{y}{1-x}\right)=f(y)+(1-y)f\left(\frac{x}{1-y}\right)$$

holds only on the open domain $D^{\circ}=\{(x,y):x,y,x+y\in]0,1[\}$ for the unknown function $f:]0,1[\to\mathbb{R}$, Maksa and Ng [68] proved that $f(x)=\varphi(x)+\varphi(1-x)+ax$ for all $x\in]0,1[$ and for some function $\varphi:[0,+\infty[\to\mathbb{R}$ satisfying functional Eq. (13) and for some $a\in\mathbb{R}$.

Obviously, if $\varphi(x)=-x\log_2 x$, $x\in[0,+\infty[$ then $\varphi\left(\frac{1}{2}\right)=\frac{1}{2}$, φ satisfies (13), and (12) implies that $f=S$. However, as it was pointed out in Aczél [4], f does not determine φ unambiguously by (12). Indeed, if $d:\mathbb{R}\to\mathbb{R}$ is a real derivation, that is, d satisfies both functional equations

$$d(x+y)=d(x)+d(y) \quad \text{and} \quad d(xy)=xd(y)+yd(x)$$

then (12) and (13) hold also with $\varphi+d$ instead of φ, moreover $(\varphi+d)\left(\frac{1}{2}\right)=\frac{1}{2}$ is valid, as well. Thus, since there are nonidentically zero real derivations, the function φ in (12) does not inherit the regularity properties of f. So even for very regular f the function φ may be very irregular. This is the main difficulty in deriving the regular solutions from the general one.

The first successful attempt in this direction is due to Daróczy [17]. By his observation, (12) and (13) imply that

$$(x+y)f\left(\frac{y}{x+y}\right)=\varphi(x)+\varphi(y)-\varphi(x+y) \qquad (x,y\in\mathbb{R}_{+},x+y\in\mathbb{R}_{++}). \tag{14}$$

If f is (say) continuous then, for all fixed $y\in\mathbb{R}_{+}$ the difference functions $x\mapsto\varphi(x+y)-\varphi(x)$, $x\in\mathbb{R}_{+}$ so are. Therefore, by a theorem of de Bruijn [23], φ is a sum of a continuous and an additive function. It is not difficult to show that the additive function is a real derivation and the other summand is a continuous solution of (13).

This is the point at which the stability idea first appeared in the investigation. Namely, supposing that the information function f is bounded by a positive real number ε, (14) implies that

$$|\varphi(x)+\varphi(y)-\varphi(x+y)|\leq\varepsilon \qquad (x,y\in\mathbb{R}_{+},x+y\leq 1),$$

that is, the Cauchy difference of φ is bounded on a triangle. While de Bruijn type theorem is not true for this case we could apply the stability theory in Maksa [60] to determine the bounded information functions by giving a new and short proof of Diderrich's theorem published in [24].

At this point, we have to highlight the problem of nonnegative information functions. First of all, we emphasis that the requirment of the nonnegativity for an information function is very natural from information theoretical point of view, since $f(x)$ is the measure of information belonging to the probability distribution $\{x, 1-x\}$, $x \in [0,1]$. On the other hand, the one-sided boundedness is important also from theoretical point of view, as well. Indeed, the solutions of the Cauchy Eq. (10) bounded below or above on a set of positive Lebesgue measure are continuous linear functions. Therefore, it was natural to expect that something similar is true for the information functions that are bounded from one side (say nonnegative on $[0,1]$). Indeed, it was conjectured in Aczél–Daróczy [7] (supported by the partial result Daróczy–Kátai [19] by which the nonnegative information functions coincide with the Shannon one at the rational points of $[0,1]$) that the only nonnegative information function is the Shannon one. The following counter example in Daróczy–Maksa [21] however disproves this conjecture since, with any nonidentically zero real derivation d, the function f_0 defined by

$$f_0(x) = \begin{cases} S(x) + \dfrac{d(x)^2}{x(1-x)}, & \text{if} \quad x \in]0,1[\\ 0, & \text{if} \quad x \in \{0,1\} \end{cases}$$

is a nonnegative information function different from S. Of course, there are positive results, as well. For example, it is also proved in [21] that $S(x) \leq f(x)$ for all nonnegative information function f and for all $x \in [0,1]$. Another one is about the set $K(f) = \{x \in [0,1] | f(x) = S(x)\}$ which was introduced by Lawrence [54] and called the Shannon kernel of the nonnegative information function f. It is proved in Gselmann–Maksa [36] that $K(f)$ has the form $[0,1] \cap L_f$ where L_f is a subfield of $\mathbb{R}$ containing the square roots of its nonnegative elements. Furthermore, if K denotes the intersection of all Shannon kernels (belonging to nonnegative information functions) then all the elements of K are algebraic over $\mathbb{Q}$ and K contains all the algebraic elements of $[0,1]$ of degree at most 3. Our first open problem is related to these latter facts.

Open Problem 1 *Prove or disprove that all algebraic elements of the closed interval* $[0,1]$ *is contained by* K*, in other words any nonnegative information function coincides with* S *at the algebraic points of the closed unit interval.*

We remark that Lawrence's conjecture in [54] is affirmative.

The last sentences of this subsection are devoted to the case $\alpha \neq 1$ which is much simpler than the case $\alpha = 1$. Indeed, in [15], Daróczy determined all the solutions $f : [0,1] \to \mathbb{R}$ of (8) satisfying the additional requirements $f(0) = f(1)$, $f\left(\frac{1}{2}\right) = 1$. Thus he characterized the entropy of degree α on Γ_n by using purely algebraic properties: semisymmetry, normalization, and α-recursivity. Since then, these results have been extended to the open domain case, as well (see, e.g., the sections about the stability).

2.2 Sum Form Equations

As we have seen earlier, the sum form Eq. (9) is the consequence of the (α, n, m)-additivity and the sum property. In connection with the characterization properties discussed above, we should remark here the following implication: the sum property follows from the symmetry (2) and α-recursivity (5), as it is shown in [7].

In several characterization theorems for the entropy of degree α based on (α, n, m)-additivity and the sum property, an additional regularity condition was supposed for the generating function f and also on the parameters α, n, and m. We list some of the results of this type in chronological order.

We begin with the Shannon case $\alpha = 1$. Chaundy–McLeod [13] proved that, if $f : [0, 1] \to \mathbb{R}$ is continuous and

$$\sum_{i=1}^{n}\sum_{j=1}^{m} f(p_i q_j) = \sum_{i=1}^{n} f(p_i) + \sum_{j=1}^{m} f(q_j) \tag{15}$$

holds for all $(p_1, \ldots, p_n) \in \Gamma_n$, $(q_1, \ldots, q_m) \in \Gamma_m$ and for all $n \geq 2, m \geq 2$ then

$$f(x) = cx \log_2(x) \qquad (x \in [0, 1]) \tag{16}$$

with some $c \in \mathbb{R}$. The same was proved by Aczél and Daróczy [6] supposing that f is continuous and (15) holds for all $n = m \geq 2$. Daróczy [16] determined the measurable solutions f supposing that $n = 3, m = 2, f(1) = 0$. Daróczy and Járai [18] found the measurable solutions of (15) in the case $n = m = 2$ discovering solutions that are not solutions when $n \geq 3$ or $m \geq 3$. This was one of the starting point of developing the regularity theory of functional equations (see Járai [42]). In Maksa [61], the solutions bounded from on a set of positive Lebesgue measure of (15) were determined. These are the same as in the continuous case (see (16)) while it was also shown that the supposition of the one-sided boundedness does not lead to the same result. Counterexample can be given by real derivations (see Maksa [64]). Connected with these investigations the following problem is still open.

Open Problem 2 *Find the general solution of Eq.* (15) *for a fixed pair* (n, m), $n \geq 2, m \geq 2$, *particularly find all functions* $f : I \to \mathbb{R}$ *satisfying the functional equation*

$$\begin{aligned} f(xy) + f((1-x)y) + f(x(1-y)) + f((1-x)(1-y)) \\ = f(x) + f(1-x) + f(y) + f(1-y) \end{aligned}$$

for all $x, y \in I$.

A partial result can be found in Losonczi–Maksa [59].

As we have already mentioned, in the characterization theorems for the entropy of degree α based on (α, n, m)-additivity and the sum property, an additional regularity condition was supposed for the generating function f. Now we present here an exceptional case (see Maksa [64]) in which all the conditions refer to the information measure itself and there is no condition on the generating function. The stability idea appears again. Indeed, suppose that the information measure (I_n) is $(1, n, m)$-additive for some $n \geq 3$, $m \geq 2$, has the sum property with generating function $f : [0, 1] \to \mathbb{R}$ and I_3 is bounded by the real number K, that is,

$$|I_3(p_1, p_2, p_3)| \leq K \qquad ((p_1, p_2, p_3) \in \Gamma_3). \tag{17}$$

Let $x, y \in [0, 1]$ such that $x+y \leq 1$ and apply (17) to the probability distributions $(x, y, 1-x-y) \in \Gamma_3$ and then to $(x+y, 1-x-y, 0) \in \Gamma_3$, respectively to get that

$$|I_3(x, y, 1-x-y)| \leq K \text{ and } |I_3(x+y, 1-x-y, 0)| \leq K.$$

Therefore, because of the triangle inequality, for the generating function f, we have that

$$|f(x+y) - f(x) - f(y) + f(0)| \leq 2\,K,$$

that is, the stability inequality holds for the function $f - f(0)$ on a triangle. The details together with the consequences are in [64].

The brief history of the case $\alpha \neq 1$ follows. The continuous solutions, supposing that (9) holds for all $n \geq 2, m \geq 2$ were determined by Behara and Nath [11], Kannappan [45], and Mittal [70] independently of each other. They found that the continuous solutions either a sum of a continuous additive function and a constant or the sum of a continuous additive function and a continuous multiplicative function (power function). The same was proved by Losonczi [56] supposing that (9) holds for a fixed pair (n, m), $n \geq 3, m \geq 2$ and the generating function f in (9) is measurable. Contrary to the case $\alpha = 1$, in the case $\alpha \neq 1$ the general solution has been determined (see Losonczi-Maksa [59] and Maksa [63]) supposing that $n \geq 3$ and $m \geq 2$ are fixed. Characterization theorems for the entropy of degree α can easily be derived from these results (see [64]).

In the case $\alpha \neq 1$, with the definition $g(p) = p+(2^{1-\alpha}-1)f(p)$, $p \in I$, Eq. (9) can be reduced to equation

$$\sum_{i=1}^{n}\sum_{j=1}^{m} g(p_i q_j) = \sum_{i=1}^{n} g(p_i) \sum_{j=1}^{m} g(q_j). \tag{18}$$

The general solution of which is not known when $n = m = 2$. Therefore we formulate the following open problem.

Open Problem 3 *Find all functions $g : I \to \mathbb{R}$ satisfying the functional equation*

$$g(xy) + g((1-x)y) + g(x(1-y)) + g((1-x)(1-y)) = (g(x) + g(1-x))\,(g(y) + g(1-y))$$

for all $x, y \in I$.

A partial result is proved in Losonczi [57].

Further investigations related to sum form equations on open domain or for functions in several variables can be found among others in Losonczi [58] and in the survey paper Ebanks–Kannappan–Sahoo–Sander [26].

3 Stability Problems

During one of his talks, held at the University of Wisconsin S. Ulam posed several problems. One of these problems has became the cornerstone of the stability theory of functional equations, see [80]. Ulam's problem reads as follows.

Let $(G, \circ)$ be a group and $(H, *)$ be a metric group with the metric d. Let $\varepsilon \geq 0$ and $f : G \to \mathbb{H}$ be a function such that

$$d\left(f(x \circ y), f(x) * f(y)\right) \leq \varepsilon$$

holds for all $x, y \in G$. Is it true that there exist $\delta \geq 0$ and a function $g : G \to \mathbb{H}$ such that

$$g(x \circ y) = g(x) * g(y), \quad (x, y \in G)$$

and

$$d\left(f(x), g(x)\right) \leq \delta$$

holds for all $x \in G$?

This question was first answered in 1941 by D. H. Hyers by proving the following theorem, see [41].

Theorem 3 *Let $\varepsilon \geq 0$, X, Y be Banach spaces and $f : X \to Y$ be a function. Suppose that*

$$\|f(x + y) - f(x) - f(y)\| \leq \varepsilon$$

holds for all $x, y \in X$. Then, for all $x \in X$, the limit

$$a(x) = \lim_{n\to\infty} \frac{f(2^n x)}{2^n}$$

does exist, the function $a : X \to \mathbb{R}$ is additive on X, i.e.,

$$a(x + y) = a(x) + a(y)$$

holds for all $x, y \in X$, furthermore,

$$\|f(x) - a(x)\| \leq \varepsilon$$

is fulfilled for arbitrary $x \in X$. Additionally, the function $a : \mathbb{R} \to \mathbb{R}$ is uniquely determined by the above formula.

The above theorem briefly expresses the following. Assume that X, Y are Banach spaces and the function $f : X \to Y$ satisfies the additive Cauchy equation only "approximatively." Then there exists a unique additive function $a : X \to Y$ which is "close" to the function f. Since 1941 this result has been extended and generalized in several ways. Furthermore, Ulam's problem can obviously be raised concerning not only the Cauchy equation but also in connection with other equations, as well. For further result consult the monograph Hyers–Isac–Rassias [40].

For instance, the stability problem of the exponential Cauchy equation highlighted a new phenomenon, which is nowadays called *superstability*. In this case the so-called stability inequality implies that the function in question is either bounded or it is the exact solution of the functional equation in question, see Baker [10].

In this work, we will meet an other notion, namely the *hyperstability*. In this case, from the stability inequality, we get that the function in question can be nothing else than the exact solution of the functional equation in question, see, e.g., Maksa–Páles [69].

Since the above result of D. H. Hyers appeared, the stability theory of functional equations became a rapidly developing area. Presently, in the theory of stability there exist several methods, e.g. the Hyers' method (c.f. Forti [28]), the method of invariant means (see Székelyhidi [74, 75]), and the method that is based on separation theorems (see Badora–Ger–Páles [9]).

As we will see in the following subsections, in case of the functional equations, we will deal with, *none of the above methods will work*. More precisely, in some cases the method of invariant means is used. However, basically we have to develop new ideas to prove stability type theorems for the functional equations, we mentioned in the introduction. Concerning topic of invariant means, we offer the expository paper Day [22]. Although the only result needed from [22] is, that on every commutative semigroup there exist an invariant mean, that is, every commutative semigroup is *amenable*.

The aim of this chapter is to investigate the stability of some functional equations that appear in the theory of information. Firstly, we will investigate the above problem concerning the parametric fundamental equation of information. The main results and also the applications will be listed in the subsequent subsections. We will prove stability, superstability, and hyperstability according to the value of the parameter α. The results, we will present can be found in Gselmann [30, 31, 32, 34] and in Gselmann–Maksa [35].

Concerning the stability of the parametric fundamental equation of information, the first result was the stability of Eq. (6) on the set D, assuming that $1 \neq \alpha > 0$ (see Maksa [67]). Furthermore, the stability constant, got in that paper is much smaller than that of ours. However, the method, used in Maksa [67], does not work if $\alpha = 1$ or $\alpha \leq 0$ or if we consider the problem on the open domain.

After that, it was proved that Eq. (6) is stable in the sense of Hyers and Ulam on the set D° as well as on D, assuming that $\alpha \leq 0$ (see [35]). After that it turned out that this method is appropriate to prove superstability in case $1 \neq \alpha > 0$. This enabled us to give a unified proof for the stability problem of Eq. (6). Finally, using a different approach, in [30] it was showed that in case $\alpha < 0$, the parametric fundamental equation of information is hyperstable on D° as well as on D.

3.1 The Cases $\alpha = 0$ and $0 < \alpha \neq 1$

In this part of the chapter, we will investigate the stability of the parametric fundamental equation of information in case for the parameter α, $\alpha = 0$, or $0 < \alpha \neq 1$ holds. The method, we will use during the proofs were firstly developed for the case $\alpha < 0$. However, it turned out that this approach works in this case also. The results we will present here can be found in [31, 32] and also in [67].

Theorem 4 *Let $\alpha, \varepsilon \in \mathbb{R}$ be fixed, $1 \neq \alpha \geq 0, \varepsilon \geq 0$. Suppose that the function $f :]0,1[\to \mathbb{R}$ satisfies the inequality*

$$\left| f(x) + (1-x)^{\alpha} f\left(\frac{y}{1-x}\right) - f(y) - (1-y)^{\alpha} f\left(\frac{x}{1-y}\right) \right| \leq \varepsilon \tag{19}$$

for all $(x, y) \in D^{\circ}$. Then, in case $\alpha = 0$, there exists a logarithmic function $l :]0,1[\to \mathbb{R}$ and $c \in \mathbb{R}$ such that

$$|f(x) - [l(1-x) + c]| \leq K(\alpha)\varepsilon, \quad (x \in]0,1[) \tag{20}$$

furthermore, if $\alpha \notin \{0,1\}$, there exist $a, b \in \mathbb{R}$ such that

$$\left| f(x) - \left[ax^{\alpha} + b(1-x)^{\alpha} - b\right] \right| \leq K(\alpha)\varepsilon \tag{21}$$

holds for all $x \in]0,1[$, where

$$K(\alpha) = \left|2^{1-\alpha} - 1\right|^{-1} \left(3 + 12 \cdot 2^{\alpha} + \frac{32 \cdot 3^{\alpha+1}}{|2^{-\alpha} - 1|}\right).$$

Proof Define the function F on $\mathbb{R}^2_{++}$ by

$$F(u, v) = (u+v)^{\alpha} f\left(\frac{v}{u+v}\right). \tag{22}$$

Then

$$F(tu, tv) = t^{\alpha} F(u, v) \quad (t, u, v \in \mathbb{R}_{++}) \tag{23}$$

and

$$f(x) = F(1-x, x), \quad (x \in]0,1[) \tag{24}$$

furthermore, with the substitutions

$$x = \frac{w}{u+v+w}, \quad y = \frac{v}{u+v+w} \quad (u, v, w \in \mathbb{R}_{++})$$

inequality (19) implies that

$$\left| f\left(\frac{w}{u+v+w}\right) + \frac{(u+v)^\alpha}{(u+v+w)^\alpha} f\left(\frac{v}{u+v}\right) \right.$$
$$\left. -f\left(\frac{v}{u+v+w}\right) - \frac{(u+w)^\alpha}{(u+v+w)^\alpha} f\left(\frac{w}{u+w}\right) \right| \leq \varepsilon \tag{25}$$

whence, by (22)

$$|F(u+v,w) + F(u,v) - F(u+w,v) - F(u,w)| \leq \varepsilon(u+v+w)^\alpha \tag{26}$$

follows for all $u, v, w \in \mathbb{R}_{++}$.

In the next step, we define the functions g and G on $\mathbb{R}_{++}$ and on $\mathbb{R}^2_{++}$, respectively by

$$g(u) = F(u,1) - F(1,u) \tag{27}$$

and

$$G(u,v) = F(u,v) + g(v). \tag{28}$$

We will show that

$$|G(u,v) - G(v,u)| \leq 3\varepsilon(u+v+1)^\alpha. \quad (u,v \in \mathbb{R}_{++}) \tag{29}$$

Indeed, with the substitution $w = 1$, inequality (26) implies that

$$|F(u+v,1) + F(u,v) - F(u+1,v) - F(u,1)| \leq \varepsilon(u+v+1)^\alpha. \tag{30}$$

Interchanging u and v, it follows from (30) that

$$|-F(u+v,1) - F(v,u) + F(v+1,u) - F(v,1)| \leq \varepsilon(u+v+1)^\alpha$$
$$(u,v \in \mathbb{R}_{++}).$$

This inequality, together with (30) and the triangle inequality imply that

$$|F(u,v) - F(v,u) - F(u+1,v) - F(u,1) + F(v+1,u) + F(v,1)|$$
$$\leq 2\varepsilon(u+v+1)^\alpha \tag{31}$$

holds for all $u, v \in \mathbb{R}_{++}$. On the other hand, with $u = 1$, we get from (26) that

$$|F(1+v,w) + F(1,v) - F(1+w,w) - F(1,w)| \leq \varepsilon(1+v+w)^\alpha.$$

Replacing here v by u and w by v, respectively, we have that

$$|F(u+1,v) + F(1,u) - F(v+1,u) - F(1,v)| \leq \varepsilon(u+v+1)^\alpha$$
$$(u,v \in \mathbb{R}_{++}).$$

Again, by the triangle inequality and the definitions (27) and (28), (31) and the last inequality imply (29).

In what follows we will investigate the function g. At this point of the proof, we have to distinguish two cases.

Case I. ($\alpha = 0$). In this case, we will show that there exists a logarithmic function $l : \mathbb{R}_{++} \to \mathbb{R}$ such that

$$|g(u) - l(u)| \leq 6\varepsilon$$

for all $u \in \mathbb{R}_{++}$. Indeed, (29) yields in this case that

$$|G(u, v) - G(v, u)| \leq 3\varepsilon. \quad (u, v \in \mathbb{R}_{++})$$

Due to (23) and (28), we obtain that

$$G(tu, tv) = F(tu, tv) + g(tv) = F(u, v) + g(tv) = G(u, v) - g(v) + g(tv)$$

that is,

$$G(tu, tv) - G(u, v) = g(tv) - g(v), \quad (t, u, v \in \mathbb{R}_{++})$$

therefore

$$\begin{aligned} |g(tv) - g(v) + g(u) - g(tu)| &= |G(tu, tv) - G(u, v) - G(tv, tu) + G(v, u)| \\ &\leq |G(tu, tv) - G(tv, tu)| + |G(v, u) - G(u, v)| \leq 6\varepsilon \end{aligned} \tag{32}$$

for all $t, u, v \in \mathbb{R}_{++}$. Now (32) with the substitution $u = 1$ implies that

$$|g(tv) - g(v) - g(t)| \leq 6\varepsilon$$

holds for all $t, v \in \mathbb{R}_{++}$, since obviously $g(1) = 0$. This means that the function g is approximately logarithmic on $\mathbb{R}_{++}$. Thus, there exists a logarithmic function $l : \mathbb{R}_{++} \to \mathbb{R}$ such that

$$|g(u) - l(u)| \leq 6\varepsilon$$

holds for all $u \in \mathbb{R}_{++}$.

Furthermore,

$$\begin{aligned} &|f(x) - l(1-x) - (f(1-x) - l(x))| \\ &= |F(1-x, x) - l(1-x) - F(x, 1-x) + l(x)| \\ &= |F(1-x, x) + g(x) - g(x) - l(1-x) \\ &\quad -F(x, 1-x) + g(1-x) - g(1-x) + l(x)\,| \\ &\leq |F(1-x, x) + g(x) - (F(x, 1-x) + g(1-x))| \\ &\quad + |g(1-x) - l(1-x)| + |l(x) - g(x)| \\ &= |G(1-x, x) - G(x, 1-x)| + |g(1-x) \end{aligned}$$

$$-l(1-x)| + |l(x) - g(x)|$$
$$\leq 3\varepsilon + 6\varepsilon + 6\varepsilon = 15\varepsilon \tag{33}$$

Define the functions f_0 and F_0 on $]0,1[$ and on $]0,1[^2$, respectively, by

$$f_0(x) = f(x) - l(1-x)$$

and

$$F_0(p,q) = f_0(p) + f_0(q) - f_0(pq) - f_0\left(\frac{1-p}{1-pq}\right)$$

Due to (3.1)

$$|f_0(x) - f_0(1-x)| \leq 15\varepsilon \tag{34}$$

holds for all $x \in\,]0,1[$. Furthermore, with the substitutions $x = 1-p$, $y = pq$ ($p,q \in\,]0,1[$) inequality (19) implies, that

$$\left| f_0(1-p) + f_0(q) - f_0(pq) - f_0\left(\frac{1-p}{1-pq}\right)\right| \leq \varepsilon \tag{35}$$

is fulfilled for all $p,q \in\,]0,1[$. Inequalities (34) and (35) and the triangle inequality imply that

$$|F_0(p,q)| \leq 16\varepsilon \tag{36}$$

for all $p,q \in\,]0,1[$. An easy calculation shows that

$$f_0(p) - f_0(q)$$

$$0 = F_0(q,p) - F_0(p,q) + F_0\left(\frac{1-p}{1-pq}, p\right) - f_0\left(1 - \frac{1-p}{1-pq}\right) + f_0\left(\frac{1-p}{1-pq}\right)$$

therefore,

$$\begin{aligned} |f_0(p) - f_0(q)| & \\ &\leq |F_0(q,p)| + |F_0(p,q)| + \left|F_0\left(\frac{1-p}{1-pq}, p\right)\right| \\ &\quad + \left| f_0\left(1 - \frac{1-p}{1-pq}\right) - f_0\left(\frac{1-p}{1-pq}\right)\right| \\ &\leq 3 \cdot 16\varepsilon + 15\varepsilon = 63\varepsilon \end{aligned} \tag{37}$$

holds for all $p,q \in\,]0,1[$. With the substitution $q = \frac{1}{2}$ inequality (37) implies that

$$\left| f_0(p) - f_0\left(\frac{1}{2}\right)\right| \leq 63\varepsilon. \quad (p \in\,]0,1[)$$

Using the definition of the function f_0, we obtain that inequality

$$|f(x) - l(1-x) - c| \leq 63\varepsilon$$

is satisfied for all $x \in\,]0,1[$, where $c = f_0\left(\frac{1}{2}\right)$. Hence, inequality (20) holds, indeed.

Case II. ($1 \neq \alpha \geq 0$). Finally, we will prove that there exists $c \in \mathbb{R}$ such that

$$|g(x) - c(x^\alpha - 1)| \leq \frac{4 \cdot 3^{\alpha+1}\varepsilon}{|2^{-\alpha} - 1|}$$

holds for all $x \in]0, 1[$.

Due to inequalities (22) and (27),

$$\begin{aligned} G(tu, tv) = F(tu, tv) + g(tv) &= t^\alpha F(u, v) + g(tv) \\ &= t^\alpha G(u, v) - t^\alpha g(v) + g(tv), \end{aligned}$$

that is,

$$G(tu, tv) - t^\alpha G(u, v) = g(tv) - t^\alpha g(v)$$

holds for all $t, v \in \mathbb{R}_{++}$. Therefore,

$$\begin{aligned} &|g(tv) - t^\alpha g(v) + t^\alpha g(u) - g(tu)| \\ &\quad = |G(tu, tv) - G(u, v) - G(tv, tu) + G(v, u)| \\ &\quad \leq |G(tu, tv) - G(tv, tu)| + |G(u, v) - G(v, u)| \\ &\quad \leq 3\varepsilon(t(u + v) + 1)^\alpha + 3\varepsilon(u + v + 1)^\alpha \end{aligned} \tag{38}$$

holds for all $t, u, v \in \mathbb{R}_{++}$, where we used (19). With the substitution $u = 1$, (38) implies that

$$|g(tv) - t^\alpha g(v) - g(t)| \leq 3\varepsilon(t(v + 1) + 1)^\alpha + 3\varepsilon(v + 2)^\alpha \quad (t, v \in \mathbb{R}_{++}) \tag{39}$$

Interchanging t and v in (39), we obtain that

$$|g(tv) - v^\alpha g(t) - g(v)| \leq 3\varepsilon(v(t + 1) + 1)^\alpha + 3\varepsilon(t + 2)^\alpha \quad (t, v \in \mathbb{R}_{++}) \tag{40}$$

Inequalities (39), (40), and the triangle inequality imply that

$$|t^\alpha g(v) + g(t) - v^\alpha g(t) - g(v)| \leq B(t, v) \tag{41}$$

is fulfilled for all $t, v \in \mathbb{R}_{++}$, where

$$B(t, v) = 3\varepsilon(t(v + 1) + 1)^\alpha + 3\varepsilon(v + 2)^\alpha + 3\varepsilon(v(t + 1) + 1)^\alpha + 3\varepsilon(t + 2)^\alpha.$$

With the substitution $t = \frac{1}{2}$ and with the definition $c = \frac{g\left(\frac{1}{2}\right)}{2^{-\alpha} - 1}$, we obtain

$$|g(v) - c(v^\alpha - 1)| \leq \frac{B\left(\frac{1}{2}, v\right)}{|2^{-\alpha} - 1|} \tag{42}$$

for all $v \in \mathbb{R}_{++}$.

Let us observe that

$$|B(t,v)| \leq 4 \cdot 3^{\alpha+1}\varepsilon$$

holds, if $t, v \in]0,1[$. Thus

$$|g(v) - c(v^{\alpha} - 1)| \leq \frac{B\left(\frac{1}{2}, v\right)}{|2^{-\alpha} - 1|} \leq \frac{4 \cdot 3^{\alpha+1}\varepsilon}{|2^{-\alpha} - 1|} \tag{43}$$

for all $v \in]0,1[$. Therefore (24), (28), (29), (43), and the triangle inequality imply that

$$\begin{aligned}
|f(x) - c(1-x)^{\alpha} + c - (f(1-x) - cx^{\alpha} + c)\,| &\\
&= |F(1-x,x) - c(1-x)^{\alpha} + c - (F(x,1-x) - cx^{\alpha} + c)\,| \\
&\leq |F(1-x,x) + g(x) - F(x,1-x) - g(1-x)| \\
&\quad + |g(x) - c(x^{\alpha} - 1)| + |g(1-x) - c((1-x)^{\alpha} - 1)| \\
&= |G(1-x,x) - G(x,1-x)| \\
&\quad + |g(x) - c(x^{\alpha} - 1)| + |g(1-x) - c((1-x)^{\alpha} - 1)| \\
&\leq 3 \cdot 2^{\alpha}\varepsilon + \frac{8 \cdot 3^{\alpha+1}\varepsilon}{|2^{-\alpha} - 1|}
\end{aligned} \tag{44}$$

holds for all $x \in]0,1[$.

As in the previous cases, we define the functions f_0 and F_0 on $]0,1[$ and on $]0,1[^2$ by

$$f_0(x) = f(x) - c(1-x)^{\alpha} \tag{45}$$

and

$$F_0(p,q) = f_0(p) + p^{\alpha} f_0(q) - f_0(pq) - (1-pq)^{\alpha} f_0\left(\frac{1-p}{1-pq}\right), \tag{46}$$

respectively. Then (19), (44), and (45) imply that

$$\left| f_0(x) + (1-x)^{\alpha} f_0\left(\frac{y}{1-x}\right) - f_0(y) - (1-y)^{\alpha} f_0\left(\frac{x}{1-y}\right) \right| \leq \varepsilon \tag{47}$$

for all $(x,y) \in D^{\circ}$ and

$$|f_0(x) - f_0(1-x)| \leq 3 \cdot 2^{\alpha}\varepsilon + \frac{8 \cdot 3^{\alpha+1}\varepsilon}{|2^{-\alpha} - 1|}. \quad (x \in]0,1[) \tag{48}$$

Furthermore, with the substitutions $x = 1-p$, $y = pq$ $(p,q \in]0,1[)$, (47) implies that

$$\left| f_0(1-p) + p^{\alpha} f_0(q) - f_0(pq) - (1-pq)^{\alpha} f_0\left(\frac{1-p}{1-pq}\right) \right| \leq \varepsilon \tag{49}$$

holds for all $p, q \in]0, 1[$. Thus (48), (49), and the triangle inequality imply that

$$|F_0(p,q)| \leq \varepsilon + 3 \cdot 2^\alpha \varepsilon + \frac{8 \cdot 3^{\alpha+1}\varepsilon}{|2^{-\alpha} - 1|}. \quad (x \in]0, 1[)$$

Similarly to the previous case, it is easy to see that the identity

$$\begin{aligned} f_0(p)\left[q^\alpha + (1-q)^\alpha - 1\right] - f_0(q)\left[p^\alpha + (1-p)^\alpha - 1\right] \\ = F_0(q,p) - F_0(p,q) \\ -(1-pq)^\alpha \left[F_0\left(\frac{1-q}{1-pq}, p\right) + f_0\left(1 - \frac{1-p}{1-pq}\right) - f_0\left(\frac{1-p}{1-pq}\right)\right] \end{aligned} \tag{50}$$

is satisfied for all $p, q \in]0, 1[$. Therefore

$$\begin{aligned} &\left| f_0(p) - \frac{f_0(q)}{q^\alpha + (1-q)^\alpha - 1}\left[p^\alpha + (1-p)^\alpha - 1\right]\right| \\ &\leq |q^\alpha + (1-q)^\alpha - 1|^{-1} \times \\ &\times \left(3\left(\varepsilon + 3 \cdot 2^\alpha \varepsilon + \frac{8 \cdot 3^{\alpha+1}\varepsilon}{|2^{-\alpha} - 1|}\right) + 3 \cdot 2^\alpha \varepsilon + \frac{8 \cdot 3^{\alpha+1}\varepsilon}{|2^{-\alpha} - 1|}\right) \end{aligned}$$

for all $p, q \in]0, 1[$. In view of (45), with $q = \frac{1}{2}$ with the definitions

$$a = f_0\left(\frac{1}{2}\right)\left(2^{1-\alpha} - 1\right)^{-1} \quad \text{and} \quad b = a + c,$$

this inequality implies that

$$\left|f(p) - \left[ap^\alpha + b(1-p)^\alpha - b\right]\right| \leq K(\alpha)\varepsilon \tag{51}$$

holds for all $p \in]0, 1[$, where

$$K(\alpha) = \left|2^{1-\alpha} - 1\right|^{-1}\left(3 + 12 \cdot 2^\alpha + \frac{32 \cdot 3^{\alpha+1}}{|2^{-\alpha} - 1|}\right),$$

which had to be proved.

In the following theorem, we shall prove that the parametric fundamental equation of information is stable not only on D° but also on D. During the proof of this theorem, the following function will be needed. For all $1 \neq \alpha > 0$, we define the function $T(\alpha)$ by

$$T(\alpha) = 3 \cdot 2^\alpha + \frac{8 \cdot 3^{\alpha+1}}{|2^{-\alpha} - 1|}.$$

Furthermore, the following relationship is fulfilled between $K(\alpha)$ and $T(\alpha)$

$$K(\alpha) = \frac{4\,T(\alpha) + 3}{|2^{1-\alpha} - 1|}$$

for all $1 \neq \alpha > 0$.

Theorem 5 *Let $\alpha, \varepsilon \in \mathbb{R}$ be fixed, $0 \leq \alpha \neq 1$, $\varepsilon \geq 0$. Suppose that the function $f : [0,1] \to \mathbb{R}$ satisfies inequality (19) for all $(x, y) \in D$. Then, in case $\alpha \neq 0$ there exist $a, b \in \mathbb{R}$ such that the function h_1 defined on $[0,1]$ by*

$$h_1(x) = \begin{cases} 0, & \text{if } x = 0 \\ ax^\alpha + b(1-x)^\alpha - b, & \text{if } x \in]0,1[\\ a - b, & \text{if } x = 1 \end{cases}$$

is a solution of (6) on D and

$$|f(x) - h_1(x)| \leq \max\{K(\alpha), T(\alpha) + 1\}\varepsilon \quad (x \in [0,1]) \tag{52}$$

holds. In case $\alpha = 0$, there exists $c \in \mathbb{R}$ such that the function h_2 defined on $[0,1]$ by

$$h_2(x) = \begin{cases} f(0), & \text{if } x = 0 \\ c, & \text{if } x \in]0,1[\\ f(1), & \text{if } x = 1 \end{cases}$$

is a solution of (6) on D and

$$|f(x) - h_2(x)| \leq K(\alpha)\varepsilon. \quad (x \in [0,1]) \tag{53}$$

is fulfilled.

Proof An easy calculation shows that the functions h_1 and h_2 are the solutions of Eq. (6) on D in case $\alpha \neq 0$ and $\alpha = 0$, respectively.

At first, we deal with the case $\alpha > 0$. Substituting $x = 0$ into (19) and with $y \to 0$, we obtain that

$$|f(0)| \leq \varepsilon \leq K(\alpha)\varepsilon,$$

that is, (52) holds for $x = 0$. If $x \in]0,1[$, then inequality (52) follows immediately from Theorem (4). Furthermore, with the substitution $y = 1 - x$ ($x \in]0,1[$) inequality (19) implies that

$$|f(x) + (1-x)^\alpha f(1) - f(1-x) - x^\alpha f(1)| \leq \varepsilon. \quad (x \in]0,1[)$$

From the proof of Theorem 4 (see definition (45)), it is known that

$$f(x) = f_0(x) + c(1-x)^\alpha, \quad (x \in]0,1[)$$

therefore, the last inequality yields that

$$|f_0(x) - f_0(1-x) + c(1-x)^\alpha - cx^\alpha + (1-x)^\alpha f(1) - x^\alpha f(1)| \leq \varepsilon \tag{54}$$

holds for all $x \in]0, 1[$. Whereas

$$|f_0(x) - f_0(1-x)| \leq T(\alpha). \quad (x \in]0, 1[)$$

Thus after rearranging (54), we get that

$$|f_0(x) - f_0(1-x) - [c + f(1)][x^\alpha - (1-x)^\alpha]| \leq \varepsilon, \quad (x \in]0, 1[)$$

that is,

$$||f_0(x) - f_0(1-x)| - |c + f(1)| \cdot |x^\alpha - (1-x)^\alpha|| \leq \varepsilon$$

holds for all $x \in]0, 1[$. Therefore

$$|c + f(1)| \cdot |x^\alpha - (1-x)^\alpha| \leq (T(\alpha) + 1)\varepsilon$$

for all $x \in]0, 1[$. Taking the limit $x \to 0+$, we obtain that

$$|c + f(1)| \leq (T(\alpha) + 1)\varepsilon.$$

However, in the proof of Theorem 4, we used the definition $c = b - a$, thus

$$|f(1) - (a - b)| \leq (T(\alpha) + 1)\varepsilon,$$

so (52) holds, indeed.

Finally, we investigate the case $\alpha = 0$. If $x = 0$ or $x = 1$, then (53) trivially holds, since

$$|f(0) - h_2(0)| = |f(0) - f(0)| = 0 \leq K(\alpha)\varepsilon$$

and

$$|f(1) - h_2(1)| = |f(1) - f(1)| = 0 \leq K(\alpha)\varepsilon.$$

Let now $x \in]0, 1[$ and $y = 1 - x$ in (19), then we obtain that

$$|f(x) - f(1-x)| \leq \varepsilon, \quad (x \in]0, 1[) \tag{55}$$

if fulfilled for all $x \in]0, 1[$.

Due to Theorem 4, there exists a logarithmic function $l :]0, 1[\to \mathbb{R}$ and $c \in \mathbb{R}$ such that

$$|f(x) - l(1-x) - c| \leq 63\varepsilon \tag{56}$$

holds for all $x \in]0, 1[$. Hence it is enough to prove that the function l is identically zero on $]0, 1[$. Indeed, due to (55) and (56)

$$\begin{aligned}
&|l(1-x) - l(x)| \\
&\quad = |l(1-x) - f(1-x) + f(1-x) + c - l(x) + f(x) - f(x) - c|
\end{aligned}$$

$$\leq |l(1-x)+c-f(x)| + |f(1-x)-l(x)-c| + |f(x)-f(1-x)|$$
$$\leq 127\varepsilon \qquad (57)$$

holds for all $x \in]0,1[$. Since the function l is uniquely extendable to $\mathbb{R}_{++}$, with the substitution $x = \frac{p}{p+q}$ $(p,q \in \mathbb{R})$, we get that

$$|l(p)-l(q)| \leq 127\varepsilon, \quad (p,q \in \mathbb{R}_{++})$$

where we used the fact that l is logarithmic, as well. This last inequality, with the substitution $q=1$ implies that

$$|l(p)| \leq 127\varepsilon$$

holds for all $p \in \mathbb{R}_{++}$, since $l(1)=0$. Thus l is bounded on $\mathbb{R}_{++}$. However, the only bounded, logarithmic function on $\mathbb{R}_{++}$ is the identically zero function. Therefore,

$$|f(x)-c| \leq 63\varepsilon$$

holds for all $x \in]0,1[$, i.e., (53) is proved.

Since

$$\lim_{\alpha \to 1} K(\alpha) = +\infty,$$

our method is inappropriate if $\alpha = 1$. Hence we cannot prove stability concerning the fundamental equation of information neither on the set D° nor on D.

The stability problem for the fundamental equation of information was raised by L. Székelyhidi (see 38. Problem in [76]) and to the best of the authors' knowledge, it is still open. Therefore, we also can formulate the following.

Open Problem 4 *Let $\varepsilon \geq 0$ be arbitrary and $f :]0,1[\to \mathbb{R}$ be a function. Suppose that*

$$\left| f(x) + (1-x)f\left(\frac{y}{1-x}\right) - f(y) - (1-y)f\left(\frac{x}{1-y}\right) \right| \leq \varepsilon$$

holds for all $(x,y) \in D^{\circ}$. Is it true that in this case there exists a solution of the fundamental equation of information $h :]0,1[\to \mathbb{R}$ and a constant $K(\varepsilon) \in \mathbb{R}$ depending only on ε such that

$$|f(x) - h(x)| \leq K(\varepsilon)$$

is fulfilled for any $x \in]0,1[$?

Concerning this problem, we remark that for the system of recursive, 3-semi-symmetric information measures, some partial results are known, see Morando [71].

Applying Theorem 4 we can prove the stability of a system of functional equations that characterizes the α-recursive, 3-semi-symmetric information measures.

Theorem 6 *Let $n \geq 2$ be a fixed positive integer and (I_n) be the sequence of functions $I_n : \Gamma_n^\circ \to \mathbb{R}$ and suppose that there exist a sequence (ε_n) of nonnegative real numbers and a real number $0 \leq \alpha \neq 1$ such that*

$$|I_n(p_1, \dots, p_n) - I_{n-1}(p_1 + p_2, p_3, \dots, p_n)$$
$$-(p_1 + p_2)^\alpha I_2 \left(\frac{p_1}{p_1 + p_2}, \frac{p_2}{p_1 + p_2}\right)\Bigg| \leq \varepsilon_{n-1} \tag{58}$$

for all $n \geq 3$ and $(p_1, \dots, p_n) \in \Gamma_n^\circ$, and

$$|I_3(p_1, p_2, p_3) - I_3(p_1, p_3, p_2)| \leq \varepsilon_1 \tag{59}$$

holds on Γ_n°. Then, in case $\alpha = 0$ there exists a logarithmic function $l :]0, 1[\to \mathbb{R}$ and $c \in \mathbb{R}$ such that

$$\left| I_n(p_1, \dots, p_n) - \left[c H_n^0(p_1, \dots, p_n) + l(p_1)\right] \right|$$
$$\leq \sum_{k=2}^{n-1} \varepsilon_k + (n-1) K(\alpha) (2\varepsilon_2 + \varepsilon_1) \tag{60}$$

for all $n \geq 2$ and $(p_1, \dots, p_n) \in \Gamma_n^\circ$. Furthermore, if $\alpha > 0$ then there exist $c, d \in \mathbb{R}$ such that

$$\left| I_n(p_1, \dots, p_n) - \left[c H_n^\alpha(p_1, \dots, p_n) + d(p_1^\alpha - 1)\right] \right|$$
$$\leq \sum_{k=2}^{n-1} \varepsilon_k + (n-1) K(\alpha)(2\varepsilon_2 + \varepsilon_1) \tag{61}$$

holds for all $n \geq 2$ and $(p_1, \dots, p_n) \in \Gamma_n^\circ$, where the convention $\sum_{k=2}^{1} \varepsilon_k = 0$ is adopted.

Proof Similarly as in the proof of Theorem 6, due to (58) and (59), it can be proved that, for the function f defined on $]0, 1[$ by $f(x) = I_2(1 - x, x)$ we get that

$$\left| f(x) + (1-x)^\alpha f\left(\frac{y}{1-x}\right) - f(y) - (1-y)^\alpha f\left(\frac{x}{1-y}\right) \right| \leq 2\varepsilon_2 + \varepsilon_1$$

for all $(x, y) \in D^\circ$, i.e., (19) holds with $\varepsilon = 2\varepsilon_2 + \varepsilon_1$. Therefore, applying Theorem 6 we obtain (20) and (21), respectively, with some $a, b, c \in \mathbb{R}$ and a logarithmic function $l :]0, 1[\to \mathbb{R}$ and $\varepsilon = 2\varepsilon_2 + \varepsilon_1$, i.e.,

$$|I_2(1-x, x) - (ax^\alpha + b(1-x)^\alpha - b)| \leq K(\alpha)(2\varepsilon_2 + \varepsilon_1), \quad (x \in]0, 1[)$$

in case $\alpha \neq 0$, and

$$|I_2(1-x, x) - (l(1-x) + c)| \leq K(\alpha)(2\varepsilon_2 + \varepsilon_1) \quad (x \in]0, 1[)$$

in case $\alpha = 0$.

Therefore, (60) holds with $c = (2^{1-\alpha} - 1)a$, $d = b - a$ in case $\alpha > 0$ and (61) holds in case $\alpha = 0$, respectively, for $n = 2$.

We continue the proof by induction on n. Suppose that (60) and (61) hold, respectively, and for the sake of brevity, introduce the notation

$$J_n(p_1, \dots, p_n) = \begin{cases} cH_n^\alpha(p_1, \dots, p_n), & \text{if } \alpha \neq 0 \\ cH_n^0(p_1, \dots, p_n) + l(p_1), & \text{if } \alpha = 0 \end{cases}$$

for all $n \geq 2$, $(p_1, \dots, p_n) \in \Gamma_n^\circ$. It can easily be seen that (60) and (61) hold on Γ_n° for J_n instead of I_n $(n \geq 3)$ with $\varepsilon_n = 0$ $(n \geq 2)$.

Therefore, if $\alpha = 0$, (58) (with $n + 1$ instead of n), (60) with $n = 2$ and the induction hypothesis (applying to $(p_1 + p_2, \dots, p_{n+1})$ instead of $(p_1, \dots, p_n)$) imply that

$$\begin{aligned} &\left| I_{n+1}(p_1, \dots, p_{n+1}) - J_{n+1}(p_1, \dots, p_{n+1}) \right| \\ &\leq \varepsilon_n + \sum_{k=2}^{n-1} \varepsilon_k + K(\alpha)(n-1)(2\varepsilon_2 + \varepsilon_1) + K(\alpha)(2\varepsilon_2 + \varepsilon_1) \\ &= \sum_{k=2}^{n} \varepsilon_k + K(\alpha)n(2\varepsilon_2 + \varepsilon_1). \end{aligned}$$

This yields that (60) holds for $n + 1$ instead of n.

Furthermore, if $\alpha > 0$, then (58) (with $n + 1$ instead of n), (61) with $n = 2$ and the induction hypothesis (applying to $(p_1 + p_2, \dots, p_{n+1})$ instead of $(p_1, \dots, p_n)$) imply that

$$\begin{aligned} &\left| I_{n+1}(p_1, \dots, p_{n+1}) - J_{n+1}(p_1, \dots, p_{n+1}) \right| \\ &\leq \varepsilon_n + \sum_{k=2}^{n-1} \varepsilon_k + K(\alpha)(n-1)(2\varepsilon_2 + \varepsilon_1) + K(\alpha)(2\varepsilon_2 + \varepsilon_1) \\ &= \sum_{k=2}^{n} \varepsilon_k + K(\alpha)n(2\varepsilon_2 + \varepsilon_1), \end{aligned}$$

that is, (61) holds for $n + 1$ instead of n.

3.2 *The Case* $\alpha < 0$

At this part of the chapter, we will turn to investigate the case $\alpha < 0$. Here it will be proved for the negative parameters, the parametric fundamental equation of information is *hyperstable* on D° as well as on D. As an application of these results, we will deduce that the system of α-recursive, 3-semi-symmetric information measures is stable.

Theorem 7 *Let $\alpha, \varepsilon \in \mathbb{R}$, $\alpha < 0$, $\varepsilon \geq 0$ and $f :]0, 1[\to \mathbb{R}$ be a function. Assume that*

$$\left| f(x) + (1-x)^{\alpha} f\left(\frac{y}{1-x}\right) - f(y) - (1-y)^{\alpha} f\left(\frac{x}{1-y}\right) \right| \leq \varepsilon \tag{62}$$

holds for all $(x, y) \in D^{\circ}$. Then, and only then, there exist $c, d \in \mathbb{R}$ such that

$$f(x) = cx^{\alpha} + d(1-x)^{\alpha} - d \tag{63}$$

for all $x \in]0, 1[$.

Proof It is easy to see that for the function f is given by formula (63) functional equation

$$f(x) + (1-x)^{\alpha} f\left(\frac{y}{1-x}\right) = f(y) + (1-y)^{\alpha} f\left(\frac{x}{1-y}\right)$$

holds for all $(x, y) \in D^{\circ}$. Thus, inequality (62) is also satisfied with arbitrary $\varepsilon \geq 0$. Therefore, it is enough to prove the converse direction.

Define the function $G : D^{\circ} \to \mathbb{R}$ by

$$G(x, y) = f(x) + (1-x)^{\alpha} f\left(\frac{y}{1-x}\right) - f(x+y). \quad ((x, y) \in D^{\circ}) \tag{64}$$

Then inequality (62) immediately implies that

$$|G(x, y) - G(y, x)| \leq \varepsilon \tag{65}$$

for all $(x, y) \in D^{\circ}$.

Let $(x, y, z) \in D_3^{\circ}$, then due to the definition of the function G,

$$G(x+y, z) = f(x+y) + (1-(x+y))^{\alpha} f\left(\frac{z}{1-(x+y)}\right) - f(x+y+z),$$

$$G(x, y+z) = f(x) + (1-x)^{\alpha} f\left(\frac{y+z}{1-x}\right) - f(x+y+z)$$

and

$$\begin{aligned}(1-x)^{\alpha} G\left(\frac{y}{1-x}, \frac{z}{1-x}\right) \\ =(1-x)^{\alpha}\left[f\left(\frac{y}{1-x}\right) + \left(1 - \frac{y}{1-x}\right)^{\alpha} f\left(\frac{\frac{z}{1-x}}{1-\frac{y}{1-x}}\right) - f\left(\frac{y+z}{1-x}\right)\right],\end{aligned}$$

therefore,

$$G(x,y) + G(x+y,z) = G(x,y+z) + (1-x)^{\alpha} G\left(\frac{y}{1-x}, \frac{z}{1-x}\right) \tag{66}$$

holds on D_3°, where we used the identity

$$\frac{z}{1-(x+y)} = \frac{\frac{z}{1-x}}{1-\frac{y}{1-x}}$$

also.

In what follows, we will show that the function G is α-homogeneous. Indeed, interchanging x and y in (66), we get

$$G(y,x) + G(x+y,z)$$
$$= G(y,x+z) + (1-y)^{\alpha} G\left(\frac{x}{1-y}, \frac{z}{1-y}\right). \quad ((x,y,z) \in D_3^\circ)$$

Furthermore, Eq. (66) with the substitution

$$(x,y,z) = (y,z,x)$$

yields that

$$G(y,z) + G(y+z,x) = G(y,x+z) + (1-y)^{\alpha} G\left(\frac{z}{1-y}, \frac{x}{1-y}\right)$$

is fulfilled for all $(x,y,z) \in D_3^\circ$.

Thus

$$G(y,z) - (1-x)^{\alpha} G\left(\frac{y}{1-x}, \frac{z}{1-x}\right)$$
$$= \left\{G(x,y) + G(x+y,z) - G(x,y+z) - (1-x)^{\alpha} G\left(\frac{y}{1-x}, \frac{z}{1-x}\right)\right\}$$
$$- G(x,y) - G(x+y,z) + G(x,y+z)$$
$$+ \left\{G(y,x) + G(x+y,z) - G(y,x+z) - (1-y)^{\alpha} G\left(\frac{x}{1-y}, \frac{z}{1-y}\right)\right\}$$
$$+ \left\{G(y,z) + G(y+z,x) - G(y,x+z) - (1-y)^{\alpha} G\left(\frac{z}{1-y}, \frac{x}{1-y}\right)\right\}$$
$$- G(y+z,x) + G(y,x+z) + (1-y)^{\alpha} G\left(\frac{z}{1-y}, \frac{x}{1-y}\right)$$
$$= G(y,x) - G(x,y) + G(x,y+z) - G(y+z,x)$$
$$+ (1-y)^{\alpha} \left(G\left(\frac{z}{1-y}, \frac{x}{1-y}\right) - G\left(\frac{x}{1-y}, \frac{z}{1-y}\right)\right) \tag{67}$$

for all $(x, y, z) \in D_3^\circ$, since the expressions in the curly brackets are zeros. Thus (67), (65), and the triangle inequality imply that

$$\left|G(y,z) - (1-x)^\alpha G\left(\frac{y}{1-x}, \frac{z}{1-x}\right)\right| \leq (2 + (1-y)^\alpha)\,\varepsilon \tag{68}$$

is fulfilled for all $(x, y, z) \in D_3^\circ$. Given any $t \in \,]0, 1[$, $(u, v) \in D^\circ$, let

$$x = 1 - t, \quad y = tu \quad \text{and} \quad z = tv.$$

Then $x, y, z \in \,]0, 1[$ and

$$x + y + z = 1 - t(1 - u - v) \in \,]0, 1[,$$

that is $(x, y, z) \in D_3^\circ$, and inequality (68) implies that

$$|G(tu, tv) - t^\alpha G(u, v)| \leq (2 + (1 - tu)^\alpha)\,\varepsilon,$$

or, after rearranging,

$$\left|\frac{G(tu, tv)}{t^\alpha} - G(u, v)\right| \leq \frac{(2 + (1 - tu)^\alpha)}{t^\alpha}\varepsilon$$

holds for arbitrary $t \in \,]0, 1[$ and $(u, v) \in D^\circ$. Taking the limit $t \to 0+$ we obtain that

$$\lim_{t\to 0+} \frac{G(tu, tv)}{t^\alpha} = G(u, v), \quad ((u, v) \in D^\circ)$$

since $\lim_{t\to 0+} (1 - tu)^\alpha = 1$ for all $u \in \,]0, 1[$ and $\lim_{t\to 0+} t^{-\alpha} = 0$, since $\alpha < 0$. This implies that the function G is α–homogeneous on D°. Indeed, for arbitrary $s \in \,]0, 1[$ and $(u, v) \in D^\circ$

$$\begin{aligned} G(su, sv) &= \lim_{t\to 0+} \frac{G(t(su), t(sv))}{t^\alpha} \\ &= s^\alpha \lim_{t\to 0+} \frac{G\left((ts)u, (ts)v\right)}{(ts)^\alpha} = s^\alpha G(u, v). \end{aligned} \tag{69}$$

At this point of the proof, we will show that inequality (65) and Eq. (69) together imply the symmetry of the function G. Indeed, due to (65)

$$|G\,(tx, ty) - G\,(ty, tx)| \leq \varepsilon$$

holds for all $(x, y) \in D^\circ$ and $t \in \,]0, 1[$. Using the α-homogeneity of the function G, we obtain that

$$|t^\alpha G\,(x, y) - t^\alpha G\,(y, x)| \leq \varepsilon, \quad ((x, y) \in D^\circ, t \in \,]0, 1[)$$

or, if we rearrange this,

$$|G(x,y) - G(y,x)| \leq \frac{\varepsilon}{t^{\alpha}}$$

holds for all $(x, y) \in D^{\circ}$ and $t \in]0, 1[$. Taking the limit $t \to 0+$, we get that

$$G(x,y) = G(y,x)$$

is fulfilled for all $(x, y) \in D^{\circ}$, since $\alpha < 0$. Therefore, the function G is symmetric. Due to definition (64), this implies that

$$f(x) + (1-x)^{\alpha} f\left(\frac{y}{1-x}\right) = f(y) + (1-y)^{\alpha} f\left(\frac{x}{1-y}\right), \quad ((x,y) \in D^{\circ})$$

i.e., the function f satisfies the parametric fundamental equation of information on D°. Thus by Theorem 3 of Maksa [62], there exist $c, d \in \mathbb{R}$ such that

$$f(x) = cx^{\alpha} + d(1-x)^{\alpha} - d$$

holds for all $x \in]0, 1[$.

In what follows, we will show that for negative α's, the parametric fundamental equation of information is stable also on the set D.

Theorem 8 *Let $\alpha, \varepsilon \in \mathbb{R}$ be fixed, $\alpha < 0$, $\varepsilon \geq 0$. Then the function $f : [0, 1] \to \mathbb{R}$ satisfies the inequality (62) for all $(x, y) \in D$ if, and only if, there exist $c, d \in \mathbb{R}$ such that*

$$f(x) = \begin{cases} 0, & \text{if } x = 0 \\ cx^{\alpha} + d(1-x)^{\alpha} - d, & \text{if } x \in]0,1[\\ c - d, & \text{if } x = 1. \end{cases} \tag{70}$$

Proof Let $y = 0$ in (62). Then, we have that

$$((1-x)^{\alpha} + 1) |f(0)| \leq \varepsilon \quad (x \in]0, 1[)$$

Since $\alpha < 0$, this yields that $f(0) = 0$. On the other hand, by Theorem 7,

$$f(x) = cx^{\alpha} + d(1-x)^{\alpha} - d \quad (x \in]0, 1[)$$

with some $c, d \in \mathbb{R}$. Finally, let $x \in]0, 1[$ and $y = 1 - x$ in (62). Then, again by Theorem 7, there exist $c, d \in \mathbb{R}$ such that

$$|c - d - f(1)| \, |x^{\alpha} - (1-x)^{\alpha}| \leq \varepsilon.$$

Since $\alpha < 0$, $f(1) = c - d$ follows.

The converse is an easy computation and it turns out that f defined by (70) is a solution of (6) on D.

Our third main result in this section says that the system of α-recursive, 3-semi-symmetric information measures is stable.

Theorem 9 *Let $n \geq 2$ be a fixed positive integer, (I_n) be the sequence of functions $I_n : \Gamma_n^\circ \to \mathbb{R}$ and suppose that there exist a sequence (ε_n) of nonnegative real numbers and a real number $\alpha < 0$ such that*

$$\left| I_n(p_1,\ldots,p_n) - I_{n-1}(p_1+p_2,p_3,\ldots,p_n) \right. \\ \left. - (p_1+p_2)^\alpha I_2\left(\frac{p_1}{p_1+p_2},\frac{p_2}{p_1+p_2}\right)\right| \leq \varepsilon_{n-1} \tag{71}$$

holds for all $n \geq 3$ and $(p_1,\ldots,p_n) \in \Gamma_n^\circ$, and

$$|I_3(p_1,p_2,p_3) - I_3(p_1,p_3,p_2)| \leq \varepsilon, \tag{72}$$

holds on D_3°. Then there exist $a, b \in \mathbb{R}$ such that

$$\left| I_n(p_1,\ldots,p_n) - \left(aH_n^\alpha(p_1,\ldots,p_n) + b\left(p_1^\alpha - 1\right)\right)\right| \leq \sum_{k=2}^{n-1} \varepsilon_k \tag{73}$$

for all $n \geq 2$ and $(p_1,\ldots,p_n) \in \Gamma_n^\circ$, where the convention $\sum_{k=2}^{1} \varepsilon_k = 0$ is adopted.

Proof As in Maksa [67], it can be proved that, due to (71) and (72), for the function f defined by $f(x) = I_2(1-x,x)$, $x \in]0,1[$ we get that

$$\left| f(x) + (1-x)^\alpha f\left(\frac{y}{1-x}\right) - f(y) - (1-y)^\alpha f\left(\frac{x}{1-y}\right)\right| \leq 2\varepsilon_2 + \varepsilon_1$$

for all $(x,y) \in D^\circ$, i.e., (62) holds with $\varepsilon = 2\varepsilon_2 + \varepsilon_1$. Therefore, applying Theorem 7, we obtain (63) with some $c, d \in \mathbb{R}$, i.e.,

$$I_2(1-x,x) = cx^\alpha + d(1-x)^\alpha - d, \quad (x \in]0,1[)$$

i.e., (73) holds for $n = 2$ with $a = (2^{1-\alpha} - 1)c$, $b = d - c$.

We continue the proof by induction on n. Suppose that (73) holds and, for the sake of brevity, introduce the notation

$$J_n(p_1,\ldots,p_n) = aH_n^\alpha(p_1,\ldots,p_n) + b\left(p_1^\alpha - 1\right)$$

for all $n \geq 2$, $(p_1,\ldots,p_n) \in \Gamma_n^\circ$. It can easily be seen that (71) and (72) hold on Γ_n° for J_n instead of I_n $(n \geq 3)$ with $\varepsilon_n = 0$ $(n \geq 2)$. Thus for all $(p_1,\ldots,p_{n+1}) \in \Gamma_{n+1}^\circ$, we get that

$$I_{n+1}(p_1,\ldots,p_{n+1}) - J_{n+1}(p_1,\ldots,p_{n+1}) \\ = I_{n+1}(p_1,\ldots,p_{n+1}) - J_n(p_1+p_2,p_3,\ldots,p_{n+1}) \\ - (p_1+p_2)^\alpha J_2\left(\frac{p_1}{p_1+p_2},\frac{p_2}{p_1+p_2}\right)$$

$$= I_{n+1}(p_1, \dots, p_{n+1}) - I_n(p_1 + p_2, p_3, \dots, p_{n+1})$$
$$- (p_1 + p_2)^{\alpha} I_2\left(\frac{p_1}{p_1 + p_2}, \frac{p_2}{p_1 + p_2}\right)$$
$$+ I_n(p_1 + p_2, p_3, \dots, p_{n+1}) - J_n(p_1 + p_2, p_3, \dots, p_{n+1})$$
$$+ (p_1 + p_2)^{\alpha} I_2\left(\frac{p_1}{p_1 + p_2}, \frac{p_2}{p_1 + p_2}\right)$$
$$- (p_1 + p_2)^{\alpha} J_2\left(\frac{p_1}{p_1 + p_2}, \frac{p_2}{p_1 + p_2}\right).$$

Therefore, (73) with $n = 2$ and the induction hypothesis imply that

$$|I_{n+1}(p_1, \dots, p_{n+1}) - J_n(p_1, \dots, p_{n+1})| \le \varepsilon_n + \sum_{k=2}^{n-1} \varepsilon_k = \sum_{k=2}^{n} \varepsilon_k,$$

that is, (73) holds for $n + 1$ instead of n.

Corollary 1.3.2.1 *Applying Theorem 9 with the choice $\varepsilon_n = 0$ for all $n \in \mathbb{N}$, we get the α-recursive, 3-semi-symmetric information measures. Hence the previous theorem says that the system of α-recursive and 3-semi-symmetric information measures is stable.*

3.3 *Related Equations*

In the previous subsections, we have investigated the stability problem of the parametric fundamental equation of information. In the remaining part of this chapter, we will discuss the stability problem of some functional equations that also have information theoretical background. Firstly, we will show that the so-called entropy equation is stable on its domain. After that some results concerning the *modified entropy equation* will follow. Finally, we will end this section with some open problems.

3.3.1 Stability of the Entropy Equation

In what follows, our aim is to prove that *the entropy equation*, i.e., equation

$$H(x, y, z) = H(x + y, 0, z) + H(x, y, 0) \tag{74}$$

is stable on the set

$$C = \left\{(x, y, z) \in \mathbb{R}^3 | x \ge 0, y \ge 0, z \ge 0, x + y + z > 0\right\}.$$

In [44], A. Kamiński and J. Mikusiński determined the continuous and 1-homogeneous solutions of Eq. (74) on the set $\mathbb{R}^3$. This result was strengthened by J. Aczél in [2]. After that, using a result of Jessen–Karpf–Thorup [43], which concerns the solution of the cocycle equation, Z. Daróczy proved the following (see [17]).

Theorem 10 *If a function $H : C \to \mathbb{R}$ is symmetric and satisfies Eq.* (74) *in the interior of C and the map $(x, y) \mapsto H(x, y, 0)$ is positively homogeneous (of order 1) for all $x, y \in \mathbb{R}_{++}$, then there exists a function $\varphi : \mathbb{R}_{++} \to \mathbb{R}$ such that*

$$\varphi(xy) = x\varphi(y) + y\varphi(x)$$

holds for all $x, y \in \mathbb{R}_{++}$ and

$$H(x, y, z) = \varphi(x + y + z) - \varphi(x) - \varphi(y) - \varphi(z)$$

for all $(x, y, z) \in C$.

During the proof of the main result the stability of the *cocycle equation* is needed. This theorem can be found in [74].

Theorem 11 *Let S be a right amenable semigroup and let $F : S \times S \to \mathbb{C}$ be a function, for which the function*

$$(x, y, z) \longmapsto F(x, y) + F(x + y, z) - F(x, y + z) - F(y, z) \tag{75}$$

is bounded on $S \times S \times S$. Then there exists a function $\Psi : S \times S \to \mathbb{C}$ satisfying the cocycle equation, i.e.,

$$\Psi(x, y) + \Psi(x + y, z) = \Psi(x, y + z) + \Psi(y, z) \tag{76}$$

for all $x, y, z \in S$ and for which the function $F - \Psi$ is bounded by the same constant as the map defined by (75).

About the symmetric, 1-homogeneous solutions of the cocycle equation one can read in [43]. Furthermore, the symmetric and α-homogeneous solutions of Eq. (76) can be found in [62], as a consequence of Theorem 3. The general solution of the cocycle equation without symmetry and homogeneity assumptions, on cancellative abelian semigroups was determined by M. Hosszú in [39].

Our main result concerning the stability of Eq. (74) is the following, see also [34].

Theorem 12 *Let $\varepsilon_1, \varepsilon_2, \varepsilon_3$ be arbitrary nonnegative real numbers, $\alpha \in \mathbb{R}$, and assume that the function $H : C \to \mathbb{R}$ satisfies the following system of inequalities.*

$$|H(x, y, z) - H(\sigma(x), \sigma(y), \sigma(z))| \leq \varepsilon_1 \tag{77}$$

for all $(x, y, z) \in C$ and for all $\sigma : \{x, y, z\} \mapsto \{x, y, z\}$ permutation;

$$|H(x, y, z) - H(x + y, 0, z) - H(x, y, 0)| \leq \varepsilon_2 \tag{78}$$

for all $(x, y, z) \in C^{\circ}$, where C° denotes the interior of the set C;

$$|H(tx, ty, 0) - t^{\alpha} H(x, y, 0)| \leq \varepsilon_3 \tag{79}$$

holds for all $t, x, y \in \mathbb{R}_{++}$. *Then, in case* $\alpha = 1$ *there exists a function* $\varphi : \mathbb{R}_{++} \to \mathbb{R}$ *which satisfies the functional equation*

$$\varphi(xy) = x\varphi(y) + y\varphi(x), \quad (x, y \in \mathbb{R}_{++})$$

and

$$|H(x, y, z) - [\varphi(x + y + z) - \varphi(x) - \varphi(y) - \varphi(z)]| \le \varepsilon_1 + \varepsilon_2 \tag{80}$$

holds for all $(x, y, z) \in C^\circ$; *in case* $\alpha = 0$ *there exists a constant* $a \in \mathbb{R}$ *such that*

$$|H(x, y, z) - a| \le 8\varepsilon_3 + 25\varepsilon_2 + 49\varepsilon_1 \tag{81}$$

for all $(x, y, z) \in C^\circ$; *finally, in all other cases there exists a constant* $c \in \mathbb{R}$ *such that*

$$\left|H(x, y, z) - c\left[(x + y + z)^\alpha - x^\alpha - y^\alpha - z^\alpha\right]\right| \le \varepsilon_1 + \varepsilon_2 \tag{82}$$

holds on C°.

Proof For the sake of brevity, here we present only the sketch of proof of the above statement. For details, the reader should consult [34].

Using inequality (79), it can be shown that the map

$$(x, y) \mapsto H(x, y, 0) \qquad (x, y \in \mathbb{R}_{++})$$

is homogeneous of degree α, assuming that $\alpha \neq 0$.

Let us consider the function $F : \mathbb{R}^2_{++} \to \mathbb{R}$ defined by

$$F(x, y) = H(x, y, 0) \qquad (x, y \in \mathbb{R}_{++}).$$

From inequalities (77) and (78), we can deduce that

$$|F(x, y) - F(y, x)| \le \varepsilon_1, \quad (x, y \in \mathbb{R}_{++}) \tag{83}$$

and

$$|F(x + y, z) + F(x, y) - F(x, y + z) - F(y, z)| \le 2\varepsilon_2 + 4\varepsilon_1. \quad (x, y, z \in \mathbb{R}_{++}). \tag{84}$$

Furthermore, in case $\alpha \neq 0$, $H(x, y, 0)$ is homogeneous of degree α, therefore

$$F(tx, ty) = t^\alpha F(x, y) \quad (\alpha \neq 0, t, x, y \in \mathbb{R}_{++}) \tag{85}$$

and if $\alpha = 0$,

$$|F(tx, ty) - F(x, y)| \le \varepsilon_3, \quad (t, x, y \in \mathbb{R}_{++}) \tag{86}$$

is fulfilled.

The set C° is a commutative semigroup with the usual addition. Thus it is amenable, as well. Therefore, by Theorem 11, there exists a function $G : \mathbb{R}^2_{++} \to \mathbb{R}$ which is a solution of the cocycle equation, and for which

$$|F(x, y) - G(x, y)| \leq 2\varepsilon_2 + 4\varepsilon_1 \tag{87}$$

holds for all $x, y \in \mathbb{R}_{++}$. Additionally, by a result of [39] there exist a function $f : \mathbb{R}_{++} \to \mathbb{R}$ and a function $B : \mathbb{R}^2_{++} \to \mathbb{R}$ which satisfies the following system

$$\begin{aligned} B(x+y,z) &= B(x,z) + B(y,z), \\ B(x,y) + B(y,x) &= 0, \end{aligned} \qquad (x, y, z \in \mathbb{R}_{++})$$

such that

$$G(x, y) = B(x, y) + f(x + y) - f(x) - f(y). \quad (x, y \in \mathbb{R}_{++})$$

All in all, this means that

$$|F(x, y) - (B(x, y) + f(x + y) - f(x) - f(y))| \leq 2\varepsilon_2 + 4\varepsilon_1 \tag{88}$$

holds for all $x, y \in \mathbb{R}_{++}$.

Using the above properties of the function B, we can show that B is identically zero on $\mathbb{R}^2_{++}$. Additionally, after some computation, we obtain that

$$F(x + y, z) + F(x, y) = F(x, y + z) + F(y, z). \quad (x, y, z \in \mathbb{R}_{++})$$

This means that also the function F satisfies the cocycle equation on $\mathbb{R}^2_{++}$. Additionally, F is homogeneous of degree α ($\alpha \neq 0$) and symmetric. Using Theorem 5 in [43], in case $\alpha = 1$, and a result of [62] in all other cases, we get that

$$F(x, y) = \begin{cases} c\left[(x+y)^{\alpha} - x^{\alpha} - y^{\alpha}\right], & \text{if} \quad \alpha \notin \{0, 1\} \\ \varphi(x+y) - \varphi(x) - \varphi(y), & \text{if} \quad \alpha = 1 \end{cases} \tag{89}$$

where the function $\varphi : \mathbb{R}_{++} \to \mathbb{R}$ satisfies the functional equation

$$\varphi(xy) = x\varphi(y) + y\varphi(x)$$

for all $x, y \in \mathbb{R}_{++}$, and $c \in \mathbb{R}$ is a constant. In view of the definition of the function F, this yields that

$$H(x, y, 0) = c\left[(x+y)^{\alpha} - x^{\alpha} - y^{\alpha}\right] \tag{90}$$

for all $x, y \in \mathbb{R}_{++}$ in case $\alpha \notin \{0, 1\}$, and

$$H(x, y, 0) = \varphi(x + y) - \varphi(x) - \varphi(y) \tag{91}$$

for all $x, y \in \mathbb{R}_{++}$ in case $\alpha = 1$.

Using this representations and inequalities (77) and (78), the statement of our theorem can be deduced.

With the choice $\varepsilon_1 = \varepsilon_2 = \varepsilon_3 = 0$, one can recognize the solutions of Eq. (74).

Corollary 1.3.3.1 *Assume that the function $H : C \to \mathbb{R}$ is symmetric, homogeneous of degree α, where $\alpha \in \mathbb{R}$ is arbitrary but fixed. Furthermore, suppose that H satisfies Eq.* (74) *on the set C°. Then, in case $\alpha = 1$ there exists a function $\varphi : \mathbb{R}_{++} \to \mathbb{R}$ which satisfies the functional equation*

$$\varphi(xy) = x\varphi(y) + y\varphi(x), \quad (x, y \in \mathbb{R}_{++})$$

and

$$H(x, y, z) = \varphi(x + y + z) - \varphi(x) - \varphi(y) - \varphi(z) \tag{92}$$

holds for all $(x, y, z) \in C^\circ$; in all other cases there exists a constant $c \in \mathbb{R}$ such that

$$H(x, y, z) = c\left[(x + y + z)^\alpha - x^\alpha - y^\alpha - z^\alpha\right] \tag{93}$$

holds on C°.

Remark 1.3.3.1 Our theorem says that the entropy equation is stable in the sense of Hyers and Ulam.

3.3.2 Stability of the Modified Entropy Equation

In this part of the chapter, we investigate the stability problem concerning the functional equation

$$f(x, y, z) = f(x, y + z, 0) + (y + z)^\alpha f\left(0, \frac{y}{y + z}, \frac{z}{y + z}\right), \tag{94}$$

where x, y, z are positive real numbers and α is a given real number. Equation (94) is a special case of the so-called modified entropy equation,

$$f(x, y, z) = f(x, y + z, 0) + \mu(y + z) f\left(0, \frac{y}{y + z}, \frac{z}{y + z}\right), \tag{95}$$

where μ is a given multiplicative function defined on the positive cone of $\mathbb{R}^k$ and (95) is supposed to hold for all elements x, y, z of the above mentioned cone and all operations on vectors are to be understood componentwise. The symmetric solutions of Eq. (95) were determined in [29] (see also [1]).

By a real interval we always mean a subinterval of positive length of $\mathbb{R}$. Furthermore, in case U and V are real intervals, then their sum

$$U + V = \{u + v \mid u \in U, v \in V\}$$

is obviously a real interval, as well.

During the proof of our main result of this subsection, the stability of a simple associativity equation should be used which is contained in the following theorem, see [33].

Theorem 13 *Let U, V, and W be real intervals, $A : (U+V) \times W \to \mathbb{R}$, $B : U \times (V+W) \to \mathbb{R}$ and suppose that*

$$|A(u+v,w) - B(u,v+w)| \leq \varepsilon \tag{96}$$

holds for all $u \in U$, $v \in V$, and $w \in W$. Then there exists a function $\varphi : U+V+W \to \mathbb{R}$ such that

$$|A(p,q) - \varphi(p+q)| \leq 2\varepsilon \quad (p \in (U+V), q \in W) \tag{97}$$

and

$$|B(t,s) - \varphi(t+s)| \leq \varepsilon \quad (t \in U, s \in (V+W)) \tag{98}$$

hold.

With the choice $\varepsilon_1 = \varepsilon_2 = 0$, we get the following theorem. Nevertheless, it was proved in Maksa [66].

Corollary 1.3.3.2 *Let U, V, and W be real intervals, $A : (U+V) \times W \to \mathbb{R}$, $B : U \times (V+W) \to \mathbb{R}$ and suppose that*

$$A(u+v,w) = B(u,v+w)$$

holds for all $u \in U, v \in V$, and $w \in W$. Then there exists a function $\varphi : U+V+W \to \mathbb{R}$ such that

$$A(p,q) = \varphi(p+q) \tag{99}$$

for all $p \in U+V$ and $q \in W$ and

$$B(t,s) = \varphi(t+s) \tag{100}$$

for all $t \in U$ and $s \in V+W$.

In view of the results of the previous sections (that is Theorems 4 and 7 and with the help of Theorem 13, the following result can be proved. For the details of the proof, see [33].

Theorem 14 *Let $\alpha, \varepsilon \in \mathbb{R}$, $\alpha \neq 1, \varepsilon \geq 0$ and $f : \mathbb{R}^3_+ \to \mathbb{R}$ be a function. Assume that*

$$\left| f(x,y,z) - f(x,y+z,0) - (y+z)^{\alpha} f\left(0, \frac{y}{y+z}, \frac{z}{y+z}\right) \right| \leq \varepsilon_1 \tag{101}$$

and

$$|f(x,y,z) - f(\sigma(x), \sigma(y), \sigma(z))| \leq \varepsilon_2 \tag{102}$$

hold for all $x, y, z \in \mathbb{R}_{++}$ *and for all permutations* $\sigma : \{x, y, z\} \to \{x, y, z\}$.

Then, in case $\alpha < 0$, *there exist* $a \in \mathbb{R}$ *and a function* $\varphi_1 : \mathbb{R}_{++} \to \mathbb{R}$ *such that*

$$\left|f(x, y, z) - \left[ax^\alpha + ay^\alpha + az^\alpha + \varphi_1(x + y + z)\right]\right| \le 2\varepsilon_1 + 3\varepsilon_2 \tag{103}$$

holds for all $x, y, z \in \mathbb{R}_{++}$.

Furthermore, if $\alpha = 0$, *then there exists a function* $\varphi_2 : \mathbb{R}_{++} \to \mathbb{R}$ *such that*

$$|f(x, y, z) - \varphi_2(x + y + z)| \le 191\varepsilon_1 + 1263\varepsilon_2 \tag{104}$$

holds for all $x, y, z \in \mathbb{R}_{++}$.

Finally, if $1 \neq \alpha > 0$, *then for all* $n \in \mathbb{N}$, *there exists a function* $\psi_n :]0, 3n] \to \mathbb{R}$ *such that*

$$\left|f(x, y, z) - \left[ax^\alpha + ay^\alpha + az^\alpha + \psi_n(x + y + z)\right]\right| \le c_n(\alpha)\varepsilon_n + d_n(\alpha)\varepsilon_2$$

holds for all $x, y, z \in]0, n]$, *where*

$$c_n(\alpha) = 2 + 7 \cdot 2^\alpha n^\alpha K(\alpha) \quad \text{and} \quad d_n(\alpha) = 4 + 7 \cdot 2^{\alpha+2} n^\alpha K(\alpha).$$

With the choice $\varepsilon_1 = \varepsilon_2 = 0$, we get the general solutions of Eq. (94), in the investigated cases.

Corollary 1.3.3.3 *Let* $\alpha \in \mathbb{R}$, $\alpha \neq 1$ *and suppose that the function* $f : \mathbb{R}^3_+ \to \mathbb{R}$ *is symmetric and satisfies functional Eq. (94) for all* $x, y, z \in \mathbb{R}_{++}$.

Then, in case $\alpha \neq 0$, *there exist* $a \in \mathbb{R}$ *and a function* $\varphi_1 : \mathbb{R}_{++} \to \mathbb{R}$ *such that*

$$f(x, y, z) = ax^\alpha + ay^\alpha + az^\alpha + \varphi_1(x + y + z)$$

holds for all $x, y, z \in \mathbb{R}_{++}$.

In case $\alpha = 0$, *there exists a function* $\varphi_2 : \mathbb{R}_{++} \to \mathbb{R}$ *such that*

$$f(x, y, z) = \varphi_2(x + y + z)$$

is fulfilled for all $x, y, z \in \mathbb{R}_{++}$.

In view of Corollary 1.3.3.3, our theorem says that the modified entropy equation is stable in the sense of Hyers and Ulam on its one-dimensional domain with the multiplicative function $\mu(x) = x^\alpha$ ($\alpha \le 0, x \in \mathbb{R}_{++}$).

In case $1 \neq \alpha > 0$ we obtain however that functional Eq. (94) is stable on every cartesian product of bounded real intervals of the form $]0, n]^3$, where $n \in \mathbb{N}$. Nevertheless, an easy computation shows that

$$\lim_{n\to+\infty} c_n(\alpha) = +\infty \qquad \lim_{n\to+\infty} d_n(\alpha) = +\infty. \quad (1 \neq \alpha > 0)$$

To the best of our knowledge, this is a new phenomenon in the stability theory of functional equations. Since we cannot prove the "standard" Hyers–Ulam stability in this case, the following problem can be raised.

Open Problem 5 *Let* $\alpha, \varepsilon_1, \varepsilon_2 \in \mathbb{R}$, $\alpha > 0, \varepsilon_1, \varepsilon_2 \geq 0$, and $f : \mathbb{R}_+^3 \to \mathbb{R}$ *be a function. Assume that*

$$\left| f(x,y,z) - f(x,y+z,0) - (y+z)^{\alpha} f\left(0, \frac{y}{y+z}, \frac{z}{y+z}\right)\right| \leq \varepsilon_1$$

and

$$|f(x,y,z) - f(\sigma(x), \sigma(y), \sigma(z))| \leq \varepsilon_2$$

holds for all $x, y, z \in \mathbb{R}_{++}$ *and for all* $\sigma : \{x, y, z\} \to \{x, y, z\}$ *permutations. Is it true that there exists a solution* $h : \mathbb{R}_{++}^3 \to \mathbb{R}$ *of equation* (94) *such that*

$$|f(x,y,z) - h(x,y,z)| \leq K_1\varepsilon_1 + K_2\varepsilon_2$$

holds for all $x, y, z \in \mathbb{R}_{++}$ *with some* $K_1, K_2 \in \mathbb{R}$? *The second open problem that can be raised is the stability problem of the modified entropy equation itself, i.e., Eq.* (95).

Open Problem 6 *Let* $\varepsilon_1, \varepsilon_2 \geq 0$, $\mu : \mathbb{R}_{++}^k \to \mathbb{R}$ *be a given multiplicative function,* $f : \mathbb{R}_+^{3k} \to \mathbb{R}$. *Assume that*

$$\left| f(x,y,z) - f(x,y+z,0) - \mu(y+z) f\left(0, \frac{y}{y+z}, \frac{z}{y+z}\right)\right| \leq \varepsilon_1$$

and

$$|f(x,y,z) - f(\sigma(x), \sigma(y), \sigma(z))| \leq \varepsilon_2$$

holds for all $x, y, z \in \mathbb{R}_{++}^k$ *and for all* $\sigma : \{x, y, z\} \to \{x, y, z\}$ *permutation.*

Is it true that there exists a solution $h : \mathbb{R}_{++}^{3k} \to \mathbb{R}$ *of Eq.* (95) such that

$$|f(x,y,z) - h(x,y,z)| \leq K_1\varepsilon_1 + K_2\varepsilon_2$$

holds for all $x, y, z \in \mathbb{R}_{++}^k$ *with certain* $K_1, K_2 \in \mathbb{R}$?

3.4 Stability of Sum Form Equations

We have to begin with an open problem since there is no stability result on Eq. (9)

$$\sum_{i=1}^{n}\sum_{j=1}^{m} f(p_i q_j) = \sum_{i=1}^{n} f(p_i) + \sum_{j=1}^{m} f(q_j) + (2^{1-\alpha} - 1)\sum_{i=1}^{n} f(p_i)\sum_{j=1}^{m} f(q_j)$$

in case $\alpha = 1$.

Open Problem 7 *Suppose that $n \geq 2, m \geq 2, 0 \leq \varepsilon \in \mathbb{R}, f : I \to \mathbb{R}$ and the stability inequality*

$$\left| \sum_{i=1}^{n} \sum_{j=1}^{m} f(p_i q_j) - \sum_{i=1}^{n} f(p_i) - \sum_{j=1}^{m} f(q_j) \right| \leq \varepsilon$$

holds for all $(p_1, \ldots, p_n) \in \mathcal{G}_n, (q_1, \ldots, q_n) \in \mathcal{G}_m$. Prove or disprove that f is the sum of a solution of (9) with $\alpha = 1$ and a bounded function.

A somewhat related result however is proved in Kocsis–Maksa [51] which reads as follows.

Theorem 15 *Let $n \geq 3, m \geq 3,\ 0 \leq \varepsilon \in \mathbb{R},\ f : [0,1] \to \mathbb{R},\ \alpha, \beta \in \mathbb{R}$ and suppose that*

$$\left| \sum_{i=1}^{n} \sum_{j=1}^{m} f(p_i q_j) - \sum_{i=1}^{n} f(p_i) \sum_{j=1}^{m} q_j^{\beta} - \sum_{j=1}^{m} f(q_j) \sum_{i=1}^{n} p_i^{\alpha} \right| \leq \varepsilon$$

holds for all $(p_1, \ldots, p_n) \in \Gamma_n,\ (q_1, \ldots, q_n) \in \Gamma_m$.

Then there exists an additive function $a : \mathbb{R} \to \mathbb{R}$, a function $\ell : \mathbb{R}_+ \to \mathbb{R}$, $\ell(0) = 0$, ℓ is logarithmic on $\mathbb{R}_{++}$, a bounded function $b : [0,1] \to \mathbb{R}$, and a real number c such that $a(1) = 0$,

$$f(p) = a(p) + c(p^{\alpha} - p^{\beta}) + b(p) \quad \text{if} \quad p \in [0,1],\ \beta \neq \alpha$$

and

$$f(p) = a(p) + p^{\alpha}\ell(p) + b(p) \quad \text{if} \quad p \in [0,1],\ \beta = \alpha \neq 1.$$

If $\varepsilon = 0$ then $b = 0$ can be chosen here, so, the above theorem is of stability type which however does not cover just with the Shannon case $\beta = \alpha = 1$.

In case $\alpha \neq 1$, the problem of the stability of Eq. (9) can easily be handled at least whenever both n and m are not less than three. First of all, introducing a new function g by $g(p) = p + (2^{1-\alpha} - 1)f(p),\ p \in I$, the stability inequality

$$\left| \sum_{i=1}^{n} \sum_{j=1}^{m} f(p_i q_j) - \sum_{i=1}^{n} f(p_i) - \sum_{j=1}^{m} f(q_j) - (2^{1-\alpha} - 1) \sum_{i=1}^{n} f(p_i) \sum_{j=1}^{m} f(q_j) \right| \leq \varepsilon$$

goes over into

$$\left| \sum_{i=1}^{n} \sum_{j=1}^{m} g(p_i q_j) - \sum_{i=1}^{n} g(p_i) \sum_{j=1}^{m} g(q_j) \right| \leq \varepsilon \cdot |2^{1-\alpha} - 1| \qquad (105)$$

and the following theorem can be proved (see Maksa [65]).

Theorem 16 *Let $n \geq 3, m \geq 3$, $0 \leq \varepsilon \in \mathbb{R}$, $g : [0,1] \to \mathbb{R}$, and suppose that* (105) *holds for all $(p_1, \ldots, p_n) \in \Gamma_n$, $(q_1, \ldots, q_n) \in \Gamma_m$. Then e*

$$g(p) = a(p) + m(p) + b(p) \qquad (p \in [0,1])$$

where $a : \mathbb{R} \to \mathbb{R}$ is an additive, $b : [0,1] \to \mathbb{R}$ is a bounded, and $m : [0,1] \to \mathbb{R}$ is a multiplicative function, respectively.

The corner point in the proofs of these theorems is the following stability result (see [65]).

Theorem 17 *Let $n \geq 3$, $0 \leq \varepsilon \in \mathbb{R}$, $\varphi : [0,1] \to \mathbb{R}$, and suppose that*

$$\left| \sum_{i=1}^{n} \varphi(p_i) \right| \leq \varepsilon \tag{106}$$

holds for all $(p_1, \ldots, p_n) \in \Gamma_n$. Then there exist an additive function $A : \mathbb{R} \to \mathbb{R}$ and a function $b : [0,1] \to \mathbb{R}$ such that $b(0) = 0$, $|b(x)| \leq \varepsilon$ for all $x \in [0,1]$ and

$$\varphi(p) - \varphi(0) = A(p) + b(p) \qquad (p \in [0,1]).$$

By an argument similar to that we used in the subsection on sum form equations in connection with the inequality (17), inequality (106) and the triangle inequality imply that

$$|\varphi(x + y) - \varphi(x) - \varphi(y) + \varphi(0)| \leq 2\varepsilon,$$

that is, the classical stability inequality holds for the function $\varphi - \varphi(0)$ on the restricted domain $\{(x, y) \in \mathbb{R}^2 \mid x, y, x+y \in [0,1]\}$. Therefore, the results (see Skof [73], Tabor and Tabor [77]) on the stability of the Cauchy equation on restricted domain can be applied to finish the proof of the above theorem.

We remark that the other basic tool for proving stability results for sum form equations was the analysis of the methods with the help of which the solutions of these equations were found. These and similar ideas proved to be fruitful in the investigations on the stability of the sum form equations in an open domain (excluding zero probabilities) and also of the several variable cases (See Kocsis [48–50]).

Acknowledgements This research has been supported by the Hungarian Scientific Research Fund (OTKA) Grant NK 814 02 and by the TÁMOP 4.2.2.C-11/1/KONV-2012-0010 project implemented through the New Hungary Development Plan cofinanced by the European Social Fund and the European Regional Development Fund.

References

1. Abbas, A.E., Gselmann, E., Maksa, Gy., Sun, Z.: General and continuous solutions of the entropy equation. In: Proceedings of the 28th International Workshop on Bayesian Inference and Maximum Entropy Methods in Science and Engineering, vol. 1073, pp. 3–7. American Institute of Physics, College Park (2008)
2. Aczél, J.: Results on the entropy equation. Bull. Acad. Pol. Sci., Sér. Sci. Math. Astron. Phys. **25**, 13–17 (1977)
3. Aczél, J.: Lectures on Functional Equations and Their Applications, vol. XIX, p. 510. Academic, New York (1966)
4. Aczél, J.: Derivations and Information Functions (A Tale of Two Surprises and a Half). Contributions to Probability, Collect. Pap. dedic. E. Lukacs, pp. 191–200 (1981)
5. Aczél, J., Daróczy, Z.: A mixed theory of information I. Symmetric, recursive and measurable entropies of randomized systems of events. RAIRO Inform. Théor. **12**, 149–155 (1978)
6. Aczél, J., Daróczy, Z.: Characterisierung der Entropien positiver Ordnung und der Shannonschen Entropie. Acta Math. Acad. Sci. Hung. **14**, 95–121 (1963)
7. Aczél, J., Daróczy, Z.: On Measures of Information and their Characterizations. Mathematics in Science and Engineering, vol. 115, XII, p. 234. Academic, New York (1975)
8. Aczél, J., Erdős, P.: The non-existence of Hamel bases and the general solution of Cauchy's equation for positive numbers. Publ. Math. Debr. **12**, 259–263 (1965)
9. Badora, R., Ger, R., Páles, Z.: Additive selections and the stability of the Cauchy functional equation. ANZIAM J. **44**(3), 323–337 (2003)
10. Baker, J.A.: The stability of the cosine equation. Proc. Am. Math. Soc. **80**, 411–416 (1980)
11. Behara, M., Nath, P.: Additive and non-additive entropies of finite measurable partitions, probability and information theory II. Lect. Notes Math. **296**, 102–138 (1973)
12. Borges, R.: Zur Herleitung der Shannonschen Information. Math. Z. **96**, 282–287 (1967) (German)
13. Chaundy, T.W., McLeod, J.B.: On a functional equation. Edinb. Math. Notes **43**, 7–8 (1960) (English)
14. Daróczy, Z.: Über das Shannonsche Maß der Information. MTA Mat. Fiz. Oszt. Közl. **19**, 9–24 (1969) (Hungarian)
15. Daróczy, Z.: Generalized information functions. Inf. Control **16**, 36–51 (1970)
16. Daróczy, Z.: On the measurable solutions of a functional equation. Acta Math, Acad. Sci. Hung. **22**, 11–14 (1971)
17. Daróczy, Z.: Remarks on the entropy equation. Symp. Quasigroupes Equat. fonct., Beograd 1974, Rec. Trav. Inst. Math., nouv. Ser., No.1(9), 31–34 (1976)
18. Daróczy, Z., Járai, A.: On measurable solutions of functional equations., Acta Math. Acad. Sci. Hungar. **34**, 105–116 (1979) (English)
19. Daróczy, Z., Kátai, I.: Äddititive zahlentheoretische Funktionen und das Maß der Information., Ann. Univ. Sci. Budapest, Eötvös, Sect. Math. **13**, (83–88) (1970) (German)
20. Daróczy, Z., Losonczi, L.: Über die Erweiterung der auf einer Punktmenge additiven Funktionen. Publ. Math. **14**, 239–245 (1967) (German)
21. Daróczy, Z., Maksa, Gy.: Nonnegative information functions. Analytic function methods in probability theory, Debrecen 1977, Colloq. Math. Soc. Janos Bolyai 21, 67–78 (1980)
22. Day, M.M.: Amenable semigroups. Ill. J. Math. **1**, 509–544 (1957)
23. de Bruijn, N.G.: Functions whose differences belong to a given class. Nieuw Arch. Wisk. **23**, 194–218 (1951) (English)
24. Diderrich, G.T.: Boundedness on a set of positive measure and the fundamental equation of information., Publ. Math. **33**, 1–7 (1986)
25. Ebanks, B.R., Sahoo, P., Sander, W., Characterization of Information Measures, vol. X, p. 281. World Scientific, Singapore (1997)
26. Kannappan, Pl., Sahoo, P.K., Ebanks, G.R., Sander, W.: Characterizations of sum form information measures on open domains. Aequationes Math. **54**, 1–30 (1997)

27. Faddeev, D.K.: On the concept of entropy of a finite probabilistic scheme. (1956) (Russian).
28. Forti, G.L.: Hyers-Ulam stability of functional equations in several variables. Aequationes Math. **50**(1–2), 143–190 (1995)
29. Gselmann, E.: On the modified entropy equation. Banach J. Math. Anal. **2**(1), 84–96 (2008)
30. Gselmann, E.: Hyperstability of a functional equation. Acta Math. Hung. **124**(1–2), 179–188 (2009)
31. Gselmann, E.: Recent results on the stability of the parametric fundamental equation of information. Acta Math. Acad. Paedagog. Nyházi. (N.S.) (2009)
32. Gselmann, E.: Stability type results concerning the fundamental equation of information of multiplicative type. Colloq. Math. **114**(1), 33–40 (2009)
33. Gselmann, E.: On the stability of the modified entropy equation. Result. Math. **58**(3–4), 255–268 (2010)
34. Gselmann, E.: Stability of the entropy equation. Publ. Math. **77**(1–2), 201–210 (2010)
35. Gselmann, E., Maksa, Gy.: Stability of the parametric fundamental equation of information for nonpositive parameters. Aequ. Math. **78**(3), 271–282 (2009)
36. Gselmann, E., Maksa, Gy.: The Shannon field of non-negative information functions. Sci. Math. Jpn. **69**(2), 241–248 (2009)
37. Hartley, R.V.: Transmission of information. Bell Syst. Tech. J. **7**, 535–563 (1928)
38. Havrda, J., Charvat, F.: Quantification method of classification processes. Concept of structural a-entropy. Kybernetika. Praha. **3**, 30–35 (1967)
39. Hosszú, M.: On the functional equation $F(x+y,z)+F(x,y) = F(x,y+z)+F(y,z)$. Period. Math. Hung. **1**, 213–216 (1971)
40. Hyers, D.H., Isac, G., Rassias, Th.M.: Stability of Functional Equations in Several Variables. Progress in Nonlinear Differential Equations and Their Applications, vol. VII, 34, p. 313. Birkhäuser, Boston (1998)
41. Hyers, D.H.: On the stability of the linear functional equation. Proc. Natl. Acad. Sci. U. S. A. **27**, 222–224 (1941)
42. Járai, A.: Regularity Properties of Functional Equations in Several Variables. Springer, Berlin (2005)
43. Jessen, B., Karpf, J., Thorup, A.: Some functional equations in groups and rings. Math. Scand. **22**, 257–265 (1968)
44. Kaminski, A., Mikusinski, J. On the entropy equation. Bull. Acad. Polon. Sci. Sér. Sci. Math. Astronom. Phys. **22**, 319–323 (1974)
45. Kannappan, Pl.: On a generalization of some measures of information theory. Glasnik Mat. **9**(29), 81–93 (1974)
46. Kendall, D.G.: Functional equations in information theory. Z. Wahrscheinlichkeitstheor. Verw. Geb. **2**, 225–229 (1964)
47. Khinchin, A. Ya.: The concept of entropy in the theory of probability. Am. Math. Soc. Transl. II. Ser. **12**, 181–197 (1959)
48. Kocsis, I.: Stability of a sum form functional equation on open domain. Publ. Math. **57** (1–2), 135–143 (2000)
49. Kocsis, I.: On the stability of a sum form functional equation of multiplicative type. Acta Acad. Paedagog. Agriensis Sect. Mat. (N.S.) **28**, 43–53 (2001)
50. Kocsis, I.: On the stability of a sum form equation in several variables. Ann. Soc. Math. Pol. Ser. I Commentat. Math. **44** (1), 67–92 (2004)
51. Kocsis, I., Maksa, Gy.: The stability of a sum form functional equation arising in information theory. Acta Math. Hung. **79**(1–2), 39–48 (1998)
52. Kuczma, M.: An introduction to the theory of functional equations and inequalities. In: Gilányi, A. (ed.) Cauchy's Equation and Jensen's Inequality, vol. XIV, 2nd ed., p. 595. Birkhäuser, Basel (2009)
53. Kullback, S.: Information Theory and Statistics. Wiley Publication in Mathematical Statistics, vol. XVII, p. 395. Wiley/Chapman, New York/London (1959)
54. Lawrence, J.: The Shannon kernel of a nonnegative information function. Aequ. Math. **23**, 233–235 (1981)

55. Lee, P.M.: On the axioms of information theory. Ann. Math. Stat. **35**, 415–418 (1964)
56. Losonczi, L.: A characterization of entropies of degree alpha. Metrika **28**, 237–244 (1981)
57. Losonczi, L.: On a functional equation of sum form. Publ. Math. Debr. **36**, 167–177 (1989)
58. Losonczi, L.: Sum form equation on an open domain I. C.R. Math. Rep. Acad. Sci. Can. **7**(1), 85–90 (1985)
59. Losonczi, L., Maksa, Gy.: On some functional equations of the information theory. Acta Math. Acad. Sci. Hung. **39**, 73–82 (1982)
60. Maksa, Gy.: Bounded symmetric information functions. C. R. Math. Acad. Sci. Soc. R. Can. **2**, 247–252 (1980)
61. Maksa, Gy.: On the bounded solutions of a functional equation. Acta Math. Acad. Sci. Hung. **37**(4) (1981), 445–450
62. Maksa, Gy.: Solution on the open triangle of the generalized fundamental equation of information with four unknown functions. Util. Math. **21C**, 267–282 (1982)
63. Maksa, Gy.: The general solution of a functional equation arising in the information theory. Acta Math. Acad, Sci. Hung. **49**, 213–217 (1987)
64. Maksa, Gy.: The role of boundedness and nonnegativity in characterizing entropies of degree α. Publ. Math. Debr. **36**, 179–185 (1989)
65. Maksa, Gy.: On the stability of a sum form equation. Result. Math. **26**(3–4), 342–347 (1994)
66. Maksa, Gy.: The generalized associativity equation revisited. Rocz. Nauk.-Dydakt. Pr. Mat. 17, 175–180 (2000)
67. Maksa, Gy.: The stability of the entropy of degree alpha. J. Math. Anal. Appl. **346**(1), 17–21 (2008)
68. Maksa, Gy., Ng, C.T.: The fundamental equation of information on open domain. Publ. Math. **33**, 9–11 (1986)
69. Maksa, Gy., Páles, Zs.: Hyperstability of a class of linear functional equations., Acta Math. Acad. Paedagog. Nyházi. (N.S.) **17**(2), 107–112 (2001)
70. Mittal, D.P.: On continuous solutions of a functional equation. Metrika. **22**, 31–40 (1970)
71. Morando, A.: A stability result concerning the Shannon entropy. Aequ. Math. **62**(3), 286–296 (2001)
72. Shannon, C.E.: A mathematical theory of communication. Bell Syst. Tech. J. **27**, 379–423 & 623–656 (1948)
73. Skof, F.: Sull'approssimazione delle applicazioni localmente delta-additive. Atti, Acad. Sc. Torino **17**, 377–389 (1983) (Italian)
74. Székelyhidi, L.: Stability properties of functional equations in several variables. Publ. Math. **47**(1–2), 95–100 (1995)
75. Székelyhidi, L.: Note on stability theorem. Can. Math. Bull. **25**, 500–501 (1982)
76. Székelyhidi, L.: 38. Problem, Report of Meeting: The 28th International Symposium on Functional Equations, 23 Aug–1 Sept, Graz-Mariatrost, Austria (1991) (Aequ. Math., vol. 41, p. 302)
77. Tabor, Ja., Tabor, Jó.: Stability of the Cauchy equation on an interval. Aequ. Math. **55**(1–2), 153–176 (1998)
78. Tsallis, C.: Possible generalization of Boltzmann-Gibbs statistics. J. Stat. Phys. **52**(1–2), 479–487 (1988)
79. Tverberg, H.: A new derivation of the information function. Math. Scand. **6**, 297–298 (1958)
80. Ulam, S.M.: Problems in Modern Mathematics. Science Editions, vol. XVII, p. 150. Wiley, New York (1964) (First published under the title 'A Collection of Mathematical Problems')

Approximate Cauchy–Jensen Type Mappings in Quasi-β-Normed Spaces

Hark-Mahn Kim, Kil-Woung Jun and Eunyoung Son

Abstract In this chapter, we find the general solution of the following Cauchy–Jensen type functional equation

$$f\left(\frac{x+y}{n}+z\right)+f\left(\frac{y+z}{n}+x\right)+f\left(\frac{z+x}{n}+y\right)=\frac{n+2}{n}[f(x)+f(y)+f(z)],$$

and then investigate the generalized Hyers–Ulam stability of the equation in quasi-β-normed spaces for any fixed nonzero integer n.

Keywords Cauchy-Jensen type mappings · Hyers-Ulam stability · Homomorphisms · Quasi-β-normed spaces

1 Introduction

In 1940, S. M. Ulam gave a talk before the Mathematics Club of the University of Wisconsin in which he discussed a number of unsolved problems. The stability problem of functional equations originated from a question of S. M. Ulam [23] concerning the stability of group homomorphisms.

We are given a group G_1 and a metric group G_2 with metric $\rho(\cdot,\cdot)$. Given $\epsilon > 0$, does there exist a number $\delta > 0$ such that if $f : G_1 \to G_2$ satisfies $\rho(f(xy), f(x)f(y)) < \delta$ for all $x, y \in G_1$, then a homomorphism $h : G_1 \to G_2$ exists with $\rho(f(x), h(x)) < \epsilon$ for all $x \in G_1$?

In 1941, D. H. Hyers [8] proved the following stability result for the case of approximate additive mappings between Banach spaces. Suppose that E_1 and E_2

E. Son (✉) · K.-W. Jun · H.-M. Kim
Department of Mathematics, Chungnam National University, 220 Yuseong-Gu, 305-764 Daejeon, Korea
e-mail: sey8405@hanmail.net

K.-W. Jun
e-mail: kwjun@cnu.ac.kr

H.-M. Kim
e-mail: hmkim@cnu.ac.kr

T. M. Rassias (ed.), *Handbook of Functional Equations*,
Springer Optimization and Its Applications 96, DOI 10.1007/978-1-4939-1286-5_11

are Banach spaces and $f : E_1 \to E_2$ satisfies the following condition: if there is a number $\epsilon \geq 0$ such that

$$\|f(x+y) - f(x) - f(y)\| \leq \epsilon$$

for all $x, y \in E_1$, then the limit $h(x) = \lim_{n\to\infty} \frac{f(2^n x)}{2^n}$ exists for all $x \in E_1$ and there exists a unique additive mapping $h : E_1 \to E_2$ such that

$$\|f(x) - h(x)\| \leq \epsilon.$$

Moreover, if $f(tx)$ is continuous in $t \in \mathbf{R}$ for each $x \in E_1$, then the mapping h is $\mathbf{R}$-linear.

The method which was provided by D. H. Hyers, and which produces the additive mapping h, is called a *direct method.* This method is the most important and most powerful tool for studying the stability of various functional equations. Hyers' theorem was generalized by T. Aoki [1] and D. G. Bourgin [3] for additive mappings by considering an unbounded Cauchy difference. In 1978, Th. M. Rassias [15] also provided a generalization of Hyers' theorem for linear mappings which allows the Cauchy difference to be unbounded. Let E_1 and E_2 be two Banach spaces and $f : E_1 \to E_2$ be a mapping such that $f(tx)$ is continuous in $t \in \mathbf{R}$ for each fixed x. Assume that there exist $\epsilon > 0$ and $0 \leq p < 1$ such that

$$\|f(x+y) - f(x) - f(y)\| \leq \epsilon(\|x\|^p + \|y\|^p), \quad \forall x, y \in E_1.$$

Then, there exists a unique $\mathbf{R}$-linear mapping $T : E_1 \to E_2$ such that

$$\|f(x) - T(x)\| \leq \frac{2\epsilon}{2 - 2^p}\|x\|^p$$

for all $x \in E_1$. In 1990, Th. M. Rassias [16] during the 27th International Symposium on Functional Equations asked the question whether such a theorem can also be proved for $p \geq 1$. In 1991, Z. Gajda [5], following the same approach as in [15], gave an affirmative solution to this question for $p > 1$. However, it was shown by Z. Gajda [5], as well as by Th. M. Rassias and P. Šemrl [20], that one cannot prove a Rassias' type theorem when $p = 1$. The counterexamples of Z. Gajda [5], as well as of Th. M. Rassias and P. Šemrl [20], have stimulated several mathematicians to invent new approximately additive or approximately linear mappings. A generalized result of Rassias' theorem was obtained by P. Găvruta in [6] and S. Jung in [10]. J. M. Rassias [18, 19] established the Hyers–Ulam stability of linear and nonlinear mappings related to Jensen and Jensen type functional equations. In 1999, P. Găvruta [7] answered a question of J. M. Rassias [14] concerning the stability of the Cauchy equation. During the last two decades, the stability problems of several functional equations have been intensively and extensively investigated by a number of authors and there are many interesting volumes containing these stability problems of several functional equations [4, 9, 11, 21].

The notion of quasi-β-normed space was introduced by J. M. Rassias and H. Kim in [17].

We fix a real number β with $0 < \beta \leq 1$ and let $\mathbf{K}$ denote either $\mathbf{R}$ or $\mathbf{C}$. Let X be a real linear space over $\mathbf{K}$. A quasi-β-norm is a real-valued function on X satisfying the following:

(1) $\|x\| \geq 0$ for all $x \in X$ and $\|x\| = 0$ if and only if $x = 0$.
(2) $\|\lambda x\| = |\lambda|^{\beta}\|x\|$ for all $\lambda \in \mathbf{K}$ and all $x \in X$.
(3) There is a constant $M \geq 1$ such that $\|x + y\| \leq M(\|x\| + \|y\|)$ for all $x, y \in X$.

The pair $(X, \|\cdot\|)$ is called a quasi-β-normed space if $\|\cdot\|$ is a quasi-β-norm on X. The smallest possible M is called the modulus of concavity of the quasi-β-norm $\|\cdot\|$. A quasi-β-Banach space is a complete quasi-β-normed space.

Let p be a real number with $(0 < p \leq 1)$. Then, the quasi-β-norm $\|\cdot\|$ on X is called a (β, p)-norm if, moreover, $\|\cdot\|^p$ satisfies the following triangle inequality

$$\|x + y\|^p \leq \|x\|^p + \|y\|^p$$

for all $x, y \in X$. In this case, a quasi-β-Banach space is called a (β, p)-Banach space. We can refer to [2, 22] for the concept of quasi-normed spaces and p-Banach spaces for $\beta = 1$ as a special case.

Given a p-norm, the formula $d(x, y) := \|x - y\|^p$ gives us a translation invariant metric on X. By the Aoki–Rolewicz theorem [22], each quasi-norm is equivalent to some p-norm (see also [2]). Since it is much easier to work with p-norms, henceforth, we restrict our attention mainly to p-norms.

We observe that if $x_1, x_2, \ldots, x_n$ are nonnegative real numbers, then

$$\left(\sum_{i=1}^{n} x_i\right)^p \leq \sum_{i=1}^{n} x_i{}^p,$$

where $0 < p \leq 1$ [13].

In 2007, A. Najati [12] has introduced the Hyers–Ulam stability of the Cauchy–Jensen type functional equation

$$f\left(\frac{x+y}{2} + z\right) + f\left(\frac{y+z}{2} + x\right) + f\left(\frac{z+x}{2} + y\right) = 2[f(x) + f(y) + f(z)] \tag{1}$$

and then has investigated homomorphisms between JB^*–triples, and derivations on JB^*–triples associated to the functional equation.

Now, we introduce a modified and generalized Cauchy–Jensen type functional equation

$$f\left(\frac{x+y}{n} + z\right) + f\left(\frac{y+z}{n} + x\right) + f\left(\frac{z+x}{n} + y\right) = \frac{n+2}{n}[f(x) + f(y) + f(z)] \tag{2}$$

for any fixed nonzero integer n. This equation reduces to functional Eq. (1) for $n = 2$. In this chapter, we establish the general solution of the functional Eq. (2).

In the sequel, we investigate the generalized Hyers–Ulam stability of (2) in (β, p)-Banach spaces.

2 Generalized Hyers–Ulam Stability of (2)

First, we present the general solution of the Eq. (2).

Lemma 1 *Let both X and Y be vector spaces. A mapping $f : X \to Y$ satisfies* (2) *if and only if f is additive.*

Proof Suppose that a mapping $f : X \to Y$ satisfies (2) for all $x, y, z \in X$. If we put x, y, z in (2) by 0, then we have $f(0) = 0$. If $n = 1$, then one can easily show that f is additive. Now, let $n \neq 1$. Setting $y, z = 0$ in (2), we get

$$f\left(\frac{x}{n}\right) = \frac{1}{n} f(x) \tag{3}$$

for all $x \in X$. Using (3), one obtains that

$$f(x + y + nz) + f(y + z + nz) + f(z + x + ny) = (n + 2)[f(x) + f(y) + f(z)] \tag{4}$$

for all $x, y, z \in X$. Putting $z = 0$ in (4), we have

$$f(x + y) + f(nx + y) + f(x + ny) = (n + 2)[f(x) + f(y)] \tag{5}$$

for all $x, y \in X$. Replacing z by $-y$ in (4) yields

$$f(x - (n - 1)y) + f(x + (n - 1)y) = 2\, f(x) + (n + 2)[f(y) + f(-y)] \tag{6}$$

for all $x, y \in X$.

Now, we claim that f is an odd mapping by showing $f_e \equiv 0$, where $f_e(x) = \frac{f(x)+f(-x)}{2}$ is the even part of f. Since the mapping $f : X \to Y$ satisfies (2), the even mapping f_e is also a solution of (2).

Thus, applying f_e to (5) and (6), we have two equations

$$f_e(x + y) + f_e(nx + y) + f_e(x + ny) = (n + 2)[f_e(x) + f_e(y)], \tag{7}$$

$$f_e(x - (n - 1)y) + f_e(x + (n - 1)y) = 2f_e(x) + 2(n + 2)f_e(y) \tag{8}$$

for all $x, y \in X$. Putting $-x + y$ instead of y in (7), one has by the evenness of f_e

$$f_e(y) + f_e((n - 1)x + y) + f_e((n - 1)x - ny) = (n + 2)[f_e(x) + f_e(-x + y)] \tag{9}$$

for all $x, y \in X$. Replacing y by $-y$ in (9) and using the evenness of f_e, one arrives at

$$f_e(y)+f_e((n - 1)x - y)+f_e((n - 1)x + ny) = (n + 2)[f_e(x)+f_e(x + y)] \tag{10}$$

for all $x, y \in X$. Adding (9) and (10), one obtains by using (8) that

$$\begin{aligned}
2f_e(y) &+ \{f_e((n-1)x+y) + f_e((n-1)x-y)\} \\
&+ \{f_e((n-1)x-ny) + f_e((n-1)x+ny)\} \\
&= 2f_e(y) + \{2f_e(y) + 2(n+2)f_e(x)\} + \{2f_e(ny) + 2(n+2)f_e(x)\} \\
&= (n+2)[2f_e(x) + f_e(x+y) + f_e(x-y)],
\end{aligned}$$

which yields

$$2(n+2)f_e(x) + 2(n+2)f_e(y) = (n+2)[f_e(x+y) + f_e(x-y)] \tag{11}$$

for all $x, y \in X$.

Thus, if $n \neq -2$, then f_e satisfies the quadratic functional equation

$$f_e(x+y) + f_e(x-y) = 2f_e(x) + 2f_e(y) \tag{12}$$

for all $x, y \in X$. Combining (3) and (12), one can easily conclude that $f_e \equiv 0$ if $n \neq -2$. If $n = -2$, then we see from (8) that f_e satisfies the equation

$$f_e(x+3y) + f_e(x-3y) = 2f_e(x)$$

for all $x, y \in X$. Thus, f_e is a Jensen and an odd mapping, and so $f_e \equiv 0$ since f_e is even. Hence, $f_e \equiv 0$ for any fixed nonzero integer $n \neq 1$. Therefore, f is an odd mapping, as claimed.

Therefore, it follows from (6) that

$$f(x-(n-1)y) + f(x+(n-1)y) = 2\,f(x)$$

for all $x, y \in X$, which is equivalent to the Cauchy–Jensen equation $f(x+y) + f(x-y) = 2\,f(x)$. Therefore, f is additive.

The proof of the converse is trivial.

From now on, we assume that X is a linear space and Y is a (β, p)-Banach space with (β, p)-norm $\|\cdot\|$ without any specific reference. For notational convenience, given a mapping $f : X \to Y$, we define the difference operator $Df : X^3 \to Y$ of the Eq. (2) by

$$\begin{aligned}
Df(x,y,z) := f\left(\frac{x+y}{n}+z\right) + f\left(\frac{y+z}{n}+x\right) + f\left(\frac{z+x}{n}+y\right) \\
- \frac{n+2}{n}[f(x) + f(y) + f(z)]
\end{aligned}$$

for all $x, y, z \in X$.

Theorem 1 *Assume that a mapping $f : X \to Y$ satisfies the functional inequality*

$$\|Df(x,y,z)\| \leq \varphi(x,y,z) \tag{13}$$

for all $x, y, z \in X$, and the perturbing function $\varphi : X^3 \to [0, \infty)$ satisfies

$$\sum_{i=0}^{\infty} \frac{1}{|k|^{\beta p i}} \varphi(k^i x, k^i y, k^i z)^p < \infty, \tag{14}$$

for all $x \in X$, where $k := \frac{n+2}{n}$. Then, there exists a unique additive mapping $A : X \to Y$ defined by $A(x) = \lim_{m\to\infty} \frac{f(k^m x)}{k^m}$ such that

$$\|f(x) - A(x)\| \le \frac{1}{|3k|^{\beta}} \left[\sum_{i=0}^{\infty} \frac{1}{|k|^{\beta p i}} \varphi(k^i x, k^i x, k^i x)^p \right]^{\frac{1}{p}} \tag{15}$$

for all $x \in X$.

Proof Letting $y = z := x$ in (1), we have

$$\left\| 3f\left(\left(\frac{n+2}{n}\right)x\right) - 3\left(\frac{n+2}{n}\right) f(x) \right\| \le \varphi(x, x, x) \tag{16}$$

for all $x \in X$. Putting $k := \frac{n+2}{n}$, we obtain

$$\|3f(kx) - 3kf(x)\| \le \varphi(x, x, x) \tag{17}$$

for all $x \in X$. If we replace x by $k^m x$ and divide both sides on (17) by $|3|^{\beta}\ |k|^{(m+1)\beta}$, we get

$$\left\| \frac{f(k^{m+1}x)}{k^{m+1}} - \frac{f(k^m x)}{k^m} \right\| \le \frac{1}{|3|^{\beta}|k|^{(m+1)\beta}} \varphi(k^m x, k^m x, k^m x) \tag{18}$$

for all $x \in X$ and all nonnegative integer m. It follows from (18) that

$$\begin{aligned} \left\| \frac{f(k^{m+1}x)}{k^{m+1}} - \frac{f(k^l x)}{k^l} \right\|^p &= \left\| \sum_{i=l}^{m} \frac{1}{k^{i+1}} f(k^{i+1}x) - \frac{1}{k^i} f(k^i x) \right\|^p \\ &\le \sum_{i=l}^{m} \left\| \frac{1}{k^{i+1}} f(k^{i+1}x) - \frac{1}{k^i} f(k^i x) \right\|^p \\ &\le \frac{1}{|3k|^{\beta p}} \sum_{i=l}^{m} \frac{1}{|k|^{\beta p i}} \varphi(k^i x, k^i x, k^i x)^p \end{aligned} \tag{19}$$

for all nonnegative integers m and l with $m > l \ge 0$ and $x \in X$. Since the right-hand side of (19) tends to zero as $l \to \infty$, we obtain that the sequence $\{\frac{f(k^m x)}{k^m}\}$ is Cauchy for all $x \in X$. Because of the fact that Y is a (β, p)-Banach space, it follows that the sequence $\{\frac{f(k^m x)}{k^m}\}$ converges in Y. Therefore, we can define a mapping $A : X \to Y$ as

$$A(x) = \lim_{m\to\infty} \frac{f(k^m x)}{k^m}, \quad x \in X.$$

Moreover, letting $l = 0$ and taking $m \to \infty$ in (19), we get

$$\|f(x) - A(x)\| \le \frac{1}{|3k|^\beta} \left[\sum_{i=0}^{\infty} \frac{1}{|k|^{\beta p i}} \varphi(k^i x, k^i x, k^i x)^p \right]^{\frac{1}{p}}$$

for all $x \in X$. It follows from (1) and (14) that

$$\begin{aligned} \|DA(x,y,z)\|^p &= \lim_{i\to\infty} \left\| \frac{Df(k^i x, k^i y, k^i z)}{k^i} \right\|^p \\ &\le \lim_{i\to\infty} \frac{1}{|k|^{\beta p i}} \varphi(k^i x, k^i y, k^i z)^p = 0 \end{aligned}$$

for all $x, y, z \in X$. Therefore, the mapping A satisfies (2) and so the mapping A is additive.

Now, to prove the uniqueness of the additive mapping A satisfying (15), let $A' : X \to Y$ be another additive mapping satisfying (15). Then, we have

$$\begin{aligned} \|A(x) - A'(x)\|^p &= \left\| \frac{1}{k^j} A(k^j x) - \frac{1}{k^j} A'(k^j x) \right\|^p \\ &\le \frac{1}{|k|^{\beta p j}} (\|A(k^j x) - f(k^j x)\|^p + \|f(k^j x) - A'(k^j x)\|^p) \\ &\le \frac{2}{|3k|^{\beta p}} \sum_{i=0}^{\infty} \frac{1}{|k|^{\beta p(i+j)}} \varphi(k^{i+j} x, k^{i+j} x, k^{i+j} x)^p \\ &\le \frac{2}{|3k|^{\beta p}} \sum_{i=j}^{\infty} \frac{1}{|k|^{\beta p i}} \varphi(k^i x, k^i x, k^i x)^p \end{aligned}$$

for all $j \in \mathbf{N}$ and all $x \in X$. Taking the limit as $j \to \infty$, we conclude that

$$A(x) = A'(x)$$

for all $x \in X$. This completes the proof.

Corollary 1 *Let X be a quasi-α-normed linear space with quasi-α-norm $\|\cdot\|$ and let r_i,θ_i be nonnegative real numbers with $0 < \alpha r_i < \beta$ for all $i = 1, 2, 3$. If a mapping $f : X \to Y$ satisfies the following functional inequality*

$$\|Df(x,y,z)\| \le \theta_1 \|x\|^{r_1} + \theta_2 \|y\|^{r_2} + \theta_3 \|z\|^{r_3}$$

for all $x, y, z \in X$, then there exists a unique additive mapping $h : X \to Y$ such that

$$\|f(x) - h(x)\| \le \frac{1}{|3|^\beta} \left[\sum_{i=1}^{3} \frac{\theta_i^p}{|k|^{\beta p} - |k|^{\alpha p r_i}} \|x\|^{r_i p} \right]^{1/p}$$

for all $x \in X$.

Proof Considering $\varphi(x, y, z) := \theta_1(\|x\|^{r_1} + \theta_2\|y\|^{r_2} + \theta_3\|z\|^{r_3})$ in Theorem 1, we lead to the desired results.

Theorem 2 *Assume that a mapping $f : X \to Y$ satisfies the functional inequality*

$$\|Df(x, y, z)\| \leq \varphi(x, y, z) \tag{20}$$

for all $x, y, z \in X$, and the perturbing function $\varphi : X^3 \to [0, \infty)$ satisfies

$$\sum_{i=1}^{\infty} |k|^{\beta p i} \varphi\left(\frac{x}{k^i}, \frac{y}{k^i}, \frac{z}{k^i}\right)^p < \infty, \tag{21}$$

for all $x \in X$, where $k := \frac{n+2}{n}$. Then, there exists a unique additive mapping $A : X \to Y$ defined by $A(x) = \lim_{m\to\infty} k^m f(\frac{x}{k^m})$ such that

$$\|f(x) - A(x)\| \leq \frac{1}{|3k|^{\beta}} \left[\sum_{i=1}^{\infty} |k|^{\beta p i} \varphi\left(\frac{x}{k^i}, \frac{x}{k^i}, \frac{x}{k^i}\right)^p\right]^{\frac{1}{p}}$$

for all $x \in X$.

Proof Dividing both sides on (17) by $|3|^{\beta}$, we obtain that

$$\|f(kx) - kf(x)\| \leq \frac{1}{|3|^{\beta}} \varphi(x, x, x) \tag{22}$$

for all $x \in X$. If we replace x in $\frac{x}{k^{m+1}}$ in (22) and multiply both sides of (22) by $|k|^{m\beta}$, we have

$$\left\|k^{m+1} f\left(\frac{x}{k^{m+1}}\right) - k^m f\left(\frac{x}{k^m}\right)\right\| \leq \frac{|k|^{m\beta}}{|3|^{\beta}} \varphi\left(\frac{x}{k^{m+1}}, \frac{x}{k^{m+1}}, \frac{x}{k^{m+1}}\right)$$

for all $x \in X$ and all nonnegative integer m. Hence

$$\begin{aligned} \left\|k^{m+1} f\left(\frac{x}{k^{m+1}}\right) - k^l f\left(\frac{x}{k^l}\right)\right\|^p &\leq \sum_{i=l}^{m} \left\|k^{i+1} f\left(\frac{x}{k^{i+1}}\right) - k^i f\left(\frac{x}{k^i}\right)\right\|^p \\ &\leq \frac{1}{|3|^{\beta p}} \sum_{i=l}^{m} |k|^{\beta p i} \varphi\left(\frac{x}{k^{i+1}}, \frac{x}{k^{i+1}}, \frac{x}{k^{i+1}}\right)^p \\ &\leq \frac{1}{|3k|^{\beta p}} \sum_{i=l}^{m} |k|^{\beta p(i+1)} \varphi\left(\frac{x}{k^{i+1}}, \frac{x}{k^{i+1}}, \frac{x}{k^{i+1}}\right)^p \end{aligned} \tag{23}$$

for all nonnegative integers m and l with $m > l \geq 0$ and all $x \in X$. The remaining proof is similar to the corresponding part of Theorem 1. This completes the proof.

Corollary 2 *Let X be a quasi-α-normed linear space with quasi-α-norm $\|\cdot\|$ and let r_i, θ_i be nonnegative real numbers with $\alpha r_i > \beta$ for all $i = 1, 2, 3$. If a mapping $f : X \to Y$ satisfies the following functional inequality*

$$\|Df(x, y, z)\| \leq \theta_1 \|x\|^{r_1} + \theta_2 \|y\|^{r_2} + \theta_3 \|z\|^{r_3}$$

for all $x, y, z \in X$, then there exists a unique additive mapping $h : X \to Y$ such that

$$\|f(x) - h(x)\| \leq \frac{1}{|3|^\beta} \left[\sum_{i=1}^{3} \frac{\theta_i^p}{|k|^{\alpha p r_i} - |k|^{\beta p}} \|x\|^{r_i p} \right]^{\frac{1}{p}}$$

for all $x \in X$.

Proof Putting $\varphi(x, y, z) := \theta_1(\|x\|^{r_1} + \theta_2 \|y\|^{r_2} + \theta_3 \|z\|^{r_3})$ in Theorem 2, we lead to the desired results.

3 Alternative Generalized Hyers–Ulam Stability of (2)

From now on, we investigate the generalized Hyers–Ulam stability of the functional inequality (2) using the contractive property of perturbing term of the inequality (2).

Theorem 3 *Assume that a mapping $f : X \to Y$ satisfies the functional inequality*

$$\|Df(x, y, z)\| \leq \varphi(x, y, z)$$

for all $x, y, z \in X$ and there exists a constant L with $0 < |k|^{1-\beta} L < 1$ for which the perturbing function $\varphi : X^3 \to [0, \infty)$ satisfies

$$\varphi(kx, ky, kz) \leq |k| L \varphi(x, y, z) \tag{24}$$

for all $x, y, z \in X$, where $k := \frac{n+2}{n}$. Then, there exists a unique additive mapping $A : X \to Y$ given by $A(x) = \lim_{m\to\infty} \frac{1}{k^m} f(k^m x)$ such that

$$\|f(x) - A(x)\| \leq \frac{1}{|3|^\beta \sqrt[p]{|k|^{\beta p} - |k|^p L^p}} \varphi(x, x, x) \tag{25}$$

for all $x \in X$.

Proof It follows from (19) and (24) that

$$\begin{aligned}
\left\| \frac{f(k^{m+1}x)}{k^{m+1}} - \frac{f(k^l x)}{k^l} \right\|^p &\leq \frac{1}{|3k|^{\beta p}} \sum_{i=l}^{m} \frac{1}{|k|^{\beta p i}} \varphi(k^i x, k^i x, k^i x)^p \\
&\leq \frac{1}{|3k|^{\beta p}} \sum_{i=l}^{m} \frac{(|k| L)^{pi}}{|k|^{\beta p i}} \varphi(x, x, x)^p \\
&= \frac{1}{|3k|^{\beta p}} \sum_{i=l}^{m} (|k|^{1-\beta} L)^{pi} \varphi(x, x, x)^p
\end{aligned}$$

for all nonnegative integers m and l with $m > l \geq 0$ and all $x \in X$. Thus, it follows that the sequence $\{\frac{f(k^m x)}{k^m}\}$ is Cauchy for all $x \in X$, and so we can define a mapping $A : X \to Y$ by

$$A(x) = \lim_{m\to\infty} \frac{f(k^m x)}{k^m}, \quad x \in X.$$

Moreover, letting $l = 0$ and $m \to \infty$ in the last inequality yields the approximation (25).

The remaining proof is similar to the corresponding part of Theorem 1. This completes the proof.

The following corollary is a generalization of the stability result of the special case $\xi(\|x\|) = \|x\|^r, r < 1$ when X is a normed space with $\alpha = 1$ and Y is a Banach space with $\beta = 1 = p$.

Corollary 3 *Let X be a quasi-α-normed linear space with quasi-α-norm $\|\cdot\|$. Let $\xi : [0,\infty) \to [0,\infty)$ be a nontrivial function satisfying*

$$\xi(|k|^\alpha t) \leq \xi(|k|)^\alpha \xi(t), \quad (t \geq 0), \qquad 0 < \xi(|k|)^\alpha < |k|^\beta,$$

where $k := \frac{n+2}{n}$. If a mapping $f : X \to Y$ satisfies the following functional inequality

$$\|Df(x,y,z)\| \leq \theta\{\xi(\|x\|) + \xi(\|y\|) + \xi(\|z\|)\}$$

for all $x,y,z \in X$ and for some $\theta \geq 0$, then there exists a unique additive mapping $A : X \to Y$ such that

$$\|f(x) - A(x)\| \leq \frac{3\theta}{|3|^\beta \sqrt[p]{|k|^{\beta p} - \xi(|k|)^{\alpha p}}}\xi(\|x\|)$$

for all $x \in X$.

Proof Letting $\varphi(x,y,z) = \theta\{\xi(\|x\|) + \xi(\|y\|) + \xi(\|z\|)\}$, and applying Theorem 3 with $L := \frac{\xi(|k|)^\alpha}{|k|}$, we obtain the desired result.

Theorem 4 *Suppose that a mapping $f : X \to Y$ satisfies the functional inequality*

$$\|Df(x,y,z)\| \leq \varphi(x,y,z)$$

for all $x,y,z \in X$ and there exists a constant L with $0 < |k|^{\beta-1}L < 1$ for which the perturbing function $\varphi : X^3 \to [0,\infty)$ satisfies

$$\varphi\left(\frac{x}{k},\frac{y}{k},\frac{z}{k}\right) \leq \frac{L}{|k|}\varphi(x,y,z) \tag{26}$$

for all $x,y,z \in X$. Then, there exists a unique additive mapping $A : X \to Y$ defined by $A(x) = \lim_{m\to\infty} k^m f(\frac{x}{k^m})$ such that

$$\|f(x) - A(x)\| \leq \frac{L}{|3|^\beta \sqrt[p]{|k|^p - |k|^{\beta p} L^p}}\varphi(x,x,x)$$

for all $x \in X$.

Proof It follows from (23) and (26) that

$$\begin{aligned}\left\|k^{m+1} f\left(\frac{x}{k^{m+1}}\right) - k^l f\left(\frac{x}{k^l}\right)\right\|^p &\leq \frac{1}{|3k|^{\beta p}} \sum_{i=l}^{m} |k|^{\beta p(i+1)} \varphi\left(\frac{x}{k^{i+1}}, \frac{x}{k^{i+1}}, \frac{x}{k^{i+1}}\right)^p \\ &= \frac{1}{|3k|^{\beta p}} \sum_{i=l+1}^{m+1} |k|^{\beta p i} \varphi\left(\frac{x}{k^i}, \frac{x}{k^i}, \frac{x}{k^i}\right)^p \\ &\leq \frac{1}{|3k|^{\beta p}} \sum_{i=l+1}^{m+1} \left(|k|^{\beta-1} L\right)^{pi} \varphi(x,x,x)^p\end{aligned}$$

for all nonnegative integers m and l with $m > l \geq 0$ and all $x \in X$.

The remaining proof is similar to the corresponding part of Theorem 1. This completes the proof.

Corollary 4 *Let X be a quasi-α-normed linear space with quasi-α-norm $\|\cdot\|$. Let $\xi : [0,\infty) \to [0,\infty)$ be a nontrivial function satisfying*

$$\xi\left(\frac{t}{|k|^\alpha}\right) \leq \xi\left(\frac{1}{|k|}\right)^\alpha \xi(t), \quad (t \geq 0), \qquad 0 < \xi\left(\frac{1}{|k|}\right)^\alpha < |k|^{-\beta},$$

where $k := \frac{n+2}{n}$. If a mapping $f : X \to Y$ satisfies the following functional inequality

$$\|Df(x,y,z)\| \leq \theta\{\xi(\|x\|) + \xi(\|y\|) + \xi(\|z\|)\}$$

for all $x, y, z \in X$ and for some $\theta \geq 0$, then there exists a unique additive mapping $A : X \to Y$ such that

$$\|f(x) - A(x)\| \leq \frac{3\theta\xi(\frac{1}{|k|})^\alpha}{|3|^\beta \sqrt[p]{1 - |k|^{\beta p}\xi(\frac{1}{|k|})^{\alpha p}}}\xi(\|x\|)$$

for all $x \in X$.

Proof Letting $\varphi(x,y,z) = \theta\{\xi(\|x\|) + \xi(\|y\|) + \xi(\|z\|)\}$ and applying Theorem 4 with $L := |k|\xi(\frac{1}{|k|})^\alpha$, we lead to the desired approximation.

References

1. Aoki, T.: On the stability of the linear transformation in Banach spaces. J. Math. Soc. Jpn. **2**, 64–66 (1950)
2. Benyamini, Y., Lindenstrauss, J.: Geometric nonlinear functional analysis, vol. 1, Colloquium Publications, vol. 48, American Mathematical Society, Providence, (2000)
3. Bourgin, D.G.: Classes of transformations and bordering transformations. Bull. Am. Math. Soc. **57**, 223–237 (1951)
4. Cho, Y.J., Rassias, Th.M. Saadati, R.: Stability of Functional Equations in Random Normed Spaces. Springer, New York (2013)

5. Gajda, Z.: On stability of additive mappings. Int. J. Math. Math. Sci. **14**, 431–434 (1991)
6. Găvruta, P.: A generalization of the Hyers–Ulam–Rassias stability of approximately additive mappings. J. Math. Anal. Appl. **184**, 431–436 (1994)
7. Găvruta, P.: An answer to a question of John M. Rassias concerning the stability of Cauchy equation, In: Advances in Equations and Inequalities, In: Hadronic Math. Ser., Hadronic Press, USA, 67–71 (1999)
8. Hyers, D.H.: On the stability of the linear functional equation. Proc. Natl. Acad. Sci. U. S. A. **27**, 222–224 (1941)
9. Hyers, D.H., Isac, G., Rassias, Th.M.: Stability of Functional Equations in Several Variables, Birkhäuser, Boston, (1998)
10. Jung, S.-M.: On the Hyers–Ulam–Rassias stability of approximately additive mappings. J. Math. Anal. Appl. **204**, 221–226 (1996)
11. Jung, S.-M., Popa, D., Rassias, M.Th.: On the stability of the linear functional equation in a single variable on complete metric groups. J. Glob. Optim. (to appear)
12. Najati, A.: Stability of homomoerphisms on JB^*–triples associated to a Cauchy–Jensen type functional equation. J. Math. Inequal. **1**, 83–103 (2007)
13. Najati, A., Moghimi, M.B.: Stability of a functional equation deriving from quadratic and additive function in quasi-Banach spaces. J. Math. Anal. Appl. **337**, 399–415 (2008)
14. Rassias, J.M.: Solution of a problem of Ulam, J. Approx. Theory **57**, 268–273 (1989)
15. Rassias Th.M.: On the stability of the linear mapping in Banach spaces, Proc. Am. Math. Soc. **72**, 297–300 (1978)
16. Rassias, Th.M.: The stability of mappings and related topics, in 'Report on the 27th ISFE'. Aequat. Math. **39**, 292–293 (1990)
17. Rassias, J.M., Kim, H.: Generalized Hyers–Ulam stability for additive functional equations in quasi-β-normed spaces. J. Math. Anal. Appl. **356**, 302–309 (2009).
18. Rassias, J.M., Rassias, M.J.: On the Ulam stability of Jensen and Jensen type mappings on restricted domains. J. Math. Anal. Appl. **281**, 516–524 (2003)
19. Rassias, J.M., Rassias, M.J.: Asymptotic behavior of alternative Jensen and Jensen type functional equations. Bull. Sci. Math. **129**, 545–558 (2005)
20. Rassias, Th.M., Šemrl, P.: On the behaviour of mappings which do not satisfy Hyers–Ulam–Rassias stability. Proc. Am. Math. Soc. **114**, 989–993 (1992)
21. Rassias, Th.M., Tabor, J., Stability of Mappings of Hyers-Ulam Type. Hadronic Press, Inc., Florida (1994)
22. Rolewicz, S.: Metric Linear Spaces. Reidel/PWN-Polish Scientific Publisher, Dordrecht (1984)
23. Ulam, S.M.: A Collection of the Mathematical Problems. Interscience Publisher, New York (1960)

An AQCQ-Functional Equation in Matrix Paranormed Spaces

Jung Rye Lee, Choonkil Park, Themistocles M. Rassias and Dong Yun Shin

Abstract In this chapter, we prove the Hyers–Ulam stability of an additive-quadratic-cubic-quartic functional equation in matrix paranormed spaces. Moreover, we prove the Hyers–Ulam stability of an additive-quadratic-cubic-quartic functional equation in matrix β-homogeneous F-spaces.

Keywords Paranormed spaces· Hyers–Ulam stability· Statistical convergence· Cauchy difference· Quadratic mapping

1 Introduction and Preliminaries

The concept of statistical convergence for sequences of real numbers was introduced by Fast [13] and Steinhaus [37] independently and since then several generalizations and applications of this notion have been investigated by various authors (see [14, 23, 26, 27, 35]). This notion was defined in normed spaces by Kolk [24].

We recall some basic facts concerning Fréchet spaces.

Definition 1 [39] Let X be a vector space. A paranorm $P(\cdot) : X \to [0, \infty)$ is a function on X such that

(1) $P(0) = 0$
(2) $P(-x) = P(x)$
(3) $P(x + y) \leq P(x) + P(y)$ (triangle inequality)

J. R. Lee (✉)
Department of Mathematics, Daejin University, Daejin, Korea
e-mail: jrlee@daejin.ac.kr

C. Park
Research Institute for Natural Sciences, Hanyang University, Hanyang, Korea
e-mail: baak@hanyang.ac.kr

T. M. Rassias
Department of Mathematics, National Technical University of Athens,
Zografou Campus, Athens, Greece
e-mail: trassias@math.ntua.gr

D. Shin
Department of Mathematics, University of Seoul, Seoul, Korea
e-mail: dyshin@uos.ac.kr

T. M. Rassias (ed.), *Handbook of Functional Equations,*
Springer Optimization and Its Applications 96, DOI 10.1007/978-1-4939-1286-5_12

(4) If $\{t_n\}$ is a sequence of scalars with $t_n \to t$ and $\{x_n\} \subset X$ with $P(x_n - x) \to 0$, then $P(t_n x_n - tx) \to 0$ (continuity of multiplication).
The pair $(X, P(\cdot))$ is called a *paranormed space* if $P(\cdot)$ is a *paranorm* on X.
The paranorm is called *total* if, in addition, we have
(5) $P(x) = 0$ implies $x = 0$.
A *Fréchet space* is a total and complete paranormed space.

Definition 2 Let X be a linear space. A nonnegative valued function $\|\cdot\|$ is an F-norm if it satisfies the following conditions:
(FN_1) $\|x\| = 0$ if and only if $x = 0$;
(FN_2) $\|\lambda x\| = \|x\|$ for all $x \in X$ and all λ with $|\lambda| = 1$;
(FN_3) $\|x + y\| \leq \|x\| + \|y\|$ for all $x, y \in X$;
(FN_4) $\|\lambda_n x\| \to 0$ provided $\lambda_n \to 0$;
(FN_5) $\|\lambda x_n\| \to 0$ provided $\|x_n\| \to 0$.
Then $(X, \|\cdot\|)$ is called an F^*-space. An F-space is a complete F^*-space.

An F-norm is called β-homogeneous ($\beta > 0$) if $\|tx\| = |t|^\beta \|x\|$ for all $x \in X$ and all $t \in \mathbf{R}$ (see [33]).

The stability problem of functional equations originated from a question of Ulam [38] concerning the stability of group homomorphisms.

The functional equation

$$f(x + y) = f(x) + f(y)$$

is called the *Cauchy additive functional equation*. In particular, every solution of the Cauchy additive functional equation is said to be an *additive mapping*. Hyers [18] gave a first affirmative partial answer to the question of Ulam for Banach spaces. Hyers' Theorem was generalized by Aoki [2] for additive mappings and by Rassias [30] for linear mappings by considering an unbounded Cauchy difference. A generalization of the Rassias theorem was obtained by Găvruta [16] by replacing the unbounded Cauchy difference by a general control function in the spirit of Rassias' approach.

In 1990, [31] during the 27th International Symposium on Functional Equations, Rassias asked the question whether such a theorem can also be proved for $p \geq 1$. In 1991, Gajda [15] following the same approach as in Rassias [30], gave an affirmative solution to this question for $p > 1$. It was shown by Gajda [15], as well as by Rassias and Šemrl [32] that one cannot prove a Rassias' type theorem when $p = 1$ (cf. the books of Czerwik [6] and Hyerset al. [19]).

The functional equation

$$f(x + y) + f(x - y) = 2f(x) + 2f(y)$$

is called a *quadratic functional equation*. In particular, every solution of the quadratic functional equation is said to be a *quadratic mapping*. A Hyers–Ulam stability problem for the quadratic functional equation was proved by Skof [36] for mappings $f : X \to Y$, where X is a normed space and Y is a Banach space. Cholewa [4] noticed that the theorem of Skof is still true if the relevant domain X is replaced by an Abelian group. Czerwik [5] proved the Hyers–Ulam stability of the quadratic functional equation.

In [21], Jun and Kim considered the following cubic functional equation

$$f(2x+y)+f(2x-y)=2f(x+y)+2f(x-y)+12f(x). \tag{1}$$

It is easy to show that the function $f(x)=x^3$ satisfies the functional Eq. (1), which is called a *cubic functional equation* and every solution of the cubic functional equation is said to be a *cubic mapping*.

In [25], Lee et al. considered the following quartic functional equation

$$f(2x+y)+f(2x-y)=4f(x+y)+4f(x-y)+24f(x)-6f(y). \tag{2}$$

It is easy to show that the function $f(x)=x^4$ satisfies the functional Eq. (2), which is called a *quartic functional equation* and every solution of the quartic functional equation is said to be a *quartic mapping*. The stability problems of several functional equations have been extensively investigated by a number of authors and there are many interesting results concerning this problem (see [1, 10, 20, 22, 28]).

The abstract characterization given for linear spaces of bounded Hilbert space operators in terms of *matricially normed spaces* [34] implies that quotients, mapping spaces and various tensor products of operator spaces may again be regarded as operator spaces. Owing in part to this result, the theory of operator spaces is having an increasingly significant effect on operator algebra theory (see [8]).

The proof given in [34] appealed to the theory of ordered operator spaces [3]. Effros and Ruan [9] showed that one can give a purely metric proof of this important theorem by using a technique of Pisier [29] and Haagerup [17] (as modified in [7]).

We will use the following notations:

$M_n(X)$ is the set of all $n\times n$-matrices in X;

$e_j\in M_{1,n}(\mathbf{C})$ is that j-th component is 1 and the other components are 0;

$E_{ij}\in M_n(\mathbf{C})$ is that (i,j)-component is 1 and the other components are 0;

$E_{ij}\otimes x\in M_n(X)$ is that (i,j)-component is x and the other components are 0;

For $x\in M_n(X)$, $y\in M_k(X)$,

$$x\oplus y=\begin{pmatrix} x & 0\\ 0 & y\end{pmatrix}.$$

Note that $(X,\{\|\cdot\|_n\})$ is a matrix normed space if and only if $(M_n(X),\|\cdot\|_n)$ is a normed space for each positive integer n and $\|AxB\|_k\le\|A\|\|B\|\|x\|_n$ holds for $A\in M_{k,n}$, $x=[x_{ij}]\in M_n(X)$ and $B\in M_{n,k}$, and that $(X,\{\|\cdot\|_n\})$ is a matrix Banach space if and only if X is a Banach space and $(X,\{\|\cdot\|_n\})$ is a matrix normed space.

Definition 3 Let $(X,P(\cdot))$ be a paranormed space.

(1) $(X,\{P_n(\cdot)\})$ is a *matrix paranormed space* if $(M_n(X),P_n(\cdot))$ is a paranormed space for each positive integer n, $P_n(E_{kl}\otimes x)=P(x)$ for $x\in X$, and $P(x_{kl})\le P_n([x_{ij}])$ for $[x_{ij}]\in M_n(X)$.

(2) $(X, \{P_n(\cdot)\})$ is a *matrix Fréchet space* if X is a Fréchet space and $(X, \{P_n(\cdot)\})$ is a *matrix paranormed space*.

Definition 4 Let $(X, \|\cdot\|)$ be an F^*-space.

(1) $(X, \{\|\cdot\|_n\})$ is a *matrix F^*-space* if $(M_n(X), \|\cdot\|_n)$ is an F^*-space for each positive integer n, $\|E_{kl} \otimes x\|_n = \|x\|$ for $x \in X$, and $\|x_{kl}\| \leq \|[x_{ij}]\|_n$ for $[x_{ij}] \in M_n(X)$.

(2) $(X, \{\|\cdot\|_n\})$ is a *matrix F-space* if X is an F-space and $(X, \{\|\cdot\|_n\})$ is a *matrix F^*-space*.

Let E, F be vector spaces. For a given mapping $h : E \to F$ and a given positive integer n, define $h_n : M_n(E) \to M_n(F)$ by

$$h_n([x_{ij}]) = [h(x_{ij})]$$

for all $[x_{ij}] \in M_n(E)$.

In this chapter, we prove the Hyers–Ulam stability of the following additive-quadratic-cubic-quartic functional equation

$$\begin{aligned} f(x+2y) + f(x-2y) = 4f(x+y) + 4f(x-y) \\ - 6f(x) + f(2y) + f(-2y) - 4f(y) - 4f(-y) \end{aligned} \tag{3}$$

in matrix normed spaces and in matrix β-homogeneous F-spaces by the direct method.

One can easily show that an odd mapping $f : X \to Y$ satisfies (3) if and only if the odd mapping $f : X \to Y$ is an additive-cubic mapping, i.e.,

$$f(x+2y) + f(x-2y) = 4f(x+y) + 4f(x-y) - 6f(x).$$

It was shown in [12, Lemma 2.2] that $g(x) := f(2x) - 2f(x)$ and $h(x) := f(2x) - 8f(x)$ are cubic and additive, respectively, and that $f(x) = \frac{1}{6}g(x) - \frac{1}{6}h(x)$.

One can easily show that an even mapping $f : X \to Y$ satisfies (3) if and only if the even mapping $f : X \to Y$ is a quadratic-quartic mapping, i.e.,

$$f(x+2y) + f(x-2y) = 4f(x+y) + 4f(x-y) - 6f(x) + 2f(2y) - 8f(y).$$

It was shown in [11, Lemma 2.1] that $g(x) := f(2x) - 4f(x)$ and $h(x) := f(2x) - 16f(x)$ are quartic and quadratic, respectively, and that $f(x) = \frac{1}{12}g(x) - \frac{1}{12}h(x)$.

2 Hyers–Ulam Stability of the AQCQ-Functional Equation (3) in Matrix Paranormed Spaces: Odd Mapping Case

In this section, we prove the Hyers–Ulam stability of the AQCQ-functional equation (3) in matrix paranormed spaces for an odd mapping case.

Throughout this section, let $(X, \{\|\cdot\|_n\})$ be a matrix Banach space and $(Y, \{P_n(\cdot)\})$ a matrix Fréchet space.

Lemma 1 *Let $(X, \{P_n(\cdot)\})$ be a matrix paranormed space. Then*

(1) $P(x_{kl}) \le P_n([x_{ij}]) \le \sum_{i,j=1}^{n} P(x_{ij})$ *for* $[x_{ij}] \in M_n(X)$.

(2) $\lim_{s\to\infty} x_s = x$ *if and only if* $\lim_{s\to\infty} x_{sij} = x_{ij}$ *for* $x_s = [x_{sij}], x = [x_{ij}] \in M_k(X)$.

Proof (1) By Definition 3, $P(x_{kl}) \le P_n([x_{ij}])$.
Since $[x_{ij}] = \sum_{i,j=1}^{n} E_{ij} \otimes x_{ij}$,

$$P_n([x_{ij}]) = P_n\left(\sum_{i,j=1}^{n} E_{ij} \otimes x_{ij}\right) \le \sum_{i,j=1}^{n} P_n(E_{ij} \otimes x_{ij}) = \sum_{i,j=1}^{n} P(x_{ij}).$$

(2) By (1), we have

$$P(x_{skl} - x_{kl}) \le P_n([x_{sij} - x_{ij}]) = P_n([x_{sij}] - [x_{ij}]) \le \sum_{i,j=1}^{n} P(x_{sij} - x_{ij}).$$

So we get the result.

Lemma 2 *Let $(X, \{\|\cdot\|_n\})$ be a matrix normed space or a matrix F^*-space. Then*

(1) $\|E_{kl} \otimes x\|_n = \|x\|$ *for* $x \in X$.

(2) $\|x_{kl}\| \le \|[x_{ij}]\|_n \le \sum_{i,j=1}^{n} \|x_{ij}\|$ *for* $[x_{ij}] \in M_n(X)$.

(3) $\lim_{n\to\infty} x_n = x$ *if and only if* $\lim_{n\to\infty} x_{ijn} = x_{ij}$ *for* $x_n = [x_{ijn}], x = [x_{ij}] \in M_k(X)$.

Proof (1) Since $E_{kl} \otimes x = e_k^* x e_l$ and $\|e_k^*\| = \|e_l\| = 1$, $\|E_{kl} \otimes x\|_n \le \|x\|$. Since $e_k(E_{kl} \otimes x)e_l^* = x$, $\|x\| \le \|E_{kl} \otimes x\|_n$. So $\|E_{kl} \otimes x\|_n = \|x\|$.

(2) Since $e_k x e_l^* = x_{kl}$ and $\|e_k\| = \|e_l^*\| = 1$, $\|x_{kl}\| \le \|[x_{ij}]\|_n$. Since $[x_{ij}] = \sum_{i,j=1}^{n} E_{ij} \otimes x_{ij}$,

$$\|[x_{ij}]\|_n = \left\|\sum_{i,j=1}^{n} E_{ij} \otimes x_{ij}\right\|_n \le \sum_{i,j=1}^{n} \|E_{ij} \otimes x_{ij}\|_n = \sum_{i,j=1}^{n} \|x_{ij}\|.$$

(3) By

$$\|x_{kln} - x_{kl}\| \le \|[x_{ijn} - x_{ij}]\|_n = \|[x_{ijn}] - [x_{ij}]\|_n \le \sum_{i,j=1}^{n} \|x_{ijn} - x_{ij}\|,$$

we get the result.

For a mapping $f : X \to Y$, define $Df : X^2 \to Y$ and $Df_n : M_n(X^2) \to M_n(Y)$ by

$$\begin{aligned} Df(a,b) := {} & f(a+2b) + f(a-2b) - 4f(a+b) - 4f(a-b) + 6f(a) \\ & - f(2b) - f(-2b) + 4f(b) + 4f(-b), \end{aligned}$$

$$Df_n([x_{ij}],[y_{ij}]) := f_n([x_{ij}]+2[y_{ij}]) + f_n([x_{ij}]-2[y_{ij}]) - 4f_n([x_{ij}]+[y_{ij}])$$
$$-4f_n([x_{ij}]-[y_{ij}])+6f_n([x_{ij}])-f_n(2[y_{ij}])-f_n(-2[y_{ij}])+4f_n([y_{ij}])+4f_n(-[y_{ij}])$$

for all $a,b \in X$ and all $x = [x_{ij}], y = [y_{ij}] \in M_n(X)$.

Theorem 1 *Let r,θ be positive real numbers with $r > 1$. Let $f : X \to Y$ be an odd mapping such that*

$$P_n\left(Df_n([x_{ij}],[y_{ij}])\right) \le \sum_{i,j=1}^{n} \theta(\|x_{ij}\|^r + \|y_{ij}\|^r) \tag{4}$$

for all $x = [x_{ij}], y = [y_{ij}] \in M_n(X)$. Then there exists a unique additive mapping $A : X \to Y$ such that

$$P_n\left(f_n(2[x_{ij}]) - 8f_n([x_{ij}]) - A_n([x_{ij}])\right) \le \sum_{i,j=1}^{n} \frac{9+2^r}{2^r-2}\theta\|x_{ij}\|^r \tag{5}$$

for all $x = [x_{ij}] \in M_n(X)$.

Proof Let $n = 1$ in (4). Then (4) is equivalent to

$$P(Df(a,b)) \le \theta(\|a\|^r + \|b\|^r) \tag{6}$$

for all $a,b \in X$.

Letting $a = b$ in (6), we get

$$P(f(3b) - 4f(2b) + 5f(b)) \le 2\theta\|b\|^r \tag{7}$$

for all $b \in X$.

Replacing a by $2b$ in (6), we get

$$P(f(4b) - 4f(3b) + 6f(2b) - 4f(b)) \le (1+2^r)\theta\|b\|^r \tag{8}$$

for all $b \in X$.

By (7) and (8),

$$\begin{aligned} &P(f(4b) - 10f(2b) + 16f(b)) \\ &\le P(4(f(3b) - 4f(2b) + 5f(b))) + P(f(4b) - 4f(3b) + 6f(2b) - 4f(b)) \\ &\le 4P(f(3b) - 4f(2b) + 5f(b)) + P(f(4b) - 4f(3b) + 6f(2b) - 4f(b)) \\ &\le (9+2^r)\theta\|b\|^r \end{aligned} \tag{9}$$

for all $b \in X$. Replacing b by $\frac{a}{2}$ and letting $g(a) := f(2a) - 8f(a)$ in (9), we get

$$P\left(g(a) - 2g\left(\frac{a}{2}\right)\right) \le \frac{9+2^r}{2^r}\theta\|a\|^r$$

for all $a \in X$. Hence

$$P\left(2^l g\left(\frac{a}{2^l}\right) - 2^m g\left(\frac{a}{2^m}\right)\right) \leq \frac{9+2^r}{2^r} \sum_{j=l}^{m-1} \frac{2^j}{2^{rj}} \theta \|a\|^r \tag{10}$$

for all nonnegative integers m and l with $m > l$ and all $a \in X$. It follows from (10) that the sequence $\{2^k g(\frac{a}{2^k})\}$ is Cauchy for all $a \in X$. Since Y is complete, the sequence $\{2^k g(\frac{a}{2^k})\}$ converges. So one can define the mapping $A : X \to Y$ by

$$A(a) := \lim_{k\to\infty} 2^k g\left(\frac{a}{2^k}\right)$$

for all $a \in X$.

By (6),

$$P(DA(a,b)) = \lim_{k\to\infty} P\left(2^k Dg\left(\frac{a}{2^k}, \frac{b}{2^k}\right)\right) \leq \lim_{k\to\infty} \frac{2^k}{2^{kr}}(2^r+8)\theta(\|a\|^r + \|b\|^r) = 0$$

for all $a, b \in X$. So $DA(a,b) = 0$. Since $g : X \to Y$ is odd, $A : X \to Y$ is odd. So the mapping $A : X \to Y$ is additive.

Moreover, letting $l = 0$ and passing the limit $m \to \infty$ in (10), we get

$$P\left(f(2a) - 8f(a) - A(a)\right) \leq \frac{9+2^r}{2^r-2} \theta \|a\|^r \tag{11}$$

for all $a \in X$.

Now, let $T : X \to Y$ be another additive mapping satisfying (11). Then we have

$$\begin{aligned} P(A(a) - T(a)) &= P\left(2^q A\left(\frac{a}{2^q}\right) - 2^q T\left(\frac{a}{2^q}\right)\right) \\ &\leq P\left(2^q \left(A\left(\frac{a}{2^q}\right) - g\left(\frac{a}{2^q}\right)\right)\right) + P\left(2^q \left(T\left(\frac{a}{2^q}\right) - g\left(\frac{a}{2^q}\right)\right)\right) \\ &\leq 2\frac{2^r+9}{2^r-2}\frac{2^q}{2^{qr}}\theta\|a\|^r, \end{aligned}$$

which tends to 0 as $q \to \infty$ for all $a \in X$. So we can conclude that $A(a) = T(a)$ for all $a \in X$. This proves the uniqueness of A. Thus the mapping $A : X \to Y$ is a unique additive mapping.

By Lemma 1 and (11),

$$\begin{aligned} P\left(f_n(2[x_{ij}]) - 8f_n([x_{ij}]) - A_n([x_{ij}])\right) &\leq \sum_{i,j=1}^{n} P(f(2x_{ij}) - 8f(x_{ij}) - A(x_{ij})) \\ &\leq \sum_{i,j=1}^{n} \frac{9+2^r}{2^r-2}\theta\|x_{ij}\|^r \end{aligned}$$

for all $x = [x_{ij}] \in M_n(X)$. Thus $A : X \to Y$ is a unique additive mapping satisfying (5), as desired.

Theorem 2 *Let r, θ be positive real numbers with $r < 1$. Let $f : Y \to X$ be an odd mapping such that*

$$\left\| Df_n([x_{ij}], [y_{ij}]) \right\|_n \leq \sum_{i,j=1}^{n} \theta(P(x_{ij})^r + P(y_{ij})^r) \tag{12}$$

for all $x = [x_{ij}], y = [y_{ij}] \in M_n(Y)$. Then there exists a unique additive mapping $A : Y \to X$ such that

$$\left\| f_n(2[x_{ij}]) - 8f_n([x_{ij}]) - A_n([x_{ij}]) \right\|_n \leq \sum_{i,j=1}^{n} \frac{9+2^r}{2-2^r} \theta P(x_{ij})^r \tag{13}$$

for all $x = [x_{ij}] \in M_n(Y)$.

Proof Let $n = 1$ in (12). Then (12) is equivalent to

$$\|Df(a,b)\| \leq \theta(P(a)^r + P(b)^r) \tag{14}$$

for all $a, b \in Y$.

Letting $b = a$ in (14), we get

$$\|f(3b) - 4f(2b) + 5f(b)\| \leq 2\theta P(b)^r \tag{15}$$

for all $a \in Y$.

Replacing a by $2b$ in (14), we get

$$\|f(4b) - 4f(3b) + 6f(2b) - 4f(b)\| \leq (1 + 2^r)\theta P(b)^r \tag{16}$$

for all $b \in Y$.

By (15) and (16),

$$\begin{aligned} &\|f(4b) - 10f(2b) + 16f(b)\| \\ &\quad \leq \|4(f(3b) - 4f(2b) + 5f(b))\| + \|f(4b) - 4f(3b) + 6f(2b) - 4f(b)\| \\ &\quad = 4\|f(3b) - 4f(2b) + 5f(b)\| + \|f(4b) - 4f(3b) + 6f(2b) - 4f(b)\| \\ &\quad \leq (9 + 2^r)\theta P(b)^r \end{aligned} \tag{17}$$

for all $b \in Y$. Replacing b by a and letting $g(a) := f(2a) - 8f(a)$ in (17), we get

$$\|2g(a) - g(2a)\| \leq (9 + 2^r)\theta P(a)^r$$

for all $a \in Y$. Hence

$$\left\| \frac{1}{2^l} g(2^l a) - \frac{1}{2^m} g(2^m a) \right\| \leq \sum_{j=l}^{m-1} \frac{9+2^r}{2} \frac{2^{jr}}{2^j} \theta P(a)^r \tag{18}$$

for all nonnegative integers m and l with $m > l$ and all $a \in Y$. It follows from (18) that the sequence $\{\frac{1}{2^k}g(2^k a)\}$ is Cauchy for all $a \in Y$. Since X is complete, the sequence $\{\frac{1}{2^k}g(2^k a)\}$ converges. So one can define the mapping $A : Y \to X$ by

$$A(a) := \lim_{k\to\infty} \frac{1}{2^k} g\left(2^k a\right)$$

for all $a \in Y$.

By (14),

$$\|DA(a,b)\| = \lim_{k\to\infty} \left\| \frac{1}{2^k} Dg\left(2^k a, 2^k b\right) \right\| \leq \lim_{k\to\infty} \frac{2^{kr}}{2^k}(2^r+8)\theta(P(a)^r + P(b)^r) = 0$$

for all $a, b \in Y$. So $DA(a,b) = 0$. Since $g : Y \to X$ is odd, $A : Y \to X$ is odd. So the mapping $A : Y \to X$ is additive.

Moreover, letting $l = 0$ and passing the limit $m \to \infty$ in (18), we get

$$\|f(2a) - 8f(a) - A(a)\| \leq \frac{9+2^r}{2-2^r}\theta P(a)^r \tag{19}$$

for all $a \in Y$.

Now, let $T : Y \to X$ be another additive mapping satisfying (19). Then we have

$$\begin{aligned}
\|A(a) - T(a)\| &= \left\| \frac{1}{2^q} A\left(2^q a\right) - \frac{1}{2^q} T\left(2^q a\right) \right\| \\
&\leq \left\| \frac{1}{2^q}\left(A\left(2^q a\right) - g\left(2^q a\right)\right) \right\| + \left\| \frac{1}{2^q}\left(T\left(2^q a\right) - g\left(2^q a\right)\right) \right\| \\
&\leq 2\frac{2^{qr}}{2^q}\frac{9+2^r}{2-2^r}\theta P(a)^r,
\end{aligned}$$

which tends to 0 as $q \to \infty$ for all $a \in Y$. So we can conclude that $A(a) = T(a)$ for all $a \in Y$. This proves the uniqueness of A. Thus the mapping $A : Y \to X$ is a unique additive mapping.

By Lemma 2 and (19),

$$\begin{aligned}
\left\| f_n(2[x_{ij}]) - 8f_n([x_{ij}]) - A_n([x_{ij}]) \right\|_n &\leq \sum_{i,j=1}^{n} \|f(2x_{ij}) - 8f(x_{ij}) - A(x_{ij})\| \\
&\leq \sum_{i,j=1}^{n} \frac{9+2^r}{2-2^r}\theta P(x_{ij})^r
\end{aligned}$$

for all $x = [x_{ij}] \in M_n(Y)$. Thus $A : Y \to X$ is a unique additive mapping satisfying (13), as desired.

Theorem 3 *Let r, θ be positive real numbers with $r > 3$. Let $f : X \to Y$ be an odd mapping satisfying* (4). *Then there exists a unique cubic mapping $C : X \to Y$ such that*

$$P_n\left(f_n(2[x_{ij}]) - 2f_n([x_{ij}]) - C_n([x_{ij}])\right) \leq \sum_{i,j=1}^{n} \frac{9+2^r}{2^r-8}\theta\|x_{ij}\|^r$$

for all $x = [x_{ij}] \in M_n(X)$.

Proof Replacing b by $\frac{a}{2}$ and letting $g(a) := f(2a) - 2f(a)$ in (9), we get

$$P\left(g(a) - 8g\left(\frac{a}{2}\right)\right) \le \frac{9+2^r}{2^r}\theta\|a\|^r$$

for all $a \in X$.

The rest of the proof is similar to the proof of Theorem 1.

Theorem 4 *Let* r, θ *be positive real numbers with* $r < 3$. *Let* $f : Y \to X$ *be an odd mapping satisfying* (12). *Then there exists a unique cubic mapping* $C : Y \to X$ *such that*

$$\left\| f_n(2[x_{ij}]) - 2f_n([x_{ij}]) - C_n([x_{ij}])\right\|_n \le \sum_{i,j=1}^{n} \frac{9+2^r}{8-2^r}\theta P(x_{ij})^r$$

for all $x = [x_{ij}] \in M_n(Y)$.

Proof Replacing b by a and letting $g(a) := f(2a) - 2f(a)$ in (17), we get

$$\|8g(a) - g\,(2a)\| \le (9+2^r)\theta P(a)^r$$

for all $a \in Y$.

The rest of the proof is similar to the proof of Theorem 2.

3 Hyers–Ulam Stability of the AQCQ-Functional Equation (3) in Matrix Paranormed Spaces: Even Mapping Case

In this section, we prove the Hyers–Ulam stability of the AQCQ-functional equation (3) in matrix paranormed spaces for an even mapping case.

Throughout this section, let $(X, \{\|\cdot\|_n\})$ be a matrix Banach space and $(Y, \{P_n(\cdot)\})$ a matrix Fréchet space.

Theorem 5 *Let* r, θ *be positive real numbers with* $r > 2$. *Let* $f : X \to Y$ *be an even mapping satisfying* $f(0) = 0$ *and* (4). *Then there exists a unique quadratic mapping* $Q : X \to Y$ *such that*

$$P_n\left(f_n(2[x_{ij}]) - 16f_n([x_{ij}]) - Q_n([x_{ij}])\right) \le \sum_{i,j=1}^{n} \frac{9+2^r}{2^r-4}\theta\|x_{ij}\|^r$$

for all $x = [x_{ij}] \in M_n(X)$.

Proof Let $n = 1$. Then (4) is equivalent to

$$P(Df(a,b)) \le \theta(\|a\|^r + \|b\|^r) \tag{20}$$

for all $a, b \in X$. Letting $a = b$ in (20), we get

$$P(f(3b) - 6f(2b) + 15f(b)) \leq 2\theta \|b\|^r \tag{21}$$

for all $b \in X$.

Replacing a by $2b$ in (20), we get

$$P(f(4b) - 4f(3b) + 4f(2b) + 4f(b)) \leq (1 + 2^r)\theta \|b\|^r \tag{22}$$

for all $b \in X$.

By (21) and (22),

$$\begin{aligned} &P(f(4b) - 20f(2b) + 64f(b)) \\ &\quad \leq P(4(f(3b) - 6f(2b) + 15f(b))) + P(f(4b) - 4f(3b) + 4f(2b) + 4f(b)) \\ &\quad \leq 4P(f(3b) - 6f(2b) + 15f(b)) + P(f(4b) - 4f(3b) + 4f(2b) + 4f(b)) \\ &\quad \leq (9 + 2^r)\theta \|b\|^r \end{aligned} \tag{23}$$

for all $b \in X$. Replacing b by $\frac{a}{2}$ and letting $g(a) := f(2a) - 16f(a)$ in (23), we get

$$P\left(g(a) - 4g\left(\frac{a}{2}\right)\right) \leq \frac{9 + 2^r}{2^r}\theta \|a\|^r$$

for all $a \in X$.

The rest of the proof is similar to the proof of Theorem 1.

Theorem 6 *Let r, θ be positive real numbers with $r < 2$. Let $f : Y \to X$ be an even mapping satisfying $f(0) = 0$ and (12). Then there exists a unique quadratic mapping $Q : Y \to X$ such that*

$$\left\| f_n(2[x_{ij}]) - 16f_n([x_{ij}]) - Q_n([x_{ij}]) \right\|_n \leq \sum_{i,j=1}^{n} \frac{9 + 2^r}{4 - 2^r}\theta P(x_{ij})^r$$

for all $x = [x_{ij}] \in M_n(Y)$.

Proof Let $n = 1$ in (12). Then (12) is equivalent to

$$\|Df(a, b)\| \leq \theta(P(a)^r + P(b)^r) \tag{24}$$

for all $a, b \in Y$.

Letting $b = a$ in (24), we get

$$\|f(3b) - 6f(2b) + 15f(b)\| \leq 2\theta P(b)^r \tag{25}$$

for all $a \in Y$.

Replacing a by $2b$ in (24), we get

$$\|f(4b) - 4f(3b) + 4f(2b) + 4f(b)\| \leq (1 + 2^r)\theta P(b)^r \tag{26}$$

for all $b \in Y$.

By (25) and (26),

$$\begin{aligned}
&\|f(4b) - 20f(2b) + 64f(b)\| \qquad (27)\\
&\quad \leq \|4(f(3b)-6f(2b)+15f(b))\| + \|f(4b)-4f(3b)+4f(2b)+4f(b)\|\\
&\quad = 4\|f(3b)-4f(2b)+5f(b)\| + \|f(4b)-4f(3b)+6f(2b)-4f(b)\|\\
&\quad \leq (9+2^r)\theta P(b)^r
\end{aligned}$$

for all $b \in Y$. Replacing b by a and letting $g(a) := f(2a) - 16f(a)$ in (27), we get

$$\|4g(a) - g(2a)\| \leq (9+2^r)\theta P(a)^r$$

for all $a \in Y$.

The rest of the proof is similar to the proof of Theorem 2.

Theorem 7 *Let r, θ be positive real numbers with $r > 4$. Let $f : X \to Y$ be an even mapping satisfying $f(0) = 0$ and (4). Then there exists a unique quartic mapping $R : X \to Y$ such that*

$$P_n\left(f_n(2[x_{ij}]) - 4f_n([x_{ij}]) - R_n([x_{ij}])\right) \leq \sum_{i,j=1}^{n} \frac{9+2^r}{2^r - 16}\theta \|x_{ij}\|^r$$

for all $x = [x_{ij}] \in M_n(X)$.

Proof Replacing b by $\frac{a}{2}$ and letting $g(a) := f(2a) - 4f(a)$ in (23), we get

$$P\left(g(a) - 16g\left(\frac{a}{2}\right)\right) \leq \frac{9+2^r}{2^r}\theta\|a\|^r$$

for all $a \in X$.

The rest of the proof is similar to the proofs of Theorems 1 and 5.

Theorem 8 *Let r, θ be positive real numbers with $r < 4$. Let $f : Y \to X$ be an even mapping satisfying $f(0) = 0$ and (12). Then there exists a unique quartic mapping $R : Y \to X$ such that*

$$\left\| f_n(2[x_{ij}]) - 4f_n([x_{ij}]) - R_n([x_{ij}])\right\|_n \leq \sum_{i,j=1}^{n} \frac{9+2^r}{16 - 2^r}\theta P(x_{ij})^r$$

for all $x = [x_{ij}] \in M_n(Y)$.

Proof Replacing b by a and letting $g(a) := f(2a) - 4f(a)$ in (27), we get

$$\|16g(a) - g(2a)\| \leq (9+2^r)\theta P(a)^r$$

for all $a \in Y$.

The rest of the proof is similar to the proofs of Theorems 2 and 6.

4 Hyers–Ulam Stability of the AQCQ-Functional Equation (3) in Matrix β-Homogeneous F^*-Spaces: Odd Mapping Case

In this section, we prove the Hyers–Ulam stability of the AQCQ-functional equation (3) in matrix β-homogeneous F^*-spaces for an odd mapping case.

From now on, we assume that $(X, \{\|\cdot\|_n\})$ is a matrix β_1-homogeneous F^*-space and $(Y, \{\|\cdot\|_n\})$ is a matrix β_2-homogeneous F-space $(0 < \beta_1, \beta_2 \leq 1)$.

Lemma 3 *Let $(X, \{\|\cdot\|_n\})$ be a matrix F^*-space. Then*

(1) $\|x_{kl}\| \leq \|[x_{ij}]\|_n \leq \sum_{i,j=1}^n \|x_{ij}\|$ *for* $[x_{ij}] \in M_n(X)$.

(2) $\lim_{s\to\infty} x_s = x$ *if and only if* $\lim_{s\to\infty} x_{sij} = x_{ij}$ *for* $x_s = [x_{sij}], x = [x_{ij}] \in M_k(X)$.

Proof (1) By Definition 4, $\|x_{kl}\| \leq \|[x_{ij}]\|_n$.

Since $[x_{ij}] = \sum_{i,j=1}^n E_{ij} \otimes x_{ij}$,

$$\|[x_{ij}]\|_n = \left\| \sum_{i,j=1}^n E_{ij} \otimes x_{ij} \right\|_n \leq \sum_{i,j=1}^n \|E_{ij} \otimes x_{ij}\|_n = \sum_{i,j=1}^n \|x_{ij}\|.$$

(2) By (1), we have

$$\|x_{skl} - x_{kl}\| \leq \|[x_{sij} - x_{ij}]\|_n = \|[x_{sij}] - [x_{ij}]\|_n \leq \sum_{i,j=1}^n \|x_{sij} - x_{ij}\|.$$

So we get the result.

Theorem 9 *Let r, θ be positive real numbers with $\beta_1 r > \beta_2$. Let $f : X \to Y$ be an odd mapping such that*

$$\left\| Df_n([x_{ij}], [y_{ij}]) \right\|_n \leq \sum_{i,j=1}^n \theta(\|x_{ij}\|^r + \|y_{ij}\|^r) \tag{28}$$

for all $x = [x_{ij}], y = [y_{ij}] \in M_n(X)$. Then there exists a unique additive mapping $A : X \to Y$ such that

$$\left\| f_n(2[x_{ij}]) - 8f_n([x_{ij}]) - A_n([x_{ij}]) \right\|_n \leq \sum_{i,j=1}^n \frac{1 + 2\cdot 4^{\beta_2} + 2^{\beta_1 r}}{2^{\beta_1 r} - 2^{\beta_2}} \theta \|x_{ij}\|^r \tag{29}$$

for all $x = [x_{ij}] \in M_n(X)$.

Proof Let $n = 1$ in (28). Then (28) is equivalent to

$$\|Df(a, b)\| \leq \theta(\|a\|^r + \|b\|^r) \tag{30}$$

for all $a, b \in X$.

Letting $a = b$ in (30), we get

$$\|f(3b) - 4f(2b) + 5f(b)\| \le 2\theta\|b\|^r \tag{31}$$

for all $b \in X$.

Replacing a by $2b$ in (30), we get

$$\|f(4b) - 4f(3b) + 6f(2b) - 4f(b)\| \le (1 + 2^{\beta_1 r})\theta\|b\|^r \tag{32}$$

for all $b \in X$.

By (31) and (32),

$$\begin{aligned}
&\|f(4b) - 10f(2b) + 16f(b)\| \\
&\quad \le \|4(f(3b) - 4f(2b) + 5f(b))\| + \|f(4b) - 4f(3b) + 6f(2b) - 4f(b)\| \\
&\quad \le 4^{\beta_2}\|f(3b) - 4f(2b) + 5f(b)\| + \|f(4b) - 4f(3b) + 6f(2b) - 4f(b)\| \\
&\quad \le (1 + 2\cdot 4^{\beta_2} + 2^{\beta_1 r})\theta\|b\|^r
\end{aligned} \tag{33}$$

for all $b \in X$. Replacing b by $\frac{a}{2}$ and letting $g(a) := f(2a) - 8f(a)$ in (33), we get

$$\left\|g(a) - 2g\left(\frac{a}{2}\right)\right\| \le \frac{1 + 2\cdot 4^{\beta_2} + 2^{\beta_1 r}}{2^{\beta_1 r}}\theta\|a\|^r$$

for all $a \in X$. Hence

$$\left\|2^l g\left(\frac{a}{2^l}\right) - 2^m g\left(\frac{a}{2^m}\right)\right\| \le \frac{1 + 2\cdot 4^{\beta_2} + 2^{\beta_1 r}}{2^{\beta_1 r}}\sum_{j=l}^{m-1}\frac{2^{\beta_2 j}}{2^{\beta_1 r j}}\theta\|a\|^r \tag{34}$$

for all nonnegative integers m and l with $m > l$ and all $a \in X$. It follows from (34) that the sequence $\{2^k g(\frac{a}{2^k})\}$ is Cauchy for all $a \in X$. Since Y is complete, the sequence $\{2^k g(\frac{a}{2^k})\}$ converges. So one can define the mapping $A : X \to Y$ by

$$A(a) := \lim_{k\to\infty} 2^k g\left(\frac{a}{2^k}\right)$$

for all $a \in X$.

By (30),

$$\|DA(a,b)\| = \lim_{k\to\infty}\left\|2^k Dg\left(\frac{a}{2^k}, \frac{b}{2^k}\right)\right\| \le \lim_{k\to\infty}\frac{2^{k\beta_2}}{2^{k\beta_1 r}}(2^{\beta_1 r} + 8^{\beta_2})\theta(\|a\|^r + \|b\|^r) = 0$$

for all $a, b \in X$. So $DA(a,b) = 0$. Since $g : X \to Y$ is odd, $A : X \to Y$ is odd. So the mapping $A : X \to Y$ is additive.

Moreover, letting $l = 0$ and passing the limit $m \to \infty$ in (34), we get

$$\|f(2a) - 8f(a) - A(a)\| \le \frac{1 + 2\cdot 4^{\beta_2} + 2^{\beta_1 r}}{2^{\beta_1 r} - 2^{\beta_2}}\theta\|a\|^r \tag{35}$$

for all $a \in X$.

Now, let $T : X \to Y$ be another additive mapping satisfying (35). Then we have

$$\begin{aligned}\|A(a) - T(a)\| &= \left\|2^q A\left(\frac{a}{2^q}\right) - 2^q T\left(\frac{a}{2^q}\right)\right\| \\ &\leq \left\|2^q \left(A\left(\frac{a}{2^q}\right) - g\left(\frac{a}{2^q}\right)\right)\right\| + \left\|2^q \left(T\left(\frac{a}{2^q}\right) - g\left(\frac{a}{2^q}\right)\right)\right\| \\ &\leq 2\frac{1 + 2 \cdot 4^{\beta_2} + 2^{\beta_1 r}}{2^{\beta_1 r} - 2^{\beta_2}} \frac{2^{\beta_2 q}}{2^{q\beta_1 r}} \theta \|a\|^r,\end{aligned}$$

which tends to 0 as $q \to \infty$ for all $a \in X$. So we can conclude that $A(a) = T(a)$ for all $a \in X$. This proves the uniqueness of A. Thus the mapping $A : X \to Y$ is a unique additive mapping.

By Lemma 3 and (35),

$$\begin{aligned}\left\| f_n(2[x_{ij}]) - 8 f_n([x_{ij}]) - A_n([x_{ij}])\right\|_n &\leq \sum_{i,j=1}^{n} \|f(2x_{ij}) - 8 f(x_{ij}) - A(x_{ij})\| \\ &\leq \sum_{i,j=1}^{n} \frac{1 + 2 \cdot 4^{\beta_2} + 2^{\beta_1 r}}{2^{\beta_1 r} - 2^{\beta_2}} \theta \|x_{ij}\|^r\end{aligned}$$

for all $x = [x_{ij}] \in M_n(X)$. Thus $A : X \to Y$ is a unique additive mapping satisfying (29), as desired.

Theorem 10 *Let r, θ be positive real numbers with $\beta_2 r < \beta_1$. Let $f : Y \to X$ be an odd mapping such that*

$$\left\| Df_n([x_{ij}], [y_{ij}])\right\|_n \leq \sum_{i,j=1}^{n} \theta(\|x_{ij}\|^r + \|y_{ij}\|^r) \tag{36}$$

for all $x = [x_{ij}], y = [y_{ij}] \in M_n(Y)$. Then there exists a unique additive mapping $A : Y \to X$ such that

$$\left\| f_n(2[x_{ij}]) - 8 f_n([x_{ij}]) - A_n([x_{ij}])\right\|_n \leq \sum_{i,j=1}^{n} \frac{1 + 2 \cdot 4^{\beta_1} + 2^{\beta_2 r}}{2^{\beta_1} - 2^{\beta_2 r}} \theta \|x_{ij}\|^r \tag{37}$$

for all $x = [x_{ij}] \in M_n(Y)$.

Proof Let $n = 1$ in (36). Then (36) is equivalent to

$$\|Df(a, b)\| \leq \theta(\|a\|^r + \|b\|^r) \tag{38}$$

for all $a, b \in Y$.

Letting $b = a$ in (38), we get

$$\|f(3b) - 4f(2b) + 5f(b)\| \leq 2\theta \|b\|^r \tag{39}$$

for all $a \in Y$.

Replacing a by $2b$ in (38), we get

$$\|f(4b) - 4f(3b) + 6f(2b) - 4f(b)\| \leq (1 + 2^{\beta_2 r})\theta\|b\|^r \quad (40)$$

for all $b \in Y$.

By (39) and (40),

$$\begin{aligned} &\|f(4b) - 10f(2b) + 16f(b)\| \quad (41)\\ &\leq \|4(f(3b) - 4f(2b) + 5f(b))\| + \|f(4b) - 4f(3b) + 6f(2b) - 4f(b)\| \\ &= 4^{\beta_1}\|f(3b) - 4f(2b) + 5f(b)\| + \|f(4b) - 4f(3b) + 6f(2b) - 4f(b)\| \\ &\leq (1 + 2\cdot 4^{\beta_1} + 2^{\beta_2 r})\theta\|b\|^r \end{aligned}$$

for all $b \in Y$. Replacing b by a and letting $g(a) := f(2a) - 8f(a)$ in (41), we get

$$\|2g(a) - g(2a)\| \leq (1 + 2\cdot 4^{\beta_1} + 2^{\beta_2 r})\theta\|a\|^r$$

for all $a \in Y$. Hence

$$\left\|\frac{1}{2^l}g\left(2^l a\right) - \frac{1}{2^m}g\left(2^m a\right)\right\| \leq \sum_{j=l}^{m-1}\frac{(1 + 2\cdot 4^{\beta_1} + 2^{\beta_2 r})}{2^{\beta_1}}\frac{2^{j\beta_2 r}}{2^{\beta_1 j}}\theta\|a\|^r \quad (42)$$

for all nonnegative integers m and l with $m > l$ and all $a \in Y$. It follows from (42) that the sequence $\{\frac{1}{2^k}g(2^k a)\}$ is Cauchy for all $a \in Y$. Since X is complete, the sequence $\{\frac{1}{2^k}g(2^k a)\}$ converges. So one can define the mapping $A : Y \to X$ by

$$A(a) := \lim_{k\to\infty}\frac{1}{2^k}g\left(2^k a\right)$$

for all $a \in Y$.

By (38),

$$\|DA(a,b)\| = \lim_{k\to\infty}\left\|\frac{1}{2^k}Dg\left(2^k a, 2^k b\right)\right\| \leq \lim_{k\to\infty}\frac{2^{k\beta_2 r}}{2^{\beta_1 k}}(2^{\beta_2 r} + 8^{\beta_1})\theta(\|a\|^r + \|b\|^r) = 0$$

for all $a, b \in Y$. So $DA(a,b) = 0$. Since $g : Y \to X$ is odd, $A : Y \to X$ is odd. So the mapping $A : Y \to X$ is additive.

Moreover, letting $l = 0$ and passing the limit $m \to \infty$ in (42), we get

$$\|f(2a) - 8f(a) - A(a)\| \leq \frac{1 + 2\cdot 4^{\beta_1} + 2^{\beta_2 r}}{2^{\beta_1} - 2^{\beta_2 r}}\theta\|a\|^r \quad (43)$$

for all $a \in Y$.

Now, let $T : Y \to X$ be another additive mapping satisfying (43). Then we have

$$\|A(a) - T(a)\| = \left\|\frac{1}{2^q}A\left(2^q a\right) - \frac{1}{2^q}T\left(2^q a\right)\right\|$$

$$\leq \left\| \frac{1}{2^q}\left(A\left(2^q a\right) - g\left(2^q a\right)\right) \right\| + \left\| \frac{1}{2^q}\left(T\left(2^q a\right) - g\left(2^q a\right)\right) \right\|$$
$$\leq 2\frac{2^{q\beta_2 r}}{2^{\beta_1 q}} \frac{1 + 2 \cdot 4^{\beta_1} + 2^{\beta_2 r}}{2^{\beta_1} - 2^{\beta_2 r}} \theta \|a\|^r,$$

which tends to 0 as $q \to \infty$ for all $a \in Y$. So we can conclude that $A(a) = T(a)$ for all $a \in Y$. This proves the uniqueness of A. Thus the mapping $A : Y \to X$ is a unique additive mapping.

By Lemma 2 and (43),

$$\left\| f_n(2[x_{ij}]) - 8f_n([x_{ij}]) - A_n([x_{ij}]) \right\|_n \leq \sum_{i,j=1}^{n} \|f(2x_{ij}) - 8f(x_{ij}) - A(x_{ij})\|$$
$$\leq \sum_{i,j=1}^{n} \frac{1 + 2 \cdot 4^{\beta_1} + 2^{\beta_2 r}}{2^{\beta_1} - 2^{\beta_2 r}} \theta \|x_{ij}\|^r$$

for all $x = [x_{ij}] \in M_n(Y)$. Thus $A : Y \to X$ is a unique additive mapping satisfying (37), as desired.

Theorem 11 *Let r, θ be positive real numbers with $\beta_1 r > 3\beta_2$. Let $f : X \to Y$ be an odd mapping satisfying* (28)*. Then there exists a unique cubic mapping $C : X \to Y$ such that*

$$\left\| f_n(2[x_{ij}]) - 2f_n([x_{ij}]) - C_n([x_{ij}]) \right\|_n \leq \sum_{i,j=1}^{n} \frac{1 + 2 \cdot 4^{\beta_2} + 2^{\beta_1 r}}{2^{\beta_1 r} - 8^{\beta_2}} \theta \|x_{ij}\|^r$$

for all $x = [x_{ij}] \in M_n(X)$.

Proof Replacing b by $\frac{a}{2}$ and letting $g(a) := f(2a) - 2f(a)$ in (33), we get

$$\left\| g(a) - 8g\left(\frac{a}{2}\right) \right\| \leq \frac{1 + 2 \cdot 4^{\beta_2} + 2^{\beta_1 r}}{2^{\beta_1 r}} \theta \|a\|^r$$

for all $a \in X$.

The rest of the proof is similar to the proof of Theorem 9.

Theorem 12 *Let r, θ be positive real numbers with $\beta_2 r < 3\beta_1$. Let $f : Y \to X$ be an odd mapping satisfying* (36)*. Then there exists a unique cubic mapping $C : Y \to X$ such that*

$$\left\| f_n(2[x_{ij}]) - 2f_n([x_{ij}]) - C_n([x_{ij}]) \right\|_n \leq \sum_{i,j=1}^{n} \frac{1 + 2 \cdot 4^{\beta_1} + 2^{\beta_2 r}}{8^{\beta_1} - 2^{\beta_2 r}} \theta \|x_{ij}\|^r$$

for all $x = [x_{ij}] \in M_n(Y)$.

Proof Replacing b by a and letting $g(a) := f(2a) - 2f(a)$ in (41), we get

$$\|8g(a) - g(2a)\| \leq (1 + 2 \cdot 4^{\beta_1} + 2^{\beta_2 r})\theta \|a\|^r$$

for all $a \in Y$.

The rest of the proof is similar to the proof of Theorem 10.

5 Hyers–Ulam Stability of the AQCQ-Functional Equation (3) in Matrix β-Homogeneous F^*-Spaces: Even Mapping Case

In this section, we prove the Hyers–Ulam stability of the AQCQ-functional equation (3) in matrix β-homogeneous F^*-spaces for an even mapping case.

Theorem 13 *Let r,θ be positive real numbers with $\beta_1 r > 2\beta_2$. Let $f : X \to Y$ be an even mapping satisfying $f(0) = 0$ and (28). Then there exists a unique quadratic mapping $Q : X \to Y$ such that*

$$\left\| f_n(2[x_{ij}]) - 16 f_n([x_{ij}]) - Q_n([x_{ij}])\right\|_n \leq \sum_{i,j=1}^{n} \frac{1+2\cdot 4^{\beta_2} + 2^{\beta_1 r}}{2^{\beta_1 r} - 4^{\beta_2}}\theta \|x_{ij}\|^r$$

for all $x = [x_{ij}] \in M_n(X)$.

Proof Let $n = 1$. Then (28) is equivalent to

$$\|Df(a,b)\| \leq \theta(\|a\|^r + \|b\|^r) \tag{44}$$

for all $a, b \in X$. Letting $a = b$ in (44), we get

$$\|f(3b) - 6f(2b) + 15f(b)\| \leq 2\theta\|b\|^r \tag{45}$$

for all $b \in X$.

Replacing a by $2b$ in (44), we get

$$\|f(4b) - 4f(3b) + 4f(2b) + 4f(b)\| \leq (1 + 2^{\beta_1 r})\theta\|b\|^r \tag{46}$$

for all $b \in X$.

By (45) and (46),

$$\begin{aligned} &\|f(4b) - 20f(2b) + 64f(b)\| \\ &\quad\leq \|4(f(3b) - 6f(2b) + 15f(b))\| + \|f(4b) - 4f(3b) + 4f(2b) + 4f(b)\| \\ &\quad\leq 4^{\beta_2}\|f(3b) - 6f(2b) + 15f(b)\| + \|f(4b) - 4f(3b) + 4f(2b) + 4f(b)\| \\ &\quad\leq (1 + 2\cdot 4^{\beta_2} + 2^{\beta_1 r})\theta\|b\|^r \end{aligned} \tag{47}$$

for all $b \in X$. Replacing b by $\frac{a}{2}$ and letting $g(a) := f(2a) - 16f(a)$ in (47), we get

$$\left\|g(a) - 4g\left(\frac{a}{2}\right)\right\| \leq \frac{1+2\cdot 4^{\beta_2} + 2^{\beta_1 r}}{2^{\beta_1 r}}\theta\|a\|^r$$

for all $a \in X$.

The rest of the proof is similar to the proof of Theorem 9.

Theorem 14 *Let r, θ be positive real numbers with $\beta_2 r < 2\beta_1$. Let $f : Y \to X$ be an even mapping satisfying $f(0) = 0$ and (36). Then there exists a unique quadratic mapping $Q : Y \to X$ such that*

$$\left\| f_n(2[x_{ij}]) - 16 f_n([x_{ij}]) - Q_n([x_{ij}]) \right\|_n \leq \sum_{i,j=1}^{n} \frac{1 + 2 \cdot 4^{\beta_1} + 2^{\beta_2 r}}{4^{\beta_1} - 2^{\beta_2 r}} \theta \|x_{ij}\|^r$$

for all $x = [x_{ij}] \in M_n(Y)$.

Proof Let $n = 1$ in (36). Then (36) is equivalent to

$$\|Df(a,b)\| \leq \theta(\|a\|^r + \|b\|^r) \tag{48}$$

for all $a, b \in Y$.

Letting $b = a$ in (48), we get

$$\|f(3b) - 6f(2b) + 15f(b)\| \leq 2\theta \|b\|^r \tag{49}$$

for all $a \in Y$.

Replacing a by $2b$ in (48), we get

$$\|f(4b) - 4f(3b) + 4f(2b) + 4f(b)\| \leq (1 + 2^{\beta_2 r})\theta \|b\|^r \tag{50}$$

for all $b \in Y$.

By (49) and (50),

$$\begin{aligned} &\|f(4b) - 20f(2b) + 64f(b)\| \\ &\quad \leq \|4(f(3b) - 6f(2b) + 15f(b))\| + \|f(4b) - 4f(3b) + 4f(2b) + 4f(b)\| \\ &\quad = 4^{\beta_1} \|f(3b) - 4f(2b) + 5f(b)\| + \|f(4b) - 4f(3b) + 6f(2b) - 4f(b)\| \\ &\quad \leq (1 + 2 \cdot 4^{\beta_1} + 2^{\beta_2 r})\theta \|b\|^r \end{aligned} \tag{51}$$

for all $b \in Y$. Replacing b by a and letting $g(a) := f(2a) - 16\, f(a)$ in (51), we get

$$\|4g(a) - g(2a)\| \leq (1 + 2 \cdot 4^{\beta_1} + 2^{\beta_2 r})\theta \|a\|^r$$

for all $a \in Y$.

The rest of the proof is similar to the proof of Theorem 10.

Theorem 15 *Let r, θ be positive real numbers with $\beta_1 r > 4\beta_2$. Let $f : X \to Y$ be an even mapping satisfying $f(0) = 0$ and (28). Then there exists a unique quartic mapping $R : X \to Y$ such that*

$$\left\| f_n(2[x_{ij}]) - 4 f_n([x_{ij}]) - R_n([x_{ij}]) \right\|_n \leq \sum_{i,j=1}^{n} \frac{1 + 2 \cdot 4^{\beta_2} + 2^{\beta_1 r}}{2^{\beta_1 r} - 16^{\beta_2}} \theta \|x_{ij}\|^r$$

for all $x = [x_{ij}] \in M_n(X)$.

Proof Replacing b by $\frac{a}{2}$ and letting $g(a) := f(2a) - 4f(a)$ in (47), we get

$$\left\| g(a) - 16g\left(\frac{a}{2}\right) \right\| \le \frac{1 + 2 \cdot 4^{\beta_2} + 2^{\beta_1 r}}{2^{\beta_1 r}} \theta \|a\|^r$$

for all $a \in X$.

The rest of the proof is similar to the proofs of Theorems 9 and 13.

Theorem 16 *Let r, θ be positive real numbers with $\beta_2 r < 4\beta_1$. Let $f : Y \to X$ be an even mapping satisfying $f(0) = 0$ and (36). Then there exists a unique quartic mapping $R : Y \to X$ such that*

$$\| f_n(2[x_{ij}]) - 4f_n([x_{ij}]) - R_n([x_{ij}]) \|_n \le \sum_{i,j=1}^{n} \frac{1 + 2 \cdot 4^{\beta_1} + 2^{\beta_2 r}}{16^{\beta_1} - 2^{\beta_2 r}} \theta \|x_{ij}\|^r$$

for all $x = [x_{ij}] \in M_n(Y)$.

Proof Replacing b by a and letting $g(a) := f(2a) - 4f(a)$ in (51), we get

$$\|16g(a) - g(2a)\| \le (1 + 2 \cdot 4^{\beta_1} + 2^{\beta_2 r})\theta \|a\|^r$$

for all $a \in Y$.

The rest of the proof is similar to the proofs of Theorems 10 and 14.

6 Conclusions

Let $f_o(x) := \frac{f(x) - f(-x)}{2}$ and $f_e(x) := \frac{f(x) + f(-x)}{2}$. Then f_o is odd and f_e is even. f_o, f_e satisfy the functional equation (3). Let $g_o(x) := f_o(2x) - 2f_o(x)$ and $h_o(x) := f_o(2x) - 8f_o(x)$. Then $f_o(x) = \frac{1}{6}g_o(x) - \frac{1}{6}h_o(x)$. Let $g_e(x) := f_e(2x) - 4f_e(x)$ and $h_e(x) := f_e(2x) - 16f_e(x)$. Then $f_e(x) = \frac{1}{12}g_e(x) - \frac{1}{12}h_e(x)$. Thus

$$f(x) = \frac{1}{6}g_o(x) - \frac{1}{6}h_o(x) + \frac{1}{12}g_e(x) - \frac{1}{12}h_e(x).$$

We summarize the above results as follows.

Let $(X, \{\|\cdot\|_n\})$ be a matrix Banach space and $(Y, \{P_n(\cdot)\})$ a matrix Fréchet space.

Theorem 17 *Let r, θ be positive real numbers with $r > 4$. Let $f : X \to Y$ be a mapping satisfying $f(0) = 0$ and (4). Then there exist an additive mapping $A : X \to Y$, a quadratic mapping $Q : X \to Y$, a cubic mapping $C : X \to Y$, and a quartic mapping $R : X \to Y$ such that*

$$P_n(24f_n(x) - 4A_n(x) - 2Q_n(x) - 4C_n(x) - 2R_n(x))$$
$$\le \left(\frac{4(9 + 2^r)}{2^r - 2} + \frac{2(9 + 2^r)}{2^r - 4} + \frac{4(9 + 2^r)}{2^r - 8} + \frac{2(9 + 2^r)}{2^r - 16} \right) \sum_{i,j=1}^{n} \theta \|x_{ij}\|^r$$

for all $x = [x_{ij}] \in M_n(X)$.

Theorem 18 *Let* r *be a positive real number with* $r < 1$. *Let* $f : Y \to X$ *be a mapping satisfying* $f(0) = 0$ *and (12). Then there exist an additive mapping* $A : Y \to X$, *a quadratic mapping* $Q : Y \to X$, *a cubic mapping* $C : Y \to X$, *and a quartic mapping* $R : Y \to X$ *such that*

$$\|24f_n(x) - 4A_n(x) - 2Q_n(x) - 4C_n(x) - 2R_n(x)\|_n \leq \left(\frac{4(9+2^r)}{2-2^r} + \frac{2(9+2^r)}{4-2^r} + \frac{4(9+2^r)}{8-2^r} + \frac{2(9+2^r)}{16-2^r}\right) \sum_{i,j=1}^{n} \theta P(x_{ij})^r$$

for all $x = [x_{ij}] \in M_n(Y)$.

From now on, we assume that $(X, \{\|\cdot\|_n\})$ is a matrix β_1-homogeneous F^*-space and $(Y, \{\|\cdot\|_n\})$ is a matrix β_2-homogeneous F-space $(0 < \beta_1, \beta_2 \leq 1)$.

Theorem 19 *Let* r, θ *be positive real numbers with* $\beta_1 r > 4\beta_2$. *Let* $f : X \to Y$ *be a mapping satisfying* $f(0) = 0$ *and (28). Then there exist an additive mapping* $A : X \to Y$, *a quadratic mapping* $Q : X \to Y$, *a cubic mapping* $C : X \to Y$, *and a quartic mapping* $R : X \to Y$ *such that*

$$\begin{aligned}\|24f_n(x) - 4A_n(x) - 2Q_n(x) - 4C_n(x) - 2R_n(x)\|_n &\leq \left(\frac{4}{2^{\beta_1 r} - 2^{\beta_2}} + \frac{2}{2^{\beta_1 r} - 4^{\beta_2}} + \frac{4}{2^{\beta_1 r} - 8^{\beta_2}} + \frac{2}{2^{\beta_1 r} - 18^{\beta_2}}\right) \\ &\quad \times (1 + 2 \cdot 4^{\beta_2} + 2^{\beta_1 r}) \sum_{i,j=1}^{n} \theta \|x_{ij}\|^r\end{aligned}$$

for all $x = [x_{ij}] \in M_n(X)$.

Theorem 20 *Let* r *be a positive real number with* $\beta_2 r < \beta_1$. *Let* $f : Y \to X$ *be a mapping satisfying* $f(0) = 0$ *and (36). Then there exist an additive mapping* $A : Y \to X$, *a quadratic mapping* $Q : Y \to X$, *a cubic mapping* $C : Y \to X$, *and a quartic mapping* $R : Y \to X$ *such that*

$$\begin{aligned}\|24f_n(x) - 4A_n(x) - 2Q_n(x) - 4C_n(x) - 2R_n(x)\|_n &\leq \left(\frac{4}{2^{\beta_1} - 2^{\beta_2 r}} + \frac{2}{4^{\beta_1} - 2^{\beta_2 r}} + \frac{4}{8^{\beta_1} - 2^{\beta_2 r}} + \frac{2}{16^{\beta_1} - 2^{\beta_2 r}}\right) \\ &\quad \times (1 + 2 \cdot 4^{\beta_1} + 2^{\beta_2 r}) \sum_{i,j=1}^{n} \theta \|x_{ij}\|^r\end{aligned}$$

for all $x = [x_{ij}] \in M_n(Y)$.

Acknowledgements C. Park was supported by Basic Science Research Program through the National Research Foundation of Korea funded by the Ministry of Education, Science and Technology (NRF-2012R1A1A2004299).

References

1. Aczel, J., Dhombres, J.: Functional Equations in Several Variables. Cambridge University Press, Cambridge (1989)
2. Aoki, T.: On the stability of the linear transformation in Banach spaces. J. Math. Soc. Jpn. **2**, 64–66 (1950)
3. Choi, M.D., Effros, E.: Injectivity and operator spaces. J. Funct. Anal. **24**, 156–209 (1977)
4. Cholewa, P.W.: Remarks on the stability of functional equations. Aequat. Math. **27**, 76–86 (1984)
5. Czerwik, S.: On the stability of the quadratic mapping in normed spaces. Abh. Math. Sem. Univ. Hamburg **62**, 59–64 (1992)
6. Czerwik, P.: Functional Equations and Inequalities in Several Variables. World Scientific Publishing Company, New Jersey (2002)
7. Effros, E.: On multilinear completely bounded module maps. Contemp. Math. **62**, Am. Math. Soc. Providence, RI, pp. 479–501 (1987)
8. Effros, E., Ruan, Z.J.: On approximation properties for operator spaces. Internat. J. Math. **1**, 163–187 (1990)
9. Effros, E., Ruan, Z.J.: On the abstract characterization of operator spaces. Proc. Am. Math. Soc. **119**, 579–584 (1993)
10. Eshaghi Gordji, M., Savadkouhi, M.B.: Stability of a mixed type cubic-quartic functional equation in non-Archimedean spaces. Appl. Math. Lett. **23**, 1198–1202 (2010)
11. Eshaghi Gordji, M., Abbaszadeh, S., Park, C.: On the stability of a generalized quadratic and quartic type functional equation in quasi-Banach spaces. J. Inequal. Appl. **2009**, Article ID 153084, (2009)
12. Eshaghi Gordji, M., Kaboli-Gharetapeh, S., Park, C., Zolfaghari, S.: Stability of an additive-cubic-quartic functional equation. Adv. Differ. Equat. **2009**, Article ID 395693 (2009)
13. Fast, H.: Sur la convergence statistique. Colloq. Math. **2**, 241–244 (1951)
14. Fridy, J.A.: On statistical convergence. Analysis **5**, 301–313 (1985)
15. Gajda, Z.: On stability of additive mappings. Internat. J. Math. Math. Sci. **14**, 431–434 (1991)
16. Găvruta, G.: A generalization of the Hyers-Ulam-Rassias stability of approximately additive mappings. J. Math. Anal. Appl. **184**, 431–436 (1994)
17. Haagerup, U.: Decomposition of completely bounded maps. unpublished manuscript.
18. Hyers, D.H.: On the stability of the linear functional equation. Proc. Natl. Acad. Sci. U. S. A. **27**, 222–224 (1941)
19. Hyers, D.H., Isac, G., Rassias, T.M.: Stability of Functional Equations in Several Variables. Birkh$\ddot{a}$user, Basel (1998)
20. Isac, G., Rassias, T.M.: On the Hyers-Ulam stability of ψ-additive mappings. J. Approx. Theory **72**, 131–137 (1993)
21. Jun, K., Kim, H.: The generalized Hyers-Ulam-Rassias stability of a cubic functional equation. J. Math. Anal. Appl. **274**, 867–878 (2002)
22. Jun, K., Lee, Y.: A generalization of the Hyers-Ulam-Rassias stability of the Pexiderized quadratic equations. J. Math. Anal. Appl. **297**, 70–86 (2004)
23. Karakus, S.: Statistical convergence on probabilistic normed spaces. Math. Commun. **12**, 11–23 (2007)
24. Kolk, E.: The statistical convergence in Banach spaces. Tartu Ul. Toime. **928**, 41–52 (1991)
25. Lee, S., Im, S., Hwang, I.: Quartic functional equations. J. Math. Anal. Appl. **307**, 387–394 (2005)
26. Mursaleen, M.: λ-statistical convergence. Math. Slovaca **50**, 111–115 (2000)
27. Mursaleen, M., Mohiuddine, S.A.: On lacunary statistical convergence with respect to the intuitionistic fuzzy normed space. J. Comput. Anal. Math. **233**, 142–149 (2009)
28. Park, C.: Homomorphisms between Poisson JC^*-algebras. Bull. Braz. Math. Soc. **36**, 79–97 (2005)

29. Pisier, G.: Grothendieck's Theorem for non-commutative C^*-algebras with an appendix on Grothendieck's constants. J. Funct. Anal. **29**, 397–415 (1978)
30. Rassias, T.M.: On the stability of the linear mapping in Banach spaces. Proc. Am. Math. Soc. **72**, 297–300 (1978)
31. Rassias, T.M.: Problem 16; 2. Report of the 27th international symposium on functional equations, Aequ. Math. **39**, 292–293; 309 (1990)
32. Rassias, T.M., Šemrl, P.: On the behaviour of mappings which do not satisfy Hyers-Ulam stability. Proc. Am. Math. Soc. **114**, 989–993 (1992)
33. Rolewicz, S.: Metric Linear Spaces. PWN-Polish Scientific Publishers, Warsaw (1972)
34. Ruan, Z.J.: Subspaces of C^*-algebras. J. Funct. Anal. **76**, 217–230 (1988)
35. Šalát, T.: On the statistically convergent sequences of real numbers. Math. Slovaca **30**, 139–150 (1980)
36. Skof, F.: Proprietà locali e approssimazione di operatori. Rend. Sem. Mat. Fis. Milano **53**, 113–129 (1983)
37. Steinhaus, H.: Sur la convergence ordinaire et la convergence asymptotique. Colloq. Math. **2**, 73–34 (1951)
38. Ulam, S.M.: A Collection of the Mathematical Problems. Interscience Publication, New York (1960)
39. Wilansky, A.: Modern Methods in Topological Vector Space. McGraw-Hill, New York (1978)

On the Generalized Hyers–Ulam Stability of the Pexider Equation on Restricted Domains

Youssef Manar, Elhoucien Elqorachi and Themistocles M. Rassias

Abstract Let $\sigma\colon E \longrightarrow E$ be an involution of the normed space E and let p, M, d be nonnegative real numbers, such that $0 < p < 1$. In this chapter, we investigate the Hyers–Ulam–Rassias stability of the Pexider functional equations

$$f(x+y) = g(x) + h(y),\ f(x+y) + g(x-y) = h(x) + k(y),$$
$$f(x+y) + g(x+\sigma(y)) = h(x) + k(y), x, y \in E$$

on restricted domains $\mathcal{B} = \{(x, y) \in E^2 : \|x\|^p + \|y\|^p \geq M^p\}$ and $\mathcal{C} = \{(x, y) \in E^2 : \|x\| \geq d \ or \ \|y\| \geq d\}$.

Keywords Hyers Ulam Rassias stability · Pexider functional equation · Metric group · Cauchy difference · Restricted domain

1 Introduction

In 1940, the following stability problem for group homomorphisms was raised by Ulam [70]. Given a group G_1 and a metric group G_2 with metric $d(.,.)$ and a positive number ϵ greater than zero, does there exist a positive number δ greater than zero such that if a function $f : G_1 \longrightarrow G_2$ satisfies the functional inequality $d(f(xy), f(x)f(y)) \leq \delta$ for all $x, y \in G_1$, then there exists a group homomorphism $h : G_1 \longrightarrow G_2$ with $d(f(x), h(x)) \leq \epsilon$ for all $x \in G_1$. The problem for the case of approximately additive mappings was solved by Hyers [25] on Banach

Y. Manar (✉)
Superior School of Technology, University Ibn Zohr, Guelmim, Morocco
e-mail: manaryoussef1984@gmail.com

E. Elqorachi
Department of Mathematics, Faculty of Sciences, University Ibn Zohr, Agadir, Morocco
e-mail: elqorachi@hotmail.com

T. M. Rassias
Department of Mathematics, National Technical University of Athens, Zografou Campus, 15780 Athens, Greece
e-mail: trassias@math.ntua.gr

T. M. Rassias (ed.), *Handbook of Functional Equations*,
Springer Optimization and Its Applications 96, DOI 10.1007/978-1-4939-1286-5_13

spaces. In 1950, Aoki [1] provided a generalization of the Hyers' theorem for additive mappings and in 1978, Rassias [57] generalized the Hyers' theorem for linear mappings by considering an unbounded Cauchy difference for sum of powers of norms $\epsilon(\|x\|^p + \|y\|^p)$. Rassias' theorem has been generalized by Găvruta [22] who permitted the Cauchy difference to be bounded by a general control function. Since then, the stability problems for several functional equations have been extensively investigated by a number mathematicians (cf. [10, 11, 16, 23, 35, 31, 41, 54, 58, 59, 61, 62, 65, and 72]). The terminology Hyers–Ulam–Rassias stability originates from these historical backgrounds. This terminology can also be applied to the case of other functional equations. For more detailed definitions of such terminologies, we can refer to [2, 3, 17, 20, 21, 26, 28, 33, 37, 42, 47–53, 60, and 63]. Concerning the stability of functional equations on a restricted domain, Skof [67] was the first author to solve Ulam problem for additive mapping on a restricted domain. Given a real normed vector spaces X and E, a function $f : X \to E$ will satisfy the functional equation

$$f(x+y) = f(x) + f(y) \quad for\ all \quad x, y \in X$$

if and only if

$$\|f(x+y) - f(x) - f(y)\| \to 0 \quad as \quad \|x\| + \|y\| \to +\infty.$$

In [27], Hyers, Isac, and Rassias considered the asymptotic aspect of Hyers–Ulam–Rassias stability that is close to the asymptotic derivability. In [29], Jung investigated the Hyers–Ulam stability for the quadratic equation

$$f(x+y) + f(x-y) = 2f(x) + 2f(y), \ x, y \in E \tag{1}$$

on a restricted domain $\mathcal{A} := \{(x, y) \in E^2 : \|x\| + \|y\| \geq d\}$. In [64], Rassias and Rassias investigated the Hyers–Ulam stability on $\mathcal{A}$ for the Jensen functional equations

$$f(x+y) + f(x-y) = 2f(x), \ x, y \in E \tag{2}$$

and

$$f(x+y) - f(x-y) = 2f(y), \ x, y \in E. \tag{3}$$

The more general equation is

$$f(x+y) + f(x+\sigma(y)) = 2f(x) + 2f(y), \ x, y \in E, \tag{4}$$

where σ is an involution and has been solved by Stetkær [69] in abelian groups. Recently, the stability theorem of Eq. (4) and the Jensen functional equations

$$f(x+y) + f(x+\sigma(y)) = 2f(x), \ x, y \in E \tag{5}$$

$$f(x+y) - f(x+\sigma(y)) = 2f(y), \ x, y \in E \tag{6}$$

has been proved (see [12, 39, 43]). In [18, 19, 44], Elqorachi, Manar, and Rassias investigated the stability of Eqs. (4), (5), and (6) on unbounded domains: $\{(x, y) \in E^2 : \|y\| \geq d\}$ and $\{(x, y) \in E^2 : \|x\| \geq d\}$, respectively. In this chapter, we consider the Pexider functional equations

$$f(x + y) = g(x) + h(y), \ x, y \in E, \tag{7}$$

$$f(x + y) + g(x - y) = h(x) + k(y), \ x, y \in E \tag{8}$$

and

$$f(x + y) + g(x + \sigma(y)) = h(x) + k(y), \ x, y \in E, \tag{9}$$

where $\sigma : E \to E$ is an involution of the normed space E, i.e., $\sigma(x + y) = \sigma(x) + \sigma(y)$ and $\sigma(\sigma(x)) = x$, for all $x, y \in E$. Jung [31], and Jung and Sahoo [36] investigated the Hyers–Ulam–Rassias stability of Eq. (8). Bouikhalene, Elqorachi, and Rassias [6] proved the Hyers–Ulam stability of Eq. (9). Recently, Pourpasha, Rassias, Saadati, and Vaezpour [55] investigated the Hyers–Ulam stability of Eq. (7) and (8) by using the fixed point method. The stability problems of several functional equations on a restricted domain have been extensively investigated by a number of authors (cf. [4, 5, 7–9, 13–15, 24, 27, 30, 32, 34, 38, 40, 45, 50, 56, 66, 68, and 71]). Chung [13] generalized the Hyers–Ulam stability of a Pexiderized logarithmic functional equation in restricted domains. In the following, we present our results as follows: In the next section, we will study the Hyers–Ulam–Rassias stability problem for equations

$$f(x + y) = g(x) + h(y), \ x, y \in E,$$

on restricted domains:

$$\mathcal{B} = \{(x, y) \in E^2 : \|x\|^p + \|y\|^p \geq M^p\}$$

and

$$\mathcal{C} = \{(x, y) \in E^2 : \|x\| \geq d \ or \ \|y\| \geq d\}.$$

In the Sect. 3, we will investigate the Hyers–Ulam–Rassias stability for the equation

$$f(x + y) + g(x - y) = k(x) + h(y), \ x, y \in E,$$

on restricted domains $\mathcal{B}$ and $\mathcal{C}$. In the last section, we study the Hyers–Ulam–Rassias stability for equation

$$f(x + y) + g(x + \sigma(y)) = k(x) + h(y), \ x, y \in E,$$

on a restricted domain $\mathcal{C}$. Throughout this paper, E denotes a normed space and F a Banach space.

2 Stability of Eq. (7) on Restricted Domains

In the present section, we prove the Hyers–Ulam–Rassias stability of the Pexider functional equation of type (7) on restricted domains $\mathcal{B}$ and $\mathcal{C}$. In the following lemma, we will apply some ideas from [41] to the proof of Hyers–Ulam–Rassias stability of Eq. (7). As an application, we study the Hyers–Ulam–Rassias stability of that equation on restricted domains $\mathcal{B}$ and $\mathcal{C}$.

Lemma 1 *Let $f_1, f_2, f_3 : E \to F$ satisfy the inequality*

$$\|f_1(x+y) - f_2(x) - f_3(y)\| \leq \delta + \epsilon(\|x\|^p + \|y\|^p) \tag{10}$$

for all $x, y \in E$, where δ, ϵ, p are given positive numbers such that $0 < p < 1$. Then, there exists a unique additive mapping $A : E \to F$ such that

$$\|f_1(x) - A(x) - f_1(0)\| \leq 6\delta + \frac{4\epsilon}{2-2^p}\|x\|^p, \tag{11}$$

$$\|f_2(x) - A(x) - f_2(0)\| \leq 8\delta + \epsilon\left(\frac{6-2^p}{2-2^p}\right)\|x\|^p \tag{12}$$

and

$$\|f_3(x) - A(x) - f_3(0)\| \leq 8\delta + \epsilon\left(\frac{6-2^p}{2-2^p}\right)\|x\|^p \tag{13}$$

for all $x \in E$.

Proof For any function $f_i : E \to F$ $(i = 1, 2, 3)$, we introduce the functions $F_i(x) = f_i(x) - f_i(0)$, $x \in E$. From (10) and the triangle inequality, we obtain

$$\|F_1(x+y) - F_2(x) - F_3(y)\| \leq 2\delta + \epsilon(\|x\|^p + \|y\|^p). \tag{14}$$

Setting $y = 0$ in (14), to obtain

$$\|F_1(x) - F_2(x)\| \leq 2\delta + \epsilon\|x\|^p \tag{15}$$

for all $x \in E$. Setting $x = 0$ in (14), we get

$$\|F_1(y) - F_3(y)\| \leq 2\delta + \epsilon\|y\|^p \tag{16}$$

for all $y \in E$. It follows from (15), (16) and the triangle inequality that

$$\begin{aligned}
&\|F_1(x+y) - F_1(x) - F_1(y)\| \\
&\quad \leq \|F_1(x+y) - F_2(x) - F_3(y)\| + \|F_1(x) - F_2(x)\| + \|F_1(y) - F_3(y)\| \\
&\quad \leq 2\delta + \epsilon(\|x\|^p + \|y\|^p) + 2\delta + \epsilon\|x\|^p + 2\delta + \epsilon\|y\|^p \\
&\quad \leq 6\delta + 2\epsilon(\|x\|^p + \|y\|^p)
\end{aligned}$$

for all $x, y \in E$. So, from [22] there exists a unique additive mapping $A : E \to F$ given by $A(x) = \lim_{n\to+\infty} 2^{-n}F_1(2^n x)$ such that

$$\|F_1(x) - A(x)\| \le 6\delta + \frac{4\epsilon}{2-2^p}\|x\|^p \tag{17}$$

for all $x \in E$. From (15) and (17), we obtain

$$\begin{aligned}\|F_2(x) - A(x)\| &\le \|F_1(x) - A(x)\| + \|F_1(x) - F_2(x)\| \\ &\le 8\delta + \epsilon\left(\frac{6-2^p}{2-2^p}\right)\|x\|^p \end{aligned}\tag{18}$$

for all $x \in E$. In a similar way, by using (16) and (17) we obtain the following inequality

$$\|F_3(x) - A(x)\| \le 8\delta + \epsilon\left(\frac{6-2^p}{2-2^p}\right)\|x\|^p \tag{19}$$

for all $x \in E$.

In the following theorem, we establish the Hyers–Ulam–Rassias stability for the Eq. (7) on restricted domains.

Theorem 1 *Let a normed vector space E and a Banach space F are given. Suppose $d \ge 0$ and $\delta \ge 0$ be given. Assume that the mappings $f_1, f_2, f_3 : E \to F$ satisfy the inequality*

$$\|f_1(x+y) - f_2(x) - f_3(y)\| \le \delta \tag{20}$$

for all $(x, y) \in \mathcal{C} = \{(x, y) \in E^2$ such that $\|x\| \ge d$ or $\|y\| \ge d\}$. Then, there exists a unique additive mapping $A : E \to F$ such that

$$\|f_1(x) - A(x) - f_1(0)\| \le 36\delta, \tag{21}$$

$$\|f_2(x) - A(x) - f_2(0)\| \le 48\delta \tag{22}$$

and

$$\|f_3(x) - A(x) - f_3(0)\| \le 48\delta \tag{23}$$

for all $x \in E$.

Proof Let $(x, y) \in E^2 \backslash \mathcal{C}$. If $x = 0$ and $y = 0$, we choose an element $z \in E$ with $\|z\| \ge d$, and we use

$$\begin{aligned}[f_1(0) - f_2(0) - f_3(0)] &= [f_1(0) - f_2(-z) - f_3(z)] + [f_1(2z) - f_2(0) - f_3(2z)] \\ &\quad + [f_1(z) - f_2(z) - f_3(0)] - [f_1(2z) - f_2(z) - f_3(z)] \\ &\quad - [f_1(z) - f_2(-z) - f_3(2z)]\end{aligned}$$

to get

$$\|f_1(0) - f_2(0) - f_3(0)\| \le 5\delta. \tag{24}$$

If $y \neq 0$ or $x \neq 0$, we choose a $z = 2^n y$ if $y \neq 0$ and we choose $z = 2^n x$ if $x \neq 0$, with $n \in \mathbf{N}$ large enough. We can easily verify that $\|z\| \ge d$, $\|y - z\| \ge d$ or $\|x + z\| \ge d$.

Therefore, from (20), the triangle inequality and the following equation:

$$\begin{aligned}
&2[f_1(x+y) - f_2(x) - f_3(y)] \\
&= [f_1(x+y) - f_2(x+z) - f_3(y-z)] + [f_1(x+y) - f_2(y-z) - f_3(x+z)] \\
&\quad + 2[f_1(y+z) - f_2(z) - f_3(y)] + 2[f_1(x+2z) - f_2(x) - f_3(2z)] \\
&\quad - [f_1(x+2z) - f_2(x+z) - f_3(z)] - [f_1(y+z) - f_2(2z) - f_3(y-z)] \\
&\quad - [f_1(x+2z) - f_2(z) - f_3(x+z)] - [f_1(y+z) - f_2(y-z) - f_3(2z)] \\
&\quad - [f_1(3z) - f_2(z) - f_3(2z)] + [f_1(3z) - f_2(2z) - f_3(z)],
\end{aligned}$$

we get

$$\|f_1(x+y) - f_2(x) - f_3(y)\| \le 6\delta. \tag{25}$$

Finally, inequality (25) holds true for all $x, y \in E$. From Lemma 1 with $\epsilon = 0$, the rest of the proof follows.

In the following theorem, we prove the Hyers–Ulam–Rassias stability of Eq. (7) on restricted domain $\mathcal{B}$.

Theorem 2 *Assume a normed vector space E and a Banach space F are given. Let $\delta, \epsilon \ge 0$ and $M, p \ge 0$ with $0 < p < 1$ be fixed. Let $f_1, f_2, f_3 : E \to F$ be mappings which satisfy the inequality*

$$\|f_1(x+y) - f_2(x) - f_3(y)\| \le \delta + \epsilon(\|x\|^p + \|y\|^p) \tag{26}$$

for all $(x, y) \in \mathcal{B} = \{(x, y) \in E^2 : \|x\|^p + \|y\|^p \ge M^p\}$. Then, there exists a unique additive mapping $A : E \to F$ such that

$$\begin{aligned}
\|f_1(x) - A(x) - f_1(0)\| &\le 36\delta + 6\epsilon(4 \times 3^p + 3 \times 2^p + 3 \times 4^p)M^p \\
&\quad + \frac{4\epsilon}{2 - 2^p}\|x\|^p,
\end{aligned} \tag{27}$$

$$\begin{aligned}
\|f_2(x) - A(x) - f_2(0)\| &\le 48\delta + 8\epsilon(4 \times 3^p + 3 \times 2^p + 3 \times 4^p)M^p \\
&\quad + \epsilon\left(\frac{6 - 2^p}{2 - 2^p}\right)\|x\|^p
\end{aligned} \tag{28}$$

and

$$\begin{aligned}
\|f_3(x) - A(x) - f_3(0)\| &\le 48\delta + 8\epsilon(4 \times 3^p + 3 \times 2^p + 3 \times 4^p)M^p \\
&\quad + \epsilon\left(\frac{6 - 2^p}{2 - 2^p}\right)\|x\|^p
\end{aligned} \tag{29}$$

for all $x \in E$.

Proof Assume $\|x\|^p + \|y\|^p < M^p$. If $x = y = 0$, we choose an element $z \in E$ with $\|z\| = M$, and we use

$$\begin{aligned}[f_1(0) - f_2(0) - f_3(0)] = {} & [f_1(0) - f_2(-z) - f_3(z)] \\ & + [f_1(2z) - f_2(0) - f_3(2z)] + [f_1(z) - f_2(z) - f_3(0)] \\ & - [f_1(2z) - f_2(z) - f_3(z)] - [f_1(z) - f_2(-z) - f_3(2z)],\end{aligned}$$

we deduce

$$\|f_1(0) - f_2(0) - f_3(0)\| \le 5\delta + 2\epsilon(3 + 2^p)M^p. \tag{30}$$

Otherwise, we take

$$z = \begin{cases} (\|x\| + M)\dfrac{x}{\|x\|}, & if\ \|x\| \ge \|y\|; \\ \\ (\|y\| + M)\dfrac{y}{\|y\|}, & if\ \|y\| \ge \|x\|. \end{cases}$$

It's clear that $\|z\| \ge M$ and

$$\|x + z\|^p + \|y - z\|^p \ge \max\{\|x + z\|^p, \|y - z\|^p\} \ge M^p,$$

$$\|x + z\|^p + \|z\|^p \ge \|z\|^p \ge M^p,$$

$$\|y - z\|^p + \|2z\|^p \ge \|2z\|^p \ge M^p,$$

$$\min\{\|z\|^p + \|y\|^p, \|x\|^p + \|2z\|^p, \|2z\|^p + \|z\|^p\} \ge \|z\|^p \ge M^p.$$

Also we have

$$\max\{\|x + z\|, \|y - z\|\} < 3M, \ \|z\| < 2M.$$

Now, by using the following equation

$$\begin{aligned} & 2[f_1(x + y) - f_2(x) - f_3(y)] \\ & = [f_1(x + y) - f_2(x + z) - f_3(y - z)] + [f_1(x + y) - f_2(y - z) - f_3(x + z)] \\ & + 2[f_1(y + z) - f_2(z) - f_3(y)] + 2[f_1(x + 2z) - f_2(x) - f_3(2z)] \\ & - [f_1(x + 2z) - f_2(x + z) - f_3(z)] - [f_1(y + z) - f_2(2z) - f_3(y - z)] \\ & - [f_1(x + 2z) - f_2(z) - f_3(x + z)] - [f_1(y + z) - f_2(y - z) - f_3(2z)] \\ & - [f_1(3z) - f_2(z) - f_3(2z)] + [f_1(3z) - f_2(2z) - f_3(z)],\end{aligned}$$

we deduce

$$\begin{aligned}\|f_1(x + y) - f_2(x) - f_3(y)\| \le {} & 6\delta + \epsilon(4 \times 3^p + 3 \times 2^p + 3 \times 4^p)M^p \\ & + \epsilon(\|x\|^p + \|y\|^p).\end{aligned} \tag{31}$$

Therefore, inequality (31) holds true for all $x, y \in E$. According to Lemma 1, the rest of the proof follows.

Now, by using ideas from [29] and Theorem 1, we provide a proof of an asymptotic behavior of that equation.

Corollary 1 *Assume a normed vector space E and a Banach space F are given. The mappings $f_1, f_2, f_3 : E \to F$ with $f_i(0) = 0$ satisfy* Eq. (7) *if and only if the asymptotic condition*

$$\|f_1(x+y) - f_2(x) - f_3(y)\| \to 0 \ \ as \ \ \|x\| + \|y\| \to +\infty$$

holds true.

Proof According to our asymptotic condition, there exists a sequence $(\delta_n)_n$ which is monotonically decreasing to zero such that

$$\|f_1(x+y) - f_2(x) - f_3(y)\| \le \delta_n \tag{32}$$

for all $x, y \in E$ with $\|x\| + \|y\| \ge n$. So

$$\|f_1(x+y) - f_2(x) - f_3(y)\| \le \delta_n \tag{33}$$

for all $(x, y) \in \mathcal{C} = \{(x, y) \in E^2 \ such \ that \ \|x\| \ge n \ or \ \|y\| \ge n\}$. By Theorem 1, there exists a unique additive mapping $A_n : E \to F$ such that

$$\|f_1(x) - A_n(x)\| \le 36\delta_n,$$
$$\|f_2(x) - A_n(x)\| \le 48\delta_n$$

and

$$\|f_3(x) - A_n(x)\| \le 48\delta_n$$

for all $x \in E$. Let n and m be integers. Since $(\delta_n)_n$ is a monotonically decreasing sequence, the additive mapping A_m satisfies

$$\|f_1(x) - A_m(x)\| \le 36\delta_m \le 36\delta_n,$$

$$\|f_2(x) - A_m(x)\| \le 48\delta_m \le 48\delta_n$$

and

$$\|f_3(x) - A_m(x)\| \le 48\delta_m \le 48\delta_n$$

for all $x \in E$. By using the uniqueness of A_n, we get $A_n = A_m$ for all $n, m \in \mathbf{N}$. By letting $n \to +\infty$, we get that f is an additive mapping. The reverse assertion is obvious.

3 Stability of Eq. (8) on Restricted Domains

In this section, we will investigate the stability of Pexider functional Eq. (8) on a restricted domain

$$\mathcal{C} = \{(x, y) \in E^2 \text{ such that } \|x\| \geq d \text{ or } \|y\| \geq d\}$$

and

$$\mathcal{B} = \{(x, y) \in E^2 : \|x\|^p + \|y\|^p \geq M^p\}.$$

First, we prove the following stability theorem.

Theorem 3 *Assume a normed vector space E and a Banach space F are given. Let* $\delta, \epsilon \geq 0$ *and p with* $0 < p < 1$ *be fixed. If the functions* $f_1, f_2, f_3, f_4 : E \to F$ *satisfy the inequality*

$$\|f_1(x + y) + f_2(x - y) - f_3(x) - f_4(y)\| \leq \delta + \epsilon(\|x\|^p + \|y\|^p) \tag{34}$$

for all $x, y \in E$*, then there exists a unique quadratic mapping* $q : E \to F$*, and exactly two additive mappings* $a_1, a_2 : E \to F$ *such that*

$$\|f_1(x) - \frac{1}{2}q(x) - \frac{1}{2}a_1(x) - \frac{1}{2}a_2(x) - f_1(0)\| \leq \frac{44}{6}\delta + \epsilon\left(\frac{1}{2} + \frac{4 + 2^{p-1}}{4 - 2^p} + \frac{3 + 2^{p-1} + 2^{2-p}}{2 - 2^p}\right)\|x\|^p, \tag{35}$$

$$\|f_2(x) - \frac{1}{2}q(x) - \frac{1}{2}a_1(x) + \frac{1}{2}a_2(x) - f_2(0)\| \leq \frac{44}{6}\delta + \epsilon\left(\frac{1}{2} + \frac{4 + 2^{p-1}}{4 - 2^p} + \frac{3 + 2^{p-1} + 2^{2-p}}{2 - 2^p}\right)\|x\|^p, \tag{36}$$

$$\|f_3(x) - q(x) - a_1(x) - f_3(0)\| \leq \frac{20}{3}\delta + \epsilon\left(\frac{2^p + 8}{4 - 2^p} + \frac{4 + 2^p}{2 - 2^p}\right)\|x\|^p \tag{37}$$

and

$$\|f_4(x) - q(x) - a_2(x) - q(x) - f_4(0)\| \leq \frac{32}{3}\delta + \epsilon\left(\frac{16 - 2^p}{4 - 2^p} + \frac{4 + 2^p}{2 - 2^p}\right)\|x\|^p \tag{38}$$

for all $x \in E$.

Proof By applying the same argument as in the proof of Theorems 3.1 [6] and 1 [72], we obtain the proof of Theorem 3.

Theorem 4 *Let a normed vector space E and a Banach space F be given. Assume that the functions $f_1, f_2, f_3, f_4 : E \to F$ satisfy the inequality*

$$\|f_1(x+y) + f_2(x-y) - f_3(x) - f_4(y)\| \le \delta \tag{39}$$

for all $(x, y) \in \mathcal{C} = \{(x, y) \in E^2$ such that $\|x\| \ge d$ or $\|y\| \ge d\}$. Then, there exists a unique quadratic mapping $q : E \to F$, and exactly two additive mappings $a_1, a_2 : E \to F$ such that

$$\|f_1(x) - \frac{1}{2}q(x) - \frac{1}{2}a_1(x) - \frac{1}{2}a_2(x) - f_1(0)\| \le \frac{308}{6}\delta, \tag{40}$$

$$\|f_2(x) - \frac{1}{2}q(x) - \frac{1}{2}a_1(x) + \frac{1}{2}a_2(x) - f_2(0)\| \le \frac{308}{6}\delta, \tag{41}$$

$$\|f_3(x) - a_1(x) - q(x) - f_3(0)\| \le \frac{140}{3}\delta \tag{42}$$

and

$$\|f_4(x) - a_2(x) - q(x) - f_4(0)\| \le \frac{224}{3}\delta \tag{43}$$

for all $x \in E$.

Proof Let $(x, y) \in E^2 \backslash \mathcal{C}$. If $x = 0$ and $y = 0$, we choose an element $z \in E$ with $\|z\| = d$, and we use

$$\begin{aligned}
&[f_1(0) + f_2(0) - f_3(0) - f_4(0)] \\
=&[f_1(0) + f_2(2z) - f_3(z) - f_4(-z)] \\
&+[f_1(-4z) + f_2(0) - f_3(-2z) - f_4(-2z)] \\
&-[f_1(-4z) + f_2(2z) - f_3(-z) - f_4(-3z)] \\
&+[f_1(-3z) + f_2(3z) - f_3(0) - f_4(-3z)] \\
&+[f_1(-z) + f_2(-z) - f_3(-z) - f_4(0)] \\
&-[f_1(-z) + f_2(3z) - f_3(z) - f_4(-2z)] \\
&-[f_1(-3z) + f_2(-z) - f_3(-2z) - f_4(-z)],
\end{aligned}$$

to get

$$\|f_1(0) + f_2(0) - f_3(0) - f_4(0\| \le 7\delta. \tag{44}$$

Otherwise, if $y \neq 0$ or $x \neq 0$, we choose a $z = 2^n y$ if $y \neq 0$ and we choose $z = 2^n x$ if $x \neq 0$ with $n \in \mathbf{N}$ large enough. We can easily verify that $\|y - z\| \ge d$,

$\|y-2z\| \geq d$, $\|z\| \geq d$ and $\|x-z\| \geq d$. Therefore, from (39), the triangle inequality and the following decomposition:

$$\begin{aligned}
&[f_1(x+y)+f_2(x-y)-f_3(x)-f_4(y)] \\
&\quad = [f_1(x+y)+f_2(x-y+2z)-f_3(x+z)-f_4(y-z)] \\
&\quad\quad + [f_1(x+y-4z)+f_2(x-y)-f_3(x-2z)-f_4(y-2z)] \\
&\quad\quad - [f_1(x+y-4z)+f_2(x-y+2z)-f_3(x-z)-f_4(y-3z)] \\
&\quad\quad + [f_1(x+y-3z)+f_2(x-y+3z)-f_3(x)-f_4(y-3z)] \\
&\quad\quad + [f_1(x+y-z)+f_2(x-y-z)-f_3(x-z)-f_4(y)] \\
&\quad\quad - [f_1(x+y-z)+f_2(x-y+3z)-f_3(x+z)-f_4(y-2z)] \\
&\quad\quad - [f_1(x+y-3z)+f_2(x-y-z)-f_3(x-2z)-f_4(y-z)],
\end{aligned}$$

we get

$$\|f_1(x+y)+f_2(x-y)-f_3(x)-f_4(y)\| \leq 7\delta. \tag{45}$$

Consequently, inequality (45) holds true for all $x, y \in E$. From Theorem 3 with $\epsilon = 0$, the rest of the proof follows.

Corollary 2 *The functions* $f_1, f_2, f_3, f_4 : E \to F$ *with* $f_i(0) = 0$, $i = 1, 2, 3, 4$ *satisfy* Eq. (8) *if and only if*

$$\|f_1(x+y)+f_2(x-y)-f_3(x)-f_4(y)\| \to 0 \;\; as \;\; \|x\|+\|y\| \to +\infty.$$

Theorem 5 *Suppose that a normed vector space E and a Banach space F are given. If the functions* $f_1, f_2, f_3, f_4 : E \to F$ *satisfy the inequality*

$$\|f_1(x+y)+f_2(x-y)-f_3(x)-f_4(y)\| \leq \delta + \epsilon(\|x\|^p + \|y\|^p) \tag{46}$$

for all $(x, y) \in \mathcal{B} = \{(x, y) \in E^2 : \|x\|^p + \|y\|^p \geq M^p\}$ *where* δ, ϵ, p *are given positive numbers such that* $0 < p < 1$, *then there exists a unique quadratic mapping* $q : E \to F$, *and exactly two additive functions* $a_1, a_2 : E \to F$ *such that*

$$\begin{aligned}
&\|f_1(x) - \frac{1}{2}q(x) - \frac{1}{2}a_1(x) - \frac{1}{2}a_2(x) - f_1(0)\| \\
&\leq \frac{308}{6}\delta + \frac{44}{6}\epsilon[12 \times 7^p]M^p + \epsilon\left(\frac{1}{2} + \frac{4+2^{p-1}}{4-2^p} + \frac{3+2^{p-1}+2^{2-p}}{2-2^p}\right)\|x\|^p,
\end{aligned} \tag{47}$$

$$\begin{aligned}
&\|f_2(x) - \frac{1}{2}q(x) - \frac{1}{2}a_1(x) + \frac{1}{2}a_2(x) - f_2(0)\| \\
&\leq \frac{308}{6}\delta + \frac{44}{6}\epsilon[12 \times 7^p]M^p + \epsilon\left(\frac{1}{2} + \frac{4+2^{p-1}}{4-2^p} + \frac{3+2^{p-1}+2^{2-p}}{2-2^p}\right)\|x\|^p,
\end{aligned} \tag{48}$$

$$\| f_3(x) - a_1(x) - q(x) - f_3(0)\| \leq \frac{140}{3}\delta + \frac{20}{3}\epsilon[12 \times 7^p]M^p + \epsilon\left(\frac{2^p+8}{4-2^p} + \frac{4+2^p}{2-2^p}\right)\|x\|^p \tag{49}$$

and

$$\| f_4(x) - a_2(x) - q(x) - f_4(0)\| \leq \frac{224}{3}\delta + \frac{32}{3}\epsilon[12 \times 7^p]M^p + \epsilon\left(\frac{16-2^p}{4-2^p} + \frac{4+2^p}{2-2^p}\right)\|x\|^p \tag{50}$$

for all $x \in E$.

Proof Assume $\|x\|^p + \|y\|^p < M^p$. If $x = y = 0$, we choose a $z \in E$ with $\|z\| = M$. Otherwise, we take

$$z = \begin{cases} (\|x\| + M)\dfrac{x}{\|x\|}, & if\ \|x\| \geq \|y\|; \\ (\|y\| + M)\dfrac{y}{\|y\|}, & if\ \|y\| \geq \|x\|. \end{cases}$$

It's clear that $\|z\| \geq M$,

$$\|x+z\|^p + \|y-z\|^p \geq \max\{\|x+z\|^p, \|y-z\|^p\} \geq M^p,$$

$$\|x-2z\|^p + \|y-2z|^p \geq \max\{\|x-2z\|^p, \|y-2z|^p\} \geq M^p,$$

$$\|x-z\|^p + \|y-3z\|^p \geq \max\{\|x-z\|^p, \|y-3z\|^p\} \geq M^p,$$

$$\|x-2z\|^p + \|y-z\|^p \geq \max\{\|x-2z\|^p, \|y-z\|^p\} \geq M^p,$$

$$\|x+z\|^p + \|y-2z\|^p \geq \max\{\|x+z\|^p, \|y-2z\|^p\} \geq M^p,$$

$$\|y\|^p + \|x-z\|^p \geq \|x-z\|^p \geq M^p,$$

$$\|x\|^p + \|y-3z\|^p \geq \|y-3z\|^p \geq M^p,$$

and

$$\max\{\|x+z\|, \|y-z\|, \|x-2z\|, \|y-2z\|, \|y-3z\|\} < 7M.$$

Consequently, we obtain

$$[f_1(x+y) + f_2(x-y) - f_3(x) - f_4(y)]$$

$$
\begin{aligned}
&= [f_1(x+y) + f_2(x-y+2z) - f_3(x+z) - f_4(y-z)] \\
&\quad + [f_1(x+y-4z) + f_2(x-y) - f_3(x-2z) - f_4(y-2z)] \\
&\quad - [f_1(x+y-4z) + f_2(x-y+2z) - f_3(x-z) - f_4(y-3z)] \\
&\quad + [f_1(x+y-3z) + f_2(x-y+3z) - f_3(x) - f_4(y-3z)] \\
&\quad + [f_1(x+y-z) + f_2(x-y-z) - f_3(x-z) - f_4(y)] \\
&\quad - [f_1(x+y-z) + f_2(x-y+3z) - f_3(x+z) - f_4(y-2z)] \\
&\quad - [f_1(x+y-3z) + f_2(x-y-z) - f_3(x-2z) - f_4(y-z)].
\end{aligned}
$$

Therefore, in view of inequality (46) and the triangle inequality, we get

$$
\begin{aligned}
\|f_1(x+y) + f_2(x-y) - f_3(x) - f_4(y)\| &\le 7\delta + \epsilon[12 \times 7^p]M^p \\
&\quad + \epsilon(\|x\|^p + \|y\|^p).
\end{aligned} \tag{51}
$$

Then, inequality (51) holds true for all $x, y \in E$. According to Theorem 3, the rest of the proof follows.

In the following corollaries, we prove the stability for the Drygas functional equation

$$f(x+y) + f(x-y) = 2f(x) + f(y) + f(-y), \quad x, y \in E \tag{52}$$

in a restricted domain

$$\mathcal{B} = \{(x,y) \in E^2 : \|x\|^p + \|y\|^p \ge M^p\}$$

and

$$\mathcal{C} = \{(x,y) \in E^2 : \|x\| \ge d \ or \ \|y\| \ge d\}.$$

As an application, we use the result for the study of an asymptotic behavior of that equation.

Corollary 3 *Suppose that a normed vector space E and a Banach space F are given. Let $\delta, \epsilon \ge 0$ and $M, p \ge 0$ with $0 < p < 1$ and let $f : E \to F$ a mapping which satisfies the inequality*

$$\|f(x+y) + f(x-y) - 2f(x) - f(y) - f(-y)\| \le \delta + \epsilon(\|x\|^p + \|y\|^p) \tag{53}$$

for all $(x,y) \in \mathcal{B} = \{(x,y) \in E^2 : \|x\|^p + \|y\|^p \ge M^p\}$. Then, there exists a unique additive mapping $a : E \to F$ and a unique quadratic mapping $q : E \to F$ such that

$$
\begin{aligned}
\|f(x) - \frac{1}{2}a(x) - \frac{1}{2}q(x)\| &\le \frac{143}{6}\delta + \frac{20}{6}\epsilon[12 \times 7^p]M^p \\
&\quad + \epsilon\left(\frac{2^p+8}{4-2^p} + \frac{2^p+4}{2-2^p}\right)\|x\|^p
\end{aligned} \tag{54}
$$

for all $x \in E$.

Corollary 4 *Suppose that a normed vector space E and a Banach space F are given. A mapping $f : E \to F$ with $f(0) = 0$ is a solution of the Drygas functional* Eq. (52) *if and only if*

$$\|f(x+y) + f(x-y) - 2f(x) - f(y) - f(-y)\| \to 0 \;\; as \;\; \|x\| + \|y\| \to +\infty.$$

4 Stability of Eq. (9) on Restricted Domain

In this section, we will investigate the Hyers–Ulam stability of the Pexider functional equation

$$f(x+y) + g(x+\sigma(y)) = k(x) + h(y), \;\; x, y \in E,$$

on a restricted domain $\mathcal{C} = \{(x, y) \in E^2 : \|x\| \geq d \; or \; \|y\| \geq d\}$.

Theorem 6 *Let a normed vector space E and a Banach space F are given. Suppose $d \geq 0$ and $\delta \geq 0$ be given. Assume that the mappings $f_1, f_2, f_3, f_4 : E \to F$ satisfy the inequality*

$$\|f_1(x+y) + f_2(x+\sigma(y)) - f_3(x) - f_4(y)\| \leq \delta \tag{55}$$

for all $(x, y) \in \mathcal{C}$. Then, there exists a unique function $q : E \to F$ solution of Eq. (4), there exists a function $v : E \to F$ solution of equation

$$v(x+y) = v(x+\sigma(y)), \;\; x, y \in E, \tag{56}$$

there exists exactly two additive functions $A_1, A_2 : E \to F$ such that $A_i \circ \sigma = -A_i$ (i=1,2)

$$\|f_1(x) - \frac{1}{2}A_1(x) - \frac{1}{2}A_2(x) - \frac{1}{2}v(x) - \frac{1}{2}q(x) - f_1(0)\| \leq 133\delta, \tag{57}$$

$$\|f_2(x) + \frac{1}{2}A_1(x) - \frac{1}{2}A_2(x) + \frac{1}{2}v(x) - \frac{1}{2}q(x) - f_2(0)\| \leq 133\delta, \tag{58}$$

$$\|f_3(x) - A_2(x) - q(x) - f_3(0)\| \leq 112\delta \tag{59}$$

and

$$\|f_4(x) - A_1(x) - q(x) - f_4(0)\| \leq 112\delta \tag{60}$$

for all $x \in E$.

Proof Let $(x, y) \in E^2 \backslash \mathcal{C}$. If $x = y = 0$, then we have

$$\begin{aligned} &2[f_1(0) + f_2(0) - f_3(0) - f_4(0)] \\ &= 2[f_1(0) + f_2(0) - f_3(-z-\sigma(z)) - f_4(z+\sigma(z))] \end{aligned}$$

$$
\begin{aligned}
&- [f_1(-\sigma(z)) + f_2(-z) - f_3(-z-\sigma(z)) - f_4(z)] \\
&+ [f_1(-\sigma(z)) + f_2(-\sigma(z)) - f_3(-\sigma(z)) - f_4(0)] \\
&- [f_1(-z) + f_2(-\sigma(z)) - f_3(-z-\sigma(z)) - f_4(\sigma(z))] \\
&+ [f_1(z) + f_2(\sigma(z)) - f_3(0) - f_4(z)] \\
&- [f_1(z) + f_2(z) - f_3(-\sigma(z)) - f_4(z+\sigma(z))] \\
&+ [f_1(\sigma(z)) + f_2(z) - f_3(0) - f_4(\sigma(z))] \\
&- [f_1(\sigma(z)) + f_2(\sigma(z)) - f_3(-z) - f_4(z+\sigma(z))] \\
&+ [f_1(-z) + f_2(-z) - f_3(-z) - f_4(0)]
\end{aligned}
$$

for all $z \in E$. Now if we choose $z = 2^n x_0$, with $x_0 \neq 0$, $x_0 + \sigma(x_0) \neq 0$ and n large enough, we obtain

$$\|f_1(0) + f_2(0) - f_3(0) - f_4(0)\| \leq 5\delta. \tag{61}$$

If $x \neq 0$ and $y \neq 0$, we choose $z = 2^n x$ or $z = 2^n y$ with $n \in \mathbf{N}$. Case 1: $\sigma(y) \neq -y$ and $\sigma(x) \neq -x$. For n large enough, we can easily verify that $\|-x-\sigma(z)\| \geq d$, $\|-x+z\| \geq d$, $\|x+z+\sigma(z)\| \geq d$, $\|x-z-\sigma(z)\| \geq d$, $\|y+z+\sigma(z)\| \geq d$, $\|y-z-\sigma(z)\| \geq d$, $\|-y-\sigma(z)\| \geq d$, and $\|-y+z\| \geq d$. Therefore, from (39), the triangle inequality and the following decomposition

$$
\begin{aligned}
&2[f_1(x+y) + f_2(x+\sigma(y)) - f_3(x) - f_4(y)] \\
&= [f_1(x+y) + f_2(x+\sigma(y)) - f_3(x-z-\sigma(z)) - f_4(y+z+\sigma(z))] \\
&\quad + [f_1(x+y) + f_2(x+\sigma(y)) - f_3(x+z+\sigma(z)) - f_4(y-z-\sigma(z))] \\
&\quad - [f_1(-\sigma(z)) + f_2(x-\sigma(x)-z) - f_3(x-z-\sigma(z)) - f_4(-x+z)] \\
&\quad + [f_1(-\sigma(z)) + f_2(-y+\sigma(y)-\sigma(z)) - f_3(-y-\sigma(z)) - f_4(y)] \\
&\quad - [f_1(z) + f_2(x-\sigma(x)+\sigma(z)) - f_3(x+z+\sigma(z)) - f_4(-x-\sigma(z))] \\
&\quad + [f_1(-\sigma(z)) + f_2(x-\sigma(x)-z) - f_3(x) - f_4(-x-\sigma(z))] \\
&\quad + [f_1(z) + f_2(x-\sigma(x)+\sigma(z)) - f_3(x) - f_4(-x+z)] \\
&\quad + [f_1(z) + f_2(-y+\sigma(y)+z) - f_3(-y+z) - f_4(y)] \\
&\quad - [f_1(-\sigma(z)) + f_2(-y+\sigma(y)-\sigma(z)) - f_3(-y+z) - f_4(y-z-\sigma(z))] \\
&\quad - [f_1(z) + f_2(-y+\sigma(y)+z) - f_3(-y-\sigma(z)) - f_4(y+z+\sigma(z))],
\end{aligned}
$$

we get

$$\|f_1(x+y) + f_2(x+\sigma(y)) - f_3(x) - f_4(y)\| \leq 5\delta. \tag{62}$$

Case 2: $\sigma(y) = -y$ or $\sigma(x) = -x$. Subcase 2.1: $\sigma(y) = -y$. By using the same decomposition in Theorem 4, we obtain

$$\|f_1(x+y) + f_2(x-y) - f_3(x) - f_4(y)\| \leq 7\delta. \tag{63}$$

Subcase 2.2: $\sigma(x) = -x$. Inequality (39) implies that

$$\|f_1(y+x) + f_2 \circ \sigma(y-x) - f_3(x) - f_4(y)\| \leq \delta. \tag{64}$$

Now, by using the decomposition in Theorem 4, we get

$$\|f_1(y+x) + f_2 \circ \sigma(y-x) - f_3(x) - f_4(y)\| \leq 7\delta. \tag{65}$$

Finally, in view of inequalities (62), (63), (64), and (65), we obtain

$$\|f_1(x+y) + f_2(x+\sigma(y)) - f_3(x) - f_4(y)\| \leq 7\delta \tag{66}$$

for all $x, y \in E$. According to Theorem 3.1 [6] one gets that there exists a unique function $q : E \to F$ solution of Eq. (4), there exists a function $v : E \to F$ solution of Eq. (56) and there exists exactly two additive functions $A_1, A_2 : E \to F$ such that $A_i \circ \sigma = -A_i$ (i=1,2), and which satisfy the inequalities (57), (58), (59), and (60). This completes the proof of theorem.

Corollary 5 *The mappings $f_1, f_2, f_3, f_4 : E \to F$ with $f_1(0) = f_2(0) = f_3(0) = f_4(0) = 0$ are solutions of* Eq. (9) *if and only if*

$$\|f_1(x+y) + f_2(x+\sigma(y)) - f_3(x) - f_4(y)\| \to 0 \ \ as \ \ \|x\| + \|y\| \to +\infty.$$

Corollary 6 *If the mappings $f_1, f_2, f_3, f_4 : E \to F$ satisfy the inequality*

$$\|f_1(x+y) + f_2(x-y) - f_3(x) - f_4(y)\| \leq \delta \tag{67}$$

for all $(x, y) \in C$, then there exists a unique function $q : E \to F$ solution of Eq. (4), *there exists $\alpha \in F$, there exist exactly two additive functions $A_1, A_2 : E \to F$ such that*

$$\|f_1(x) - \frac{1}{2}A_1(x) - \frac{1}{2}A_2(x) - \frac{1}{2}q(x) - f_1(0) - \alpha\| \leq 133\delta, \tag{68}$$

$$\|f_2(x) + \frac{1}{2}A_1(x) - \frac{1}{2}A_2(x) - \frac{1}{2}q(x) - f_2(0) + \alpha\| \leq 133\delta, \tag{69}$$

$$\|f_3(x) - A_2(x) - q(x) - f_3(0)\| \leq 112\delta \tag{70}$$

and

$$\|f_4(x) - A_1(x) - q(x) - f_4(0)\| \leq 112\delta \tag{71}$$

for all $x \in E$.

Corollary 7 *The mappings* $f_1, f_2, f_3, f_4 : E \to F$ *with* $f_1(0) = f_2(0) = f_3(0) = f_4(0) = 0$ *are solutions of* Eq. (8) *if and only if*

$$\|f_1(x+y) + f_2(x-y) - f_3(x) - f_4(y)\| \to 0 \ \text{as} \ \|x\| + \|y\| \to +\infty.$$

Corollary 8 *If the mappings* $f, g, h : E \to F$ *satisfy the inequality*

$$\|f(x+y) - g(x) - h(y)\| \leq \delta \tag{72}$$

for all $(x, y) \in C$, *then there exists a unique additive function* $A : E \to F$ *such that*

$$\|f(x) - A(x) - f(0)\| \leq 266\delta, \tag{73}$$

$$\|g(x) - A(x) - g(0)\| \leq 112\delta \tag{74}$$

and

$$\|h(x) - A(x) - h(0)\| \leq 112\delta \tag{75}$$

for all $x \in E$.

Corollary 9 *The mappings* $f, g, h : E \to F$ *with* $f(0) = g(0) = h(0) = 0$ *are solutions of* Eq. (7) *if and only if*

$$\|f(x+y) - g(x) - h(y)\| \to 0 \ \text{as} \ \|x\| + \|y\| \to +\infty.$$

Corollary 10 *The mapping* $f : E \to F$ *with* $f(0) = 0$, *is additive if and only if*

$$\|f(x+y) - f(x) - f(y)\| \to 0 \ \text{as} \ \|x\| + \|y\| \to +\infty.$$

Corollary 11 *The mapping* $f : E \to F$ *with* $f(0) = 0$ *is a quadratic function if and only if*

$$\|f(x+y) + f(x-y) - 2f(x) - 2f(y)\| \to 0 \ \text{as} \ \|x\| + \|y\| \to +\infty.$$

Corollary 12 *The mapping* $f : E \to F$ *with* $f(0) = 0$ *is a solution of the quadratic functional* Eq. (4) *if and only if*

$$\|f(x+y) + f(x+\sigma(y)) - 2f(x) - 2f(y)\| \to 0 \ \text{as} \ \|x\| + \|y\| \to +\infty.$$

In the following corollaries, we state the Hyers–Ulam stability for Drygas functional equation

$$f(x+y) + f(x+\sigma(y)) = 2f(x) + f(y) + f(\sigma(y)), \tag{76}$$

in the restricted domain C.

Corollary 13 *If the mapping $f : E \to F$ satisfies the inequality*

$$\|f(x+y) + f(x+\sigma(y)) - 2f(x) - f(y) - f(\sigma(y))\| \leq \delta \tag{77}$$

for all $(x, y) \in C$, then there exists a unique additive mapping $A : E \to F$ and a unique quadratic mapping $q : E \to F$ such that $A \circ \sigma = -A$ and

$$\|f(x) - q(x) - A(x) - f(0)\| \leq 112\delta \tag{78}$$

for all $x \in E$.

Corollary 14 *The mapping $f : E \to F$ with $f(0) = 0$ is a solution of Drygas functional* Eq. (76) *if and only if*

$$\|f(x+y)+f(x+\sigma(y))-2f(x)-f(y)-f(\sigma(y))\| \to 0 \quad as \quad \|x\|+\|y\| \to +\infty.$$

References

1. Aoki, T.: On the stability of the linear transformation in Banach spaces. J. Math. Soc. Japan. **2**, 64–66 (1950)
2. Akkouchi, M.: Stability of certain functional equations via a fixed point of Ćirić. Filomat **25**, 121–127 (2011)
3. Akkouchi, M.: Hyers–Ulam–Rassias stability of Nonlinear Volterra integral equation via a fixed point approach. Acta. Univ. Apulensis Math. Inform. **26**, 257–266 (2011)
4. Bae, J.-H.: On the stability of 3-dimensional quadratic functional equation. Bull. Korean Math. Soc. **37**(3), 477–486 (2000)
5. Bae, J.-H., Jun, K.-W.: On the generalized Hyers–Ulam–Rassias stability of a quadratic functional equation. Bull. Korean Math. Soc. **38**(2), 325–336 (2001)
6. Bouikhalene, B., Elqorachi, E., Rassias, Th.M.: On the Hyers–Ulam stability of approximately Pexider mappings. Math. Inequal. Appl. **11**, 805–818 (2008)
7. Brzdęk, J.: On a method of proving the Hyers–Ulam stability of functional equations on restricted domains. Aust. J. of Math. Anal. Appl. **6**(1), 1–10, Article 4, (2009)
8. Brzdęk, J.: A note on stability of the Popoviciu functional equation on restricted domain. Dem. Math. **XLIII**(3), 635–641 (2010)
9. Brzdęk, J., Pietrzyk, A.: A note on stability of the general linear equation. Aequ. Math. **75**, 267–270 (2008)
10. Cădariu, L., Radu, V.:Fixed points and the stability of Jensens functional equation. J. Inequal. Pure Appl. Math. **4**(1), Article 4 (2003)
11. Cădariu, L., Radu, V.: On the stability of the Cauchy functional equation: A fixed point approach. Grazer Math. Berichte. **346**, 43–52 (2003)
12. Charifi, A., Bouikhalene, B., Elqorachi, E.: Hyers–Ulam–Rassias stability of a generalized Pexider functional equation. Banach J. Math. Anal. **1**(2), 176–185 (2007)
13. Chung, J.-Y.: A generalized Hyers–Ulam stability of a Pexiderized logarithmic functional equation in restricted domains. J. Ineq. Appl. **2012**, 15 (2012)
14. Chung, J.-Y.: Stability of functional equations on restricted domains in a group and their asymptotic behaviors. Comp. Math. Appl. **60**(9), 2653–2665 (2010)
15. Chung, J., Sahoo, P.K..: Stability of a logarithmic functional equation in distributions on a restricted domain. Abstr. Appl. Anal. 9 p, Article ID 751680, (2013)
16. Cieplisński, K.: Applications of fixed point theorems to the Hyers–Ulam stability of functional equations. A survey. Ann. Funct. Anal. **3**, 151–164 (2012)

17. Dales, H.G., Moslehian, M.S.: Stability of mappings on multi-normed spaces. Glasgow Math. J. **49**, 321–332 (2007)
18. Elqorachi, E., Manar, Y., Rassias, Th.M: Hyers–Ulam stability of the quadratic functional equation. Int. J. Nonlin. Anal. Appl. **1**(2), 11–20 (2010)
19. Elqorachi, E., Manar, Y., Rassias, Th.M: Hyers-Ulam stability of the quadratic and Jensen functional equations on unbounded domains. J. Math. Sci. Adv. Appl. **4**(2), 287–301 (2010)
20. Forti, G.L.: Hyers-Ulam stability of functional equations in several variables. Aequ. Math. **50**, 143–190 (1995)
21. Forti, G.-L., Sikorska, J.: Variations on the Drygas equation and its stability. Nonlin. Anal.: Theory Meth. Appl. **74**, 343–350 (2011)
22. Găvruta, P.: A generalization of the Hyers–Ulam–Rassias stability of approximately additive mappings. J. Math. Anal. Appl. **184**, 431–436 (1994)
23. Gajda, Z.: On stability of additive mappings. Int. J. Math. Math. Sci. **14**, 431–434 (1991)
24. Ger, R.: On a factorization of mappings with a prescribed behaviour of the Cauchy difference. Ann. Math. Sil. **8** 141–155 (1994)
25. Hyers, D.H.: On the stability of the linear functional equation. Proc. Nat. Acad. Sci. U. S. A. **27**, 222–224 (1941)
26. Hyers, D.H., Rassias, Th.M.: Approximate homomorphisms. Aequ. Math. **44**, 125–153 (1992)
27. Hyers, D.H., Isac, G., Rassias, T.M.: On the asymptoticity aspect of Hyers–Ulam stability of mappings. Proc. Amer. Math. Soc. **126**, 425–430 (1998)
28. Hyers, D.H., Isac, G., Rassias, T.M.: Stability of Functional Equations in Several Variables. Birkhäuser, Basel (1998)
29. Jung, S.-M.: On the Hyers-Ulam stability of the functional equation that have the quadratic property. J. Math. Anal. Appl. **222**, 126–137 (1998)
30. Jung, S.-M.: Hyers–Ulam–Rassias stability of Jensen's equation and its application. Proc. Amer. Math. Soc. **126**(11), 3137–3143 (1998)
31. Jung, S.-M.: Stability of the quadratic equation of Pexider type. Abh. Math. Sem. Univ. Hamburg **70**, 175–190 (2000)
32. Jung, S.-M.: Hyers–Ulam–Rassias stability of isometries on restricted domains. Nonlin. Stud. **8**(1), 125 (2001)
33. Jung, S.-M.: Hyers-Ulam-Rassias Stability of Functional Equations in Nonlinear Analysis. Springer, New York (2011)
34. Jung, S.-M., Kim, B.: Local stability of the additive functional equation and its applications. Int. J. Math. Math. Sci. **2003**, 15–26 (2003)
35. Jun, K.-W., Lee, Y.-H.: A generalization of the Hyers-Ulam-Rassias stability of Jensen's equation. J. Math. Anal. Appl. **238**, 305–315 (1999)
36. Jung, S.-M., Sahoo, P.K.: Hyers–Ulam stability of the quadratic equation of Pexider type. J. Korean Math. Soc. **38**(3), 645–656 (2001)
37. Kannappan, Pl.: Functional Equations and Inequalities with Applications. Springer, New York, 2009
38. Kim, B.: Local stability of the additive equation. Int. J. Pure Appl. Math., **5** (1), 65–74 (2003)
39. Kim, G.H., Lee, S.H.: Stability of the d'Alembert type functional equations. Nonlin. Funct. Anal. Appl. **9**, 593–604 (2004)
40. Kim, G.H., Lee, Y.-H.: Hyers–Ulam stability of a Bi-Jensen functional equation on a punctured domain. J. Ineq. Appl. **2010,** 15 p, Article ID 476249, (2010)
41. Lee, Y.H., Jung, K.W.: A generalization of the Hyers-Ulam-Rassias stability of the pexider equation. J. Math. Anal. Appl. **246**, 627–638 (2000)
42. Lee, J.R., Shin, D.Y., Park C.: Hyers-Ulam stability of functional equations in matrix normed spaces. J. Inequal. Appl. **2013**, 11 p (2013)
43. Manar, Y., Elqorachi, E., Rassias, Th.M.: Hyers–Ulam stability of the Jensen functional equation in quasi-Banach spaces. Nonlin. Funct. Anal. Appl. **15**(4), 581–603 (2010)
44. Manar, Y., Elqorachi, E., Rassias, Th.M: On the Hyers–Ulam stability of the quadratic and Jensen functional equations on a restricted domain. Nonlin. Funct. Anal. Appl. **15**(4), 647–655 (2010)

45. Moghimi, M.B., Najati, A., Park, C.: A functional inequality in restricted domains of Banach modules. Adv. Diff. Eq. **2009**, 14 p, Article ID 973709, doi:10.1155/2009/973709
46. Moslehian, M.S.: The Jensen functional equation in non-Archimedean normed spaces. J. Funct. Spaces Appl. **7**, 13–24 (2009)
47. Moslehian, M.S., Najati, A.: Application of a fixed point theorem to a functional inequality. Fixed Point Theory **10**, 141–149 (2009)
48. Moslehian, M.S., Sadeghi, G.: Stability of linear mappings in quasi-Banach modules. Math. Inequal. Appl. **11**, 549–557 (2008)
49. Najati, A.: On the stability of a quartic functional equation. J. Math. Anal. Appl. **340**, 569–574 (2008)
50. Najati, A., Jung, S.M.: Approximately quadratic mappings on restricted domains, J. Ineq. Appl. 10 p, Article ID 503458, (2010)
51. Najati, A., Moghimi, M.B.: Stability of a functional equation deriving from quadratic and additive functions in quasi-Banach spaces. J. Math. Anal. Appl. **337**, 399–415 (2008)
52. Najati, A., Park, C.: Hyers–Ulam–Rassias stability of homomorphisms in quasi-Banach algebras associated to the Pexiderized Cauchy functional equation. J. Math. Anal. Appl. **335**, 763–778 (2007)
53. Pardalos, P.M., Georgiev, P.G., Srivastava, H.M. (eds.): Nonlinear Analysis - Stability, Approximation, and Inequalities. In honor of Th.M. Rassias on the occasion of his 60th birthday, Springer, Berlin (2012)
54. Park, C.G.: On the stability of the linear mapping in Banach modules. J. Math. Anal. Appl. **275**, 711–720 (2002)
55. Pourpasha, M.M., Rassias, J.M., Saadati, R., Vaezpour, S.M.: A fixed point approach to the stability of Pexider quadratic functional equation with involution, J. Ineq. Appl. Article ID 839639, (2010). doi:10.1155/2010/839639
56. Rahimi, A., Najati, A., Bae J.-H.: On the asymptoticity aspect of Hyers–Ulam stability of quadratic mappings. J. Ineq. Appl. **2010**, 14 p (2010)
57. Rassias, Th.M.: On the stability of linear mapping in Banach spaces. Proc. Amer. Math. Soc. **72**, 297–300 (1978)
58. Rassias, J.M.: On approximation of approximately linear mappings by linear mappings. J. Funct. Anal. **46**, 126–130 (1982)
59. Rassias, J.M.: Solution of a problem of Ulam. J. Approx. Theory **57**, 268–273 (1989)
60. Rassias, Th.M.: On the stability of functional equations in Banach spaces. J. Math. Anal. Appl. **251**(1), 264–284 (2000)
61. Rassias, Th.M.: The problem of S. M. Ulam for approximately multiplicative mappings. J. Math. Anal. Appl. **246**, 352–378 (2000)
62. Rassias, Th.M.: On the stability of the functional equations and a problem of Ulam. Acta Appl. Math. **62**, 23–130 (2000)
63. Rassias, Th.M., Brzdęk, J.: Functional Equations in Mathematical Analysis. Springer, New York (2011)
64. Rassias, J.M., Rassias, M.J.: On the Ulam stability of Jensen type mappings on restricted domains. J. Math. Anal. Appl. **281**, 516–524 (2003)
65. Rassias, Th.M., Šemrl, P.: On the behavior of mappings which do not satisfy Hyers-Ulam stability. Proc. Amer. Math. Soc. **114**, 989–993 (1992)
66. Sikorska, J.: Exponential functional equation on spheres. Appl. Math. Lett. **23**, 156–160 (2010)
67. Skof, F.: Approssimazione di funzioni δ-quadratic su dominio restretto. Atti. Accad. Sci. Torino Cl. Sci. Fis. Mat. Natur. **118**, 58–70 (1984)
68. Skof, F.: Sull'approssimazione delle applicazioni localmente δ-additive. Atti. Accad. Sci. Torino Cl. Sci. Fis. Mat. Natur. **117**, 377–389 (1983)
69. Stetkær, H.: Functional equations on abelian groups with involution. Aequ. Math. **54**, 144–172 (1997)
70. Ulam, S.M.: A Collection of Mathematical Problems. Interscience, New York, (1961); Problems in Modern Mathematics, *Wiley, New York*, (1964)

71. Xiang, S.-H.: Hyers-Ulam-Rassias stability of approximate isometries on restricted domains. J. Cent. S. Univ. Tech. **9**(4), 289–292 (2002)
72. Yang, D.: Remarks on the stability of Drygas' equation and the Pexider-quadratic equation. Aequ. Math. **68**, 108–116 (2004)

Hyers-Ulam Stability of Some Differential Equations and Differential Operators

Dorian Popa and Ioan Raşa

Abstract This chapter contains results on generalized Hyers–Ulam stability, obtained by the authors, for linear differential equations, linear differential operators and partial differential equations in Banach spaces. As a consequence we improve some known estimates of the difference between the perturbed and the exact solution.

Keywords Hyers–Ulam stability · Differential operators · Linear differential equations · Partial differential equations

1 Introduction

In 1940, on a talk given at Wisconsin University, S. M. Ulam posed the following problem: "Under what conditions does there exist an homomorphism near an approximately homomorphism of a complete metric group?", more precisely: "*Given a metric group* $(G, \cdot, \rho)$, *a number* $\varepsilon > 0$ *and a mapping* $f : G \to G$ *which satisfies the inequality* $\rho(f(xy), f(x)f(y)) < \varepsilon$ *for all* $x, y \in G$, *does there exist a homomorphism* a *of* G *and a constant* $k > 0$, *depending only on* G, *such that* $\rho(a(x), f(x)) \le k\varepsilon$ *for all* $x \in G$?"

If the answer is affirmative, the equation $a(xy) = a(x)a(y)$ of the homomorphism is called **stable** (for more details see [39]). A year later, D. H. Hyers in [12] gave an answer to the problem of Ulam for the Cauchy functional equation in Banach spaces. "Let E_1, E_2 be two real Banach spaces and $\varepsilon > 0$. Then, for every mapping $f : E_1 \to E_2$ satisfying

$$\|f(x+y) - f(x) - f(y)\| \le \varepsilon \qquad (1)$$

D. Popa (✉) · I. Raşa
Technical University of Cluj-Napoca, Department of Mathematics,
28 Memorandumului Street, 400114, Cluj-Napoca, Romania,
e-mail: Popa.Dorian@math.utcluj.ro

I. Raşa
e-mail: Ioan.Rasa@math.utcluj.ro

T. M. Rassias (ed.), *Handbook of Functional Equations*,
Springer Optimization and Its Applications 96, DOI 10.1007/978-1-4939-1286-5_14

for all $x, y \in E_1$, there exists a unique additive mapping $g : E_1 \to E_2$ with the property

$$\|f(x) - g(x)\| \leq \varepsilon, \ \forall x \in E_1." \tag{2}$$

After Hyers' result many papers dedicated to this topic, extending Ulam's problem to other functional equations and generalizing Hyers' result in various directions, were published (see e.g., [3–7, 10, 11, 13, 14, 18, 28, 29]. A new direction of research in the stability theory of functional equations, called today Hyers–Ulam stability, was opened by the papers of Aoki and Rassias by considering instead of ε in (1) a function depending on x and y ([2, 33]). Obłoza seems to be the first author who investigated Hyers–Ulam stability of differential equations [26, 27]. Later Alsina and Ger proved that for every differentiable mapping $f : I \to \mathbb{R}$ satisfying $|f'(x) - f(x)| \leq \varepsilon$ for every $x \in I$, where $\varepsilon > 0$ is a given number and I is an open interval of $\mathbb{R}$, there exists a differentiable function $g : I \to \mathbb{R}$ with the property $g'(x) = g(x)$ and $|f(x) - g(x)| \leq 3\varepsilon$ for all $x \in I$. The result of Alsina and Ger [1] was extended by Miura, Miyajima and Takahasi [24, 25, 37] and by Takahasi, Takagi, Miura and Miyajima [38] to the Hyers–Ulam stability of the first order linear differential equations and linear differential equations of higher order with constant coefficients. Furthermore, S.-M. Jung [15, 16, 17, 19] obtained results on the stability of linear differential equations extending the results of Takahasi, Takagi and Miura. I. A. Rus obtained some results on the stability of differential and integral equations using Gronwall lemma and the technique of weakly Picard operators [35, 36]. Recently, G. Wang, M. Zhou and L. Sun [40] and Y. Liand Y. Shen [20] proved the Hyers–Ulam stability of the linear differential equation of the first order and the linear differential equation of the second order with constant coefficients by using the method of integral factor.

An extension of the results given in [16, 20, 25] was obtained by D. S. Cîmpean and D. Popa, and by D. Popa and I. Raşa for the linear differential equation of nth order with constant coefficients and the linear differential operator of nth order with nonconstant coefficients [8, 30, 31]. It seems that the first paper on Hyers–Ulam stability of partial differential equations was written by Prastaro and Rassias [32]. For recent results on this subject we refer the reader to [9, 21–23, 34].

Throughout this paper by $(X, \|\cdot\|)$ we denote a Banach space over the field K (K is one of the fields $\mathbb{R}$ or $\mathbb{C}$). In what follows by $\Re z$, we denote the real part of the complex number z.

2 Stability of the Linear Differential Equation of Order One

In what follows, $I = (a, b)$, $a, b \in \mathbb{R} \cup \{\pm\infty\}$ is an open interval in $\mathbb{R}$, $c \in [a, b]$, $C \in \overline{\mathbb{R}}$, $f \in C(I, X)$, $\lambda \in C(I, K)$ and $\varepsilon \in C(I, \mathbb{R})$ with $\varepsilon \geq 0$. We deal with the stability of the linear differential equation (see [30])

$$y'(x) - \lambda(x)y(x) = f(x), \quad x \in I. \tag{3}$$

For a function $g : (a,b) \to X$, define $g(b) := \lim_{x\to b} g(x)$ and $g(a) := \lim_{x\to a} g(x)$, if the limits exist. Let $L \in C^1(I,K)$ be an antiderivative of λ, i.e., $L' = \lambda$ on I. Define $\psi_c : I \to \mathbb{R}$ by

$$\psi_c(x) := e^{\Re L(x)} \left| \int_c^x e^{-\Re L(t)} \varepsilon(t) dt \right|. \tag{4}$$

If $c \in \{\pm\infty\}$ then we suppose that the integral which defines ψ_c is convergent for every $x \in I$. Therefore, $\psi_c(c) = 0$ for all $c \in I$.

The following well-known lemma is useful in the proof of our stability results.

Lemma 1 *The general solution of the equation*

$$y'(x) - \lambda(x)y(x) = f(x), \quad x \in I \tag{5}$$

is given by

$$y(x) = e^{L(x)} \left(\int_{x_0}^x f(t)e^{-L(t)} dt + k \right) \tag{6}$$

where $x_0 \in I$ and $k \in X$ is an arbitrary constant.

The first result on Aoki–Rassias stability for a first order linear differential equation is contained in the next theorem.

Theorem 1 *For every $y \in C^1(I,X)$ satisfying*

$$\|y'(x) - \lambda(x)y(x) - f(x)\| \le \varepsilon(x), \quad x \in I \tag{7}$$

there exists a unique solution $u \in C^1(I,X)$ of the Eq. (5) with the property

$$\|y(x) - u(x)\| \le \psi_c(x), \quad x \in I. \tag{8}$$

Proof *Existence.* Let $y \in C^1(I,X)$ satisfying (7) and define

$$g(x) := y'(x) - \lambda(x)y(x) - f(x), \quad x \in I. \tag{9}$$

Then, according to Lemma 1, it follows

$$y(x) = e^{L(x)} \left(\int_{x_0}^x e^{-L(t)} f(t) dt + \int_{x_0}^x e^{-L(t)} g(t) dt + k \right), \quad x_0 \in I,\ k \in X.$$

Let $G : I \to X$ be given by

$$G(x) := \int_c^x e^{-L(t)} g(t) dt, \quad x \in I. \tag{10}$$

If $c \in \{\pm\infty\}$, the integral which defines G is convergent since

$$\|g(t)\| \le \varepsilon(t) \quad \text{for all} \quad t \in I.$$

(See the remark after (4)).

Now let u be defined by

$$u(x) := e^{L(x)} \left(\int_{x_0}^{x} f(t) e^{-L(t)} dt + k - G(x_0) \right).$$

Then, obviously u satisfies the Eq. (5) and we get

$$\begin{aligned} \|y(x) - u(x)\| &= e^{\Re L(x)} \left\| \int_{x_0}^{x} g(t) e^{-L(t)} dt + G(x_0) \right\| = e^{\Re L(x)} \|G(x)\| \\ &\le e^{\Re L(x)} \left| \int_{c}^{x} \|e^{-L(t)} g(t)\| \right| dt \\ &\le e^{\Re L(x)} \left| \int_{c}^{x} e^{-\Re L(t)} \varepsilon(t) dt \right| \\ &= \psi_c(x),\ x \in I. \end{aligned}$$

Therefore, the existence is proved.
Uniqueness. Suppose that for a y satisfying (7) there exist $u_1, u_2, u_1 \neq u_2$, satisfying (5) and (8). Then

$$u_j(x) = e^{L(x)} \left(\int_{x_0}^{x} f(t) e^{-L(t)} dt + k_j \right), \quad k_j \in X,\ j = 1, 2,\ k_1 \neq k_2$$

and

$$\begin{aligned} e^{\Re L(x)} \|k_1 - k_2\| &= \|u_1(x) - u_2(x)\| \\ &\le \|u_1(x) - y(x)\| + \|y(x) - u_2(x)\| \\ &\le 2 e^{\Re L(x)} \left| \int_{c}^{x} e^{-\Re L(t)} \varepsilon(t) dt \right| \end{aligned}$$

for all $x \in I$. Therefore,

$$\|k_1 - k_2\| \le 2 \left| \int_{c}^{x} e^{-\Re L(t)} \varepsilon(t) dt \right|, \quad x \in I. \tag{11}$$

Now letting $x \to c$ in (11) it follows $k_1 = k_2$, contradiction.
Theorem 1 leads to the following result for the Cauchy problem of Eq. (5).

Corollary 1 *For every $y \in C^1(I, X)$ satisfying*

$$\begin{cases} \|y'(x) - \lambda(x) y(x) - f(x)\| \le \varepsilon(x) \\ \\ y(c) = C \end{cases}, \quad x \in I$$

there exists a unique solution $u \in C^1(I, X)$ *of the Cauchy problem*

$$\begin{cases} u'(x) - \lambda(x)u(x) - f(x) = 0 \\ \\ u(c) = C \end{cases}, \quad x \in I,$$

with the property

$$\|y(x) - u(x)\| \leq \psi_c(x), \quad x \in I.$$

The result obtained in Theorem 1 is more general than the result of [8, [Lemma 2.2]] and [16, Theorem 1] since it gives a better estimation of the difference between the approximate solution and the exact solution of Eq. (5). This is obvious in the cases $c = a$ and $c = b$, but for $c \in (a, b)$ this better approximation is not always valid on the entire interval (a, b). We will show in the next example that in some cases, this estimation is global for $c \in (a, b)$ and we will find the optimal ψ_c.

Example 1 Let $\theta \in \mathbb{R} \setminus \{0\}$ and $\varepsilon(x) = \theta \Re\lambda(x)$, $x \in I$. Then

$$\psi_c(x) = |\theta| \cdot |1 - e^{\Re(L(x) - L(c))}|, \quad x \in I.$$

First we consider the case $\theta > 0$, i.e., $\Re\lambda(x) \geq 0$ for all $x \in I$. Then,

$$\Re L'(x) \geq 0, \quad x \in I,$$

hence, $\Re L$ is increasing on I,

$$\psi_c(x) = \theta \cdot \begin{cases} e^{\Re L(x) - \Re L(c)} - 1, & x \in [c, b), \\ \\ 1 - e^{\Re L(x) - \Re L(c)}, & x \in (a, c), \end{cases}$$

and

$$\|\psi_c\|_\infty = \theta \max\left\{e^{\Re(L(b) - L(c))} - 1, 1 - e^{\Re(L(a) - L(c))}\right\}.$$

Obviously $\|\psi_c\|_\infty$ is minimum for $e^{\Re(L(b) - L(c))} - 1 = 1 - e^{\Re(L(a) - L(c))}$, i.e.,

$$e^{\Re L(c)} = \frac{e^{\Re L(a)} + e^{\Re L(b)}}{2}.$$

The relation from above gives $\widetilde{c}$ optimal, therefore, the following estimation holds

$$\|y - u\|_\infty \leq \|\psi_{\widetilde{c}}\|_\infty,$$

where

$$\|\psi_{\widetilde{c}}\|_\infty = \theta(e^{\Re(L(b) - L(\widetilde{c}))} - 1) = \theta \cdot \frac{e^{\Re L(b)} - e^{\Re L(a)}}{e^{\Re L(b)} + e^{\Re L(a)}}$$

i.e.,

$$\min_{c \in I} \|\psi_c\|_\infty = \theta \cdot \frac{e^{\Re L(b)} - e^{\Re L(a)}}{e^{\Re L(b)} + e^{\Re L(a)}}. \tag{12}$$

The case $\theta < 0$ leads analogously to

$$\min_{c\in I}\|\psi_c\|_\infty = -\theta \cdot \frac{e^{\Re L(a)} - e^{\Re L(b)}}{e^{\Re L(b)} + e^{\Re L(a)}};$$

therefore for all $\theta \in \mathbb{R}\setminus\{0\}$, we have

$$\min_{c\in I}\|\psi_c\|_\infty = |\theta| \cdot \frac{|e^{\Re L(b)} - e^{\Re L(a)}|}{e^{\Re L(b)} + e^{\Re L(a)}}.$$

Remark 1 Now let λ be constant with $\Re\lambda \neq 0$. Then, $L(x) = \lambda x$ and

$$\min_{c\in I}\|\psi_c\|_\infty = |\theta| \cdot \frac{|e^{b\Re\lambda} - e^{a\Re\lambda}|}{e^{b\Re\lambda} + e^{a\Re\lambda}}. \tag{13}$$

Taking now an arbitrary $\delta > 0$ and $\theta = \dfrac{\delta}{|\Re\lambda|}$ it is easy to check that

$$\min_{c}\|\psi_c\|_\infty < \frac{\delta}{|\Re\lambda|}(1 - e^{-|\Re(\lambda)|(b-a)})$$

if $a, b \in \mathbb{R}$ and

$$\min_{c}\|\psi_c\|_\infty = \frac{\delta}{|\Re\lambda|}$$

if $a = -\infty$ or $b = +\infty$, therefore, we improve the result obtained in [[8], [Corollary 2.4]], along all interval I in the case of classical Hyers–Ulam stability.

More precisely we have the following result.

Corollary 2 *Suppose that $\lambda \in \mathbb{C}\setminus\{0\}$ and $\delta \geq 0$. Then, for every $y \in C^1(I, X)$ satisfying*

$$\|y'(x) - \lambda y(x) - f(x)\| \leq \delta, \quad x \in I$$

there exists a unique solution of (5) such that

$$\|y(x) - u(x)\| \leq \begin{cases} \dfrac{\delta}{|\Re\lambda|} \cdot \dfrac{|e^{b\Re\lambda} - e^{a\Re\lambda}|}{e^{b\Re\lambda} + e^{a\Re\lambda}}, & \text{if} \quad a, b \in \mathbb{R} \\[2ex] \dfrac{\delta}{|\Re\lambda|}, & \text{if} \quad a = -\infty \quad \text{or} \quad b = +\infty. \end{cases}$$

3 Stability of the Linear Differential Equation of Higher Order with Constant Coefficients

The results proved in the previous theorems and corollaries lead to stability of the linear differential equation with constant coefficients (see [31]). We will improve in what follows the results obtained in [8] and [25] for this equation. Suppose that

$(X, \|\cdot\|)$ is a Banach space over $\mathbb{C}$ and $a_0, a_1, \ldots, a_{n-1} \in \mathbb{C}$, $n \geq 1$, are given numbers. We study the stability of the linear differential equation

$$y^{(n)}(x) - \sum_{j=0}^{n-1} a_j y^{(j)}(x) = f(x), \quad x \in I. \tag{14}$$

Let

$$P(z) = z^n - \sum_{j=0}^{n-1} a_j z^j \tag{15}$$

be the characteristic polynomial of the Eq. (14) and denote by $r_1, r_2, \ldots, r_n$ the complex roots of (15). For $\lambda \in \mathbb{C}$ and $c \in [a, b]$ define

$$\phi_\lambda(h)(x) := e^{\Re(\lambda)x} \left| \int_c^x e^{-\Re(\lambda)t} h(t) dt \right|, \quad x \in I \tag{16}$$

for all h with the property that the integral from the right hand side of (16) is convergent. We suppose that $\phi_{r_k} \circ \phi_{r_{k-1}} \circ \ldots \circ \phi_{r_1}(\varepsilon)$ exist for every $k \in \{1, 2, \ldots, n\}$ if $c = a$ or $c = b$.

Theorem 2 *For every $y \in C^n(I, X)$ with the property*

$$\left\| y^{(n)}(x) - \sum_{j=0}^{n-1} a_j y^{(j)}(x) - f(x) \right\| \leq \varepsilon(x), \quad x \in I \tag{17}$$

there exists a solution of the Eq. (14) such that

$$\|y(x) - u(x)\| \leq \phi_{r_n} \circ \phi_{r_{n-1}} \circ \ldots \circ \phi_{r_1}(\varepsilon)(x), \quad x \in I. \tag{18}$$

Proof The proof by induction is analogous to the proof of [8, Theorem 2.3].

For $n = 1$, Theorem 2 holds in virtue of Theorem 1.

Now suppose that Theorem 2 holds for an $n \in \mathbb{N}$. We have to prove that for all $y \in C^{n+1}(I, X)$ satisfying the relation

$$\left\| y^{(n+1)}(x) - \sum_{j=0}^{n} a_j y^{(j)}(x) - f(x) \right\| \leq \varepsilon(x), \ \forall\, x \in I \tag{19}$$

there exists a unique solution $u \in C^{n+1}(I, X)$ satisfying

$$u^{(n+1)}(x) - \sum_{j=0}^{n} a_j u^{(j)}(x) - f(x) = 0, \ \forall\, x \in I \tag{20}$$

such that

$$\|y(x) - u(x)\| \leq \phi_{r_{n+1}} \circ \phi_{r_n} \circ \ldots \circ \phi_{r_1}(\varepsilon)(x),\ \forall\, x \in I. \tag{21}$$

Let $y \in C^{n+1}(I, X)$ be a mapping satisfying (19). According to Vieta's relations we get

$$\|y^{(n+1)}(x) - (r_1 + \ldots + r_{n+1})y^{(n)}(x) + \ldots + (-1)^{n+1} r_1 r_2 \ldots r_{n+1} y(x) - f(x)\| \leq \varepsilon(x)$$

or

$$\|(y^{(n+1)}(x) - r_{n+1} y^{(n)}(x)) - (r_1 + \ldots + r_n)(y^{(n)}(x) - r_{n+1} y^{(n-1)}(x)) + \ldots +$$

$$+(-1)^n r_1 \ldots r_n (y'(x) - r_{n+1} y(x)) - f(x)\| \leq \varepsilon(x), \quad x \in I. \tag{22}$$

Let z be given by

$$z := y' - r_{n+1} y.$$

Then, (22) becomes

$$\|z^{(n)}(x) - (r_1 + \ldots + r_n) z^{(n-1)}(x) + \ldots + (-1)^n r_1 \ldots r_n z(x) - f(x)\| \leq \varepsilon(x)$$

for all $x \in I$. Therefore, in virtue of the induction hypothesis, there exists a unique v such that

$$v^{(n)}(x) - (r_1 + \ldots + r_n) v^{(n-1)}(x) + \ldots + (-1)^n r_1 \ldots r_n v(x) = f(x), \quad x \in I$$

and

$$\|z(x) - v(x)\| \leq \phi_{r_n} \circ \ldots \circ \phi_{r_1}(\varepsilon)(x), \quad x \in I$$

which is equivalent to

$$\|y'(x) - r_{n+1}(x) y(x) - v(x)\| \leq \phi_{r_n} \circ \ldots \circ \phi_{r_1}(\varepsilon)(x).$$

Taking account of Theorem 1, it follows that there exists a unique mapping $u \in C^1(I, X)$ such that

$$u'(x) - r_{n+1} u(x) - v(x) = 0, \quad x \in I, \tag{23}$$

and

$$\|y(x) - u(x)\| \leq \phi_{r_{n+1}} \circ \phi_{r_n} \circ \ldots \circ \phi_{r_1}(\varepsilon)(x), \quad x \in I.$$

Finally taking into account the properties of u and v, it follows that u satisfies (20). The theorem is proved.

Theorem 3 *Let δ be a positive number and suppose that all the roots of the characteristic Eq. (15) have the property $\Re r_k \neq 0$, $1 \leq k \leq n$. Then, for every mapping $y \in C^n(I, X)$ satisfying the relation*

$$\left\| y^{(n)}(x) - \sum_{j=0}^{n-1} a_j y^{(j)}(x) - f(x) \right\| \leq \delta, \quad x \in I$$

there exists a solution $u \in C^n(I, X)$ of the equation

$$y^{(n)}(x) - \sum_{j=0}^{n-1} a_j y^{(j)}(x) - f(x) = 0, \quad x \in I,$$

such that

$$\|y(x) - u(x)\| \leq L$$

where

$$L = \begin{cases} \delta \cdot \displaystyle\prod_{k=1}^{n} \frac{1}{|\Re r_k|} \cdot \frac{|e^{b\Re r_k} - e^{a\Re r_k}|}{e^{b\Re r_k} + e^{a\Re r_k}}, & \text{if } a, b \in \mathbb{R}, \\ \dfrac{\delta}{\displaystyle\prod_{k=1}^{n} |\Re r_k|}, & \text{if } a = -\infty \text{ or } b = +\infty. \end{cases}$$

Proof The proof follows analogously to the proof of Theorem 2 taking account of Corollary 2.

Remark 2 The uniqueness of the solution u in Theorem 3 holds if its characteristic polynomial P has no pure imaginary roots and $I = \mathbb{R}$ (see [16]).

4 Stability of First Order Linear Differential Operator

In what follows, $I = (a, b)$, $a, b \in \mathbb{R} \cup \{\pm\infty\}$ is an open interval, $c \in (a, b)$, $(X, \|\cdot\|)$ is a Banach space over $\mathbb{C}$, $C^n(I, X)$ is the set of all n-times strongly differentiable functions $f : I \to X$ with $f^{(n)}$ continuous on I, $n \in \mathbb{N}$, and $C(I, X)$ is the set of all continuous functions $f : I \to X$.

Let also $\lambda, a_1, \ldots, a_n \in C(I, \mathbb{C})$ be given. We deal with the Hyers–Ulam stability of the linear differential operator $D^n : C^n(I, X) \to C(I, X)$ defined by

$$D^n(y) = y^{(n)} + a_1 y^{(n-1)} + \ldots + a_n y, \quad y \in C^n(I, X). \tag{24}$$

For every $h \in C^n(I, X)$, define $\|h\|_\infty$ by

$$\|h\|_\infty = \sup\{\|h(t)\| : t \in I\}. \tag{25}$$

Then, $\|\cdot\|_\infty$ is a gauge function on $C^n(I, X)$. (Recall that $\rho : Y \to [0, \infty]$ is a gauge function on the linear complex space Y if $\rho(\alpha x) = |\alpha|\rho(x)$ for all $x \in Y$ and all $\alpha \in \mathbb{C}$, see [25]). For an arbitrary function $f : A \to B$ we denote by $R(f)$ the range of f, i.e., $R(f) = \{y \mid y = f(x),\ x \in A\}$.

Definition 1 The operator D^n is said to be stable in Hyers–Ulam sense if for every $\varepsilon \geq 0$ there exists $\delta \geq 0$ such that for every $f \in R(D^n)$ and every $y \in C^n(I, X)$ satisfying

$$\|D^n(y) - f\|_\infty \leq \varepsilon \tag{26}$$

there exists $u \in C^n(I, X)$ such that $D^n(u) = f$ and

$$\|y - u\|_\infty \leq \delta. \tag{27}$$

Let ε be a non-negative number. As in Sect. 2, for a function $g : (a, b) \to X$ define $g(a) := \lim_{x \to a} g(x)$, $g(b) := \lim_{x \to b} g(x)$, if the limits exist. Let L be an antiderivative of λ, i.e., $L \in C^1(I, \mathbb{C})$ and $L' = \lambda$ on I. For $n = 1$ and $a_1 = \lambda$, denote D^1 by D_λ, i.e.

$$D_\lambda = y' + \lambda y, \quad y \in C^1(I, X).$$

The next results concern the Hyers–Ulam stability of the first order linear differential operator and improve some results obtained in [24].

Theorem 4 *Suppose that*

$$\inf_{x \in I} |\Re\lambda(x)| := m > 0. \tag{28}$$

Then, for every $f \in C(I, X)$ and every $y \in C^1(I, X)$ satisfying

$$\|D_\lambda(y) - f\|_\infty \leq \varepsilon \tag{29}$$

there exists $u \in C^1(I, X)$ with the properties $D_\lambda(u) = f$ and

$$\|y - u\|_\infty \leq \frac{\varepsilon}{m} \cdot \delta_\lambda \tag{30}$$

where

$$\delta_\lambda = \begin{cases} 1 - e^{\Re L(a) - \Re L(b)}, & \text{if} \quad \Re\lambda > 0 \quad \text{on} \quad I, \\ 1 - e^{\Re L(b) - \Re L(a)}, & \text{if} \quad \Re\lambda < 0 \quad \text{on} \quad I. \end{cases}$$

Moreover, if one of the following conditions

$$\begin{aligned} &i) \quad \Re L(a) = -\infty, \quad \text{if} \quad \Re\lambda > 0 \quad \text{on} \quad I, \\ &ii) \quad \Re L(b) = -\infty, \quad \text{if} \quad \Re\lambda < 0 \quad \text{on} \quad I, \end{aligned} \tag{31}$$

is satisfied, then u is uniquely determined.

Proof From (28), it follows that $\Re\lambda \neq 0$ on I, therefore, $\Re\lambda$ has constant sign on I, in view of its continuity. We conclude that $\Re L$ is strictly monotone on I, hence, there exist $\Re L(a)$ and $\Re L(b)$, finite or infinite.
Existence. Let $y \in C^1(I, X)$ satisfying (29) and define

$$g(x) := y'(x) + \lambda(x)y(x) - f(x), \quad x \in I.$$

Then, according to Lemma 1, we get

$$y(x) = e^{-L(x)}\left(\int_{x_0}^{x} e^{L(t)} f(t)dt + \int_{x_0}^{x} e^{L(t)} g(t)dt + k\right), \; x_0 \in I, \; k \in X. \tag{32}$$

1° Suppose first that $\Re\lambda > 0$ on I. Define

$$G(x) := \int_{a}^{x} e^{L(t)} g(t)dt, \quad x \in I.$$

Since I is an open interval we have to prove that $G(x)$ is defined for all $x \in I$. We get

$$\|e^{L(t)} g(t)\| \leq \varepsilon \cdot e^{\Re L(t)}, \quad t \in I \tag{33}$$

and

$$\begin{aligned}\int_{a}^{x} e^{\Re L(t)}dt &= \int_{a}^{x} \frac{1}{\Re\lambda(t)} \cdot \Re\lambda(t) \cdot e^{\Re L(t)}dt \leq \frac{1}{m}\int_{a}^{x} (e^{\Re L(t)})'dt \\ &= \frac{1}{m}(e^{\Re L(x)} - e^{\Re L(a)}) \leq \frac{1}{m}e^{\Re L(x)}, \; x \in I.\end{aligned} \tag{34}$$

($\Re L(a) < \Re L(x)$ for all $x \in I$, since $\Re L$ is increasing).

From (33) and (34), it follows that $G(x)$ is absolutely convergent for all $x \in I$. Now defining

$$u(x) := e^{-L(x)}\left(\int_{x_0}^{x} e^{L(t)} f(t)dt + k - G(x_0)\right), \quad x \in I,$$

and using (34) we get

$$\begin{aligned}\|y(x) - u(x)\| &\leq \varepsilon e^{-\Re L(x)} \int_{a}^{x} e^{\Re L(t)}dt \\ &\leq \frac{\varepsilon}{m} e^{-\Re L(x)} \left(e^{\Re L(x)} - e^{\Re L(a)}\right) \\ &\leq \frac{\varepsilon}{m}\left(1 - e^{\Re L(a) - \Re L(b)}\right), \; x \in I.\end{aligned}$$

2° The case $\Re\lambda < 0$ can be treated analogously, setting

$$G(x) := -\int_x^b e^{L(t)} g(t)dt, \quad x \in I.$$

The existence is proved.

Uniqueness. Suppose that one of the conditions (i), (ii) is satisfied and for a function $y \in C^1(I, X)$ satisfying (29) there exist two solutions u_1, u_2 of (5), $u_1 \neq u_2$, with the property (30). Then

$$u_j(x) = e^{-L(x)} \left(\int_{x_0}^x e^{L(t)} f(t)dt + k_j \right), \quad k_j \in X, \ j = 1, 2,$$

with $k_1 \neq k_2$, according to Lemma 1, and

$$\begin{aligned} e^{-\Re L(x)} \|k_1 - k_2\| &= \|u_1(x) - u_2(x)\| \\ &\leq \|u_1(x) - y(x)\| + \|y(x) - u_2(x)\| \\ &\leq \frac{2\varepsilon}{m} \delta_\lambda, \ x \in I. \end{aligned} \tag{35}$$

Letting in (35) $x \to a$ if (i) is satisfied, or $x \to b$ if (ii) is satisfied, it follows $\infty \leq \frac{2\varepsilon}{m}$, contradiction.

The theorem is proved.

Theorem 5 *Suppose that* $\inf_{x \in I} |\Re\lambda(x)| := m > 0$ *and let* $f \in C(I, X)$. *Then, for every* $y \in C^1(I, X)$ *satisfying*

$$\|D_\lambda(y) - f\|_\infty \leq \varepsilon \tag{36}$$

there exists $u \in C^1(I, X)$ *with the properties* $D_\lambda(u) = f$ *and*

$$\|y - u\|_\infty \leq \frac{\varepsilon}{m} \delta_\lambda(c), \tag{37}$$

where

$$\delta_\lambda(c) = \begin{cases} \max\{|1 - e^{\Re L(c) - \Re L(b)}|, |e^{\Re L(c) - \Re L(a)} - 1|\}, & \text{if } \Re L(a) > -\infty \\ & \text{and } \Re L(b) > -\infty, \\ 1, \text{if } \Re L(a) = -\infty \quad \text{or} \quad \Re L(b) = -\infty. & \end{cases} \tag{38}$$

Proof If $\Re L(a) = -\infty$ or $\Re L(b) = -\infty$, the statement follows from Theorem 1. Suppose now that $\Re L(a) > -\infty$ and $\Re L(b) > -\infty$.

Similarly to the proof of Theorem 1, we get that $\Re\lambda$ has constant sign on I and $\Re L$ is strictly monotone on I.

Let $y \in C^1(I, X)$ satisfying (36) and

$$g(x) := y'(x) + \lambda(x)y(x) - f(x), \quad x \in I.$$

Then, y is given by (32). Define G and u by

$$G(x) := \int_c^x e^{L(t)} g(t)dt$$

$$u(x) := e^{-L(x)} \left(\int_{x_0}^x e^{L(t)} f(t)dt + k - G(x_0) \right), \quad x \in I.$$

We get, analogously to the proof of Theorem 1,

$$\begin{aligned} \|y(x) - u(x)\| &= e^{-\Re L(x)} \|g(x)\| \\ &\le e^{-\Re L(x)} \left| \int_c^x \|e^{L(t)} g(t)\| dt \right| \\ &\le \varepsilon e^{-\Re L(x)} \left| \int_c^x e^{\Re L(t)} dt \right|, \ x \in I. \end{aligned} \tag{39}$$

On the other hand, since $\Re\lambda \cdot e^{\Re L}$ has constant sign on I, it follows

$$\begin{aligned} \left| \int_c^x e^{\Re L(t)} dt \right| &= \left| \int_c^x \frac{1}{\Re\lambda(t)} \cdot \Re\lambda(t) \cdot e^{\Re L(t)} dt \right| \\ &\le \frac{1}{m} |e^{\Re L(x)} - e^{\Re L(c)}|, \ x \in I. \end{aligned} \tag{40}$$

The relations (39) and (40) lead to

$$\|y(x) - u(x)\| \le \frac{\varepsilon}{m} |1 - e^{\Re L(c) - \Re L(x)}|, \quad x \in I. \tag{41}$$

The relation (37) follows from (41) taking account of the monotonicity of $\Re L$. The theorem is proved.

Remark 3 If $L(a) > -\infty$ and $L(b) > -\infty$ it is easy to verify that $\delta_\lambda(c)$ is minimal in Theorem 5 for

$$1 - e^{\Re L(c) - \Re L(b)} = e^{\Re L(c) - \Re L(a)} - 1$$

or

$$e^{-\Re L(c)} = \frac{e^{-\Re L(a)} + e^{-\Re L(b)}}{2}. \tag{42}$$

The relation (42) gives $\widetilde{c}$ optimal. Note that since $\Re L$ is strictly increasing and continuous on I, $\widetilde{c}$ exists and is unique.

Choosing $c = \widetilde{c}$ in Theorem 5, we get

$$\delta_\lambda(\widetilde{c}) = \frac{|e^{-\Re L(a)} - e^{-\Re L(b)}|}{e^{-\Re L(a)} + e^{-\Re L(b)}}, \tag{43}$$

therefore, the relation (37) becomes

$$\|y - u\|_\infty \le \frac{\varepsilon}{m} \delta_\lambda(\widetilde{c}). \tag{44}$$

5 Stability of Higher Order Linear Differential Operator

The results expressed in Theorems 4 and 5 lead to the Hyers–Ulam stability of the operator D^n defined by (24), in appropriate conditions. We suppose that there exist $r_1, r_2, \dots, r_n \in C(I, \mathbb{C})$ such that

$$D^n = D_{r_1} \circ D_{r_2} \circ \dots \circ D_{r_n}. \tag{45}$$

We remark that D^n is a surjective operator as a composition of surjective operators (D_λ is a surjective operator in view of Lemma 1).

Let R_k be an antiderivative of r_k, $1 \le k \le n$ and $f \in C(I, X)$ an arbitrary function.

Theorem 6 *Suppose that* $\inf_{x\in I} |r_k(x)| := m_k > 0$ *for every* $k \in \{1, 2, \dots, n\}$. *Then, for every* $y \in C^n(I, X)$ *satisfying the relation*

$$\|D^n(y) - f\|_\infty \le \varepsilon \tag{46}$$

there exists $u \in C^n(I, X)$ *with the properties* $D^n(u) = f$ *and*

$$\|y - u\|_\infty \le \frac{\varepsilon}{m_1 m_2 \dots m_n} \delta_{r_1} \delta_{r_2} \dots \delta_{r_n}. \tag{47}$$

Proof We prove the theorem by induction on n.

For $n = 1$, Theorem 6 holds in virtue of Theorem 4.

Now suppose that Theorem 6 holds for an $n \in \mathbb{N}$. We have to prove that for all $y \in C^{n+1}(I, X)$ satisfying

$$\|D^{n+1}(y) - f\|_\infty \le \varepsilon \tag{48}$$

there exists $u \in C^{n+1}(I, X)$, $D^{n+1}(u) = f$, such that

$$\|y - u\|_\infty \le \frac{\varepsilon}{m_1 m_2 \dots m_{n+1}} \delta_{r_1} \delta_{r_2} \dots \delta_{r_{n+1}}. \tag{49}$$

Let $y \in C^{n+1}(I, X)$ satisfying (48). Then

$$\|D^n(z) - f\|_\infty \le \varepsilon$$

with $z := D_{r_{n+1}}(y)$. Hence, in virtue of the induction hypothesis, there exists $v \in C^n(I, X)$, $D^n(v) = f$, and

$$\|z - v\|_\infty \le \frac{\varepsilon}{m_1 m_2 \dots m_n} \delta_{r_1} \delta_{r_2} \dots \delta_{r_n}$$

which is equivalent to

$$\|D_{r_{n+1}}(y) - v\|_\infty \le \frac{\varepsilon}{m_1 m_2 \dots m_n} \delta_{r_1} \delta_{r_2} \dots \delta_{r_n}. \tag{50}$$

Then, according to Theorem 4, from (50), it follows that there exists a mapping $u \in C^1(I, X)$, $D_{r_{n+1}}(u) = v$, and

$$\|y - u\|_\infty \leq \frac{\varepsilon}{m_1 m_2 \ldots m_{n+1}} \delta_{r_1} \delta_{r_2} \ldots \delta_{r_{n+1}}.$$

Finally the relations $D^n(v) = f$ and $D_{r_{n+1}}(u) = v$ lead to $D^{n+1}(u) = f$. The theorem is proved.

An analogous result follows from Theorem 5.

Theorem 7 *Suppose that* $\inf_{x \in I} |r_k(x)| = m_k > 0$ *for every* $k \in \{1, 2, \ldots, n\}$. *Then, for every* $y \in C^n(I, X)$ *satisfying*

$$\|D^n(y) - f\|_\infty \leq \varepsilon$$

there exists $u \in C^n(I, X)$, $D^n(u) = f$, *such that*

$$\|y - u\|_\infty \leq \frac{\varepsilon}{m_1 m_2 \ldots m_n} \delta_{r_1}(c) \ldots \delta_{r_n}(c). \tag{51}$$

Proof Analogous to the proof of Theorem 6.

Remark 4 If $\Re(R_k(a)) > -\infty$ for all $k \in \{1, 2, \ldots, n\}$, choosing $c = \widetilde{c}$ in Theorem 7 the estimate (51) can be improved to

$$\|y - u\|_\infty \leq \frac{\varepsilon}{m_1 m_2 \ldots m_n} \prod_{k=1}^{n} \frac{|e^{-\Re R_k(a)} - e^{\Re R_k(b)}|}{e^{-\Re R_k(a)} + e^{-\Re R_k(b)}}.$$

Proof Follows from Theorem 7 and Remark 3.

The results obtained in Theorems 6, 7 and their consequences improve and extend the estimates given in [8, Theorem 1.1], on the Hyers–Ulam stability for the linear differential operator with constant coefficients. The estimates obtained in (47) and (51), concerning the difference between the perturbed and the exact solution, improve also the results on stability in Aoki–Rassias sense for systems of differential equations [17, Theorem 2], and for linear differential equations with constant coefficients in Banach spaces, given in [8, Theorem 2.3] and [30, Theorem 3.2].

6 Stability of Partial Differential Equations

In what follows, let $D = [a, b) \times \mathbb{R}$, $a \in \mathbb{R}$, $b \in \mathbb{R} \cup \{+\infty\}$ be a subset of $\mathbb{R}^2$. We deal with the Hyers–Ulam stability of the linear partial differential equation

$$p(x, y)\frac{\partial u}{\partial x} + q(x, y)\frac{\partial u}{\partial y} = p(x, y)r(x)u + f(x, y) \tag{52}$$

where $p, q \in C(D, K)$, $f \in C(D, X)$, $r \in C([a, b), \mathbb{R})$ are given functions and $u \in C^1(D, X)$ is the unknown function (see [21]). We suppose that $p(x, y) \neq 0$ for every $(x, y) \in D$.

Let $\varepsilon \geq 0$ be a given number. The Eq. (52) is said to be stable in Hyers–Ulam sense if there exists $\delta \geq 0$ such that for every function $u \in C^1(D, X)$ satisfying

$$\left\| p(x, y)\frac{\partial u}{\partial x}(x, y) + q(x, y)\frac{\partial u}{\partial y}(x, y) - p(x, y)r(x)u(x, y) - f(x, y) \right\| \leq \varepsilon \quad (53)$$

for all $(x, y) \in D$, there exists a solution $v \in C^1(D, X)$ of (52) with the property

$$\|u(x, y) - v(x, y)\| \leq \delta, \quad \forall\, (x, y) \in D. \quad (54)$$

We will prove in what follows that the existence of a global prime integral $\varphi : [a, b) \to \mathbb{R}$ of the Eq. (52) leads, in appropriate conditions, to the stability of the Eq. (52). The following lemma is a useful tool in the proof of the main result of this section.

Lemma 2 *Let $\varphi : [a, b) \to \mathbb{R}$ be a solution of the differential equation*

$$y' = \frac{q(x, y)}{p(x, y)}.$$

Then, u is a solution of the Eq. (52) if and only if there exists a function $F \in C^1(I, X)$ such that

$$u(x, y) = e^{-L(x)}\left(\int_a^x \frac{f(\theta, \varphi(\theta) + y - \varphi(x))}{p(\theta, \varphi(\theta) + y - \varphi(x))} e^{L(\theta)} d\theta + F(y - \varphi(x))\right) \quad (55)$$

for every $(x, y) \in D$, where $L(x) = -\int_a^x r(\theta)d\theta$, $x \in [a, b)$ and $I = \{y - \varphi(x) : (x, y) \in D\}$.

Proof Let u be a solution of the Eq. (52) and consider the change of coordinates

$$\begin{cases} s = x \\ t = y - \varphi(x) \end{cases} \Leftrightarrow \begin{cases} x = s \\ y = \varphi(s) + t \end{cases} \quad (56)$$

Define the function v by

$$v(s, t) = u(s, \varphi(s) + t) \Leftrightarrow u(x, y) = v(x, y - \varphi(x)). \quad (57)$$

Then

$$\frac{\partial u}{\partial x} = \frac{\partial v}{\partial s} - \varphi'(s) \cdot \frac{\partial v}{\partial t}, \quad \frac{\partial u}{\partial y} = \frac{\partial v}{\partial t}$$

and replacing in (52) it follows

$$\frac{\partial v}{\partial s} - r(s) \cdot v = \frac{f(s, \varphi(s) + t)}{p(s, \varphi(s) + t)}. \quad (58)$$

Equation (58) is equivalent to

$$\frac{\partial}{\partial s}(v \cdot e^{L(s)}) = \frac{f(s, \varphi(s) + t)}{p(s, \varphi(s) + t)} \cdot e^{L(s)}. \tag{59}$$

An integration on the interval $[a, s)$, $s \in [a, b)$, leads to

$$v(s, t) = e^{-L(s)} \left(\int_a^s \frac{f(\theta, \varphi(\theta) + t)}{p(\theta, \varphi(\theta) + t)} e^{L(\theta)} d\theta + F(t) \right) \tag{60}$$

where F is an arbitrary function of class C^1.

Replacing s, t from (56) in (60) the relation (55) is obtained.

Now let u be given by (55), we have to prove that u is a solution of (52). Taking account of the change of coordinates (56), it is sufficient to prove that v, given by (60), satisfies (58). A simple calculation shows that v is a solution of (58).

The main result of this section is given in the next theorem.

Theorem 8 *Let $\varepsilon \geq 0$ be a given number. Suppose that the equation $y' = \dfrac{q(x, y)}{p(x, y)}$ admits a solution $\varphi : [a, b) \to \mathbb{R}$ and $\inf\limits_{(x,y) \in D} |p(x, y)| \cdot r(x) =: m > 0$. Then, for every solution u of (53) there exists a solution v of (52) with the property*

$$\|u(x, y) - v(x, y)\| \leq \frac{\varepsilon}{m}, \quad (x, y) \in D. \tag{61}$$

Moreover, if $L(b) =: \lim\limits_{x \to b} L(x) = -\infty$ then v is uniquely determined.

Proof *Existence.* Let u be a solution of (53) and put

$$p(x, y)\frac{\partial u}{\partial x}(x, y) + q(x, y)\frac{\partial u}{\partial y}(x, y) - p(x, y)r(x)u(x, y) - f(x, y) =: g(x, y)$$

for every $(x, y) \in D$. Then, according to Lemma 2, we have:

$$u(x, y) = e^{-L(x)} \left(\int_a^x \frac{f(\theta, \varphi(\theta) + y - \varphi(x)) + g(\theta, \varphi(\theta) + y - \varphi(x))}{p(\theta, \varphi(\theta) + y - \varphi(x))} e^{L(\theta)} d\theta \right.$$
$$\left. + F(y - \varphi(x)) \right)$$

where $F \in C^1(I, X)$ is an arbitrary function.

Let v be defined by

$$v(x, y) = e^{-L(x)} \left(\int_a^x \frac{f(\theta, \varphi(\theta) + y - \varphi(x))}{p(\theta, \varphi(\theta) + y - \varphi(x))} e^{L(\theta)} d\theta \right.$$
$$\left. + \int_a^b \frac{g(\theta, \varphi(\theta) + y - \varphi(x))}{p(\theta, \varphi(\theta) + y - \varphi(x))} e^{L(\theta)} d\theta + F(y - \varphi(x)) \right), \quad (x, y) \in D.$$

The function v is well defined since the integral

$$G(t) := \int_a^b \frac{g(\theta, \varphi(\theta) + t)}{p(\theta, \varphi(\theta) + t)} e^{L(\theta)} d\theta, \quad t \in I$$

is convergent. Indeed,

$$\begin{aligned} \| G(t) \| &\leq \int_a^b \left\| \frac{g(\theta, \varphi(\theta) + t)}{p(\theta, \varphi(\theta) + t) \cdot r(\theta)} \cdot r(\theta) e^{L(\theta)} \right\| d\theta \\ &\leq \frac{\varepsilon}{m} \int_a^b r(\theta) e^{L(\theta)} d\theta \\ &= -\frac{\varepsilon}{m} \int_a^b (e^{L(\theta)})' d\theta = \frac{\varepsilon}{m}(1 - e^{L(b)}) \leq \frac{\varepsilon}{m}, \quad t \in I, \end{aligned}$$

therefore $G(t)$ is absolutely convergent.

(Since r is positive on $[a, b)$ it follows that the function L is decreasing on $[a, b)$, a monotone function has left and right limits at every point, therefore $L(b) = -\lim_{x \to b} \int_a^x r(\theta) d\theta$ exists and is negative).

On the other hand v is a solution of (52) being of the form (55). We have:

$$\begin{aligned} \|u(x, y) - v(x, y)\| &= \left\| e^{-L(x)} \left(- \int_x^b \frac{g(\theta, \varphi(\theta) + y - \varphi(x))}{p(\theta, \varphi(\theta) + y - \varphi(x))} \cdot e^{L(\theta)} d\theta \right) \right\| \\ &\leq e^{-L(x)} \int_x^b \frac{\varepsilon}{|p(\theta, \varphi(\theta) + y - \varphi(x))|} e^{L(\theta)} d\theta \\ &= e^{-L(x)} \int_x^b \frac{\varepsilon}{|p(\theta, \varphi(\theta) + y - \varphi(x))| r(\theta)} r(\theta) e^{L(\theta)} d\theta \\ &\leq \frac{\varepsilon}{m} e^{-L(x)} \int_x^b (-e^{L(\theta)})' d\theta \\ &= \frac{\varepsilon}{m}(1 - e^{L(b) - L(x)}) \leq \frac{\varepsilon}{m}, \ (x, y) \in D. \end{aligned}$$

Uniqueness. Suppose that $L(b) = -\infty$ and for a solution u of (53), there exist two solutions v_1, v_2 of (52), $v_1 \neq v_2$, with the property (61), given by

$$v_k(x, y) = e^{-L(x)} \left(\int_a^x \frac{f(\theta, \varphi(\theta) + y - \varphi(x))}{p(\theta, \varphi(\theta) + y - \varphi(x))} e^{L(\theta)} d\theta + F_k(y - \varphi(x)) \right)$$

$(x, y) \in D$, $k \in \{1, 2\}$. We have

$$\begin{aligned} \|v_1(x, y) - v_2(x, y)\| &\leq \|v_1(x, y) - u(x, y)\| + \|u(x, y) - v_2(x, y)\| \\ &\leq \frac{2\varepsilon}{m}, \ (x, y) \in D \end{aligned}$$

which is equivalent to

$$e^{-L(x)}\|F_1(y-\varphi(x))-F_2(y-\varphi(x))\| \le \frac{2\varepsilon}{m}, \quad (x,y)\in D. \tag{62}$$

Since $v_1 \neq v_2$ it follows that there exists x_0 such that $F_1(x_0) \neq F_2(x_0)$. For $y=\varphi(x)+x_0$, the relation (62) becomes

$$e^{-L(x)}\|F_1(x_0)-F_2(x_0)\| \le \frac{2\varepsilon}{m}, \quad x\in[a,b). \tag{63}$$

Now letting $x \to b$ in (63), it follows $\infty \le \frac{2\varepsilon}{m}$, contradiction. Uniqueness is proved.

Corollary 3 *Let $D=(0,\infty)\times\mathbb{R}$ and $p,q \in C(D,\mathbb{R})$, $r\in C([0,\infty),\mathbb{R})$, $f\in C(D,X)$. Suppose that p,q are homogeneous functions of the same degree, $\frac{q(x,y)}{p(x,y)} \neq \frac{y}{x}$ on D and $\inf_{(x,y)\in D}|p(x,y)|\cdot r(x)=m>0$. Then, for every $\varepsilon\ge 0$ and every solution u of (53) there exists a solution v of (52) with the property (54). If $\int_0^\infty r(\theta)d\theta=\infty$, then v is uniquely determined.*

Proof Suppose that p,q are homogeneous functions of nth degree. First, we prove that the equation

$$y'=\frac{q(x,y)}{p(x,y)} \tag{64}$$

admits a solution $\varphi:(0,\infty)\to\mathbb{R}$.

Taking account of the homogeneity of p and q, it follows

$$\frac{q(x,y)}{p(x,y)}=\frac{q\left(x\cdot 1, x\cdot\frac{y}{x}\right)}{p\left(x\cdot 1, x\cdot\frac{y}{x}\right)}=\frac{x^n q\left(1,\frac{y}{x}\right)}{x^n p\left(1,\frac{y}{x}\right)}=\frac{q\left(1,\frac{y}{x}\right)}{p\left(1,\frac{y}{x}\right)}=:h\left(\frac{y}{x}\right)$$

for all $(x,y)\in D$, therefore the Eq. (64) is equivalent to the homogeneous differential equation

$$y'=h\left(\frac{y}{x}\right). \tag{65}$$

Let $H:\mathbb{R}\to\mathbb{R}$ be given by

$$H(z)=\int_0^z \frac{d\theta}{h(\theta)-\theta}, \quad z\in\mathbb{R}. \tag{66}$$

Obviously H is well defined since $h(\theta)\neq\theta$ for all $\theta\in\mathbb{R}$.

The change of variable in (65) given by

$$y(x) = xz(x), \quad x \in (0, \infty)$$

leads to the equation with separate variables

$$\frac{dz}{h(z) - z} = \frac{dx}{x}$$

with a solution given by

$$H(z) = \ln x, \quad x \in (0, \infty). \tag{67}$$

By the condition $h(\theta) \neq \theta, \theta \in \mathbb{R}$ and the continuity of h it follows that $h(\theta) - \theta$ has constant sign on $\mathbb{R}$, therefore H is strictly monotone.

In this case, there exists $H^{-1} : H(\mathbb{R}) \to \mathbb{R}$. From (67), we get the explicit solution of the Eq. (65) given by

$$z(x) = H^{-1}(\ln x)$$

and finally the prime integral

$$\varphi(x) = x \cdot H^{-1}(\ln x), \quad x \in (0, \infty).$$

Now the conclusion follows from Theorem 8.

Remark 5 If $m = 0$, then the result obtained in Theorem 8 is not generally true.

Indeed consider the equation

$$x\frac{\partial u}{\partial x} + y\frac{\partial u}{\partial y} = 0, \; x, y \in [a, \infty), a > 0. \tag{68}$$

and let $\varepsilon > 0$. A solution of the equation $x\frac{\partial u}{\partial x} + y\frac{\partial u}{\partial y} = \varepsilon$ is of the form $u(x, y) = \varepsilon \ln x + \varphi(\frac{y}{x})$ where $\varphi : (0, \infty) \longrightarrow X$ is an arbitrary function of class C^1, according to Lemma 2.

Let $v(x, y) = \psi(\frac{y}{x})$ be an arbitrary solution of (68), $\psi \in C^1((0, \infty), X)$. The condition

$$\left\| x\frac{\partial u}{\partial x}(x, y) + y\frac{\partial u}{\partial y}(x, y) \right\| \leq \varepsilon$$

is satisfied for all $x, y \in (0, \infty)$, but

$$\sup_{x \in [a,\infty)} \|u(x, x) - v(x, x)\| = +\infty,$$

and therefore, the Eq. (68) is not stable.

References

1. Alsina, C., Ger, R. *On some inequalities and stability results related to the exponential function*, J. Inequal. Appl., 2(1998), 373–380.
2. Aoki, T.: On the stability of the linear transformations in Banach spaces. J. Math. Soc. Japan. **2**, 64–66 (1950)
3. Brillouet-Beluot, N., Brzdek, J., Cieplinski, K. *On some recent developments in Ulam's type stability*, Abstract and Applied Analysis, vol. 2012, article ID716936, 41 pag.
4. Brzdek, J.: On the quotient stability of a family of functional equations. Nonlinear Analysis. **71**, 4396–4404 (2009)
5. Brzdek, J., Popa, D., Xu, B.: The Hyers-Ulam stability of nonlinear recurrences. J. Math. Anal. Appl. **335**, 443–449 (2007)
6. Brzdek, J., Th., M.. Rassias: *Functional Equations in Mathematical Analysis*, Springer, 2011
7. Cadariu, L., Radu, V. *Fixed point methods for the generalized stability of functional equations in a single variable*, Fixed Point Theory A., vol. 2008, article ID749392, 15 pag.
8. Cîmpean, D.S., Popa, D.: On the stability of the linear differential equation of higher order with constant coefficients. Appl. Math. Comput. **217**, 4141–4146 (2010)
9. Cîmpean, D.S., Popa, D.: Hyers-Ulam stability of Euler's equation. Appl. Math. Lett. **24**, 1539–1543 (2011)
10. Czerwik, S., *Functional Equations and Inequalities in Several Variables*, World Scientific, 2002
11. Forti, G.-L.: Comments on the core of the direct method for proving Hyers-Ulam stability of functional equations. J. Math. Anal. Appl. **295**, 127–133 (2004)
12. Hyers, D.H.: *On the stability of the linear functional equation*, Proc. Nat. Acad. Sci. U. S.A. **27**, 222–224 (1941)
13. Hyers, D.H., Isac, G., Th., M. Rassias: Stability of Functional Equations in Several Variables. Birkhäuser, Basel (1998)
14. Jung, S.-M., *Hyers-Ulam Rassias Stability of Functional Equations in Mathematical Analysis*, Hadronic Press, Palm Harbour, 2001
15. Jung, S.-M.: Hyers-Ulam stability of linear differential equation of the first order (III). J. Math. Anal. Appl. **311**, 139–146 (2005)
16. Jung, S.-M.: Hyers-Ulam stability of linear differential equations of first order (II). Appl. Math. Lett. **19**, 854–858 (2006)
17. Jung, S.-M.: Hyers-Ulam stability of a system of first order linear differential equations with constant coefficients. J. Math. Anal. Appl. **320**, 549–561 (2006)
18. Jung, S.M., Popa, D., Rassias, M.Th, *On the stability of the linear functional equation in a single variable on complete metric groups*, J. Glob. Optim. DOI 10.1007/s10898-013-0083-9.
19. S.M. Jung, H.: Şevli, *Power series method and approximate linear differential equations of second order*, Advances Diff. Equations. **2013**, 76 (2013)
20. Li, Y., Shen, Y.: Hyers-Ulam stability of linear differential equations of second order. Appl. Math. Lett. **23**, 306–309 (2010)
21. Lungu, N., Popa, D.: Hyers-Ulam stability of a first order partial differential equation. J. Math. Anal. Appl. **385**, 86–91 (2012)
22. Lungu, N., Popa, D. *On the Hyers-Ulam stability of a first order partial differential equation*, Carpathian J. Math., 28(2012), 77–82.
23. Lungu, N., Rus, I.A. *Ulam stability of nonlinear hyperbolic differential equations*, Carpathian J. Math., 24(2008), 403–408.
24. Miura, T., Miyajima, S., Takahasi, S.E.: A characterization of Hyers-Ulam stability of first order linear differential operators. J. Math. Anal. Appl. **286**, 136–146 (2003)
25. Miura, T., Miyajima, S., Takahasi, S.E.: Hyers-Ulam stability of linear differential operator with constant coefficients. Math. Nachr. **258**, 90–96 (2003)
26. Obłoza, M. *Hyers-Ulam stability of the linear differential equations*, Rocznik Nauk, Dydakt. Prace. Mat., 13(1993), 259–270.

27. Obłoza, M. *Connections between Hyers and Lyapunov stability of the ordinary differential equations*, Rocznik Naukm Dydakt. Prace Mat., 14(1997), 141–146.
28. Petru, T.P., Petruşel, A., Yao, J.-C.: Ulam-Hyers stability for operatorial equations and inclusions via nonself operators. Taiwanese J. Math. **15**, 2195–2212 (2011)
29. Popa, D.: Hyers-Ulam-Rassias stability of a linear recurrence. J. Math. Anal. Appl. **309**, 591–597 (2005)
30. Popa, D., Raşa, I.: On the Hyers-Ulam stability of the linear differential equation. J. Math. Anal. Appl. **381**, 530–537 (2011)
31. Popa, D., Raşa, I.: Hyers-Ulam stability of the linear differential operator with nonconstant coefficients. Appl. Math. Comput. **219**, 1562–1568 (2012)
32. Prastaro, A., Th., M. Rassias: *Ulam stability in geometry PDE's*, Nonlinear Funct. Anal. Appl., 8(2)(2003), 259–278.
33. Th. M. Rassias: On the stability of the linear mapping in Banach spaces. Proc. Amer. Math. Soc. **72**, 297–300 (1978)
34. Rezaei, H., Jung, S.M., Th., M. Rassias: Laplace transform and Hyers-Ulam stability of the linear differential equations. J. Math. Anal. Appl. **403**, 244–251 (2013)
35. Rus, I.A.: Remarks on Ulam stability of the operatorial equations. Fixed Point Theory. **10**, 305–320 (2009)
36. Rus, I.A. *Ulam stability of ordinary differential equations*, Studia Univ. Babeş-Bolyai Math., 54(2009), 125–134.
37. Takahasi, S.E., Miura, T., Miyajima, S. *On the Hyers-Ulam stability of the Banach space-valued differential equation* $y' = \lambda y$, Bull. Korean Math. Soc., 39(2002), 309–315.
38. Takahasi, S.E., Takagi, H., Miura, T., Miyajima, S.: The Hyers-Ulam stability constants of first order linear differential operators. J. Math. Anal. Appl. **296**, 403–409 (2004)
39. Ulam, S.M., *A Collection of Mathematical Problems*, Interscience, New York, 1960
40. Wang, G., Zhou, M., Sun, L.: Hyers-Ulam stability of linear differential equations of first order. Appl. Math. Lett. **21**, 1024–1028 (2008)

Results and Problems in Ulam Stability of Operatorial Equations and Inclusions

Ioan A. Rus

Abstract In this chapter we survey some results and problems in Ulam stability of fixed point equations, coincidence point equations, operatorial inclusions, integral equations, ordinary differential equations, partial differential equations and functional inclusions. Some new results and problems are also presented.

Keywords Ulam stability · Inclusions · Operational equations · Coincidence point · Differential equations

1 Introduction

In the general theory of differential equations, integral equations, operatorial equations and operatorial inclusions the data dependence (monotony, continuity, differentiability, stability, ...) is a crucial part ([3, 8, 11, 17, 22, 28, 33, 35, 43–46, 68, 69, 73, 74, 88, 93, 107, 108, 110, 115, 125, 126, 127, 134, 135, 136, ...]). On the other hand, in the theory of functional equations ([12, 30, 42, 103, 104, ...]) there are some special kind of data dependence (Ulam (1940; [133]), Hyers (1941; [48]), Hyers-Ulam (1945; [53]), Aoki (1950; [4]), Bourgin (1951; [15]), Gruber (1978; [45]), Rassias (1978: [95]), Hyers (1983; [49]), Baker (1951; [9]), Găvruţă (1994; [39]), Radu (2003, [94]); see also: [2, 7, 16, 18–20, 23–27, 37, 38, 40, 41, 47, 50–66, 67, 77–79, 81–83, 85, 86, 89–91, 96–101, 102, 103–106, 129–132, 137, ...]). With these results in mind, we introduced in [117] and [122] six types of Ulam stability for operatorial equations in metric and generalized metric spaces.

The aim of this paper is to revisit these results and to present some open problems.

Throughout this paper we shall use the terminology and the notations in [117] and [122]. We shall specify some of them along the paper.

I. A. Rus (✉)
Department of Mathematics, Babeş-Bolyai University, Kogălniceanu Street No. 1, 400084 Cluj-Napoca, Romania
e-mail: iarus@math.ubbcluj.ro

T. M. Rassias (ed.), *Handbook of Functional Equations*,
Springer Optimization and Its Applications 96, DOI 10.1007/978-1-4939-1286-5_15

2 Operatorial Equations and Inclusions

Let X be a nonempty set, $f : X \to X$ be a singlevalued operator and $T : X \multimap X$ be a multivalued operator. Then we denote:

$$P(X) := \{Y \subset X \mid Y \neq \emptyset\}$$

$$f^0 := 1_X, f^1 := f, f^2 := f \circ f, \ldots, f^n := f \circ f^{n-1} \text{ - the iterates of } f$$

$$T(Y) := \bigcup_{y \in Y} T(y), \text{for} Y \subset X$$

$$T^1(Y) := T(Y), T^2(Y) := T(T(Y)), \ldots, T^n(Y) := T(T^{n-1}(Y)) \text{ - the iterates of } T.$$

We consider the following operatorial equations:

(a) $x = f(x)$

A solution of this equation is by definition a fixed point of the operator f and we denote by F_f the solution set of this equation. By $F_f = \{x^*\}$ we mean that the operator f has a unique fixed point and we denote this fixed point by x^*.

(b) $x \in T(x)$

A solution of this equation is by definition a fixed point of the multivalued operator T and we denote by F_T the solution set of this equation. We name equation (b), operatorial inclusion.

(c) $\{x\} = T(x)$

By definition, a solution of this equation is a strict fixed point of the multivalued operator T and we denote the solution set of equation (c) by $(SF)_T$.

Let X and Y be two nonempty sets, $f, g : X \to Y$ be two singlevalued operators and $T, S : X \to P(Y)$ be two multivalued operators. In this case we consider the following operatorial equations:

(d) $f(x) = g(x)$

A solution $x \in X$ of this equation is by definition a coincidence point of the pair f, g. We denote by $C(f, g) := \{x \in X \mid f(x) = g(x)\}$ the solution set of equation (d).

(e) $T(x) \cap S(x) \neq \emptyset$

A solution $x \in X$ of this equation is by definition a coincidence point of the pair T, S. We denote by $C(T, S) := \{x \in X \mid T(x) \cap S(x) \neq \emptyset\}$ the solution set of equation (e).

The basic problems of the operatorial equations and inclusions are the following: In which conditions we have:

Problem 1 $F_f \neq \emptyset$?

Problem 2 $F_f = \{x^*\}$?

Problem 3 $F_T \neq \emptyset$?

Problem 4 $F_T = \{x^*\}$?

Problem 5 $(SF)_T \neq \emptyset$?

Problem 6 $(SF)_T = \{x^*\}$?

Problem 7 $F_T = (SF)_T$?

Problem 8 $F_T = (SF)_T = \{x^*\}$?

Problem 9 $(SF)_T \neq \emptyset \Rightarrow F_T = (SF)_T = \{x^*\}$?

Problem 10 $C(f,g) \neq \emptyset$?

Problem 11 $C(f,g) = \{x^*\}$?

Problem 12 $C(T,S) \neq \emptyset$?

Problem 13 $C(T,S) = \{x^*\}$?

Other problems of the theory of operatorial equations and inclusions are in connection with data dependence of solutions. In what follows we shall present some of them.

Let (X,d) be a metric space and $f,g : X \to X$ be two operators. Let us consider, for example, the fixed point equations

$$x = f(x), \tag{1}$$

$$x = g(x) \tag{2}$$

Problem 14 We suppose that:

(i) $F_f = \{x_f^*\}$;
(ii) $F_g \neq \emptyset$.

In which conditions there exists a function $\theta : \mathbb{R}_+ \to \mathbb{R}_+$ such that we have the following implication

$$\eta > 0 \quad \text{and} \quad d(f(x), g(x)) \leq \eta,\ \forall\, x \in X \Rightarrow d(x_f^*, x_g^*) \leq \theta(\eta),\ \forall\, x_g^* \in F_g \text{ ?}$$

For example if f is an α-contraction, i.e., $0 \leq \alpha < 1$ and

$$d(f(x), f(y)) \leq \alpha d(x,y),\ \forall\, x, y \in X,$$

then, $F_f = \{x_f^*\}$ and $\theta(\eta) = (1-\alpha)^{-1}\eta$.

Indeed, let x_g^* be a fixed point of g. We have

$$\begin{aligned} d(x_f^*, x_g^*) = d(f(x_f^*), g(x_g^*)) &\leq d(f(x_f^*), f(x_g^*)) + d(f(x_g^*), g(x_g^*)) \\ &\leq \alpha d(x_f^*, x_g^*) + \eta. \end{aligned}$$

So, $d(x_f^*, x_g^*) \leq (1-\alpha)^{-1}\eta$.

Problem 15 (Ulam problem) Let (X, d) be a metric space and $f : X \to X$ be an operator. We consider the fixed point equation, (1), and for each $\varepsilon > 0$ the inequation

$$d(y, f(y)) \leq \varepsilon. \tag{3}$$

In which conditions there exists a function $\theta : \mathbb{R}_+ \to \mathbb{R}_+$ such that for each solution y^* of (3) there exists a solution x^* of (2) with the following property, $d(x^*, y^*) \leq \theta(\varepsilon)$?

Since the problem is suggested by the well-known Ulam problem (see section 3 of this paper) we call it *Ulam Problem of Data Dependence of a Fixed Point Equation.*

For the theory of fixed point equations, i.e., the fixed point theory see: [13, 17, 33, 44, 68, 69, 110, 113, 124], ...

For the theory of coincidence equations, i.e., the coincidence theory, see: [17, 21, 33, 68, 69, 113, 124], ...

For the theory of fixed point equations with multivalued operator, i.e., the inclusion theory, see: [33, 68, 88, 110, 113, 124], ...

The aim of this paper is to study the Ulam Problem of Data Dependence.

3 From the Ulam Problem to the Notion of Ulam-Hyers Stability of an Operatorial Equation

In 1940, S.M. Ulam proposed the following problem (see [133]; see also [23, 51, 56, 63], ...):

Let $(G_1, +)$ be a group and $(G_2, \oplus, d)$ be a metric group. For each $\varepsilon > 0$ find a positive number $\delta(\varepsilon)$ such that for every mapping $f : G_1 \to G_2$ satisfying

$$d(f(x + y), f(x) \oplus f(y)) \leq \delta(\varepsilon)$$

there exists a group homomorphism $h : G_1 \to G_2$ with

$$d(f(x), h(x)) \leq \varepsilon, \ \forall\, x \in G_1.$$

In 1941, D.H. Hyers [48] gave the following answer to the Ulam Problem:

Hyers Theorem. *Let* $(E_1, +, \mathbb{R}, \|\cdot\|, (E_2, +, \mathbb{R}, \|\cdot\|$ *be two Banach spaces and let* $f : E_1 \to E_2$ *be a mapping satisfying:*

$$\|f(x + y) - f(x) - f(y)\| \leq \varepsilon, \ \forall\, x, y \in E_1$$

with $\varepsilon > 0$*. Then, there exists a unique additive mapping* $h : E_1 \to E_2$ *which satisfies*

$$\|f(x) - h(x)\| \leq \varepsilon, \ \forall\, x \in E_1.$$

In 1945, D.R. Hyers and S.M. Ulam [53] considered the following problem:

Ulam-Hyers Problem. *Let (X, d) and (Y, ρ) be two metric spaces. In which conditions there exists a constant $k(X, Y) > 0$ such that for each $\varepsilon > 0$ and for each mapping $f : X \to Y$ with*

$$|d(x, y) - \rho(f(x), f(y))| \leq \varepsilon, \ \forall \, x, y \in X,$$

there exists an isometry $h : X \to Y$ such that

$$\rho(f(x), h(x)) \leq k(X, Y)\varepsilon?$$

In 1978, P.M. Gruber presented the following problem ([45]):
General Stability Problem of Gruber. *Suppose that a mathematical object satisfies a certain property approximately. Is then possible to approximate this object by objects satisfying the property exactly?*

It is not a problem to remark that the Ulam problem has generated a lot of research directions in the theory of functional equations, operatorial equations and inclusions. With these results in mind, we introduced in [117] and [122] some types of Ulam stability for the operatorial equations in a metric space. In what follows we shall present some of them, in the case of a fixed point equation.

Definition 1 Let (X, d) be a metric space and $f : X \to X$ be an operator. By definition, the fixed point equation

$$x = f(x) \tag{4}$$

is Ulam-Hyers stable if there exists a constant $c_f > 0$ such that: for each $\varepsilon > 0$ and each solution $y^* \in X$ of the inequation

$$d(y, f(y)) \leq \varepsilon \tag{5}$$

there exists a solution x^* of the Eq. (4) such that

$$d(y^*, x^*) \leq c_f \varepsilon.$$

Definition 2 The Eq. (4) is generalized Ulam-Hyers stable if there exists $\theta : \mathbb{R}_+ \to \mathbb{R}_+$ increasing and continuous in 0 with $\theta(0) = 0$ such that: for each $\varepsilon > 0$ and for each solution y^* of (5) there exists a solution x^* of (4) such that

$$d(y^*, x^*) \leq \theta(\varepsilon).$$

Remark 1 A solution of the inequation (5) is called an ε-solution of the Eq. (4).

Let us denote by S_ε the ε-solution set of (4) and by H_d the Pompeiu-Hausdorff functional (see [10, 33, 88, 124]). Then from Definition 2 we have that if the Eq. 4 is generalized Ulam-Hyers stable, then

$$H_d(F_f, S_\varepsilon) \leq \theta(\varepsilon).$$

Indeed, this follows from the following property of the functional H_d (see [110], p. 76).

Lemma 1 *Let (X, d) be a metric space and $A, B \in P(X)$. Then, if $\eta > 0$ is such that:*

(1) for each $a \in A$, there exists $b \in B$ such that $d(a, b) \leq \eta$;
(2) for each $b \in B$, there exists $a \in A$ such that $d(a, b) \leq \eta$;

then, $H_d(A, B) \leq \eta$.

Remark 2 Let d and ρ be two metrics on a nonempty set X and $f : X \to X$ be an operator. We suppose that the metrics d and ρ are metric equivalent, i.e., there exists $c_1, c_2 > 0$ such that

$$c_1 d(x, y) \leq \rho(x, y) \leq c_2 d(x, y), \ \forall\, x, y \in X.$$

Then the following statements are equivalent:

(1) the Eq. (4) is Ulam-Hyers stable in (X, d);
(2) the Eq. (4) is Ulam-Hyers stable in (X, ρ).

We have a similar result for the generalized Ulam-Hyers stability.

For more considerations on the role of metric in Ulam-Hyers stability see [117] and [122].

4 ψ-Weakly Picard Operators and Ulam-Hyers Stability of a Fixed Point Equation

Let (X, d) be a metric space. Following [112] we shall present some notions and examples from weakly Picard operatory theory.

Definition 3 An operator $f : X \to X$ is weakly Picard operator (WPO) if the sequence $(f^n(x))_{n \in \mathbb{N}}$ of successive approximations converges for all $x \in X$ and the limit (which may depend of x) is a fixed point of f. If f is WPO and $F_f = \{x^*\}$, then by definition f is Picard operator (PO).

Definition 4 If $f : X \to X$ is WPO, then we define the operator $f^\infty : X \to X$ by $f^\infty(x) = \lim\limits_{n \to \infty} f^n(x)$.

From the definition of f^∞ it follows that $f^\infty(x) \in F_f$ and $f^\infty(X) = F_f$.

Definition 5 A WPO $f : X \to X$ is c-WPO if c is a positive constant and

$$d(x, f^\infty(x)) \leq c d(x, f(x)), \ \forall\, x \in X.$$

Definition 6 A WPO is ψ-WPO if $\psi : \mathbb{R}_+ \to \mathbb{R}_+$ is increasing, continuous in 0 with $\psi(0) = 0$ and

$$d(x, f^\infty(x)) \leq \psi(d(x, f(x))), \ \forall\, x \in X.$$

It is clear that if f is ψ-WPO and $\psi(t) = ct$ then f is c-WPO.

Example 1 Let (X, d) be a complete metric space and $f : X \to X$ be an α -contraction. Then, f is $(1-\alpha)^{-1}$-PO.

Indeed, by the contraction principle f is PO. Let $F_f = \{x^*\}$. Then, $f^\infty(x) = x^*$, $\forall\, x \in X$ and we have

$$d(x, f^\infty(x)) = d(x, x^*) \leq d(x, f(x)) + d(f(x), f(x^*)) \leq$$
$$\leq d(x, f(x)) + \alpha d(x, x^*).$$

So,

$$d(x, f^\infty(x)) \leq (1-\alpha)^{-1} d(x, f(x)),\ \forall\, x \in X.$$

Example 2 Let (X, d) be a complete metric space and $\varphi : \mathbb{R}_+ \to \mathbb{R}_+$ be a strict comparison function, i.e.,

(1) φ is increasing;
(2) $\varphi^n(t) \to 0$ as $n \to \infty$, $\forall\, t \in \mathbb{R}_+$;
(3) $t - \varphi(t) \to \infty$ as $t \to \infty$.

Let $f : X \to X$ satisfying the following condition

$$d(f(x), f(y)) \leq \varphi(d(x, y)),\ \forall\, x, y \in X.$$

Then, f is ψ_φ-PO, where ψ_φ is defined by

$$\psi_\varphi(t) := \sup\{s \in \mathbb{R}_+ \mid s - \varphi(s) \leq t\}.$$

Indeed, by the Matkowski fixed point theorem (see [68, 110, 124]) f is PO. Let $F_f = \{x^*\}$. We have

$$d(x, f^\infty(x)) = d(x, x^*) \leq d(x, f(x)) + d(f(x), f(x^*)) \leq$$
$$\leq d(x, f(x)) + \varphi(d(x, x^*)),\ \forall\, x \in X.$$

Hence

$$d(x, x^*) - \varphi(d(x, x^*)) \leq d(x, f(x)).$$

So,

$$d(x, x^*) \leq \psi_\varphi(d(x, f(x))),\ \forall\, x \in X.$$

Example 3 Let (X, d) be a complete metric space, $\alpha \in [0, 1[$ and $f : X \to X$ be an operator. We suppose that:

(i) $d(f^2(x), f(x)) \leq \alpha d(x, f(x))$, $\forall\, x \in X$;
(ii) f is with closed graphic.

Then, f is $(1-\alpha)^{-1}$-WPO.

Indeed, by the graphic contraction principle (see [110, 124]) f is WPO. Let $x \in X$. We have

$$d(x, f^{\infty}(x)) \leq d(x, f^n(x)) + d(f^n(x), f^{\infty}(x)) \leq$$
$$\leq d(x, f(x)) + \ldots + d(f^{n-1}(x), f^n(x)) + d(f^n(x), f^{\infty}(x)) \leq$$
$$\leq \frac{1-\alpha^n}{1-\alpha} d(x, f(x)) + d(f^n(x), f^{\infty}(x)), \ \forall\, n \in \mathbb{N}^*.$$

So,

$$d(x, f^{\infty}(x)) \leq (1-\alpha)^{-1} d(x, f(x)), \ \forall\, x \in X.$$

For other examples of ψ-WPO see [13, 29, 110, 124], ...

The basic results of this section are the following:

Theorem 1 *Let (X, d) be a metric space. If $f : X \to X$ is c-WPO, then the equation*

$$x = f(x) \tag{6}$$

is Ulam-Hyers stable.

Proof For $\varepsilon > 0$ we consider the inequation

$$d(y, f(y)) \leq \varepsilon. \tag{7}$$

Let $y^* \in X$ be a solution of the inequation (7). Then, $x^* := f^{\infty}(y^*)$ is a solution of (6). Since f is c-WPO we have

$$d(y^*, x^*) \leq c d(y^*, f(y^*)) \leq c\varepsilon.$$

So, the Eq. (6) is Ulam-Hyers stable. □

In a similar way we have

Theorem 2 *Let (X, d) be a metric space. If $f : X \to X$ is ψ-WPO, then the Eq. (6) is generalized Ulam-Hyers stable.*

Now, some applications of Theorems 1 and 2.

Example 4 Let $X := \mathbb{R}$, $d(x, y) := |x - y|$, $0 < m < 1$ and $M > 0$. We consider the Kepler equation

$$x = m \sin x + M$$

where $f(x) := m \sin x + M$.

It is clear that f is a m-contraction and by Example 1 and Theorem 1, the Kepler equation is Ulam-Hyers stable.

Example 5 Let $X := \{x : \mathbb{R} \to \mathbb{R} \mid x \text{is bounded}\}$ and the metric $d(x, y) := \sup_{t \in \mathbb{R}} |x(t) - y(t)|$. Then (X, d) is a complete metric space. Let $\varphi : \mathbb{R} \to \mathbb{R}$ be a function and $0 < \lambda < 1$. Let us consider the Schröder functional equation

$$x(t) = \lambda x(\varphi(t)), \ t \in \mathbb{R}. \tag{8}$$

In this case f is defined by

$$f(x)(t) := \lambda x(\varphi(t)).$$

We have that

$$|f(x)(t) - f(y)(t)| \leq \lambda |x(\varphi(t)) - y(\varphi(t))| \leq \lambda d(x, y), \ \forall\, x, y \in X.$$

Hence,

$$d(f(x), f(y)) \leq \lambda d(x, y), \text{ i.e., } f \text{ is a } \lambda\text{-contraction.}$$

By Example 1 and Theorem 1, the Schröder equation is Ulam-Hyers stable. Moreover, if $y^* \in X$ is a solution of the inequation

$$d(y, \lambda y(\varphi)) \leq \varepsilon \tag{9}$$

then there exists a solution $x^* \in X$ of (8) such that

$$d(y^*, x^*) \leq (1 - \lambda)^{-1}\varepsilon.$$

Here x^* is the unique solution of (8).

Remark 3 Let $y^* \in X$ be a solution of the inequation

$$|y(t) - \lambda y(\varphi(t))| \leq \varepsilon, \ \forall\, t \in \mathbb{R} \tag{10}$$

then

$$|y^*(t) - x^*(t)| \leq \varepsilon, \ \forall\, t \in \mathbb{R}.$$

Indeed, we observe that if y^* is a solution of (10), then y^* is a solution of (9).

Example 6 Let $X := C(\overline{\Omega}) := \{x : \overline{\Omega} \to \mathbb{R} \mid x \text{is continuous}\}$, where Ω is a bounded domain in $\mathbb{R}^m$. We consider on $C(\overline{\Omega})$ the Chebyshev metric, $d(x, y) := \max_{t\in\overline{\Omega}}|x(t) - y(t)|$. With this metric $C(\overline{\Omega})$ is a complete metric space. We consider on $C(\overline{\Omega})$ the following integral equation of Fredholm type

$$x(t) = \int_{\Omega} K(t, s, x(s))ds + k(t), \ t \in \Omega. \tag{11}$$

We have

Theorem 3 *We suppose that:*

(*i*) $K \in C(\overline{\Omega} \times \overline{\Omega} \times \mathbb{R})$ *and* $k \in C(\overline{\Omega})$;
(*ii*) *there exists* $L_K > 0$ *such that*

$$|K(t, s, u) - K(t, s, v)| \leq L_K |u - v|, \ \forall\, t, s \in \Omega, \ \forall\, u, v \in \mathbb{R};$$

(*iii*) $L_K mes(\Omega) < 1$.

Then, the integral Eq. (11) is Ulam-Hyers stable. Moreover, if $y^* \in C(\overline{\Omega})$ *is a solution of the inequation*

$$d\left(y, \int_{\Omega} K((\cdot), s, y(s))ds + k\right) \leq \varepsilon \tag{12}$$

and x^* *is a unique solution of (11) then*

$$d(y^*, x^*) \leq (1 - L_K mes(\Omega))^{-1}\varepsilon.$$

Proof Let $f : C(\overline{\Omega}) \to C(\overline{\Omega})$ be defined by

$$f(x)(t) := \int_{\Omega} K(t, s, x(s))ds + k(t),\ t \in \Omega.$$

We have

$$\begin{aligned} |f(x)(t) - f(y)(t)| &\leq \int_{\Omega} |K(t, s, x(s)) - K(t, s, y(s))|ds \leq \\ &\leq L_K \int_{\Omega} |x(s) - y(s)|ds \leq \\ &\leq L_K d(x, y) \int_{\Omega} ds = L_K mes(\Omega) d(x, y). \end{aligned}$$

Hence,

$$d(f(x).f(y)) \leq L_K mes(\Omega) d(x, y),\ \forall\, x, y \in C(\overline{\Omega}).$$

From the condition (iii) it follows that f is a contraction. By Example 1 and Theorem 1, the integral Eq. (11) is Ulam-Hyers stable. □

Remark 4 If $y^* \in C(\overline{\Omega})$ is a solution of the inequation

$$|y(t) - \int_{\Omega} K(t, s, y(s))ds - k(t)| \leq \varepsilon,\ \forall\, t \in \overline{\Omega} \tag{13}$$

then

$$|y^*(t) - x^*(t)| \leq (1 - L_K mes(\Omega))^{-1}\varepsilon,\ \forall\, t \in \overline{\Omega}.$$

Indeed, we observe that if y^* is a solution of (13), then y^* is a solution of (12).

Example 7 We consider the Fredholm integral Eq. (11) on $C(\overline{\Omega})$, but we shall endow the set $C(\overline{\Omega})$ with two metrics, the metric d of Chebyshev and the metric ρ defined by

$$\rho(x, y) := \left(\int_{\Omega} |x(s) - y(s)|^2 ds \right)^{\frac{1}{2}} =: \|x - y\|_{L^2(\Omega)}.$$

We have

Theorem 4 *We suppose that:*

(i) $K \in C(\overline{\Omega} \times \overline{\Omega} \times \mathbb{R})$ *and* $k \in C(\overline{\Omega})$;
(ii) there exists $L \in C(\overline{\Omega} \times \overline{\Omega})$ *such that*

$$|K(t, s, u) - K(t, s, v)| \leq L(t, s)|u - v|, \ \forall\, t, s \in \overline{\Omega},\ u, v \in \mathbb{R};$$

(iii) $\int_{\Omega \times \Omega} |L(t, s)|^2 dtds < 1.$

Then, the integral Eq. (11) is Ulam-Hyers stable with respect to the metric ρ. *Moreover, if* $y^* \in C(\overline{\Omega})$ *is a solution of the inequation*

$$\rho\left(y, \int_{\Omega} K((\,\cdot\,), s, y(s))ds + k \right) \leq \varepsilon$$

and x^* *is the unique solution of (11) then*

$$\rho(y^*, x^*) \leq c\varepsilon$$

where $c := \left(1 - \left(\int_{\Omega \times \Omega} |L(t, s)|^2 dtds \right)^{\frac{1}{2}} \right)^{-1}.$

Proof Let f be defined as in Example 7. The conditions (i)-(iii) imply that f satisfies the conditions of the fixed point theorem of Maia (see [110], pp. 28–29; [124], pp. 39–40). From this theorem it follows that f is c-PO with respect to the metric ρ, where $c = (1 - (\int_{\Omega \times \Omega} |L(t, s)|^2 dtds)^{\frac{1}{2}})^{-1}$. So, by Theorem 1, the Eq. (11) is Ulam-Hyers stable with respect to the metric ρ. □

Example 8 Let $X := C[a, b]$ with Chebyshev metric d. We consider on $C[a, b]$ the following Volterra integral equation

$$x(t) = \int_a^t K(t, s, x(s))ds + k(t),\ t \in [a, b] \tag{14}$$

In a similar way as in Example 6 we have

Theorem 5 *We suppose that:*

(*i*) $K \in C([a,b] \times [a,b] \times \mathbb{R})$ *and* $k \in C[a,b]$*;*
(*ii*) *there exists* $L_K > 0$ *such that*

$$|K(t,s,u) - K(t,s,v)| \leq L_K|u-v|, \ \forall\, t,s \in \Omega, \ \forall\, u,v \in \mathbb{R};$$

(*iii*) $L_K(b-a) < 1$.

Then, the integral Eq. (14) *is Ulam-Hyers stable. Moreover, if* $y^* \in C[a,b]$ *is a solution of the inequation*

$$d\left(x, \int_a^{(\cdot)} K((\cdot), s, x(s))ds + k\right) \leq \varepsilon \tag{15}$$

and x^* *is a unique solution of* (14)*, then*

$$d(y^*, x^*) \leq (1 - L_K(b-a))^{-1}\varepsilon.$$

Remark 5 If $y^* \in C[a,b]$ is a solution of the inequation

$$\left| x(t) - \int_a^t K(t,s,x(s))ds - k(t) \right| \leq \varepsilon, \ \forall\, t \in [a,b] \tag{16}$$

then

$$|y^*(t) - x^*(t)| \leq (1 - L_K(b-a))^{-1}\varepsilon.$$

Example 9 Let $X := C[a,b]$ and we consider on $C[a,b]$ the Bielecki metric (for $\tau > 0$)

$$d_\tau(x,y) := \max_{a \leq t \leq b}(|x(t) - y(t)|e^{-\tau(t-a)}).$$

With respect to d_τ, $C[a,b]$ is a complete metric space.
We consider on $C[a,b]$ the Volterra integral Eq. (14). We have

Theorem 6 *We suppose that:*

(*i*) $K \in C([a,b] \times [a,b] \times \mathbb{R})$ *and* $k \in C[a,b]$*;*
(*ii*) *there exists* $L_K > 0$ *such that*

$$|K(t,s,u) - K(t,s,v)| \leq L_K|u-v|, \ \forall\, t,s \in [a,b], \ u,v \in \mathbb{R}.$$

If $\tau > 0$ *is such that* $\frac{L_K}{\tau} < 1$*, then the Eq.* (14) *Is Ulam-Hyers stable with respect to Bielecki metric,* d_τ*.*

Proof Let $f : C[a,b] \to C[a,b]$ be defined by

$$f(x)(t) := \int_a^t K(t,s,x(s))ds + k(t), \ t \in [a,b].$$

Then we have

$$|f(x)(t) - f(y)(t)| \leq L_K \int_a^t |x(s) - y(s)|ds \leq$$

$$\leq L_K \int_a^t |x(s) - y(s)|e^{-\tau(s-a)}e^{\tau(s-a)}ds \leq \frac{L_K}{\tau} d_\tau(x,y)e^{\tau(t-a)}.$$

Hence,

$$d_\tau(f(x), f(y)) \leq \frac{L_K}{\tau} d_\tau(x,y), \ \forall \, x,y \in C[a,b].$$

Since $\frac{L_K}{\tau} < 1$, from Example 1 and Theorem 1 it follows that the Volterra integral Eq. (14) is Ulam-Hyers stable. Moreover, if y^* is a solution of the inequation

$$d_\tau\left(y, \int_a^{(\cdot)} K((\cdot), s, y(s))ds + k\right) \leq \varepsilon$$

and x^* is the unique solution of (14), then

$$d_\tau(y^*, x^*) \leq \left(1 - \frac{L_K}{\tau}\right)\varepsilon.$$

□

For the theory of integral equations see: [46, 61, 69, 108, 119] . . .

5 Ulam-Hyers Stability of a Coincidence Equation

Let (X,d) and (Y,ρ) be two metric spaces and $f, g : X \to X$ be two operators.

Definition 7 Let $\psi : \mathbb{R}_+ \to \mathbb{R}_+$ be increasing, continuous in 0 with $\psi(0) = 0$. By definition, the pair f, g is ψ-weakly Picard pair if there exists an operator $h : X \to X$ such that

(*i*) h is WPO;
(*ii*) $F_h = C(f,g)$;
(*iii*) $d(x, h^\infty(x)) \leq \psi(\rho(f(x), g(x)))$, $\forall \, x \in X$.

If the pair is ψ-weakly Picard and $\psi(t) = ct$, then the pair f, g is called c-weakly Picard.

Definition 8 The coincidence equation

$$f(x) = g(x) \tag{17}$$

is Ulam-Hyers stable if there exists $c > 0$ such that: for each $\varepsilon > 0$ and for each solution y^* of the inequation

$$\rho(f(y), g(y)) \leq \varepsilon \tag{18}$$

there exists a solution x^* of (17) such that

$$d(y^*, x^*) \leq c\varepsilon.$$

Definition 9 The coincidence Eq. (17) is generalized Ulam-Hyers stable if there exists an increasing function $\psi : \mathbb{R}_+ \to \mathbb{R}_+$, continuous in 0 with $\psi(0) = 0$, such that: for each $\varepsilon > 0$ and for each solution y^* of the coincidence inequation (18) there exists a solution $x*$ of (17) such that

$$d(y^*, x^*) \leq \psi(\varepsilon).$$

The following results are very useful to study Ulam-Hyers stability of coincidence equations.

Theorem 7 *If a pair $f, g : X \to Y$ is c-weakly Picard pair, then the Eq.* (17) *is Ulam-Hyers stable.*

Proof Let y^* be a solution of (18). Let $h : X \to X$ be the operator which appears in Definition 7. We take $x^* := h^\infty(y^*)$. For this solution of (17) we have

$$d(y^*, x^*) \leq c\rho(f(y^*), g(y^*)) \leq c\varepsilon. \qquad \square$$

In a similar way we have

Theorem 8 *If a pair $f, g : X \to Y$ is ψ-weakly Picard pair, then the coincidence Eq. (17) is generalized Ulam-Hyers stable.*

For some examples of c-weakly Picard pairs see [21], pp. 37–40.

For the coincidence point theory see [17, 21, 33, 44, 110, 113, 124].

Problem 16 To construct a theory of ψ-weakly Picard pairs.

Problem 17 To give some relevant applications of Theorem 7 and Theorem 8.

References: [21, 117, 122].

6 The Case of Spaces of Functions: Ulam-Hyers and Ulam-Hyers-Rassias Stability

Let $\Omega \subset \mathbb{R}^m$ be a nonempty subset of $\mathbb{R}^m$, X be a set of functions $x : \Omega \to \mathbb{R}$ and $f, g : X \to X$. If we have on X a metric then we have for the coincidence equation

$$f(x) = g(x) \tag{19}$$

the notions of Ulam-Hyers stability given by Definitions 8 and 9.

Now, we consider on X the generalized metric $d : X \times X \to X_+ := \{x : \Omega \to \mathbb{R} \mid x \geq 0\}$, defined by

$$d(x, y)(t) := |x(t) - y(t)|, \ \forall\, t \in \Omega.$$

With respect to this generalized metric we have the following notions of Ulam stability.

Definition 10 The Eq. (19) is Ulam-Hyers stable with respect to the generalized metric d if there exists a real number $c > 0$ such that: for each $\varepsilon > 0$ and for each $y^* \in X$ solution of the inequation

$$|f(y)(t) - g(y)(t)| \leq \varepsilon, \ \forall\, t \in \Omega \tag{20}$$

there exists a solution x^* of (19) such that

$$|y^*(t) - x^*(t)| \leq c\varepsilon, \ \forall\, t \in \Omega.$$

Definition 11 The Eq. (19) is generalized Ulam-Hyers stable with respect to the generalized metric d if there exists an increasing function $\psi : \mathbb{R}_+ \to \mathbb{R}_+$ continuous in 0 with $\psi(0) = 0$, such that: for each $\varepsilon > 0$ and for each solution y^* of (20) there exists a solution x^* of (19) such that

$$|y^*(t) - x^*(t)| \leq \psi(\varepsilon), \ \forall\, t \in \Omega.$$

Definition 12 Let $\varphi : \Omega \to \mathbb{R}_+$ be a function. The Eq. (19) is Ulam-Hyers-Rassias stable with respect to φ and to the generalized metric d if there exists $c > 0$ such that: for each $\varepsilon > 0$ and for each solution y^* of the inequation

$$|f(y)(t) - g(y)(t)| \leq \varepsilon\varphi(t), \ \forall\, t \in \Omega \tag{21}$$

there exists a solution x^* of (19) with

$$|y^*(t) - x^*(t)| \leq c\varepsilon\varphi(t), \ \forall\, t \in \Omega.$$

Definition 13 Let $\varphi : \Omega \to \mathbb{R}_+$ be a function. The Eq. (19) is generalized Ulam-Hyers-Rassias stable with respect to φ and to the generalized metric d if there exists $c > 0$ such that: for each solution y^* of the inequation

$$|f(y(t)) - g(y(t))| \leq \varphi(t), \ \forall\, t \in \Omega \tag{22}$$

there exists a solution x^* of (19) with

$$|y^*(t) - x^*(t)| \leq c\varphi(t), \ \forall\, t \in \Omega.$$

Example 10 Let $\Omega := [a, b]$ and $X := C^1[a, b]$. Let us consider the differential equation

$$x'(t) = h(t, x(t)), \ t \in [a, b] \tag{23}$$

and the differential inequation

$$|y'(t) - h(t, y(t))| \leq \varphi(t),\ t \in [a, b]. \tag{24}$$

We have

Theorem 9 *We suppose that:*

(*i*) $h \in C([a, b] \times \mathbb{R})$;
(*ii*) $\varphi \in C([a, b], \mathbb{R}_+)$ *is increasing;*
(*iii*) *there exists* $l_h \in L^1[a, b]$ *such that*

$$|h(t, u) - h(t, v)| \leq l_h(t)|u - v|,\ \forall\, t \in [a, b],\ \forall\, u, v \in \mathbb{R}.$$

Then, the differential Eq. (23) *is generalized Ulam-Hyers-Rassias stable.*

Proof Let $y \in C^1[a, b]$ be a solution of the differential inequation (24). Let x be the unique solution of the Cauchy problem (conditions (i)-(iii) imply the existence and uniqueness of Cauchy problem!)

$$x'(t) = h(t, x(t)),\ t \in [a, b]$$
$$x(a) = y(a).$$

For such y and x we have

$$x(t) = y(a) + \int_a^t h(s, x(s))ds,\ t \in [a, b]$$

and

$$\left|y(t) - y(a) - \int_a^t h(s, y(s))ds\right| \leq \int_a^t \varphi(s)ds \leq \varphi(t),\ t \in [a, b].$$

From these relations it follows

$$|y(t) - x(t)| \leq \left|y(t) - y(a) - \int_a^t h(s, y(s))ds\right| + \int_a^t |h(s, y(s)) - h(s, x(s)|ds \leq$$

$$\leq \varphi(t) + \int_a^t l_h(s)|y(s) - x(s)|ds.$$

By a well-known Gronwall lemma (see [3, 28, 134], ...) we have

$$|y(t) - x(t)| \leq c\varphi(t),\ t \in [a, b]$$

where $c := \exp \int_a^b l_h(s)ds$. □

For other results on Ulam stability of integral equations, of differential equations and of partial differential equations see: [2, 36, 41, 57–59, 61, 62, 64–68, 71, 72, 75, 76, 80, 92, 117–120, 122], ...

For the basic theory of differential and integral equations see: [3, 11, 28, 43, 46, 93, 108, 116, 127, 134–136], ...

Problem 18 To give some abstract results for Ulam stabilities as in Definitions 10–13.

References: [21, 56, 63, 82, 87, 104, 107, 112, 117, 119–122, 137], ...

7 Equations with Multivalued Operators

Let (X, d) be a metric space. Let us denote

$P_{cl}(X) := \{Y \in P(X) \mid Y \text{is closed}\}$,
$P_{cp}(X) := \{Y \in P(X) \mid Y \text{is compact}\}$,
$P_b(X) := \{Y \in P(X) \mid Y \text{is bounded}\}$.

In what follows we need the following functionals:

- $\delta_d : P(X) \times P(X) \to \mathbb{R}_+ \cup \{+\infty\}$,

  $\delta_d(Y, Z) := \sup\{d(y, z) \mid y \in Y,\ z \in Z\}$- the diameter functional,

- $D_d : P(X) \times P(X) \to \mathbb{R}_+$,

  $D_d(Y, Z) := \inf\{d(y, z) \mid y \in Y,\ z \in Z\}$- the gap functional,

- $H_d : P(X) \times P(X) \to \mathbb{R}_+ \cup \{+\infty\}$,

$$H_d(Y, Z) := \max\left\{\sup_{y \in Y} D_d(y, Z),\ \sup_{z \in Z} D_d(Y, z)\right\}$$

  - the generalized Pompeiu-Hausdorff functional.

Following [123] we shall present some notions from multivalued weakly Picard operator theory.

Definition 14 Let (X, d) be a metric space and $T : X \to P(X)$ be a multivalued operator. By definition T is WPO if for each $x \in X$ and each $y \in T(x)$ there exists a sequence of successive approximations $(x_n)_{n \in \mathbb{N}}$, $x_{n+1} \in T(x_n)$, $n \in \mathbb{N}$, such that $x_0 = x$, $x_1 = y$ and $x_n \overset{d}{\to} x^* \in F_T$.

Definition 15 Let $T : X \to P(X)$ be a multivalued WPO. Then we define the multivalued operator $T^\infty : G(T) \to P(F_T)$ by,

$$T^\infty(x, y) := \{z \in F_T \mid \text{there exists a sequence of successive approximations of } T \text{starting from} (x, y) \text{that converges to} z\}.$$

Here $G(T)$ denotes the graphic of T.

Definition 16 Let $\psi : \mathbb{R}_+ \to \mathbb{R}_+$ be an increasing function, continuous in 0 with $\psi(0) = 0$. An WPO $T : X \to P(X)$ is ψ-weakly Picard multivalued operator if there exists a selection t^∞ of T^∞ such that

$$d(x, t^\infty(x, y)) \leq \psi(d(x, y)), \ \forall\, x, y \in G(T).$$

If $\psi(t) = ct$, then T is called c-multivalued WPO.

Definition 17 Let us consider the multivalued fixed point equation

$$x \in T(x) \tag{25}$$

and the multivalued inequation

$$D_d(u, Tu) \leq \varepsilon \tag{26}$$

for $\varepsilon > 0$.

The Eq. (25) is Ulam-Hyers stable if there exists $c > 0$ such that: for each $\varepsilon > 0$ and for each solution u^* of (26) there exists a solution x^* of (25) such that

$$d(u^*, x^*) \leq c\varepsilon.$$

The Eq. (25) is generalized Ulam-Hyers stable if there exists an increasing function $\psi : \mathbb{R}_+ \to \mathbb{R}_+$, continuous in 0 with $\psi(0) = 0$ such that: for each $\varepsilon > 0$ and for each solution u^* of (26) there exists a solution x^* of (25) such that

$$d(u^*, x^*) \leq \psi(\varepsilon).$$

Now let us consider the strict fixed point equation

$$\{x\} = T(x) \tag{27}$$

and the strict fixed point inequation

$$H_d(\{u\}, T(u)) \leq \varepsilon. \tag{28}$$

We observe that $H_d(\{u\}, T(u)) = \delta_d(\{u\}, T(u))$.

Definition 18 The Eq. (27) is Ulam-Hyers stable if there exists $c > 0$ such that: for each $\varepsilon > 0$ and for each solution u^* of (28) there exists a solution x^* of (27) such that

$$d(u^*, x^*) \leq c\varepsilon.$$

The Eq. (27) is generalized Ulam-Hyers stable if there exists an increasing function $\psi : \mathbb{R}_+ \to \mathbb{R}_+$, continuous in 0 with $\psi(0) = 0$ such that: for each $\varepsilon > 0$ and each solution u^* of (28) there exists a solution x^* of (27) such that

$$d(u^*, x^*) \leq \psi(\varepsilon).$$

We have

Theorem 10 *Let (X, d) be a metric space and $T : X \to P_{cp}(X)$ be a multivalued ψ-WPO. Then, the inclusion* (25) *is generalized Ulam-Hyers stable.*

Proof Let u^* be a solution of (25). Let $y^* \in T(u^*)$ be such that $D_d(u^*, T(u^*)) = d(u^*, y^*)$. If we take $x^* := t^\infty(u^*, y^*)$, then we have

$$d(u^*, x^*) = d(u^*, t^\infty(u^*, y^*)) \le \psi(d(u^*, y^*)) \le \psi(\varepsilon). \qquad \square$$

For other results for Ulam stabilities in the case of multivalued operators see [14, 87, 117, 128].

Problem 19 To study the Ulam-Hyers stability of a strict fixed point equation.

References: [87, 111, 124].

Another operatorial equation with multivalued operators is the coincidence equation.

Let (X, d) and (Y, ρ) be two metric spaces and $T, S : X \to P(Y)$ be two multivalued operators from X to Y. Let us consider the coincidence equation

$$T(x) \cap S(x) \neq \emptyset \tag{29}$$

and the inequation

$$D_\rho(T(u), S(u)) \le \varepsilon \tag{30}$$

for $\varepsilon > 0$.

By definition the Eq. (29) is generalized Ulam-Hyers stable if there exists an increasing function $\psi : \mathbb{R}_+ \to \mathbb{R}_+$, continuous in 0 with $\psi(0) = 0$ such that: for each $\varepsilon > 0$ and for each solution $u^* \in X$ of (30) there exists a solution x^* of (29) such that

$$d(u^*, x^*) \le \psi(\varepsilon).$$

Problem 20 To study the Ulam-Hyers stability of a multivalued coincidence equation.

References: [14] and the references therein.

8 Other Problems

8.1 *Ulam Stability in the Case of a Generalized Metric Space ($d(x, y) \in \mathbb{R}_+$)*

There are several concepts of generalized metric of type $d : X \times X \to \mathbb{R}_+$. The following axioms appear in different definitions of such metrics:

(i) $d(x, y) = 0 \Leftrightarrow x = y, \ \forall\, x, y \in X$;
(i_1) $d(x, x) = 0, \ \forall\, x \in X$;
(i_2) $d(x, y) = 0 \Rightarrow x = y, \ \forall\, x, y \in X$;

(i_3) $d(x,y) = d(y,x) = 0 \Leftrightarrow x = y,\ \forall\, x, y \in X$;
(i_4) $d(x,y) = d(y,x) = 0 \Rightarrow x = y,\ \forall\, x, y \in X$;
(i_5) $d(x,x) = d(y,y) = d(x,y) \Leftrightarrow x = y,\ \forall\, x, y \in X$;
(i_6) $d(x,x) \leq d(x,y),\ \forall\, x, y \in X$;
(i_7) $d(y,y) \leq d(x,y),\ \forall\, x, y \in X$;
(ii) $d(x,y) = d(y,x),\ \forall\, x, y \in X$;
(ii_1) there exists $c > 0$ such that $d(x,y) \leq cd(y,x),\ \forall\, x, y \in X$;
(iii) $d(x,y) \leq d(x,z) + d(z,y),\ \forall\, x, y, z \in X$;
(iii_1) $d(x,y) \leq d(x,z) + d(y,z),\ \forall\, x, y, z \in X$;
(iii_2) $d(x,y) \leq \max(d(x,z), d(z,y)),\ \forall\, x, y, z \in X$;
(iii_3) $\forall\, \varepsilon > 0\ d(x,z) \leq \varepsilon,\ d(z,y) \leq \varepsilon \Rightarrow d(x,y) \leq \varepsilon,\ \forall\, x, y, z \in X$;
(iii_4) there exists $a \geq 1$ such that: $d(x,y) \leq a(d(x,z) + d(z,y)),\ \forall\, x, y, z \in X$;
(iii_5) there exists $a \geq 1$ such that:

$$d(x,y) \leq a\max(d(x,y), d(y,z)),\ \forall\, x, y, z \in X;$$

(iii_6) $d(x,y) \leq d(x,z) + d(z,y) - d(z,z),\ \forall\, x, y, z \in X$.

By definition d is a:

1) *premetric* if it satisfies: $(i_1) + (iii)$;
2) *pseudometric* if it satisfies: $(i_1) + (ii) + (iii)$;
3) *quasimetric* if it satisfies: $(i_3) + (iii)$;
4) *semimetric* if it satisfies: $(i) + (ii)$;
5) *symmetric* if it satisfies: $(i_2) + (ii)$;
6) *dislocated metric* if it satisfies: $(i_4) + (ii) + (iii)$;
7) *ultrametric* if it satisfies: $(i) + (ii) + (iii_2)$ or $(i) + (ii) + (iii_3)$;
8) *quasiultrametric* if it satisfies: $(i) + (ii_1) + (iii_5)$;
9) *b-metric* if it satisfies: $(i) + (ii) + (iii_4)$;
10) *partial metric* if it satisfies: $(i_5) + (i_6) + (ii) + (iii_6)$.

Problem 21 To study Ulam stability of operatorial equations and inclusions in each of the above generalized metric spaces.

References: [34, 114, 124] and the references therein.

Commentaries: Let X be a nonempty set and $f : X \to X$ be an operator. Let us consider the fixed point equation

$$x = f(x)$$

and the functional $d : X \times X \to \mathbb{R}_+$. If d is a semimetric, an ultrametric, a quasiultrametric or a b-metric then the definitions of Ulam-Hyers stability and of gneralized Ulam-Hyers stability can be given as in Definitions 1 and 2. If the functional d do not satisfies axiom (i) then it is necessary to take instead of inequation (5) another inequation. Let us consider, for example, that d is a partial metric. In this case a good candidate for (5) is the following inequation

$$2d(x, f(x)) - d(x,x) - d(f(x), f(x)) \leq \varepsilon,$$

since the functional $\rho : X \times X \to \mathbb{R}_+$ definded by, $\rho(x,y) := 2d(x,y) - d(x,x) - d(y,y)$ is a metric (see [114]).

8.2 Ulam Stability in the Case of a Generalized Metric Space ($d(x, y) \in E_+$)

Let $(E, +, \mathbb{R}, \leq, \to)$ be an ordered linear L-space (see [121] and [122]). Let

$$E_+ := \{e \in E \mid e \geq 0\}$$

and

$$E_+^* := \{e \in E \mid e \geq 0 \text{and} e \neq 0\}.$$

Let X be a nonempty set and $d : X \times X \to E_+$ be a generalized metric on X, i.e., d satisfies the following axioms:

(i) $d(x, y) = 0 \Leftrightarrow x = y,\ \forall\, x, y \in X;$
(ii) $d(x, y) = d(y, x),\ \forall\, x, y \in X;$
(iii) $d(x, y) \leq d(x, z) + d(z, y),\ \forall\, x, y, z \in X.$

Following [122] we present the following definitions and results.

Definition 19 Let (X, d) be a generalized metric space with $d(x, y) \in E_+, \forall\, x, y \in X$. Let $f : X \to X$ be an operator. By definition the equation

$$x = f(x) \tag{31}$$

is Ulam-Hyers stable if there exists a linear increasing operator $c : E \to E$ such that: for each $\varepsilon \in E_+^*$ and each solution $y^* \in X$ of the inequation

$$d(y, f(y)) \leq \varepsilon \tag{32}$$

there exists a solution $x^* \in X$ of (31) with

$$d(y^*, x^*) \leq c(\varepsilon).$$

Definition 20 The Eq. (31) is generalized Ulam-Hyers stable if there exists an increasing operator $\psi : E_+ \to E_+$, continuous in 0 with $\psi(0) = 0$, such that: for each $\varepsilon \in E_+^*$ and for each solution $y^* \in X$ of (32) there exists a solution $x^* \in X$ of (31) such that

$$d(y^*, x^*) \leq \psi(\varepsilon).$$

Example 11 Let $E := \mathbb{R}^m$, $X := C(\overline{\Omega}, \mathbb{R}^m)$, $\Omega \subset \mathbb{R}^p$ is a bounded domain and

$$d(x, y) := \begin{pmatrix} \max\limits_{t\in\Omega}|x_1(t) - y_1(t)| \\ \vdots \\ \max\limits_{t\in\Omega}|x_m(t) - y_m(t)| \end{pmatrix} \in \mathbb{R}_+^m$$

where $x = (x_1, \ldots, x_m),\ y = (y_1, \ldots, y_m)$ are from $C(\overline{\Omega}, \mathbb{R}^m)$.

Let $f : C(\overline{\Omega}, \mathbb{R}^m) \to C(\overline{\Omega}, \mathbb{R}^m)$ be defined by

$$f(x)(t) := \int_{\Omega} K(t, s, x(s))ds + k(t),\ t \in \overline{\Omega}$$

and we consider the fixed point equation

$$x(t) = f(x)(t),\ t \in \overline{\Omega}. \tag{33}$$

We have

Theorem 11 *We suppose that:*

(i) $K \in C(\overline{\Omega} \times \overline{\Omega} \times \mathbb{R}^m, \mathbb{R}^m)$ *and* $k \in C(\overline{\Omega}, \mathbb{R}^m)$*;*
(ii) there exists a matrix $L_K \in \mathbb{R}_+^{m \times m}$ *such that*

$$\begin{pmatrix} |K_1(t,s,u) - K_1(t,s,v)| \\ \vdots \\ |K_m(t,s,u) - K_m(t,s,v)| \end{pmatrix} \leq L_K \begin{pmatrix} |u_1 - v_1| \\ \vdots \\ |u_m - v_m| \end{pmatrix},$$

$\forall\, t, s \in \overline{\Omega},\ \forall\, u, v \in \mathbb{R}^m.$
(iii) the matrix $mes(\Omega)L_K$ *is such that*

$$(mes(\Omega)L_K)^n \to 0 as n \to \infty.$$

Then, the Eq. (33) *has a unique solution and is Ulam-Hyers stable.*

Proof
First of all, we observe that

$$d(f(x), f(y)) \leq mes(\Omega)L_K d(x, y),\ \forall\, x, y \in C(\overline{\Omega}, \mathbb{R}^m). \tag{34}$$

From the Perov fixed point theorem (see [110], pp. 96–97; [124], pp. 83) the Eq. (33) has a unique solution x^*. Let y^* be a solution of the inequation ($\varepsilon \in \mathbb{R}_+^*$)

$$d(y, f(y)) \leq \varepsilon,\ \forall\, x \in C(\overline{\Omega}, \mathbb{R}^m). \tag{35}$$

From (34) we have that

$$d(y^*, x^*) \leq (I_m - mes(\Omega)L_K)^{-1}\varepsilon.$$

So, the Eq. (33) is Ulam-Hyers stable. □

For more considerations on Ulam-Hyers stability in a generalized metric space with $d(x, y) \in E_+$ see [122].

For a fixed point theory in a such generalized metric space see [31, 32, 124, 139].

Problem 22 To construct a theory of WPO in a generalized metric space ($d(x, y) \in E_+$) and to apply this theory to Ulam-Hyers stability of a fixed point equation in a such space.

References: [122].

8.3 *Ulam Stability in the Case of Equations with Set-To-Point Operators*

Let (X, d) and (Y, ρ) be two metric spaces, $Z \subset P(X)$, $Z \neq \emptyset$ and $T, S : Z \to Y$ be two set-to-point operators. We consider on Z the equation

$$T(A) = S(A). \tag{36}$$

Definition 21 The Eq. (36) is Ulam-Hyers stable if there exists $c > 0$ such that: for each $\varepsilon > 0$ and each solution $B^* \in Z$ of the inequation

$$\rho(T(B), S(B)) \leq \varepsilon \tag{37}$$

there exists a solution $A^* \in Z$ of (36) such that

$$H_d(B^*, A^*) \leq c\varepsilon.$$

In a similar way we define the generalized Ulam-Hyers stability of the set-to-point Eq. (36).

Problem 23 To study Ulam stability of (36).

Problem 24 Let (X, d) be a metric space, $Z \subset P(X)$, $Z \neq \emptyset$ and $T : Z \to \mathbb{R}_+$ be a point-to-set functional. Let $r > 0$ be a given positive real number ($S(A) := r$, $\forall\, A \in Z$). The problem is to study Ulam stability of the equation

$$T(A) = r \tag{38}$$

We can take in Problem 8.3, $T := \delta_d$, $T := \alpha$ - an abstract measure of noncompactness, $T := \beta$ - an abstract measure of nonconvexity, ...

References: [122]. For the measures of noncompactness see: [5, 6, 10, 33, 44, 68, 113, 124]. For the measures of nonconvexity see: [10, 113] and the references therein.

8.4 *Difference Equations as Operatorial Equations*

Let $k \in \mathbb{N}^*$ and $f_n : \mathbb{R}^k \to \mathbb{R}$, $n \in \mathbb{N}^*$ be some given functions. We consider, for example, the following difference equation

$$x_n = f_n(x_{n-k}, x_{n-k+1}, \ldots, x_{n-1}). \tag{39}$$

Let us denote

$$s(\mathbb{R}) := \{(x_n)_{n\in\mathbb{N}^*} \mid x_n \in \mathbb{R}\}$$

and

$$M(\mathbb{R}) := \{(x_{ij})_1^\infty \mid x_{ij} \in \mathbb{R},\ i, j \in \mathbb{N}^*\}.$$

We consider on $s(\mathbb{R})$ the generalized metric

$$d(x, y) := (|x_n - y_n|)_{n\in\mathbb{N}^*}$$

and on $M(\mathbb{R})$ the generalized metric

$$d(A, B) := \sup_{i\in\mathbb{N}^*} \sum_{j=1}^{\infty} |a_{ij}|.$$

Let us consider the operator

$$T : \mathbb{R}^k \times s(\mathbb{R}) \to \mathbb{R}^k \times s(\mathbb{R})$$

defined by

$$(x_{-k+1}, \dots, x_0, x_1, \dots, x_n, \dots) \mapsto (x_{-k+1}, \dots, x_0, f_1(x_{-k+1}, \dots, x_0), \dots,$$
$$f_n(x_{n-k}, \dots, x_{n-1}), \dots).$$

In terms of the operator T, the difference Eq. (39) takes the following form:

$$x = T(x). \tag{40}$$

The Eq. (40) is a fixed point equation on the generalized metric space, $\mathbb{R}^k \times s(\mathbb{R})$.

Problem 25 To study the Ulam stability of the Eq. (39) by operatorial equations tehniques.

References: [1, 70, 92, 117, 122], ...

For the Ulam stability of the difference equations see [18–20, 90, 91, 138], ...

8.5 *Ulam Stability of Fractal Equations*

Let (X, d) be a metric space and $T : X \to P_{cp}(X)$ be an upper semicontinuous operator. Let $\hat{T}$ be the fractal operator corresponding to T, i.e., $\hat{T} : P_{cp}(X) \to P_{cp}(X)$, defined by $\hat{T}(A) := \bigcup_{a\in A} T(a)$. Let us consider the equations:

$$x \in T(x) \tag{41}$$

and

$$A = \hat{T}(A). \tag{42}$$

Equation (41) is an operatorial inclusion on X and Eq. (42) is a fixed point equation with singlevalued operator on $P_{cp}(X)$.

Problem 26 In which conditions we have that: If the Eq. (41) is Ulam-Hyers stable on (X, d), then the Eq. (42) is Ulam-Hyers stable on $(P_{cp}(X), H_d)$?

For fractal operators see: [113], pp. 19–20; [124], pp. 275–277.

References

1. Agarwal, R.P.: Difference Equations and Inequalities. Dekker, New York (1992)
2. Alsina, C., Ger, R.: On some inequalities results related to the exponential function. J. Inequal. Appl. **2**, 373–380 (1998)
3. Amann, H.: Ordinary Differential Equations. Walter de Gruyter, Berlin (1990)
4. Aoki, T.: On the stability of the linear transformation in Banach spaces. J. Math. Soc. Jpn. **2**, 64–66 (1950)
5. Appell, J.: Measure of noncompactness, condensing operators and fixed points: an application-oriented survey. Fixed Point Theory **6**, 157–229 (2005)
6. Ayerbe Toledane, J.M., Domingues Benavides, T., López, G.: Measures of Noncompactness in Metric Fixed Point Theory. Birkhäuser , Basel, (1997)
7. Baak, C., Boo, D.-H., Rassias, Th.M.: Generalized additive mapping in Banach modules and isomorphisms between C^*-algebras. J. Math. Anal. Appl. **314**(1), 150–161 (2006)
8. Bailey, P.B., Shampine, L.F., Waltman, P.E.: Nonlinear Two Point Boundary Value Problems. Academic, New York (1968)
9. Baker, J.A.: The stability of certain functional equations. Proc. Am. Math. Soc. **112**, 729–732 (1991)
10. Ban, A.I., Gal, S.G.: Defects of Properties in Mathematics. World Scientific, New Jersey (2002)
11. Barbu, V.: Nonlinear Semigroups and Differential Equations in Banach Spaces. Nordhoff, Leyden (1976)
12. Belitskii, G., Tkachenko, V.: One-dimensional Functional Equations. Birkhäuser , Basel (2003)
13. Berinde, V.: Iterative Approximation of Fixed Points. Springer, New York (2007)
14. Bota-Boriceanu, M.F., Petruşel, A.: Ulam-Hyers stability for operatorial equations. Analele Ştiinţifice ale Univ. "Al. I. Cuza" din Iaşi, Matematica **97**, 65–74 (2011)
15. Bourgin, D.G.: Classes of transformations and bordering transformations. Bull. Am. Math. Soc. **57**, 223–237 (1951)
16. Breckner, W.W., Trif, T.: Convex Functions and Related Functional Equations. Cluj University Press, Cluj-Napoca (2008)
17. Brown, R.F., Furi, M., Gorniewicz, L., Jiang, B.: Handbook of Topological Fixed Point Theory. Springer, New York (2005)
18. Brzdek, J., Popa, D., Xu, B.: Note on nonstability of the linear recurrence. Abh. Math. Sem. Univ. Hamburg **76**, 183–189 (2006)
19. Brzdek, J., Popa, D., Xu, B.: The Hyers-Ulam stability of nonlinear recurrences. J. Math. Anal. Appl. **335**, 443–449 (2007)
20. Brzdek, J., Popa, D., Xu, B.: Hyers-Ulam stability for linear equations of higher orders. Acta Math. Hungar. **1–2**, 1–8 (2008)
21. Buică, A.: Principii de coincidenţă şi aplicaţii. Presa Univ. Clujeană, Cluj-Napoca (2001)
22. Burton, T.A.: Stability by Fixed Point Theory for Functional Differential Equations. Dover Publications, Mineola (2006)
23. Cădariu, L.: Stabilitatea Ulam-Hyers-Bourgin pentru ecuaţii funcţionale. Editura Univ. de Vest Timişoara (2007)
24. Cădariu, L., Radu, V.: The stability of the Cauchy functional equation: a fixed point approach. In: Iteration Theory (ECIT 02), pp. 43–52, Grazer Math. Ber., 346, Karl-Franzens-Univ. Graz, Graz (2004)

25. Cădariu, L., Radu, V.: A general fixed point method for the stability of Jensen functional equation. Bul. Şt. Univ. Politehnica din Timişoara **51**(2), 63–72 (2006)
26. Cădariu, L., Radu, V.: The alternative of fixed point and stability results for functional equations. Int. J. Appl. Math. Stat **7**(Fe 07), 40–58 (2007)
27. Cădariu, L., Radu, V.: A general fixed point method for stability of the monomial functional equation. Carpathian J. Math. **28**(1), 25–36 (2012)
28. Chicone, C.: Ordinary Differential Equations with Applications. Springer, New York (2006)
29. Chiş-Novac, A., Precup, R., Rus, I.A.: Data dependence of fixed points for non-self generalized contractions. Fixed Point Theory **10**(1), 73–87 (2009)
30. Czerwik, S.: Functional Equations and Inequalities in Several Variable. World Scientific, New Jersey (2002)
31. De Pascale, E., De Pascale, L.: Fixed points for some nonobviously contractive operators. Proc. Am. Math. Soc. **130**, 3249–3254 (2002)
32. De Pascale, E., Marino, G., Pietramala, P.: The use of the E-metric spaces in the search for fixed points. Le Matematiche **48**(II), 367–376 (1993)
33. Deimling, K.: Nonlinear Functional Analysis. Springer, Berlin (1985)
34. Deza, M.M., Deza, E.: Encyclopedia of Distances. Springer, Heidelberg (2009)
35. Dontchev, A.L., Rockafellar, R.T.: Implicit Functions and Solution Mappings. Springer , New York (2009)
36. Egri, E.: Ulam stabilities of a first order iterative functional-differential equation. Fixed Point Theory **12**(2), 321–328 (2011)
37. Eshaghi Gordji, M., Khodaei, H., Rassias, Th.M.: Fixed point methods for the stability of general quadratic functional equation. Fixed Point Theory **12**(1), 71–82 (2011)
38. Faiziev, V.A., Rassias, Th.M., Sahoo, P.K.: The space of (ψ, γ)-additive mappings on semigroups. Trans. Am. Math. Soc. **354**(11), 4455—4472 (2002)
39. Găvruţă, P.: A generalization of the Hyers-Ulam-Rassias stability of approximately additive mappings. J. Math. Anal. Appl. **184**, 431–436 (1994)
40. Găvruţă, P.: On a problem of G. Isac and Th. Rassias concerning the stability of mappings. J. Math. Anal. Appl. **261**, 543–553 (2001)
41. Găvruţă, P., Găvruţă, L.: A new method for the generalized Hyers-Ulam-Rassias stability. Int. J. Nonlinear Anal. Appl. **1**(2), 11–18 (2010)
42. Ghermănescu, M.: Ecuaţii funcţionale. Editura Academiei, Bucureşti (1960)
43. Goldstein, J.A.: Semigroups of Linear Operators and Applications. Oxford University Press, Oxford (1985)
44. Granas, A., Dugundji, J.: Fixed Point Theory. Springer, New York (2003)
45. Gruber, P.M.: Stability of isometries. Trans. Am. Math. Soc. **245**, 263–277 (1978)
46. Guo, D., Lakshmikantham, V., Liu, X.: Integral Equations in Abstract Spaces. Kluwer, Dordrecht (1996)
47. Hirasawa, G., Miura, T.: Hyers-Ulam stability of a closed operator in Hilbert space. Bull. Korean Math. **43**(1), 107–117 (2006)
48. Hyers, D.H.: On the stability of the linear functional equation. Proc. Natl. Acad. Sci. U. S. A. **27**, 222–224 (1941)
49. Hyers, D.H.: The stability of homomorphism and related topics. In: Global Analysis—Analysis on Manifolds, (Th.M. Rassias, ed.), pp. 140–153. Teubner, Leipzig (1983)
50. Hyers, D.H., Isac, G., Rassias, Th.M: On the asymptoticity aspect of Hyers-Ulam stability of mappings. Proc. Am. Math. Soc. **126**, 425–430 (1998)
51. Hyers, D.H., Isac, G., Rassias, Th.M.: Stability of Functional Equations in Several Variables. Birkhäuser, Basel (1998)
52. Hyers, D.H., Rassias, Th.M.: Approximate homomorphisms. Aequationes Math. **44**, 125–153 (1992)
53. Hyers, D.R., Ulam, S.: On approximate isometries. Bull. Am. Math. Soc. **51**, 288–292 (1945)
54. Isac, G., Rassias, Th.M.: On the Hyers-Ulam stability of ψ-additive mappings. J. Approx. Theory **72**, 131–137 (1993)

55. Isac, G., Rassias, Th.M.: Stability of additive mappings: application to nonlinear analysis. Int. J. Math. Math. Sci. **19**, 219–228 (1996)
56. Jung, S.-M.: Hyers-Ulam-Rassias Stability of Functional Equations in Mathematical Analysis. Hadronic Press, Palm Harbor (2001)
57. Jung, S.-M.: Hyers-Ulam stability of linear differential equations of first order. Appl. Math. Lett. **17**(10), 1135–1140 (2004)
58. Jung, S.-M.: Hyers-Ulam stability of linear differential equations of first order, III. J. Math. Anal. Appl. **311**(1), 139–146 (2005)
59. Jung, S.-M.: Legendre's differential equations and its Hyers-Ulam stability. Abstr. Appl. Anal. Art. ID56419, 14 p (2007)
60. Jung, S.-M.: A fixed point approach to the stability of isometries. J. Math. Anal. Appl. **329**, 879–890 (2007)
61. Jung, S.-M.: A fixed point approach to the stability of a Volterra integral equation. Fixed Point Theory Appl., Art. ID57064, 9 p. (2007)
62. Jung, S.-M.: Hyers-Ulam stability of linear partial differential equations of first order. Appl. Math. Lett. **22**, 70–74 (2009)
63. Jung, S.-M.: Hyers-Ulam-Rassias Stability of Functional Equations in Nonlinear Analysis. Springer , New York (2011)
64. Jung, S.-M., Lee, F: Hyers-Ulam-Rassias stability of linear differential equations of second order. J. Comput. Math. Optim. **3**(3), 193–200 (2007)
65. Jung, S.-M., Lee, K.-S.: Hyers-Ulam stability of first order linear partial differential equations with constant coefficients. Math. Ineq. Appl. **10**(2), 261–266 (2007)
66. Jung, S.-M., Rassias, Th.M.: Generalized Hyers-Ulam stability of Riccati differential equation. Math. Ineq. Appl. **11**(4), 777–782 (2008)
67. Jung, S.-M., Popa, D., Rassias, Th.M.: On the stability of the linear functional equation in a single variable on complete metric groups. J Glob Optimization (to appear).
68. Kirk, W.A., Sims, B. (eds.): Handbook of Metric Fixed Point Theory. Kluwer, Dordrecht (2001)
69. Krasnoselskii, M.A.: Topological Methods in the Theory of Nonlinear Integral Equations. Pergamon, Oxford (1964)
70. Lakshmikantham, V., Trigiante, D.: Theory of Difference Equations: Numerical Methods and Applications. Academic, New York (1988)
71. Lungu, N., Popa, D.: On the Hyers-Ulam stability of a first order partial differential equations. Carpathian J. Math. **28**(1), 77–82 (2012)
72. Lungu, N., Rus, I.A.: Ulam stability of nonlinear hyperbolic partial differential equations. Carpathian J. Math. **24**, 403–408 (2008)
73. Măruşter, Şt.: The stability of gradient-like methods. Appl. Math. Comput. **117**, 103–115 (2001)
74. Mauldin, R.D., Ulam, S.M.: Mathematical problems and games. Adv. Appl. Math. **8**(3), 281–344 (1987)
75. Mészáros, A.R.: Ulam-Hyers stability of elliptic partial differential equations in Sobolev spaces. International Conference on fixed point theory and its applications, Cluj-Napoca (2012)
76. Miura, T., Jung, S.-M., Takahasi, S.-E.: Hyers-Ulam-Rassias stability of the Banach space valued linear differential equations $y' = \lambda y$. J. Korean Math. Soc. **41**(6), 995–1005 (2004)
77. Moslehian, M.S., Najati, A.: An application of a fixed point theorem to a functional inequality. Fixed Point Theory **10**(1), 141–149 (2009)
78. Moslehian, M.S., Rassias, Th.M.: Stability of functional equations in non-Archimedian spaces. Appl. Anal. Discrete Math. **1**(2), 325–334 (2007)
79. Nikodem, K., Popa, D.: On selection of general linear inclusions. Publ. Math. Debrecen **75**(1–2), 239–249 (2009)
80. Otrocol, D.: Ulam stabilities of differential equation with abstract Volterra operator in a Banach space. Nonlinear Funct. Anal. Appl. **15**(4), 613–619 (2010)

81. Páles, Zs.: Hyers-Ulam stability of the Cauchy functional equation on square-symmetric groupoids. Publ. Math. Debrecen **58**, 651–666 (2001)
82. Paradalos, P., Rassias, Th., Khan, A.A. (eds.): Nonlinear Analysis and Variational Problems. Springer, New York (2009)
83. Park, C.: Fixed points and Hyers-Ulam-Rassias stability of Cauchy-Jensen functional equations in Banach algebras. Fixed Point Theory Appl. Art. ID50175, 15 p. (2007)
84. Park, C.: A fixed point approach to stability of additive functional inequalities in RN-spaces. Fixed Point Theory **12**(2), 429–442 (2011)
85. Park, C., Rassias, Th.M.: Hyers-Ulam stability of a generalized Apollonius type quadratic mapping. J. Math. Anal. Appl. **322**(1), 371–381 (2006)
86. Park, C., Rassias, Th.M.: Homomorphisms and derivations in proper JCQ^*-triples. J. Math. Anal. Appl. **337**(2) 1404–1414 (2008)
87. Petru, T.P., Petruşel, A., Yao, J.-C.: Ulam-Hyers stability for operatorial equations and inclusions via nonself operators. Taiwanese J. Math. **15**(5), 2195–2212 (2011)
88. Petruşel, A.: Operatorial Inclusions. House of the Book of Science, Cluj-Napoca (2002)
89. Pop, V.: Stability of some functional equations defined by quasiarithmetic means. Carpathian J. Math. **28**(1), 151–156 (2012)
90. Popa, D.: Hyers-Ulam stability of the linear recurrence with constant coefficients. Adv. Diff. Eq. **2**, 101–107 (2005)
91. Popa, D.: Hyers-Ulam-Rassias stability of linear recurrence. J. Math. Anal. Appl. **309**, 591–597 (2005)
92. Prastaro, A., Rassias, Th.M.: Ulam stability in geometry of PDEs. Nonlinear Funct. Anal. Appl. **8**, 259–278 2003)
93. Precup, R.: Lecţii de ecuaţii cu derivate parţiale. Presa Univ. Clujeană, Cluj-Napoca (2004)
94. Radu, V.: The fixed point alternative and the stability of functional equations. Fixed Point Theory **4**, 91–96 (2003)
95. Rassias, Th.M.: On the stability of the linear mapping in Banach spaces. Proc. Am. Math. Soc. **72**(2) 297–300 (1978)
96. Rassias, Th.M.: On a modified Hyers–Ulam sequence. J. Math. Anal. Appl. **158**, 106–113 (1991)
97. Rassias, Th.M.: On the stability of the quadratic functional equation and its applications. Studia Univ. "Babeş-Bolyai", Math. **43**(3), 89–124 1998
98. Rassias, Th.M.: The problem of S. M. Ulam for approximately multiplicative mappings. J. Math. Anal. Appl. **246**(2), 352–378 (2000)
99. Rassias, Th.M.: On the stability of functional equations and a problem of Ulam. Acta Appl. Math. **62**(1), 23–130 (2000)
100. Rassias, Th.M.: On the stability of functional equations in Banach spaces. J. Math. Anal. Appl. **251**(1), 264–284 (2000)
101. Rassias, Th.M.: On the stability of functional equations originated by a problem of Ulam. Mathematica **44**(67), no. 1, 39–75 (2002)
102. Rassias, Th.M.: On the stability of minimum points. Mathematica **45**(68), no. 1, 93–104 (2003)
103. Rassias, Th.M. (eds.): Functional Equations, Inequalities and Applications. Kluwer, Dordrecht (2003)
104. Rassias, Th.M., Brzdek, J. (eds.): Functional Equations in Mathematical Analysis. Springer, New York (2012)
105. Rassias, Th.M., Šemrl, P. : On the Hyers-Ulam stability of linear mappings. J. Math. Anal. Appl. **173**, 325–338 (1993)
106. Rassias, Th.M., Tabor, J. (eds.): Stability of Mappings of Hyers-Ulam Type. Handronic Press, Palm Harbor (1994)
107. Reich, S., Zaslawski, A.J.: A stability result in fixed point theory. Fixed Point Theory **6**(1), 113–118 (2005)

108. Rus, I.A.: Ecuaţii diferenţiale, ecuaţii integrale şi sisteme dinamice. Transilvania Press, Cluj-Napoca (1996)
109. Rus, I.A.: An abstract point of view in the nonlinear difference equations. Conference on analysis, functional equations, approximation and convexity in honour of professor Elena Popoviviu, Cluj-Napoca, 272–276 (1999)
110. Rus, I.A.: Generalized Contractions and Applications. Cluj University Press, Cluj-Napoca (2001)
111. Rus, I.A.: Strict fixed point theory. Fixed Point Theory **4**(2), 177–183 (2003)
112. Rus, I.A.: Picard operators and applications. Sci. Math. Japonicae. **58**, 191–219 (2003)
113. Rus, I.A.: Fixed Point Structure Theory. Cluj University Press, Cluj-Napoca (2006)
114. Rus, I.A.: Fixed point theory in partial metric spaces. An. Univ. Vest Timişoara, Mat.-Inform. **46**(2), 149–160 (2008)
115. Rus, I.A.: The theory of a metrical fixed point theorem: theoretical and applicative relevance. Fixed Point Theory **9**(2), 541–559 (2008)
116. Rus, I.A.: Gronwall lemmas: ten open problems. Sci. Math. Japonicae **70**(2), 221–228 (2009)
117. Rus, I.A.: Remarks on Ulam stability of the operatorial equations. Fixed Point Theory **10**(2), 305–320 (2009)
118. Rus, I.A.: Ulam stability of ordinary differential equations. Studia Univ. "Babeş-Bolyai", Math. **54**(4), 125–133 (2009)
119. Rus, I.A.: Gronwall lemma approach to the Hyers-Ulam-Rassias stability of an integral equation. In: Pardalos, P., Rassias, Th.M., Khan, A.A. (eds.) Nonlinear Analysis and Variational Problems. Springer optimization and its applications, vol. 35, pp. 147–152. Springer, New York (2010)
120. Rus, I.A.: Ulam stabilities of ordinary differential equations in a Banach space. Carpathian J. Math. **26**(1), 103–107 (2010)
121. Rus, I.A.: Some nonlinear functional differential and integral equations, via weakly Picard operator theory: a survey. Carpathian J. Math. **26**(2), 230–258 (2010)
122. Rus, I.A.: Ulam stability of the operatorial equations. In: Rassias, Th.M., Brzdek, J. (eds.) Functional Equations in Mathematical Analysis. Springer optimization and its applications, vol. 52, pp. 287–305. Springer, New York (2012)
123. Rus, I.A., Petruşel, A., Sîntămărian, A.: Data dependence of the fixed point set of some multivalued weakly Picard operators. Nonlinear Anal. **52**, 1947–1959 (2003)
124. Rus, I.A., Petruşel, A., Petruşel, G.: Fixed Point Theory. Cluj University Press, Cluj-Napoca (2008)
125. Talpalaru, P.: Quelques problèmes concernant l'équivalence asymptotique des systèmes différentiels. Boll. U.M.I. **4**, 164–186 (1971)
126. Talpalaru, P.: Asymptotic relationship between solutions of two systems of difference equations. Bulet. Inst. Politehnic din Iaşi **21**, fas. 3–4, 49–58 (1975)
127. Terrel, W.J.: Stability and Stabilization. Princeton University Press, Princeton (2009)
128. Tişe, F.A., Tişe, I.C.: Ulam-Hyers-Rassias stability for set integral equations. Fixed Point Theory **13**(2), 659–668 (2012)
129. Trif, T.: On the stability of a functional equation deriving from an inequality of Popoviciu for convex functions. J. Math. Anal. Appl. **272**, 604–616 (2002)
130. Trif, T.: On the stability of a general gamma-type functional equation. Publ. Math. Debrecen **60**, fasc. 1–2, 47–61 (2002)
131. Trif, T.: Hyers-Ulam-Rassias stability of a quadratic functional equation. Bull. Korean Math. Soc. **40**(2), 253–267 (2003)
132. Trif, T.: Hyers-Ulam-Rassias stability of a linear functional equation with constant coefficients. Nonlinear Funct. Anal. Appl. **11**(5), 881–889 (2006)
133. Ulam, S.M.: Problems in modern mathematics. Wiley, New York (1964)
134. Ver Eecke, P.: Applications du calcul différentiel. Presses Univ. de France, Paris (1985)
135. Vrabie, I.I.: C_0-Semigroups and Applications. North-Holland, Amsterdam (2003)

136. Vrabie, I.I.: Differential Equations. World Scientific, New Jersey (2004)
137. Wang, J.: Some further generalizations of the Ulam-Hyers-Rassias stability of functional equations. J. Math. Anal. Appl. **263**, 406–423 (2001)
138. Xu, M.: Hyers-Ulam-Rassias stability of a system of first order linear recurrences. Bull. Korean Math. Soc. **44**(4), 841–849 (2007)
139. Zabrejko, P.: K-metric and K-normed linear spaces: survey. Collect. Math. **48**, 825–859 (1997)

Superstability of Generalized Module Left Higher Derivations on a Multi-Banach Module

T. L. Shateri and Z. Afshari

Abstract The problem of stability of functional equations was originally raised by Ulam in 1940. During the last decades, several stability problems for various functional equations have been investigated by several authors. In this chapter, by defining a multi-Banach space, we introduce a multi-Banach module. Also, we define the notion of generalized module left higher derivations and approximate generalized module left higher derivations. Then, we discuss the superstability of an approximate generalized module left higher derivation on a multi-Banach module. In fact, we show that an approximate generalized module left higher derivation on a multi-Banach module is a generalized module left higher derivation. Finally, we get the similar result for a linear generalized module left higher derivation.

Keywords Superstability · Multi-Banach module · Derivation · Normed space · Group homomorphisms · Bimodule · Multi-Banach space

1 Introduction and Preliminaries

The problem of stability of functional equations was originally raised by Ulam [26] in 1940 concerning the stability of group homomorphisms. Hyers [12] provided an affirmative partial solution to the question of Ulam for the case of approximate additive mappings between banach spaces. Superstability, the result of Hyers was generalized by Aoki [1], Bourgin [4] and Rassias [22]. During the last decades several stability problems for various functional equations have been investigated by several authors. We refer the reader to the monographs [2, 6, 13, 15, 16, 18, 21, 23, 24].

Let $(E, \|.\|)$ be a complex-normed space, and let $k \in \mathbb{N}$. We denote by E^k the linear space $E \oplus \cdots \oplus E$ consisting of k-tuples $(x_1, \cdots, x_k)$, where $x_1, \cdots, x_k \in E$. The linear operations on E^k are defined coordinatewise. The zero element of either

T. L. Shateri (✉) · Z. Afshari
Department of Mathematics and Computer Sciences, P. O. Box 397, Sabzevar, Iran
e-mail: t.shateri@hsu.ac.ir

Z. Afshari
e-mail: afshari1002@yahoo.com

T. M. Rassias (ed.), *Handbook of Functional Equations*,
Springer Optimization and Its Applications 96, DOI 10.1007/978-1-4939-1286-5_16

E or E^k is denoted by 0. We denote by $\mathbb{N}_k$ the set $\{1, 2, \cdots, k\}$ and by C_k the group of permutations on k symbols.

Definition 1 A multi-norm on $\{E^k : k \in \mathbb{N}\}$ is a sequence $(\|.\|_k) = (\|.\|_k : k \in \mathbb{N})$ such that $\|.\|_k$ is a norm on E^k for each $k \in \mathbb{N}$, $\|x\|_1 = \|x\|$ for each $x \in E$, and the following axioms are satisfied for each $k \in \mathbb{N}$ with $k \geq 2$
$(M1)$ $\|(x_{\sigma(1)}, \cdots, x_{\sigma(k)})\|_k = \|(x_1, \cdots, x_k)\|_k \quad (\sigma \in \mathrm{C}_k, x_1, \cdots, x_k \in E)$
$(M2)$ $\|(\alpha_1 x_1, \cdots, \alpha_k x_k)\|_k \leq (\max_{i \in \mathbb{N}_k} |\alpha_i|)\|(x_1, \cdots, x_k)\|_k$
$(\alpha_1, \cdots, \alpha_k \in \vec{C},\ x_1, \cdots, x_k \in E)$
$(M3)$ $\|(x_1, \cdots, x_{k-1}, 0)\|_k = \|(x_1, \cdots, x_{k-1})\|_{k-1} \quad (x_1, \cdots, x_k \in E)$
$(M4)$ $\|(x_1, \cdots, x_{k-1}, x_{k-1})\|_k = \|(x_1, \cdots, x_{k-1})\|_{k-1} \quad (x_1, \cdots, x_k \in E)$.
In this case, we say that $((E^k, \|.\|), k \in \mathbb{N})$ is a multi-normed space.

We recall that the notion of multi-normed space was introduced by H. G. Dales and M. E. Polyakov in [7]. Motivations for the study of multi-normed spaces and many examples are given in [7].

Suppose that $((E^k, \|.\|_k) : k \in \mathbb{N})$ is a multi-normed space, and $k \in \mathbb{N}$. The following properties are almost immediate consequences of the axioms.

(i) $\|(x, \cdots, x)\|_k = \|x\| \quad (x \in E)$
(ii) $max_{i \in \mathbb{N}_k} \|x_i\| \leq \|(x_1, \cdots, x_k)\|_k \leq \sum_{i=1}^{k} \|x_i\| \leq k\, max_{i \in \mathbb{N}_k} \|x_i\| (x_1, \cdots, x_k \in E)$.

It follows from (ii) that, if $(E, \|.\|)$ is a Banach space, then $(E^k, \|.\|_k)$ is a Banach space for each $k \in \mathbb{N}$. In this case, $((E^k, \|.\|_k) : k \in \mathbb{N})$ is a multi-Banach space. By (ii), we get the following lemma.

Lemma 1 *Suppose that* $k \in \mathbb{N}$ *and* $(x_1, \cdots, x_k) \in E^k$. *For each* $j \in \mathbb{N}_k$, *let* $\{x_n^j\}_{n \in \mathbb{N}}$ *be a sequence in* E *such that* $\lim_{n \to \infty} x_n^j = x_j$. *Then, for each* $(y_1, \cdots, y_k) \in E^k$, *we have*

$$\lim_{n \to \infty} (x_n^1 - y_1, \cdots, x_n^k - y_k) = (x_1 - y_1, \cdots, x_k - y_k).$$

Definition 2 Let $((E^k, \|.\|_k) : k \in \mathbb{N})$ be a multi-normed space. A sequence $\{x_n\}$ in E is a multi-null sequence if, for each $\epsilon > 0$, there exists $n_0 \in \mathbb{N}$ such that

$$\sup_{k \in \mathbb{N}} \|(x_n, \cdots, x_{n+k-1})\|_k < \epsilon \quad (n \geq n_0).$$

Let $x \in E$. We say that $\lim_{n \to \infty} x_n = x$ if $\{x_n - x\}$ is a multi-null sequence.

Definition 3 Let $(\mathcal{A}, \|.\|)$ be a normed algebra such that $((\mathcal{A}^k, \|.\|_k) : k \in \mathbb{N})$ is a multi-normed space. Then, $((\mathcal{A}^k, \|.\|_k) : k \in \mathbb{N})$ is a multi-normed algebra if

$$\|(a_1 b_1, \cdots, a_k b_k)\|_k \leq \|(a_1, \cdots, a_k)\|_k \|(b_1, \cdots, b_k)\|_k \tag{1}$$

for $k \in \mathbb{N}$ and $x_1, \cdots, x_k, y_1, \cdots, y_k \in \mathcal{A}$. Furthermore, if $((\mathcal{A}^k, \|.\|_k) : k \in \mathbb{N})$ is a multi-Banach space, then $((\mathcal{A}^k, \|.\|_k) : k \in \mathbb{N})$ is a multi-Banach algebra. Let $\mathcal{X}$ be a Banach $\mathcal{A}$-bimodule such that $((\mathcal{X}^k, \|.\|_k) : k \in \mathbb{N})$ is a multi-normed space, then

$((\mathcal{A}^k, \|.\|_k)k \in \mathbb{N})$ is said to be a multi-Banach $\mathcal{A}$-bimodule if there is a non-negative number M

$$\|(a_1x_1, \cdots, a_kx_k)\|_k \leq M\|(a_1, \cdots, a_k)\|_k\|(x_1, \cdots, x_k)\|_k, \tag{2}$$

$$\|(x_1a_1, \cdots, x_ka_k)\|_k \leq M\|(a_1, \cdots, a_k)\|_k\|(x_1, \cdots, x_k)\|_k \tag{3}$$

for $k \in \mathbb{N}$ and $a_1, \cdots, a_k \in \mathcal{A}, x_1, \cdots, x_k \in \mathcal{X}$.

Let $\mathcal{A}$ be an algebra and $k_0 \in \{0, 1, \cdots, \} \cup \{\infty\}$. A family $\{D_j\}_{j=0}^{k_0}$ of linear mappings on $\mathcal{A}$ is said to be a *higher derivation* of rank k_0 if the functional equation $D_j(xy) = \sum_{i=0}^{j} D_i(x)D_{j-i}(y)$, holds for all $x, y \in \mathcal{A}$, $j = 0, 1, 2, \ldots, k_0$. If $D_0 = id_{\mathcal{A}}$, where $id_{\mathcal{A}}$ is the identity map on $\mathcal{A}$, then D_1 is a derivation and $\{D_j\}_{j=0}^{k_0}$ is called a *strongly* higher derivation. A standard example of a higher derivation of rank k_0 is $\{\frac{D^j}{j!}\}_{j=0}^{k_0}$ where $D : \mathcal{A} \to \mathcal{A}$ is a derivation. The reader may find more information about higher derivations in [3, 8–11, 14, 27].

Let $\mathcal{A}$ be an algebra over the real or complex field $\vec{F}$ and $\mathcal{X}$ be an $\mathcal{A}$-bimodule.

Definition 4 A family $\{\delta_j\}_{j=0}^{k_0}$ of mappings from $\mathcal{A}$ into $\mathcal{A}$ is called a *module-$\mathcal{X}$ additive* if

$$x\delta_j(a + b) = x\delta_j(a) + x\delta_j(b) \quad (a, b \in \mathcal{A}, x \in \mathcal{X}, j = 0, 1, 2, \cdots, k_0). \tag{4}$$

A module-$\mathcal{X}$ additive family $\{\delta_j\}_{j=0}^{k_0}$ is called a *module-$\mathcal{X}$* left higher derivation (resp., *module-$\mathcal{X}$* higher derivation) if $\delta_0 = id_{\mathcal{A}}$ and for all $1 \leq j \leq k_0$

$$x\delta_j(ab) = ax\delta_j(b) + bx\delta_j(a) + \sum_{i=1}^{j-1} x\delta_i(b)\delta_{j-i}(a) \quad (a, b \in \mathcal{A}, x \in \mathcal{X}) \tag{5}$$

(resp.,

$$x\delta_j(ab) = \sum_{i=0}^{j} \delta_i(a)x\delta_{j-i}(b) \quad (a, b \in \mathcal{A}, x \in \mathcal{X})) \tag{6}$$

holds.

Definition 5 A family $\{f_j\}_{j=0}^{k_0}$ of mappings from $\mathcal{X}$ into $\mathcal{X}$ is called a *module-$\mathcal{A}$-additive* if

$$af_j(x + y) = af_j(x) + af_j(y) \quad (a \in \mathcal{A}, x, y \in \mathcal{X}, j = 0, 1, 2, \cdots, k_0) \tag{7}$$

A module-$\mathcal{A}$ additive family $\{f_j\}_{j=0}^{k_0}$ is called a *generalized module-$\mathcal{A}$* left higher derivation (resp., *generalized module-$\mathcal{A}$* higher derivation) if $f_0 = id_{\mathcal{A}}$ and there exists a module-$\mathcal{X}$ left higher derivation $\{\delta_j\}_{j=0}^{k_0}$ such that for all $1 \leq j \leq k_0$

$$af_j(bx) = abf_j(x) + a\sum_{i=1}^{j} f_{j-i}(x)\delta_i(b) \quad (a, b \in \mathcal{A}, x \in \mathcal{X}). \tag{8}$$

(resp.,

$$af_j(bx) = abf_j(x) + a\sum_{i=1}^{j} \delta_i(b)f_{j-i}(x) \quad (a, b \in \mathcal{A}, x \in \mathcal{X})). \tag{9}$$

Remark 1 If $\mathcal{X} = \mathcal{A}$ is a unital algebra or a Banach algebra with an approximate unit, then module-$\mathcal{A}$ left higher derivations, module-$\mathcal{A}$ higher derivations, generalized module-$\mathcal{A}$ left higher derivations and generalized module-$\mathcal{A}$ higher derivations on $\mathcal{A}$ become left higher derivations, higher derivations, generalized left higher derivations and generalized higher derivations on $\mathcal{A}$. Superstability of generalized higher derivations discussed in [25].

The stability of derivations was studied by C.-G.Park [19, 20]. In this chapter, using some ideas from [17, 5], we investigate the superstability of generalized module left higher derivations in multi-Banach algebras.

2 Main Results

In this section, we define the notion of an approximate generalized module-$\mathcal{A}$ left higher derivation. Then, we show that an approximate generalized left higher derivation on a multi-Banach algebra is a generalized module-$\mathcal{A}$ left higher derivation.

Lemma 2 *Let $((E^k, \|.\|_k)k \in \mathbb{N})$ be a multi-Banach space. Let $\psi : E \times E \to [0, \infty)$ satisfies the following conditions*

(i) $\lim_{n\to\infty} t^{-n}\psi(\sum_{i=1}^{k} t^n x_i, \sum_{i=1}^{k} t^n y_i) = 0,$
(ii) $\tilde{\psi}(x_1, \cdots, x_k) = \sum_{n=0}^{\infty} t^{1-n}\psi(\sum_{i=1}^{k} t^n x_i, 0) < \infty$
for $x_1, \cdots, x_k, y_1, \cdots, y_k \in E$. Suppose that $f : E \to E$ is a mapping satisfying $f(0) = 0$ and

$$\sup_{k\in\bar{N}} \left\| \left(f\left(\frac{x_1}{t} + \frac{y_1}{l}\right) + f\left(\frac{x_1}{t} - \frac{y_1}{l}\right) - \frac{2f(x_1)}{t}, \cdots, f\left(\frac{x_k}{t} + \frac{y_k}{l}\right) \right.\right.$$
$$\left.\left. + f\left(\frac{x_k}{t} - \frac{y_k}{l}\right) - \frac{2f(x_k)}{t} \right) \right\|_k$$
$$\leq \psi\left(\sum_{i=1}^{k} x_i, \sum_{i=1}^{k} y_i\right), \tag{10}$$

for all integer $t, l > 1$ and all $x_1, \cdots, x_k, y_1, \cdots, y_k \in E$. Then, there exists an additive mapping $d : E \to E$ such that

$$\|(f(x_1) - d(x_1), \cdots, f(x_k) - d(x_k))\| \leq \tilde{\psi}(x_1, \cdots, x_k) \tag{11}$$

for $x_1, \cdots, x_k \in E$.

Proof Substituting $y_i = 0$ for $i = 1, \cdots, k$ and replacing x_i by $tx_1, \cdots, tx_k$ in (10), we get

$$\sup_{k\in\mathbb{N}} \left\| \left(f(x_1) - \frac{f(tx_1)}{t}, \cdots, f(x_k) - \frac{f(tx_k)}{t} \right) \right\|_k \leq \frac{1}{2}\psi(tx_1 + \cdots + tx_k, 0). \quad (12)$$

From (12), we have that

$$\sup_{k\in\mathbb{N}} \left\| \left(f(x_1) - \frac{f(t^2x_1)}{t^2}, \cdots, f(x_k) - \frac{f(t^2x_k)}{t^2} \right) \right\|_k \leq \frac{1}{2}\psi\left(\sum_{i=1}^{k} tx_i, 0\right)$$
$$+ \frac{1}{2}t^{-1}\psi\left(\sum_{i=1}^{k} t^2x_i, 0\right). \quad (13)$$

An induction argument implies that

$$\sup_{k\in\mathbb{N}} \left\| \left(f(x_1) - \frac{f(t^nx_1)}{t^n}, \cdots, f(x_k) - \frac{f(t^nx_k)}{t^n} \right) \right\|_k \leq \frac{1}{2}\sum_{j=1}^{n} t^{1-j}\psi\left(\sum_{i=1}^{k} t^jx_i, 0\right) \quad (14)$$

for $x_1, \cdots, x_k \in E$ and $n \in \mathbb{N}$. Let $n > m$, then by (14) and condition (ii), we obtain that

$$\sup_{k\in\vec{N}} \left\| \left(\frac{f(t^nx_1)}{t^n} - \frac{f(t^mx_1)}{t^m}, \cdots, \frac{f(t^nx_k)}{t^n} - \frac{f(t^mx_k)}{t^m} \right) \right\|_k$$
$$\leq \frac{1}{t^m}\sup_{k\in\vec{N}} \left\| \left(\frac{f(t^{n-m}t^mx_1)}{t^{n-m}} - f(t^mx_1), \cdots, \frac{f(t^{n-m}t^mx_k)}{t^{n-m}} - f(t^mx_k) \right) \right\|_k$$
$$\leq \frac{1}{t^m}\frac{1}{2}\sum_{j=1}^{n-m} t^{1-j}\psi\left(\sum_{i=1}^{k} t^j.t^mx_i, 0\right)$$
$$\leq \frac{1}{2}\sum_{j=m}^{\infty} t^{1-j}\psi\left(\sum_{i=1}^{k} t^jx_i, 0\right) \to 0\ (m \to \infty).$$

Hence, the sequence $\{\frac{f(t^nx)}{t^n}\}$ is a cauchy sequence in the multi-Banach space E and therefore converges for all $x \in E$. Put $d(x) = \lim_{n\to\infty}\frac{f(t^nx)}{t^n}$ $(x \in E)$, so $d(0) = f(0) = 0$. By (14), we get

$$\sup_{k\in\mathbb{N}} \left\| \left(\frac{f(t^nx_1)}{t^n} - d(x_1), \cdots, \frac{f(t^nx_k)}{t^n} - d(x_k) \right) \right\|_k \leq \tilde{\psi}(x_1, \cdots, x_k). \quad (15)$$

In particular, the property (ii) of multi-norm implies that

$$\lim_{n\to\infty} \|\frac{f(t^nx)}{t^n} - d(x)\| = 0 \quad (x \in E). \quad (16)$$

Now, we show that d is additive. To do this, let $x, y \in E$ put $x_1 = \cdots = x_k = t^n x, y_1 = \cdots = y_k = t^n y$ in (10). Therefore

$$\left\| t^{-n} f\left(\frac{t^n x}{t} + \frac{t^n y}{l}\right) + t^{-n} f\left(\frac{t^n x}{t} - \frac{t^n y}{l}\right) - \frac{1}{t} 2\frac{f(t^n x)}{t^n} \right\| \leq t^{-n} \psi\left(\sum_{i=1}^{k} t^n x, \sum_{i=1}^{k} t^n y\right). \tag{17}$$

By letting $n \to \infty$, the condition (i) yields that

$$d\left(\frac{x}{t} + \frac{y}{l}\right) + d\left(\frac{x}{t} - \frac{y}{l}\right) = \frac{2}{t} d(x) \tag{18}$$

for all $x, y \in E$. Since $d(0) = 0$, taking $y = 0$ and $y = \frac{l}{t} x$, respectively, we get $d(\frac{2x}{t}) = 2d(\frac{x}{t})$, and hence, $d(2x) = 2d(x)$ for all $x \in E$. We obtain that $d(x + y) + d(x - y) = 2d(x)$ for all $x, y \in E$. Now, for all $z, w \in E$, put $x = \frac{t}{2}(z + w), y = \frac{l}{2}(z - w)$. Then by (18), we see that

$$d(z) + d(w) = d\left(\frac{x}{t} + \frac{y}{l}\right) + d\left(\frac{x}{t} - \frac{y}{l}\right) = \frac{2}{t} d(x) = \frac{2}{t} d\left(\frac{t}{2}(z + w)\right) = d(z + w). \tag{19}$$

Therefore, d is additive.

Definition 6 Let $((\mathcal{A}^k, \|.\|_k) : k \in \mathbb{N})$ be a multi-Banach algebra and $((\mathcal{X}^k, \|.\|_k)k \in \mathbb{N})$ a multi-Banach $\mathcal{A}$-bimodule. Suppose that $t, l > 1$ are integers, $0 < \alpha < t$ and $\psi : \mathcal{X} \times \mathcal{X} \times \mathcal{A} \times \mathcal{X} \to [0, \infty)$ satisfy the following conditions
(i) $\lim_{n\to\infty} t^{-n}[\psi(\sum_{i=1}^k t^n x_i, \sum_{i=1}^k t^n y_i, 0, 0) + \psi(0, 0, \sum_{i=1}^k t^n a_i, \sum_{i=1}^k z_i)] = 0$
(ii) $\tilde{\psi}(x_1, \cdots, x_k) = \sum_{n=0}^{\infty} t^{-n+1} \psi(\sum_{i=1}^k t^n x_i, 0, 0, 0) < \infty$
(iii) $\lim_{n\to\infty} t^{-2n} \psi(0, 0, \sum_{i=1}^k t^n a_i, \sum_{i=1}^k t^n z_i)] = 0$
for all $x_1, \cdots, x_k, y_1, \cdots, y_k, z_1, \cdots, z_k \in \mathcal{X}, a_1, \cdots, a_k \in \mathcal{A}$ and
(iv) $\psi(0, 0, t^n a, t^m x) \leq \alpha^{n+m} \psi(0, 0, a, x) \quad (a \in \mathcal{A}, x \in \mathcal{X}, m, n \in \mathbb{N})$.

Let $\{f_j : \mathcal{X} \to \mathcal{X}\}_{j=0}^{k_0}$ and $\{g_j : \mathcal{A} \to \mathcal{A}\}_{j=0}^{k_0}$ be two families of mappings such that $f_j(0) = 0$ and $\delta_j(a) := \lim_{n\to\infty} \frac{1}{t^n} g_j(t^n a)$ exists for all $a \in \mathcal{A}$ and $0 < j \leq k_0$. If for each $0 < j \leq k_0$

$$\begin{aligned} \sup_{k\in\vec{N}} \Big\| \Big(f_j &\left(\frac{x_1}{t} + \frac{y_1}{l} + a_1 z_1\right) + f_j\left(\frac{x_1}{t} - \frac{y_1}{l} + a_1 z_1\right) - 2\frac{f_j(x_1)}{t} - 2a_1 f_j(z_1) \\ &- 2z_1 g_j(a_1), \cdots, f_j\left(\frac{x_k}{t} + \frac{y_k}{l} + a_k z_k\right) + f_j\left(\frac{x_k}{t} - \frac{y_k}{l} + a_k z_k\right) - 2\frac{f_j(x_k)}{t} \\ &- 2a_k f_j(z_k) - 2z_k g_j(a_k)\Big)\Big\|_k \leq \psi\left(\sum_{i=1}^k x_i, \sum_{i=1}^k y_i, \sum_{i=1}^k a_i, \sum_{i=1}^k z_i\right) \end{aligned} \tag{20}$$

for all $x_1, \cdots, x_k, y_1, \cdots, y_k, z_1, \cdots, z_k \in \mathcal{X}, a_1, \cdots, a_k \in \mathcal{A}$ and

$$\left\| a f_j(bx) - ab f_j(x) - a \sum_{i=1}^{j} f_{j-i}(x) g_i(b) \right\| \leq \psi(0, 0, b, x) \tag{21}$$

for all $x \in \mathcal{X}, a, b \in \mathcal{A}$. Then, $\{f_j\}_0^{k_0}$ and $\{g_j\}_0^{k_0}$ are called (ψ, α)-approximate generalized module-$\mathcal{A}$ left higher derivation and (ψ, α)-approximate module-$\mathcal{X}$ left higher derivation, respectively.

In the following theorem, we show that a (ψ, α)-approximate generalized module-$\mathcal{A}$ left higher derivation on a multi-Banach $\mathcal{A}$-bimodule is a generalized module left higher derivation.

Theorem 1 *Let $\mathcal{A}$ be a Banach algebra with unit e and $\mathcal{X}$ a Banach $\mathcal{A}$-bimodule. Suppose that $\{f_j\}_0^{k_0}$ is a (ψ, α)-approximate generalized module-$\mathcal{A}$ left higher derivation on the multi-Banach $\mathcal{A}$-bimodule $((\mathcal{X}^k, \|.\|_k) : k \in \mathbb{N})$ and $\{g_j\}_0^{k_0}$ is a (ψ, α)-approximate module-$\mathcal{X}$ left higher derivation on the multi-Banach algebra $((\mathcal{A}^{||}, \|.\|_{||}) : || \in \mathbb{N})$. Then, $\{f_j\}_0^{k_0}$ is a generalized module-$\mathcal{A}$ left higher derivation and $\{g_j\}_0^{k_0}$ is a module-$\mathcal{X}$ left higher derivation.*

Proof Letting $a_i = z_i = 0$ for $i = 1, \cdots k$ in (20), Lemma 2 implies that for each $0 < j \leq k_0$, there exists an additive mapping d_j on $\mathcal{X}$ defined by $d_j(x) = \lim_{n\to\infty} \frac{f_j(t^n x)}{t^n}$ such that

$$\|(f(x_1) - d(x_1), \cdots, f(x_k) - d(x_k))\| \leq \tilde{\psi}(x_1, \cdots, x_k) \tag{22}$$

for $x_1, \cdots, x_k \in E$ and $d_j(0) = f_j(0) = 0$. If $j = 1$ [5, Theorem 2.1] implies that f_1 is a generalized module-$\mathcal{A}$ left derivation and δ_1, g_1 are module-$\mathcal{X}$ left derivation on $\mathcal{A}$ and for all $a, b \in \mathcal{A}$ and $x \in \mathcal{X}$

$$af_1(bx) = abf_1(x) + ax\delta_1(b), \quad d_1(x) = f_1(x), \tag{23}$$
$$xg_1(ab) = axg_1(b) + bxg_1(a), \quad xg_1(a) = x\delta_1(a).$$

By induction for $1 \leq j \leq k_0 - 1$, we assume that

$$af_j(bx) = abf_j(x) + a\sum_{i=1}^{j} f_{j-i}(x)\delta_i(b), \; d_j(x) = f_j(x)$$

$$x\delta_j(ab) = ax\delta_j(b) + bx\delta_j(a) + \sum_{i=1}^{j-1} x\delta_i(b)\delta_{j-i}(a)$$

$$xg_j(ab) = axg_j(b) + bxg_j(a) + x\sum_{i=1}^{j-1} g_{j-i}(b)g_i(a), \; xg_j(a) = x\delta_j(a) \tag{24}$$

and each δ_j is module-$\mathcal{X}$ additive. Then, we prove that

$$af_{k_0}(bx) = abf_{k_0}(x) + a\sum_{i=1}^{k_0} f_{k_0-i}(x)\delta_i(b), \; d_{k_0}(x) = f_{k_0}(x) \tag{25}$$

$$x\delta_{k_0}(ab) = ax\delta_{k_0}(b) + bx\delta_{k_0}(a) + \sum_{i=1}^{k_0-1} x\delta_i(b)\delta_{k_0-i}(a)$$

$$xg_{k_0}(ab) = axg_{k_0}(b) + bxg_{k_0}(a) + x\sum_{i=1}^{k_0-1} g_{k_0-i}(b)g_i(a), \; xg_{k_0}(a) = x\delta_{k_0}(a).$$

Let $b \in \mathcal{A}, x \in \mathcal{X}$, by replacing b with $t^n b$, x with $t^n x$ in (21), respectively and $a = e$, we obtain

$$\left\| \frac{1}{t^{2n}} f_{k_0}(t^{2n}bx) - \frac{b}{t^n} f_{k_0}(t^n x) - \frac{1}{t^{2n}} \sum_{i=1}^{k_0} f_{k_0 - i}(t^n x) g_{k_0}(t^n b) \right\|$$

$$\leq \frac{1}{t^{2n}} \psi(0,0,t^n b, t^n x) \leq \left(\frac{\alpha}{t}\right)^{2n} \psi(0,0,b,x) \to 0 \quad (n \to \infty). \tag{26}$$

Therefore,

$$d_{k_0}(bx) = b d_{k_0}(x) + \sum_{i=1}^{k_0} d_{k_0 - i}(x) \delta_i(b) \tag{27}$$

for all $b \in \mathcal{A}$ and $x \in \mathcal{X}$. Since d_{k_0} is additive and δ_s $(1 \leq s \leq k_0 - 1)$ are module-$\mathcal{X}$ additive, δ_{k_0} is module-$\mathcal{X}$ additive. Put $F_{k_0}(b,x) = f_{k_0}(bx) - b f_{k_0}(x) - \sum_{i=1}^{k_0} f_{k_0-i}(x) g_i(b)$, then we see that

$$\frac{1}{t^n} \| F_{k_0}(t^n b, x) \| \leq \left(\frac{\alpha}{t}\right)^n \psi(0,0,b,x) \to 0 \quad (n \to \infty) \tag{28}$$

for all $b \in \mathcal{A}$ and $x \in \mathcal{X}$. Hence

$$d_{k_0}(bx) = \lim_{n\to\infty} \frac{f_{k_0}(t^n bx)}{t^n} = \lim_{n\to\infty} \left(\frac{F_{k_0}(t^n b, x) + t^n b f_{k_0}(x) + \sum_{i=1}^{k_0} f_{k_0-i}(x) g_i(t^n b)}{t^n} \right)$$

$$= b f_{k_0}(x) + \sum_{i=1}^{k_0} f_{k_0-i}(x) \delta_i(b) = b f_{k_0}(x) + \sum_{i=1}^{k_0} d_{k_0-i}(x) \delta_i(b).$$

It follows from (27) that $b d_{k_0}(x) = b f_{k_0}(x)$ for all $b \in \mathcal{A}$ and $x \in \mathcal{X}$, and hence, $d_{k_0}(x) = f_{k_0}(x)$ for all $x \in \mathcal{X}$. Since d_{k_0} is additive, f_{k_0} is module-$\mathcal{A}$ additive. Therefore, for all $a, b \in \mathcal{A}$ and $x \in \mathcal{X}$,

$$a f_{k_0}(bx) = a d_{k_0}(bx) = ab f_{k_0}(x) + a \sum_{i=1}^{k_0} f_{k_0-i}(x) \delta_i(b)$$

and then (24) implies that

$$x \delta_{k_0}(ab) = f_{k_0}(abx) - ab f_{k_0}(x) - \sum_{i=1}^{k_0-1} f_{k_0-i}(x) \delta_i(ab)$$

$$= a f_{k_0}(bx) + bx \delta_{k_0}(a) + \sum_{i=1}^{k_0-1} f_{k_0-i}(bx) \delta_i(a) - ab f_{k_0}(x) - \sum_{i=1}^{k_0-1} f_{k_0-i}(x) \delta_i(ab)$$

$$= ab f_{k_0}(x) + ax \delta_{k_0}(b) + a \sum_{i=1}^{k_0-1} f_{k_0-i}(x) \delta_i(b) + bx \delta_{k_0}(a)$$

$$+\sum_{i=1}^{k_0-1}\left[bf_{k_0-i}(x)+x\delta_{k_0-i}(b)+\sum_{k=1}^{k_0-i-1}f_{k_0-i-k}(x)\delta_k(b)\right]\delta_i(a)$$

$$-abf_{k_0}(x)-\sum_{i=1}^{k_0-1}[af_{k_0-i}(x)\delta_i(b)+bf_{k_0-i}(x)\delta_i(a)$$

$$+sum_{k=1}^{i-1}f_{k_0-i}(x)\delta_k(b)\delta_{i-k}(b)]$$

$$=ax\delta_{k_0}(b)+bx\delta_{k_0}(a)+\sum_{i=1}^{k_0-1}x\delta_{k_0-i}(b)\delta_i(a)+\sum_{i=1}^{k_0-1}\sum_{k=1}^{k_0-i-1}f_{k_0-k-i}(x)\delta_k(b)\delta_i(a)$$

$$-\sum_{i=1}^{k_0-1}\sum_{k=1}^{i-1}f_{k_0-i}(x)\delta_k(b)\delta_{i-k}(a)=ax\delta_{k_0}(b)+bx\delta_{k_0}(a)+\sum_{i=1}^{k_0-1}x\delta_{k_0-i}(b)\delta_i(a).$$

This shows that $\{\delta_j\}_{j=0}^{k_0}$ is a module-$\mathcal{X}$ left higher derivation on $\mathcal{A}$ and then $\{f_j\}_{j=0}^{k_0}$ is a generalized module-$\mathcal{A}$ left higher derivation on $\mathcal{X}$. Finally, we prove that $\{g_j\}_{j=0}^{k_0}$ is a module-$\mathcal{X}$ left higher derivation on $\mathcal{A}$. We conclude from (21) and the condition (i) that

$$\left\|\frac{1}{t^n}f_{k_0}(t^nbx)-\frac{b}{t^n}f_{k_0}(t^nx)-\frac{1}{t^n}\sum_{i=1}^{k_0}f_{k_0-i}(t^nx)g_{k_0}(b)\right\|$$

$$\leq\frac{1}{t^n}\psi(0,0,b,t^nx)\leq(\frac{\alpha}{t})^n\psi(0,0,b,x)\to 0,$$

as $n\to\infty$. Therefore,

$$d_{k_0}(bx)=bd_{k_0}(x)+\sum_{i=1}^{k_0}d_{k_0-i}(x)g_i(b)$$

for all $b\in\mathcal{A}$ and $x\in\mathcal{X}$. Now, the induction assumptions in (24) and Eq. (27) implies that $xg_{k_0}(a)=x\delta_{k_0}(a)$, for all $b\in\mathcal{A}$ and $x\in\mathcal{X}$. Therefore, $\{g_j\}_{j=0}^{k_0}$ is a module-$\mathcal{X}$ left higher derivation on $\mathcal{A}$, and this completes the proof.

Remark 2 A typical example of the function ψ in Theorem 1 is

$$\psi(x,y,a,z)=\epsilon(\|x^p\|+\|y\|^q+\|a\|^r\|z\|^s)$$

in which $\epsilon\geq 0$ and $p,q,r,s\in[0,1)$.

If we put $\mathcal{X}=\mathcal{A}$ in Theorem 1, we get the next corollary.

Corollary 1 *Let $\mathcal{A}$ be a Banach algebra with unit $e,\epsilon\geq 0$ and $t,l\geq 1$ be integers. Suppose that $\{f_j\}_{j=0}^{k_0}$ and $\{g_j\}_{j=0}^{k_0}$ are two family mappings from $\mathcal{A}$ into $\mathcal{A}$ with $f_j(0)=0$ $(0\leq j\leq k_0)$ such that*

$$\sup_{k\in\vec{N}}\Big\|\Big(f_j\Big(\frac{x_1}{t}+\frac{y_1}{l}+a_1z_1\Big)+f_j\Big(\frac{x_1}{t}-\frac{y_1}{l}+a_1z_1\Big)-2\frac{f_j(x_1)}{t}-2a_1f_j(z_1)-2z_1g_j(a_1)$$
$$,\cdots,f_j\Big(\frac{x_k}{t}+\frac{y_k}{l}+a_kz_k\Big)+f_j\Big(\frac{x_k}{t}-\frac{y_k}{l}+a_kz_k\Big)-2\frac{f_j(x_k)}{t}$$
$$-2a_kf_j(z_k)-2z_kg_j(a_k)\Big)\Big\|_k\le\epsilon, \tag{29}$$

for all $x_1,\cdots,x_k,y_1,\cdots,y_k,z_1,\cdots,z_k,a_1,\cdots,a_k\in\mathcal{A}$ *and*

$$\Big\|af_j(bx)-abf_j(x)-a\sum_{i=1}^{j}f_{j-i}(x)g_i(b)\Big\|\le\epsilon \tag{30}$$

for all $x,a,b\in\mathcal{A}$. *Then,* $\{f_j\}_0^{k_0}$ *is a generalized left higher derivation and* $\{g_j\}_0^{k_0}$ *is a left higher derivation.*

Proof [5, Corollary 2.3] implies that f_1 is a generalized left derivation and g_1 is a left derivation. By induction, let the result holds for $\{f_j\}_0^{k_0-1}$ and $\{g_j\}_0^{k_0-1}$. If we put $\psi(x,y,a,z)=\epsilon$ in Theorem 1 and $a=e$ in (25), then we have

$$f_j(bx)=bf_j(x)+\sum_{i=1}^{j}f_{j-i}(x)\delta_i(b),\quad g_j(ab)=ag_j(b)+bg_j(a)+\sum_{i=1}^{j-1}g_i(b)g_{j-i}(a)$$

for all $a,b,x\in\mathcal{A}$. This completes the proof.

With the help of Theorem 1, we get the following result for a linear generalized module left higher derivation.

Theorem 2 *Let* $\mathcal{A}$ *be a Banach algebra with unit e and* $\mathcal{X}$ *a Banach* $\mathcal{A}$*-bimodule. Suppose that* $\{f_j\}_0^{k_0}$ *is a* (ψ,α)*-approximate generalized module-*$\mathcal{A}$ *left higher derivation on the multi-Banach* $\mathcal{A}$*-bimodule* $((\mathcal{X}^k,\|.\|_k)k\in\mathbb{N})$ *and* $\{g_j\}_0^{k_0}$ *is a* (ψ,α)*-approximate module-*$\mathcal{X}$ *left higher derivation on the multi-Banach algebra* $((\mathcal{A}^{||},\|.\|_{||})||\in\mathbb{N})$ *such that*

$$\sup_{k\in\mathbb{N}}\Big\|\Big(f_j\Big(\frac{\beta x_1}{t}+\frac{\gamma y_1}{l}+a_1z_1\Big)+f_j\Big(\frac{\beta x_1}{t}-\frac{\gamma y_1}{l}+a_1z_1\Big)-\frac{2\beta f_j(x_1)}{t}$$
$$-2a_1f_j(z_1)-2z_1g_j(a_1)$$
$$,\cdots,f_j\Big(\frac{\beta x_k}{t}+\frac{\gamma y_k}{l}+a_kz_k\Big)+f_j\Big(\frac{\beta x_k}{t}-\frac{\gamma y_k}{l}+a_kz_k\Big)-\frac{2\beta f_j(x_k)}{t}$$
$$-2a_kf_j(z_k)-2z_kg_j(a_k)\Big)\Big\|_k$$
$$\le\psi\Big(\sum_{i=1}^{k}x_i,\sum_{i=1}^{k}y_i,\sum_{i=1}^{k}a_i,\sum_{i=1}^{k}z_i\Big) \tag{31}$$

for all $x_1,\cdots,x_k,y_1,\cdots,y_k,z_1,\cdots,z_k\in\mathcal{X},a_1,\cdots,a_k\in\mathcal{A}$ *and all* $\beta,\gamma\in\vec{T}=\{z\in\vec{C}:|z|=1\}$. *Then,* $\{f_j\}_0^{k_0}$ *is a linear generalized module-*$\mathcal{A}$ *left higher derivation and* $\{g_j\}_0^{k_0}$ *is a linear module-*$\mathcal{X}$ *left higher derivation.*

Proof It is clear that the inequality (20) is satisfied. Theorem (1) shows that $\{f_j\}_0^{k_0}$ is a generalized module-$\mathcal{A}$ left higher derivation and $\{g_j\}_0^{k_0}$ is a module-$\mathcal{X}$ left higher derivation with

$$f_j(x) = \lim_{n\to\infty} \frac{f_j(t^n x)}{t^n}, \quad g_j(b) = f_j(b) - bf_j(e) - \sum_{i=1}^{j-1} f_{j-i}(e)g_i(b) \tag{32}$$

for all $x \in \mathcal{X}, b \in \mathcal{A}$ and $j = 1, \cdots, k_0$. Taking $a_i = z_i = 0$ and replacing x_i and y_i with $t^n x$ and $t^n y$ for $i = 1, \cdots k$ and $x, y \in \mathcal{X}$ in (31), respectively, then we see that

$$\left\| \frac{1}{t^n} f_j\left(\frac{\beta t^n x}{t} + \frac{\gamma t^n y}{l}\right) + \frac{1}{t^n} f_j\left(\frac{\beta t^n x}{t} - \frac{\gamma t^n y}{l}\right) - \frac{1}{t^n}\frac{2\beta f_j(t^n x)}{t} \right\|$$
$$\leq \frac{1}{t^n}\psi\left(kt^n x, kt^n y, 0, 0\right) \to 0$$

as $n \to \infty$, for all $x, y \in \mathcal{X}$ and all $\beta, \gamma \in \vec{T}$. Therefore,

$$f_j\left(\frac{\beta x}{t} + \frac{\gamma y}{l}\right) + f_j\left(\frac{\beta x}{t} - \frac{\gamma y}{l}\right) = \frac{2\beta f_j(x)}{t} \tag{33}$$

for all $x, y \in \mathcal{X}$ and all $\beta, \gamma \in \vec{T}$. Taking $y = 0$ in (33) implies that $f_j(\beta x) = \beta f_j(x)$ for all $x \in \mathcal{X}$ and all $\beta \in \vec{T}$. Since f_j is additive, [5, Lemma 2.4] shows that f_j is linear and (32) yields that g_j is linear.

Employing the similar way as in the proof of Theorem 2, we get the next corollary for a linear generalized left higher derivation.

Corollary 2 *Let $\mathcal{A}$ be a Banach algebra with unit e, $\epsilon \geq 0$ and $t, l \geq 1$ be integers. Suppose that $\{f_j\}_{j=0}^{k_0}$ and $\{g_j\}_{j=0}^{k_0}$ are two family mappings from $\mathcal{A}$ into $\mathcal{A}$ with $f_j(0) = 0$ $(0 \leq j \leq k_0)$ such that*

$$\sup_{k\in\mathbb{N}} \Bigg\| \left(f_j\left(\frac{\beta x_1}{t} + \frac{\gamma y_1}{l} + a_1 z_1\right) + f_j\left(\frac{\beta x_1}{t} - \frac{\gamma y_1}{l} + a_1 z_1\right) - \frac{2\beta f_j(x_1)}{t}\right.$$
$$- 2a_1 f_j(z_1) - 2z_1 g_j(a_1)$$
$$, \cdots, f_j\left(\frac{\beta x_k}{t} + \frac{\gamma y_k}{l} + a_k z_k\right) + f_j\left(\frac{\beta x_k}{t} - \frac{\gamma y_k}{l} + a_k z_k\right) - \frac{2\beta f_j(x_k)}{t}$$
$$\left. - 2a_k f_j(z_k) - 2z_k g_j(a_k)\right)\Bigg\|_k \leq \epsilon, \tag{34}$$

for all $x_1, \cdots, x_k, y_1, \cdots, y_k, z_1, \cdots, z_k, a_1, \cdots, a_k \in \mathcal{A}$ and $\beta, \gamma \in \vec{T}$. If

$$\|af_j(bx) - abf_j(x) - a\sum_{i=1}^{j} f_{j-i}(x)g_i(b)\| \leq \epsilon \tag{35}$$

for all $x, a, b \in \mathcal{A}$, then $\{f_j\}_0^{k_0}$ is a linear generalized left higher derivation and $\{g_j\}_0^{k_0}$ is a linear left higher derivation.

References

1. Aoki, T.: On the stability of the linear transformations in Banach spaces. J. Math. Soc. Jpn. **2**, 64–66 (1950)
2. Baak, C., Boo, D.-H., Rassias, Th.M.: Generalized additive mapping in Banach modules and isomorphisms between C* -algebras. J. Math. Anal. Appl. **314**(1), 150–161 (2006)
3. Bland, P.E.: Higher derivations on rings and modules. Int. J. Math. Math. Sci. **15**, 2373–2387 (2005)
4. Bourgin, D.G.: Classes of transformations and bordering transformations. Bull. Am. Math. Soc. **57**, 223–237 (1951)
5. Cao, H.-X., Lv, J.-R., Rassias, J.M.: Superstability for generalized module left derivations and generalized module derivations on a Banach module (II). J. Inequal. Pure Appl. Math. **10**(2), Art. 85 (2009)
6. Czerwik, S.: Stability of Functional Equations of Ulam–Hyers–Rassias Type. Hadronic, Palm Harbor (2003)
7. Dales, H.G., Polyakov, M.E.: Multi-normed spaces and multi-Banach algebras. (preprint)
8. Hasse, H., Schmidt, F.K.: Noch eine Begrüdung der theorie der höheren differential quotienten in einem algebraischen funtionenkörper einer unbestimmeten. J. Reine Angew. Math. **177**, 215–237 (1937)
9. Hejazian, S., Shatery, T.L.: Automatic continuity of higher derivations on JB^*-algebras. Bull. Iran. Math. Soc. **33**(1), 11–23 (2007)
10. Hejazian, S., Shatery, T.L.: Higher derivations on Banach algebras. J. Anal. Appl. **6**, 1–15 (2008)
11. Hejazian, S., Shateri, T.L.: A characterization of higher derivations, to appear in Italian. J. Pure Appl. Math. (2015)
12. Hyers, D.H.: On the stability of the linear functional equation. Proc. Nat. Acad. Sci. U. S. A. **27**, 222–224 (1941)
13. Hyers, D.H., Isac, G., Rassias, Th.M.: Stability of Functional Equations in Several Variables. Birkhäuser, Basel (1998)
14. Jewell, N.P.: Continuity of module and higher derivations. Pac. J. Math. **68**, 91–98 (1977)
15. Jung, S.-M.: Hyers–Ulam–Rassias Stability of Functional Equations in Mathematical Analysis. Hadronic Press, Palm Harbor (2001)
16. Jung, S.-M., Popa, D., Rassias, M.Th: On the stability of the linear functional equation in a single variable on complete metric groups. J. Glob. Optim. (to appear)
17. Kang, S.-Y., Chang, I.-S.: Approximation of generalized left derivations. J. Abst. Appl. Anal. **2008**, 1–8 (2008)
18. Lee, Y.-H., Jung, S.-M., Rassias, M.Th: On an n-dimensional mixed type additive and quadratic functional equation, Appl. Math. Comput. (to appear)
19. Park, C.-G.: Linear derivations on Banach algebras. Nonlinear Funct. Anal. Appl. **9**, 359–368 (2004)
20. Park, C.-G.: Lie *-homomorphisms between Lie C*-algebras and Lie *- derivations on Lie C*-algebras. J. Math. Anal. Appl. **293**, 419–434 (2004)
21. Park, C.-G., Rassias, Th.M.: Hyers-Ulam stability of a generalized Apollonius type quadratic mapping. J. Math. Anal. Appl. **322**(1), 371–381 (2006)
22. Rassias, Th.M.: On the stability of the linear mappings in Banach space. Proc. Am. Math. Soc. **72**, 297–300 (1978)
23. Rassias, Th.M.: Functional Equations, Inequalities and Applications. Kluwer, Dordrecht (2003)
24. Rassias, Th.M., Tabor, J.: Stability of Mappings of Hyers - Ulam Type. Hadronic Press, Florida (1994)
25. Shateri, T.L.: Superstability of generalized higher derivations. Abstr. Appl. Anal. (2011). doi:10.1155/2011/239849

26. Ulam, S.M.: A Collection of Mathematical Problems. Problems in Modern Mathematics. Wiley, New York (1964)
27. Uchino, Y., Satoh, T.: Function field modular forms and higher derivations. Math. Ann. **311**, 439–466 (1998)

D'Alembert's Functional Equation and Superstability Problem in Hypergroups

D. Zeglami, A. Roukbi and Themistocles M. Rassias

Abstract Our main goal is to determine the continuous and bounded complex valued solutions of the functional equation

$$\langle \delta_x * \delta_y, g\rangle + \langle \delta_x * \delta_{\check{y}}, g\rangle = 2g(x)g(y), \ x, y \in X,$$

where X is a hypergroup. The solutions are expressed in terms of 2 -dimensional representations of X. The papers of Davison [10] and Stetkær [25, 26] are the essential motivation for this first part of the present work and the methods used here are closely related to and inspired by those in [10, 25, 26]. In addition, superstability problem for this functional equation on any hypergroup and without any condition on f is considered.

Keywords Superstability · Hypergroup · D'Alembert's functional equation · Involution · Wilson's functional equation

1 Introduction

A number of results have been obtained for the d'Alembert's functional equation (1) and the corresponding Wilson's functional equation (2) on groups

$$g(xy) + g(x\sigma(y)) = 2g(x)g(y), \ x, y \in G, \tag{1}$$

D. Zeglami (✉)
Department of Mathematics, E.N.S.A.M., Moulay Ismail University, B.P.: 15290, Al Mansour, Meknes, Morocco
e-mail: zeglamidriss@yahoo.fr

A. Roukbi
Department of Mathematics, Ibn Tofail University, 14000 Kenitra, Morocco
e-mail: rroukbi.a2000@gmail.com

T. M. Rassias
Department of Mathematics, National Technical University of Athens, Zofrafou Campus, 15780 Athens, Greece
e-mail: trassias@math.ntua.gr

T. M. Rassias (ed.), *Handbook of Functional Equations*,
Springer Optimization and Its Applications 96, DOI 10.1007/978-1-4939-1286-5_17

$$f(xy) + f(x\sigma(y)) = 2f(x)g(y), \; x, y \in G, \tag{2}$$

where f, g are complex-valued functions on a group G and $\sigma : G \to G$ be an involution of G, i.e., $\sigma(xy) = \sigma(y)\sigma(x)$ and $\sigma(\sigma(x)) = x$, for all $x, y \in G$. In [15], from 1968, Kannappan proved that the non-zero complex valued functions g satisfying

$$g(x + y) + g(x - y) = 2g(x)g(y), \; x, y \in G$$

are the functions of the form

$$g(x) = \frac{m(x) + m(-x)}{2}, \; x \in G$$

where m is an homomorphism of $(G, +)$. Mathematicians extended Kannappan's result to an even more general setting; the group inversion is below replaced by a general involution, and the group by a semigroup (See Baker [3], Sinopoulos [21] and Stetkaer [22]). In this context, the most general result is the following:

Theorem 1 *Let S be a semigroup, and let $\sigma : S \to S$ be an involution of S. Assume that $g : S \to \mathbb{C}$ satisfies Kannappan's condition, i.e., that $g(xyz) = g(xzy)$ for all $x, y, z \in S$. Then, g is a solution of the functional* Eq. (1) *if and only if there exists a multiplicative homomorphism $m : S \to \mathbb{C}$, such that*

$$g = \frac{m + m \circ \sigma}{2}.$$

and m is unique, except that it can be replaced by $m \circ \sigma$.

The problem was how to find solutions of (1) in the non-abelian case. I. Corovei (see, e.g., [8, 9]) discussed them on certain nilpotent groups. H. Stetkær [23, 24] solved d'Alembert's and Wilson's functional equations on Step 2 nilpotent groups and derived many properties of d'Alembert functions on groups (see[26]). In 2008, T. Davison [10] proved, with algebraic methods only, the following structure theorem, which encompasses both abelian and non-abelian d'Alembert functions.

Theorem 2 [10] *Let G be a topological group and $f : G \to \mathbb{C}$ a continuous function with $f(e) = 1$ satisfying*

$$f(xy) + f(xy^{-1}) = 2f(x)f(y) \tag{3}$$

for all x, y in G. Then, there is a continuous (group) homomorphism $h : G \to SL_2(\mathbb{C})$ such that

$$f(x) = \frac{1}{2}tr(h(x)), \; x \in G.$$

In [25], H. Stetkær gave solutions of (3) introducing the theory of representation. Precisely, he proved that *the non-zero continuous solutions f of the* Eq. (3) *are the functions of the form $f = \frac{1}{2}\chi_\pi$, where π ranges over the 2-dimensional continuous representations of G for which $\pi(x) \in SL_2(\mathbb{C})$ for all $x \in G$.*

The study of functional equations on hypergroups started with some recent results. Székelyhidy [27, 28] and Orosz and Székelyhidi [16] described moment functions, additive functions and multiplicative functions in special cases of hypergroups. In

[17], sine and cosine functional equations are considered and solved on arbitrary polynomial hypergroups in a single variable and the method of solution is based on spectral synthesis. Recently, Roukbi and Zeglami [19] studied the abelian solutions of the d'Alembert's functional equation

$$\langle \delta_x * \delta_y, g\rangle + \langle \delta_x * \delta_{\check{y}}, g\rangle = 2g(x)g(y),\ x, y \in X, \tag{4}$$

where g is an unknown complex-valued function to be determined on a hypergroup $(X, *)$, and obtains the following result.

Theorem 3 *Let $(X, *)$ be a hypergroup and $g \in C_b(X)$. Assume that g is abelian (i.e., satisfies Kannappan's type condition defined below).*

Then, g is a solution of the functional equation (4), if and only if there exists a continuous multiplicative function $\chi : X \longrightarrow \mathbb{C}$, such that $g = \frac{\chi+\check{\chi}}{2}$ and χ is unique, except that it can be replaced by $\check{\chi}$.

The purpose of this chapter is to develop a coherent theory for d'Alembert's functional equation (4) on hypergroups that includes most of the results just mentioned. Precisely, we determine the continuous and bounded solutions of the functional equation (4). In this context, the papers [10] of Davison and [25, 26] of Steatkær just mentioned are the essential motivation for the present work and the methods used here are closely related to and inspired by those in [10, 25, 26].

In the last section of this chapter, we shall extend the investigation given by J. A. Baker [3], L. Székelyhidi [27, 29], R. Badora [2], and E. Elqorachi and M. Akkouchi [11] to Eq. (4). We consider the superstability of the Eq. (4) on any hypergroup. This chapter has the following content. In Sect. 2, we give some preliminaries on hypergroups and some notations and definitions which will be used in this work. In Sect. 3, we derive a series of elementary but useful properties of d'Alembert functions on hypergroups. In Sect. 4, for a non-abelian d'Alembert function g, we study the space $W(g)$ of Wilson functions corresponding to g. In particular, we prove that $W(g)$ is a finite-dimensional subspace of $C_b(X)$, invariant under the left representation L of X and $\dim W(g) = 4$. The main results (Theorem 6) will be proved in Sect. 5. In Sect. 6, superstability problem for the Eq. (4) is considered. On the stability problem, the interested reader should refer to [1–5, 11–13, 18, 20, 27, 29–36].

2 Preliminaries and Notations

Our notations and definitions are described in this section. We will, without further mentioning, keep them in rest of the chapter.

2.1 Hypergroups

We start with some notations. For a locally compact Hausdorff space X, let $M(X)$ denote the complex space of all bounded Borel measures on X, if $\mu \in M(X)$,

$supp(\mu)$ is the support of μ. The unit point mass concentrated at x is indicated by δ_x. Let $K(X)$ be the complex algebra of all continuous complex-valued functions on X with compact support and $C(X)$ (resp. $C_b(X)$) the complex algebra of all continuous (resp. continuous and bounded) complex-valued functions on X. Now, recall some basic notions and used notation from the hypergroup theory.

Definition 1 If $M(X)$ is a Banach algebra with an associative multiplication $*$ (called a convolution), then $(X, *)$ is a hypergroup if the following axioms are satisfied:

X1. If μ and ν are probability measures, then so is $\mu * \nu$.
X2. The mapping $(\mu, \nu) \longrightarrow \mu * \nu$ is continuous from $M(X) \times M(X)$ into $M(X)$, where $M(X)$ is endowed with the weak topology with respect to $K(X)$.
X3. There is an element $e \in X$ such that $\delta_e * \mu = \mu * \delta_e = \mu$ for all $\mu \in M(X)$.
X4. There is a homeomorphic mapping $x \longrightarrow \check{x}$ of X into itself such that $(\check{x})\check{} = x$ and $e \in supp(\delta_x * \delta_y)$ if and only if $y = \check{x}$.
X5. For all $\mu, \nu \in M(X)$, $(\mu * \nu)\check{} = \check{\nu} * \check{\mu}$, where $\check{\mu}$ is defined by

$$\langle \check{\mu}, f \rangle = \langle \mu, \check{f} \rangle = \int_X f(\check{t}) d\mu(t); \quad f \in C_b(X).$$

X6. The mapping $(x, y) \longrightarrow supp(\delta_x * \delta_y)$ is continuous from $X \times X$ into the space of compact subsets of X with the topology described in ([14], Sect. 2.5).

The definitive set of axioms was given first by Jewett in his encyclopedic article [14]. A hypergroup $(X, *)$ is called commutative if its convolution is commutative.

We review some notations: Let $f \in C_b(X)$, for all $x \in X$ and $\mu \in M(X)$, we put

$$\langle \delta_x, f \rangle = f(x)$$

$$\langle \mu, f \rangle = \int_X \langle \delta_x, f \rangle \, d\mu(x)$$

$$\check{f}(x) = f(\check{x}), \; x \in X.$$

If $\mu, \nu \in M(X)$, we define the convolution measure $\mu * \nu$ by

$$\langle \mu * \nu, f \rangle = \int_X \int_X \left\langle \delta_x * \delta_y, f \right\rangle d\mu(x) d\mu(y).$$

f is said to be even or invariant, (resp. odd), if $\check{f} = f$, (resp. $\check{f} = -f$).

Definition 2 [6] Let $(X, *)$ be a hypergroup and $\chi : X \longrightarrow \mathbb{C}$ be a function, we say that

(i) χ is a multiplicative function of $(X, *)$ if it has the the property

$$\left\langle \delta_x * \delta_y, \chi \right\rangle = \chi(x)\chi(y) \text{ for all } x, y \in X.$$

(ii) χ is a hermitian function if $\chi(\check{x}) = \overline{\chi(x)}$, for all $x \in X$.
(iii) χ is a hypergroup character of $(X, *)$ if it is bounded, continuous, multiplicative, and hermitian function.

Definition 3 **([14], Sect. 11.3)** Let $\mathcal{H}$ be a Hilbert space, let I be the identity operator, and let $B(\mathcal{H})$ be the space of all bounded operators on $\mathcal{H}$. We say that π is a representation of X on $\mathcal{H}$ if the following four conditions are satisfied:

i) The mapping $\mu \longmapsto \pi(\mu)$ is a $*$-homomorphism from $M(X)$ into $B(\mathcal{H})$.
ii) If $\mu \in M(X)$ then $\|\pi(\mu)\| \leq \|\mu\|$.
iii) $\pi(\delta_e) = I$.
iv) If $a, b \in \mathcal{H}$, then the mapping $\mu \longmapsto \langle \pi(\mu)a, b\rangle$ is bounded and continuous.

Let L and R denote, respectively, the left and right representation of X on $C_b(X)$, i.e., $[L(y)f](x) = \langle \delta_{\check{y}} * \delta_x, f\rangle$ and $[R(y)f](x) = \langle \delta_x * \delta_y, f\rangle$ for all $x, y \in X$ and $f \in C_b(X)$. Note that $L(a)$ and $R(b)$ commute for all $a, b \in X$ as it is well known and also easy to check.

2.2 D'Alembert Function on Hypergroups

Definition 4 Let $f : X \longrightarrow \mathbb{C}$ be a continuous and bounded function on X.

(i) f is said to satisfy Kannappan's type condition if

$$\langle \mu * \nu * w, f\rangle = \langle \mu * w * \nu, f\rangle, \text{ for all } \mu, \nu, w \in M(X) \tag{5}$$

(ii) We say that f is abelian if it satisfies (5).

Definition 5 Let $(X, *)$ be a hypergroup.

(i) A d'Alembert function on X is a continuous and bounded non-zero solution $g : X \longrightarrow \mathbb{C}$ of d'Alembert's functional equation (4).
(ii) A solution of Wilson functional equation is a pair $\{f, g\}$ of functions in $C_b(X)$ satisfying

$$\langle \delta_x * \delta_y, f\rangle + \langle \delta_x * \delta_{\check{y}}, f\rangle = 2f(x)g(y),\ x, y \in X. \tag{6}$$

We say that the function f in the solution $\{f, g\}$ is a Wilson function corresponding to g.

3 Properties of D'Alembert Functions

In this section, let $(X, *)$ denotes a hypergroup with neutral element e. First, the notation of some pertinent functions.

Definition 6 Let $g \in C_b(X)$. We define

(i) $d = d_g : M(X) \longrightarrow \mathbb{C}$ by $\langle \mu, d\rangle := 2\langle \mu, g\rangle^2 - \langle \mu * \mu, g\rangle$ for all $\mu \in M(X)$. In particular, $d(x) = 2g(x)^2 - \langle \delta_x * \delta_x, g\rangle$ for all $x \in X$.

(ii) $g_\mu : X \longrightarrow \mathbb{C}$ by $g_\mu(y) = \langle \mu * \delta_y, g\rangle - \langle \mu, g\rangle g(y)$ for all $\mu \in M(X),\ y \in X$. In particular, $g_x(y) = \langle \delta_x * \delta_y, g\rangle - g(x)g(y)$ for all $x,\ y \in X$.

(iii) $\Delta : M(X) \times M(X) \longrightarrow \mathbb{C}$ by $\langle \mu \otimes \nu, \Delta\rangle := \langle \mu, g_\mu\rangle\langle \nu, g_\nu\rangle - \langle \nu, g_\mu\rangle^2$ for all $\mu, \nu \in M(X)$. In particular, $\Delta(x, y) = g_x(x)g_y(y) - g_x(y)^2$ for all $x,\ y \in X$.

The function Δ was introduced by Davison in [10]. Let us start by collecting a number of results for d'Alembert function.

Proposition 1 *Let $g : X \longrightarrow \mathbb{C}$ be a d'Alembert function on X and g_μ be as in Definition 6, then*

(i) g is invariant, that is $\check{g} = g$.
(ii) g is central, that is $\langle \mu * \nu, g\rangle = \langle \nu * \mu, g\rangle$ for all $\mu, \nu \in M(X)$.
(iii) If $e \in X$ is the neutral element, then $g(e) = 1$.
(iv) For all $\mu, \nu, w \in M(X)$, we have the following equalities

$$\langle \mu, g_\nu\rangle = \langle \nu, g_\mu\rangle. \tag{7}$$

$$\langle \check{\nu}, g_\mu\rangle = -\langle \nu, g_\mu\rangle. \qquad (i.e., g_\mu \text{ is odd}) \tag{8}$$

$$\langle \nu * w, g_\mu\rangle + \langle \nu * \check{w}, g_\mu\rangle = 2\langle \nu, g_\mu\rangle\langle w, g\rangle. \tag{9}$$

$$\langle \nu * w, g_\mu\rangle + \langle w * \nu, g_\mu\rangle = 2\langle \nu, g_\mu\rangle\langle w, g\rangle + 2\langle \nu, g\rangle\langle w, g_\mu\rangle \tag{10}$$

$$\langle \mu * \nu, g_\mu\rangle = \langle \nu * \mu, g_\mu\rangle = \langle \mu, g\rangle\langle \nu, g_\mu\rangle + \langle \nu, g\rangle\langle \mu, g_\mu\rangle. \tag{11}$$

Proof (i) The left hand side of (4) does not change if ν is replaced by $\check{\nu}$. Applying this to the right hand side, we infer that $\langle \check{\nu}, g\rangle = \langle \nu, g\rangle$ for all $\nu \in M(X)$, proving (i). (ii) If we interchange μ and ν in (4), we obtain

$$\langle \nu * \mu, g\rangle + \langle \nu * \check{\mu}, g\rangle = 2\langle \mu, g\rangle\langle \nu, g\rangle.$$

Since g is an invariant, we get that

$$\langle \nu * \mu, g\rangle + \langle \mu * \check{\nu}, g\rangle = 2\langle \mu, g\rangle\langle \nu, g\rangle,$$

for all $\mu, \nu \in M(X)$. By (4), we conclude that

$$\langle \mu * \nu, g\rangle = \langle \nu * \mu, g\rangle \text{for all } \mu, \nu \in M(X).$$

(iii) Noting that $\check{e} = e$, we get with $\nu = \delta_e$ in (4) that

$$\langle \mu, g\rangle + \langle \mu, g\rangle = 2\langle \mu, g\rangle g(e) \text{ for all } \mu \in M(X).$$

This implies, since $g \neq 0$, that $g(e) = 1$. (iv) The first property follows from the definition of g_μ and the fact that g is central. The rest of the proof consists of straightforward computations except possibly for (10); to get (10), we interchange ν and w in (9) and add the identities. We get the result by using (8).

The following results are also hold.

Proposition 2 *Let $g \in C_b(X)$ be a d'Alembert function on X and d be as in Definition 6.*

(i) If $w \in M(X)$ such that $w = \check{w}$, then $\langle \mu * w, g\rangle = \langle \mu, g\rangle\langle w, g\rangle$ for all $\mu \in M(X)$. In particular,

$$\langle \mu * \nu * \check{\nu}, g\rangle = \langle \mu, g\rangle\langle \nu, d\rangle, \text{ for all } \mu, \nu \in M(X).$$

(ii) $d : X \longrightarrow \mathbb{C}$ is a multiplicative invariant function satisfying $\langle \mu, d\rangle = \langle \mu * \check{\mu}, g\rangle$ for all $\mu \in M(X)$, and $d(e) = 1$.

(iii) For all $\mu, \nu, w \in M(X)$, we have

$$\begin{aligned}\langle \mu * \nu * w, g\rangle + \langle \mu * w * \nu, g\rangle = 2\langle \mu, g\rangle\langle \nu * w, g\rangle + 2\langle \nu, g\rangle\langle w * \mu, g\rangle \\ + 2\langle w, g\rangle\langle \mu * \nu, g\rangle - 4\langle \mu, g\rangle\langle \nu, g\rangle\langle w, g\rangle.\end{aligned} \tag{12}$$

Proof (i) From the functional equation (4), we get that

$$\begin{aligned}2\langle \mu, g\rangle\langle w, g\rangle &= \langle \mu * w, g\rangle + \langle \mu * \check{w}, g\rangle \\ &= \langle \mu * w, g\rangle + \langle \mu * w, g\rangle \\ &= 2\langle \mu * w, g\rangle,\end{aligned}$$

which implies (i). (ii) By definition of d and the fact that g is central, we have

$$\begin{aligned}\langle \mu * \nu, d\rangle = \langle \mu * \nu * (\mu * \nu)\check{}, g\rangle &= \langle \check{\mu} * \mu * \nu * \check{\nu}, g\rangle \\ = \langle \check{\mu} * \mu, g\rangle\langle \nu * \check{\nu}, g\rangle &= \langle \mu, d\rangle\langle \nu, d\rangle.\end{aligned}$$

(iii) Since

$$\langle \mu * \nu, g\rangle = \langle \nu, g_\mu\rangle + \langle \mu, g\rangle\langle \nu, g\rangle,$$

we may write

$$\begin{aligned}\langle \mu * \nu * w, g\rangle + \langle \mu * w * \nu, g\rangle &= \langle \nu * w, g_\mu\rangle + \langle \mu, g\rangle\langle \nu * w, g\rangle \\ &+ \langle w * \nu, g_\mu\rangle + \langle \mu, g\rangle\langle w * \nu, g\rangle,\end{aligned}$$

from which we, using (10) and the fact that g is central, derive that

$$\begin{aligned}\langle \mu * \nu * w, g\rangle + \langle \mu * w * \nu, g\rangle &= 2\langle w, g\rangle\langle \nu, g_\mu\rangle + 2\langle \nu, g\rangle\langle w, g_\mu\rangle + 2\langle \mu, g\rangle\langle \nu * w, g\rangle \\ &= 2\langle w, g\rangle\langle \mu * \nu, g\rangle - 2\langle \mu, g\rangle\langle \nu, g\rangle\langle w, g\rangle \\ &\quad + 2\langle \nu, g\rangle\langle \mu * w, g\rangle \\ &\quad - 2\langle \mu, g\rangle\langle \nu, g\rangle\langle w, g\rangle + 2\langle \mu, g\rangle\langle \nu * w, g\rangle,\end{aligned}$$

which is the stated result.

We now derive some properties of the function Δ introduced in Definition 6.

Proposition 3 *Let $g \in C_b(X)$ be a d'Alembert function on X and Δ be as in Definition 6. Then*

(i) For all $\mu, \nu \in M(X)$, we have

$$\langle \mu \otimes \nu, \Delta \rangle = \langle \nu \otimes \mu, \Delta \rangle = \langle \mu \otimes \check{\nu}, \Delta \rangle. \tag{13}$$

$$\begin{aligned} \langle \mu \otimes \nu, \Delta \rangle = & \left[\langle \mu, g \rangle^2 - \langle \mu, d \rangle \right] \left[\langle \nu, g \rangle^2 - \langle \nu, d \rangle \right] \\ & - [\langle \mu * \nu, g \rangle - \langle \mu, g \rangle \langle \nu, g \rangle]^2 \end{aligned} \tag{14}$$

$$\langle \mu \otimes \nu, \Delta \rangle = \frac{1}{2}(\langle \mu, d \rangle \langle \nu, d \rangle - \langle \mu * \nu * \check{\mu} * \check{\nu}, g \rangle). \tag{15}$$

$$\langle \mu \otimes \nu, \Delta \rangle = \frac{1}{2}(\langle \mu * \mu * \nu * \nu, g \rangle - \langle \mu * \nu * \mu * \nu, g \rangle. \tag{16}$$

(ii) The right representation R of X on $C_b(X)$ has the property

$$\left(\frac{R(\mu * \nu) - R(\nu * \mu)}{2} \right)^2 g = -\langle \mu \otimes \nu, \Delta \rangle g, \text{ for all } \mu, \nu \in M(X). \tag{17}$$

Proof (i) The equalities (13) and (14) sign come immediately from the definition of Δ and the equality

$$\langle \mu, g_\mu \rangle = \langle \mu, g \rangle^2 - \langle \mu, d \rangle, \text{ for all } \mu \in M(X).$$

For the proof of (15), substituting $w = \check{\mu} * \check{\nu}$ in (12), we get that

$$\begin{aligned} & \langle \mu * \nu * (\check{\mu} * \check{\nu}), g \rangle + \langle \mu * \check{\mu} * \check{\nu} * \nu), g \rangle \\ & = 2\langle \mu, g \rangle \langle \nu * \check{\mu} * \check{\nu}), g \rangle + 2\langle \nu, g \rangle \langle \check{\mu} * \check{\nu} * \mu, g \rangle \\ & \quad + 2\langle \check{\mu} * \check{\nu}, g \rangle \langle \mu * \nu, g \rangle - 4\langle \mu, g \rangle \langle \nu, g \rangle \langle \check{\mu} * \check{\nu}, g \rangle \\ & = 2\langle \mu, g \rangle \langle \check{\mu} * \check{\nu} * \nu, g \rangle + 2\langle \nu, g \rangle \langle \check{\nu} * \mu * \check{\mu}, g \rangle + 2\langle \mu * \nu, g \rangle^2 \\ & \quad - 4\langle \mu, g \rangle \langle \nu, g \rangle \langle \check{\mu} * \check{\nu}, g \rangle \\ & = 2\langle \mu, g \rangle \langle \check{\mu}, g \rangle \langle \check{\nu} * \nu, g \rangle + 2\langle \nu, g \rangle \langle \check{\nu}, g \rangle \langle \mu * \check{\mu}, g \rangle \\ & \quad + 2\langle \mu * \nu, g \rangle^2 - 4\langle \mu, g \rangle \langle \nu, g \rangle \langle \check{\mu} * \check{\nu}, g \rangle \\ & = 2\langle \mu, g \rangle^2 \langle \nu, d \rangle + 2\langle \nu, g \rangle^2 \langle \mu, d \rangle) + 2\langle \mu * \nu, g \rangle^2 \\ & \quad - 4\langle \mu, g \rangle \langle \nu, g \rangle \langle \mu * \nu, g \rangle. \end{aligned}$$

Using the fact that d is multiplicative, we get the second equality sign. To get (16), we compute $\langle \mu * \nu * (\check{\mu} * \check{\nu}), g \rangle$ as follows:

$$\begin{aligned}\langle \mu * \nu * (\check{\mu} * \check{\nu}), g\rangle &= \langle \mu * \nu * (\nu \check{*} \mu), g\rangle + \langle \mu * \nu * \nu * \mu, g\rangle - \langle \mu * \nu * \nu * \mu, g\rangle \\ &= 2\langle \mu * \nu, g\rangle\langle \nu * \mu), g\rangle - \langle \mu * \nu * \nu * \mu), g\rangle \\ &= 2\langle \mu * \nu, g\rangle^2 - \langle \mu * \mu * \nu * \nu), g\rangle \\ &= \langle \mu * \nu * \mu * \nu, g\rangle + \langle \mu * \nu * (\nu \check{*} \mu), g\rangle - \langle \mu * \mu * \nu * \nu, g\rangle \\ &= \langle \mu * \nu, g\rangle + \langle \mu * \nu, d\rangle - \langle \mu * \mu * \nu * \nu), g\rangle,\end{aligned}$$

from which the result follows.

(ii) By definition, for all $\mu_1, \nu_1 \in M(X)$, we have

$$\begin{aligned}(R(\mu_1 * \nu_1) - R(\nu_1 * \mu_1))^2 &= R(\mu_1 * \nu_1 * \mu_1 * \nu_1) - R(\mu_1 * \nu_1 * \nu_1 * \mu_1) \\ &\quad - R(\nu_1 * \mu_1 * \mu_1 * \nu_1) + R(\nu_1 * \mu_1 * \nu_1 * \mu_1),\end{aligned}$$

we get for any $w \in M(X)$ that

$$\begin{aligned}\langle w, (R(\mu_1 * \nu_1) - R(\nu_1 * \mu_1))^2 g\rangle &= \langle w * \mu_1 * \nu_1 * \mu_1 * \nu_1,\ g\rangle \\ &\quad - \langle w * \mu_1 * \nu_1 * \nu_1 * \mu_1,\ g\rangle \\ &\quad + \langle w * \nu_1 * \mu_1 * \mu_1 * \nu_1,\ g\rangle \\ &\quad - \langle w * \nu_1 * \mu_1 * \nu_1 * \mu_1,\ g\rangle.\end{aligned}$$

Using the definition of g_w, we have

$$\begin{aligned}&\langle w, (R(\mu_1 * \nu_1) - R(\nu_1 * \mu_1))^2 g\rangle = \langle \mu_1 * \nu_1 * \mu_1 * \nu_1, g_w\rangle + \langle \nu_1 * \mu_1 * \nu_1 * \mu_1, g_w\rangle \\ &- \langle \mu_1 * \nu_1 * \nu_1 * \mu_1, g_w\rangle - \langle \nu_1 * \mu_1 * \mu_1 * \nu_1, g_w\rangle + \langle \mu, g\rangle \times \{\langle \mu_1 * \nu_1 * \mu_1 * \nu_1, g\rangle \\ &+ \langle \nu_1 * \mu_1 * \nu_1 * \mu_1, g\rangle - \langle \mu_1 * \nu_1 * \nu_1 * \mu_1, g\rangle - \langle \nu_1 * \mu_1 * \mu_1 * \nu_1, g\rangle\}.\end{aligned}$$

The first four terms on the right hand side cancel. Indeed, on the two first terms we apply the property (10) with $\nu = \mu_1 * \nu_1$ and $w = \mu_1 * \nu_1$, then we get

$$\langle \mu * \nu * \mu * \nu, g_w\rangle = 2\langle \mu * \nu, g_w\rangle\langle \mu * \nu, g\rangle,$$

and on the two next terms we apply (10) with $\mu = \mu_1 * \nu_1$ and $w = \nu_1 * \mu_1$. Using freely that g is central, we get that

$$\langle w, (R(\mu * \nu) - R(\nu * \mu))^2 g\rangle = 2\langle w, g\rangle(\langle \mu * \nu * \mu * \nu, g\rangle - \langle \mu * \mu * \nu * \nu, g\rangle),$$

we get (ii) by referring to (16).

We now characterize the abelian solution of d'Alembert's equation (4) by using the function Δ.

Theorem 4 *Let $(X, *)$ be a hypergroup and $g : X \longrightarrow \mathbb{C}$ be a d'Alembert function, then*

(i) g is abelian if and only if $\Delta = 0$.
(ii) If g is non-abelian then there are elements $\mu_0, \nu_0 \in M(X)$ such that $g_{\mu_0} \neq 0$ and simultaneously $\langle \mu_0 \otimes \nu_0, \Delta\rangle \neq 0$.

Proof The proof of Theorem 4 consists of modifications of the corresponding computations of [25] and [26]. (i) That $\Delta = 0$ if g satisfies Kannappan's type condition, which is an immediate consequence of the formula (16).

Let us conversely assume that $\Delta = 0$. We get from (15) that

$$\langle \mu * \nu * \check{\mu} * \check{\nu}, g\rangle = \langle \mu, d\rangle\langle \nu, d\rangle for\ all \mu, \nu \in M(X).$$

But more is true when $\Delta = 0$, viz. that the following holds for all $\mu, \nu, w \in M(X)$:

$$\langle w * \mu * \nu * \check{\mu} * \check{\nu}, g\rangle = \langle w, g\rangle\langle \mu * \nu * \check{\mu} * \check{\nu}, g\rangle = \langle w, g\rangle\langle \mu, d\rangle\langle \nu, d\rangle. \quad (18)$$

Indeed, writing w instead of μ and $\mu * \nu * \check{\mu} * \check{\nu}$ instead of ν in (14), we find

$$\begin{aligned}
&[\langle w * \mu * \nu * \check{\mu} * \check{\nu}, g\rangle - \langle w, g\rangle\langle \mu * \nu * \check{\mu} * \check{\nu}, g\rangle]^2 \\
&= [\langle w, g\rangle^2 - \langle w, d\rangle][\langle \mu * \nu * \check{\mu} * \check{\nu}, g\rangle^2 - \langle \mu * \nu * \check{\mu} * \check{\nu}, d\rangle] \\
&= [\langle w, g\rangle^2 - \langle w, d\rangle][\langle \mu * \nu * \check{\mu} * \check{\nu}, g\rangle^2 - \langle \mu, d\rangle^2\langle \nu, d\rangle^2] \\
&= [\langle w, g\rangle^2 - \langle w, d\rangle][\langle \mu * \nu * \check{\mu} * \check{\nu}, g\rangle + \langle \mu, d\rangle\langle \nu, d\rangle] \\
&\quad \times [\langle \mu * \nu * \check{\mu} * \check{\nu}, g\rangle - \langle \mu, d\rangle\langle \nu, d\rangle] \\
&= [\langle w, g\rangle^2 - \langle w, d\rangle][\langle \mu * \nu * \check{\mu} * \check{\nu}, g\rangle + \langle \mu, d\rangle\langle \nu, d\rangle] \times 0 = 0.
\end{aligned}$$

When we replace w by $\mu * \nu * w$ in (18), we find using the fact that g is central and Proposition 2(i) that

$$\begin{aligned}
\langle \mu * \nu * w, g\rangle\langle \mu, d\rangle\langle \nu, d\rangle &= \langle \mu * \nu * w, g\rangle\langle \nu, d\rangle\langle \mu, d\rangle \\
&= \langle (\mu * \nu * w) * \nu * \mu * \check{\nu} * \check{\mu}, g\rangle \\
&= \langle (\mu * \nu) * w * \nu * \mu * \check{\nu} * \check{\mu}, g\rangle \\
&= \langle w * \nu * \mu * \check{\nu} * \check{\mu} * (\mu * \nu), g\rangle \\
&= \langle w * \nu * \mu * (\mu * \nu)^{\vee} * (\mu * \nu), g\rangle \\
&= \langle w * \nu * \mu, g\rangle\langle \mu * \nu, d\rangle \\
&= \langle w * \nu * \mu, g\rangle\langle \mu, d\rangle\langle \nu, d\rangle,
\end{aligned}$$

so

$$[\langle \mu * \nu * w, g\rangle - \langle \mu * w * \nu, g\rangle]\langle \mu, d\rangle\langle \nu, d\rangle = 0, \quad (19)$$

for all $\mu, \nu, w \in M(X)$. If d vanishes nowhere, we get that $\langle \mu * \nu * w, g\rangle = \langle \mu * w * \nu, g\rangle$, i.e., that g satisfies Kannappan's type condition. In general case, we must work a bit longer to get the desired conclusion:

$$\langle \mu * \nu * w, g\rangle = \langle \mu * w * \nu, g\rangle, \ \textit{for all } \mu, \nu, w \in M(X).$$

Case 1 Assume that $\langle \mu, g_\mu\rangle = 0$, then for any $\nu \in M(X)$, we find that

$$0 = \langle \nu \otimes \mu, \Delta\rangle = \langle \nu, g_\nu\rangle\langle \mu, g_\mu\rangle - \langle \mu, g_\nu\rangle^2 = 0 - [\langle \nu * \mu, g\rangle - \langle \nu, g\rangle\langle \mu, g\rangle]^2,$$

so $\langle \nu * \mu, g \rangle = \langle \nu, g \rangle \langle \mu, g \rangle$. For any $\nu, w \in M(X)$, we now get

$$\langle \nu * w * \mu, g \rangle = \langle \nu * w, g \rangle \langle \mu, g \rangle = \langle w * \nu, g \rangle \langle \mu, g \rangle = \langle w * \nu * \mu, g \rangle = \langle \nu * \mu * w, g \rangle,$$

so that the desired conclusion holds in this case.

Case 2 Assume that $\langle \nu, g_\mu \rangle = 0$. We have also the desired conclusion. In this case, $0 = \langle \mu \otimes \nu, \Delta \rangle = \langle \mu, g_\mu \rangle \langle \nu, g_\nu \rangle - \langle \nu, g_\mu \rangle^2 = \langle \mu, g_\mu \rangle \langle \nu, g_\nu \rangle$, so either

$$\langle \mu, g_\mu \rangle = 0 \quad \text{or} \quad \langle \nu, g_\nu \rangle = 0.$$

We are thus back in the Case 1.

General Case Let $\mu, \nu, w \in M(X)$, since g is central we get that

$$\begin{aligned}\langle \nu * w, g_{\nu * w} \rangle &= \langle \nu * w * \nu * w, g \rangle - \langle \nu * w, g \rangle^2 \\ &= \langle w * \nu * w * \nu, g \rangle - \langle w * \nu, g \rangle^2 = \langle w * \nu, g_{w * \nu} \rangle,\end{aligned}$$

and so (using $\langle \mu \otimes (\nu * w), \Delta \rangle = 0$) that

$$\langle \nu * w, g_\mu \rangle^2 = \langle \mu, g_\mu \rangle \langle \nu * w, g_{\nu * w} \rangle = \langle \mu, g_\mu \rangle \langle w * \nu, g_{w * \nu} \rangle = \langle w * \nu, g_\mu \rangle^2,$$

which means that

$$[\langle \nu * w, g_\mu \rangle - \langle w * \nu, g_\mu \rangle][\langle \nu * w, g_\mu \rangle + \langle w * \nu, g_\mu \rangle] = 0.$$

Rewriting the individual factors by using the definition of g_μ on the first factor and (10) on the second one, we get that

$$[\langle \mu * \nu * w, g \rangle - \langle \mu * w * \nu, g \rangle][\langle \nu, g_\mu \rangle \langle w, g \rangle + \langle \nu, g \rangle \langle w, g_\mu \rangle] = 0,$$

for all $\mu, \nu, w \in M(X)$. Replacing w by $\check{w}$ only changes the sign of the first factor, while we in the second factor get that $\langle \check{w}, g_\mu \rangle = -\langle w, g_\mu \rangle$, so that

$$[\langle \mu * \nu * w, g \rangle - \langle \mu * w * \nu, g \rangle][\langle \nu, g_\mu \rangle \langle w, g \rangle - \langle \nu, g \rangle \langle w, g_\mu \rangle] = 0.$$

Adding the two identities yields

$$[\langle \mu * \nu * w, g \rangle - \langle \mu * w * \nu, g \rangle] \langle \nu, g_\mu \rangle \langle w, g \rangle = 0. \tag{20}$$

If $\langle \nu, g_\mu \rangle = 0$ then $\langle \mu * \nu * w, g \rangle = \langle \mu * w * \nu, g \rangle$ from the second case above. In particular, $[\langle \mu * \nu * w, g \rangle - \langle \mu * w * \nu, g \rangle] \langle w, g \rangle = 0$, an identity which also holds if $\langle \nu, g_\mu \rangle \neq 0$ as we see from (20). Renaming letters we have in any case that

$$[\langle \mu * \nu * w, g \rangle - \langle \mu * w * \nu, g \rangle] \langle \mu, g \rangle = 0 \text{ for all } \mu, \nu, w \in M(X). \tag{21}$$

Writing $\langle \mu, g_\mu \rangle = \langle \mu, g \rangle^2 - \langle \mu, d \rangle$ and similarly for g_ν we get for any $\mu, \nu, w \in M(X)$ by help of (20) and (21) that

$$[\langle \mu * \nu * w, g \rangle - \langle \mu * w * \nu, g \rangle] \langle \mu, g_\mu \rangle \langle \nu, g_\nu \rangle]$$

$$= [\langle \mu * \nu * w, g\rangle - \langle \mu * w * \nu, g\rangle](\langle \mu, g\rangle^2 - \langle \mu, d\rangle)(\langle \nu, g\rangle^2 - \langle \nu, d\rangle)$$
$$= 0.$$

Applying the first case we get that $\langle \mu * \nu * w, g\rangle - \langle \mu * w * \nu, g\rangle = 0$. i.e., the function g is abelian.

(ii) By (i) there exist $\sigma, \upsilon \in M(X)$ such that

$$\langle \sigma \otimes \upsilon, \Delta\rangle = \langle \sigma, g_\sigma\rangle\langle \upsilon, g_\upsilon\rangle - \langle \upsilon, g_\sigma\rangle^2 \neq 0. \tag{22}$$

If $\langle \sigma, g_\sigma\rangle \neq 0$, we have the desired conclusion with $\mu_0 = \sigma$ and $\nu_0 = \upsilon$. Similarly if $\langle \upsilon, g_\upsilon\rangle \neq 0$. So, we may in the remainder of the proof assume that $\langle \sigma, g_\sigma\rangle = \langle \upsilon, g_\upsilon\rangle = 0$. We get from (22) that $\langle \upsilon, g_\sigma\rangle \neq 0$. From $\langle \sigma, g_\sigma\rangle = 0$, we get that

$$\langle \sigma * \sigma, g\rangle = \langle \sigma, g\rangle^2 = \langle \sigma * \check{\sigma}, g\rangle \neq 0,$$

where the inequality sign was established in Proposition 2(ii). By the formula (11) in Proposition 2(iv), we get that

$$\langle \sigma * \upsilon, g_\sigma\rangle = \langle \sigma, g\rangle\langle \upsilon, g_\sigma\rangle + \langle \sigma, g_\sigma\rangle\langle \upsilon, g\rangle = \langle \sigma, g\rangle\langle \upsilon, g_\sigma\rangle \neq 0.$$

So the pair $\{\sigma, \sigma * \upsilon\}$ also satisfies the inequality (22), i.e., $\langle \sigma \otimes (\sigma * \upsilon), \Delta\rangle \neq 0$.

We are through if $\langle \sigma * \upsilon, g_{\sigma*\upsilon}\rangle \neq 0$, So we may from now assume that $\langle \sigma * \upsilon, g_{\sigma*\upsilon}\rangle = 0$. This means that

$$\begin{aligned}\langle (\sigma * \upsilon) * (\sigma * \upsilon), g\rangle &= \langle \sigma * \upsilon, g\rangle^2 = \langle (\sigma * \upsilon) * (\sigma * \upsilon)\check{}, g\rangle \\ &= \langle \sigma * \upsilon * \check{\upsilon} * \check{\sigma}, g\rangle \\ &= \langle \sigma * \check{\sigma}, g\rangle\langle \upsilon * \check{\upsilon}, g\rangle = \langle \sigma, g\rangle^2\langle \upsilon, g\rangle^2\end{aligned}$$

Now,

$$\begin{aligned}&[\langle \sigma * \upsilon, g\rangle + \langle \sigma, g\rangle\langle \upsilon, g\rangle]\langle \upsilon, g_\sigma\rangle \\ &= [\langle \sigma * \upsilon, g\rangle + \langle \sigma, g\rangle\langle \upsilon, g\rangle][\langle \sigma * \upsilon, g\rangle + \langle \sigma, g\rangle\langle \upsilon, g\rangle] \\ &= \langle \sigma * \upsilon, g\rangle^2 - \langle \sigma, g\rangle^2\langle \upsilon, g\rangle^2 = 0,\end{aligned}$$

which, since $\langle \upsilon, g_\sigma\rangle \neq 0$, implies that $\langle \sigma * \upsilon, g\rangle + \langle \sigma, g\rangle\langle \upsilon, g\rangle = 0$.

Since $\langle \sigma \otimes \upsilon, \Delta\rangle = \langle \sigma \otimes \check{\upsilon}, \Delta\rangle \neq 0$ we may go through the considerations above with υ replaced by $\check{\upsilon}$. This will give us $\langle \sigma * \check{\upsilon}, g\rangle + \langle \sigma, g\rangle\langle \check{\upsilon}, g\rangle = 0$, i.e., that $\langle \sigma * \check{\upsilon}, g\rangle + \langle \sigma, g\rangle\langle \upsilon, g\rangle = 0$, unless we on the way get a pair $\{\mu_0, \nu_0\}$ of measures in $M(X)$ such that $\langle \mu_0, g_{\mu_0}\rangle\langle \mu_0 \otimes \nu_0, \Delta\rangle \neq 0$. Adding this to $\langle \sigma * \upsilon, g\rangle + \langle \sigma, g\rangle\langle \upsilon, g\rangle = 0$ gives $4\langle \sigma, g\rangle\langle \upsilon, g\rangle = 0$, contradicting that $\langle \sigma, g\rangle \neq 0$ and $\langle \upsilon, g\rangle \neq 0$.

4 The Space $W(g)$ of Wilson Functions

We discuss Wilson functions corresponding to a non-abelian d'Alembert function. Throughout this section, let g denote a non-abelian d'Alembert function on a hypergroup X. That g is non-abelian means that there exist $\mu_0, \nu_0 \in M(X)$ such that $\langle \mu_0 \otimes \nu_0, \Delta \rangle \neq 0$ (Theorem 4(i)). We choose such element $\mu_0, \nu_0 \in M(X)$, keep them fixed during the remainder of this section.

Definition 7 We let $W(g)$ denote the space of functions $f \in C_b(X)$ satisfying

$$\langle \mu * \nu, f \rangle + \langle \mu * \check{\nu}, f \rangle = 2\langle \mu, f \rangle \langle \nu, g \rangle, \ \mu, \nu \in M(X). \tag{23}$$

The elements of $W(g)$ are called Wilson functions on X. The set of even (resp. odd) Wilson functions is denoted $W(g)_e$ (resp. $W(g)_o$). Even/odd is meant with respect to the involution of X.

A consequence that we will derive below, is that $W(g)$ is finite-dimensional and invariant under the right and the left representation of X. We generalize the result for groups obtained recently by Davison in [10] and Stetkær in [25]. For the reader's convenience we include proofs. It is an adaptation of Davison's and Stetkær's proofs to the hypergroup status.

Proposition 4 *Let $g \in C_b(X)$ be a d'Alembert function on X.*

(i) A Wilson function f is odd if and only if $f(e) = 0$.
(ii) $g \in W(g)_e$ while $g_\mu \in W(g)_o$ for all $\mu \in M(X)$.
(iii) $W(g)$ is a vector subspace of $C_b(X)$, and it is invariant under the left representation L of X.
(iv) $W(g)_e$ and $W(g)_o$ are vector subspaces of $W(g)$ such that $W(g) = W(g)_e \oplus W(g)_o$. Furthermore, $W(g)_e = \mathbb{C}g$. Finally, if $f \in W(g)_o$ then for all $\mu, \nu \in M(X)$

$$\langle \mu * \nu, f \rangle + \langle \nu * \mu, f \rangle = 2\langle \mu, f \rangle \langle \nu, g \rangle + 2\langle \nu, f \rangle \langle \mu, g \rangle. \tag{24}$$

Proof (i) Assume that $f(e) = 0$. In (23), we replace μ by δ_e then we get that $\langle \nu, f \rangle + \langle \check{\nu}, f \rangle = 0$. The other implication is obvious.

(ii) It follows from Proposition 1.
(iii) Using the definition of the left regular representation of X, we get that

$$\begin{aligned}\langle \mu * \nu, L(w)f \rangle + \langle \mu * \check{\nu}, L(w)f \rangle &= \langle \check{w} * \mu * \nu, f \rangle + \langle \check{w} * \mu * \check{\nu}, f \rangle \\ &= 2\langle \check{w} * \mu, f \rangle \langle \nu, g \rangle \\ &= 2\langle \mu, L(w)f \rangle \langle \nu, g \rangle\end{aligned}$$

which shows that $L(w)f \in W(g)$ for all $w \in M(X)$.

(iv) $f \in W(g)$ is by

$$f = \frac{f+\check{f}}{2} + \frac{f-\check{f}}{2},$$

decomposed into its even and odd parts in $C_b(X)$. The only problem is to show that the parts belong to $W(g)$. Taking $\mu = \delta_e$ in (23), we see that $\frac{f+\check{f}}{2} = f(e)g \in W(g)$. It follows that $\frac{f-\check{f}}{2} = f - \frac{f+\check{f}}{2} = f - f(e)g \in W(g)$ as well. To get (24), we interchange μ, ν in (23) and add the result to (23).

We have so far in this section not used that g is non-abelian. But the assumption will be essential from now on. We need a pair $\{\mu_0, \nu_0\}$ of measures in $M(X)$ such that $\Delta_0 = \langle \mu_0 \otimes \nu_0, \Delta \rangle \neq 0$.

Lemma 1 *If $f \in C_b(X)$ is an odd Wilson function such that $\langle \mu_0, f\rangle = \langle \nu_0, f\rangle = \langle \mu_0 * \nu_0, f\rangle = 0$, then $f = 0$.*

Proof Let $f \in W(g)_o$ satisfies the conditions of the lemma, and let $\mu, \nu, w \in M(X)$ be arbitrary. We take our point of departure in (24) which gives us the two identities

$$\langle \mu * \nu * w, f\rangle + \langle \nu * w * \mu, f\rangle = 2\langle \mu, f\rangle\langle \nu * w, f\rangle + 2\langle \mu, g\rangle\langle \nu * w, f\rangle,$$
$$\langle \nu * w * \mu, f\rangle + \langle w * \mu * \nu, f\rangle = 2\langle \nu, f\rangle\langle w * \mu, g\rangle + 2\langle \nu, g\rangle\langle w * \mu, f\rangle.$$

We get two more identities by interchanging μ and ν in these two identities. Adding all four identities we get, using (24) and that g is central to get reductions on the left and right hand sides, that

$$\langle \mu * w * \nu, f\rangle + \langle \nu * w * \mu, f\rangle = 2\langle \mu, f\rangle\langle \nu * w, g\rangle + 2\langle \nu, f\rangle\langle \mu * w, g\rangle + 2\langle w, f\rangle[2\langle \mu, f\rangle\langle \nu, f\rangle - \langle \mu * \nu, f\rangle]. \tag{25}$$

In (25), we put $\nu = \mu$ and get by the definition of d that

$$\langle \mu * w * \mu, f\rangle = 2\langle \mu, f\rangle\langle \mu * w, g\rangle + \langle w, f\rangle\langle \mu, d\rangle. \tag{26}$$

In (26), we replace w by $\nu * w * \nu$, and add it to the same identity with μ and ν interchanged. Using (26), we get

$$\langle \mu * \nu * w * \nu * \mu, f\rangle + \langle \nu * \mu * w * \mu * \nu, f\rangle = 2\langle \mu, f\rangle[\langle \mu * \nu * w * \nu, g\rangle + \langle \nu, d\rangle\langle \mu * w, f\rangle] + 2\langle \nu, f\rangle[\langle \nu * \mu * w * \mu, f\rangle + \langle \mu, d\rangle\langle \nu * w, f\rangle] + 2\langle w, f\rangle\langle \mu, d\rangle\langle \nu, d\rangle.$$

We note that the left hand side is the left hand side of (25) if we in (25) replace μ by $\nu * \mu$ and ν by $\mu * \nu$, so equating the right hand sides we obtain, using (16) that

$$2\langle \mu, f\rangle[\langle \mu * \nu * w * \nu, g\rangle + \langle \nu, d\rangle\langle \mu * w, g\rangle] + 2\langle \nu, f\rangle[\langle \nu * \mu * w * \mu, g\rangle + \langle \mu, d\rangle\langle w * \nu, g\rangle] + 2\langle w, f\rangle\langle \mu, d\rangle\langle \nu, d\rangle$$
$$= 2\langle \nu * \mu, f\rangle\langle \mu * \nu * w, g\rangle + 2\langle \mu * \nu, f\rangle\langle \nu * \mu * w, g\rangle + 2\langle w, f\rangle [2\langle \nu * \mu, g\rangle^2 - \langle \nu * \mu * \mu * \nu, g\rangle]$$

Using (14), (23), and

$$2\langle \nu * \mu, g\rangle^2 = 2\langle \nu * \mu, g\rangle\langle \mu * \nu, g\rangle = \langle \nu * \mu * \mu * \nu, g\rangle + \langle \nu * \mu * \check{\nu} * \check{\mu}, g\rangle$$

we obtain

$$\begin{aligned}
&2\langle \mu, f\rangle[\langle \mu * \nu * w * \nu, g\rangle + \langle \nu, d\rangle\langle \mu * w, g\rangle + 2\langle \nu, f\rangle[\langle \nu * \mu * w * \mu, g\rangle \\
&\qquad + \langle \mu, d\rangle\langle \nu * w, g\rangle] + 2\langle w, f\rangle\langle \mu, d\rangle\langle \nu, d\rangle \\
&= 2[-\langle \mu * \nu, f\rangle + 2\langle \mu, f\rangle\langle \nu, g\rangle + 2\langle \nu, f\rangle\langle \mu, g\rangle]\langle \mu * \nu * w, g\rangle \\
&\qquad + 2\langle \mu * \nu, g\rangle\langle \nu * \mu * w, g\rangle + 2\langle w, f\rangle[\langle \mu, d\rangle\langle \nu, d\rangle - 2\langle \mu \otimes \nu, \Delta\rangle].
\end{aligned}$$

This simplifies enormously for $\mu = \mu_0$ and $\nu = \nu_0$ because $\langle \mu_0, f\rangle = \langle \nu_0, f\rangle = \langle \mu_0 * \nu_0, f\rangle = 0$. Indeed, we get

$$2\langle w, f\rangle\langle \mu_0, d\rangle\langle \nu_0, d\rangle = 2\langle w, f\rangle[\langle \mu_0, d\rangle\langle \nu_0, d\rangle - 4\langle \mu_0 \otimes \nu_0, \Delta\rangle],$$

which implies that $\langle w, f\rangle = 0$, since $\langle \mu_0 \otimes \nu_0, \Delta\rangle \neq 0$.

Corollary 1 *The central Wilson functions corresponding to a non-abelian d'Alembert function g are the complex multiples of g.*

Proof $g \in W(g)$ is central, so any complex multiple of g is a central Wilson function. To get the converse, let $f \in W(g)$ be a central function. According to Proposition 4(iv) we may write it in the form $f = cg + f_0$ where $c \in \mathbb{C}$ and $f_0 \in W(g)_o$. It suffices to prove that $f_0 = 0$. Noting that $f_0 = f - cg$ is central, we get from (24) that

$$\langle \mu * \nu, f_0\rangle = \langle \mu, f_0\rangle\langle \nu, g\rangle + \langle \nu, f_0\rangle\langle \mu, g\rangle \text{ for all } \mu, \nu \in M(X).$$

That we use a number of times in the following computation

$$\begin{aligned}
\langle \nu, g_\mu\rangle\langle \mu, f_0\rangle &= \langle \mu * \nu, g\rangle\langle \mu, f_0\rangle - \langle \mu, g\rangle\langle \nu, g\rangle\langle \mu, f_0\rangle \\
&= \langle \mu * \nu, g\rangle\langle \mu, f_0\rangle + \langle \mu * \nu, f_0\rangle\langle \mu, g\rangle - \langle \mu, g\rangle[\langle \mu, f_0\rangle\langle \nu, g\rangle \\
&\qquad + \langle \nu, f_0\rangle\langle \mu, g\rangle] - \langle \mu, g\rangle\langle \nu, g\rangle\langle \mu, f_0\rangle \\
&= \langle \mu * \mu * \nu, f_0\rangle - \langle \mu, g\rangle^2\langle \nu, f_0\rangle - 2\langle \mu, g\rangle\langle \nu, g\rangle\langle \mu, f_0\rangle \\
&= \langle \mu * \mu, f_0\rangle\langle \nu, g\rangle + \langle \mu * \mu, g\rangle\langle \nu, f_0\rangle - \langle \mu, g\rangle^2\langle \nu, f_0\rangle \\
&\qquad - 2\langle \mu, g\rangle\langle \nu, g\rangle\langle \mu, f_0\rangle \\
&= 2\langle \mu, f_0\rangle\langle \mu, g\rangle\langle \nu, g\rangle + [\langle \mu * \mu, g\rangle \\
&\qquad - \langle \mu, g\rangle^2]\langle \nu, f_0\rangle - 2\langle \mu, g\rangle\langle \nu, g\rangle\langle \mu, f_0\rangle \\
&= [\langle \mu * \mu, g\rangle - \langle \mu, g\rangle^2]\langle \nu, f_0\rangle \\
&= \langle \mu, g_\mu\rangle\langle \nu, f_0\rangle
\end{aligned}$$

Taking first $\mu = \mu_0$ and $\nu = \nu_0$, and then $\mu = \nu_0$ and $\nu = \mu_0$ we find that

$$\langle \mu_0, g_{\mu_0} \rangle \langle \nu_0, f_0 \rangle - \langle \nu_0, g_{\mu_0} \rangle \langle \mu_0, f_0 \rangle = 0,$$

and

$$\langle \mu_0, g_{\nu_0} \rangle \langle \nu_0, f_0 \rangle - \langle \nu_0, g_{\nu_0} \rangle \langle \mu_0, f_0 \rangle = 0.$$

This is a system of equations in the unknowns $\langle \nu_0, f_0 \rangle$ and $\langle \mu_0, f_0 \rangle$. Its determinant

$$\langle \mu_0, g_{\mu_0} \rangle \langle \nu_0, g_{\nu_0} \rangle - \langle \nu_0, g_{\mu_0} \rangle^2 = \langle \mu_0 \otimes \nu_0, \Delta \rangle \neq 0,$$

then $\langle \nu_0, f_0 \rangle = \langle \mu_0, f_0 \rangle = 0$ and

$$\langle \mu_0 * \nu_0, f_0 \rangle = \langle \mu_0, f_0 \rangle \langle \nu_0, g \rangle + \langle \nu_0, f_0 \rangle \langle \mu_0, g \rangle = 0.$$

By Lemma 1, we conclude that $f_0 = 0$.

Definition 8 Let $g \in C_b(X)$ be a non-abelian d'Alembert function on X, let

$$f_1 := g_{\mu_0}, \quad f_2 := g_{\nu_0} \text{and } \ f_3 := (\frac{R(\mu_0 * \nu_0) - R(\nu_0 * \mu_0)}{2})g,$$

and define $\alpha, \beta, \gamma \in \mathbb{C}$ by

$$\alpha := \frac{1}{\Delta_0} \langle \mu_0, g_{\mu_0} \rangle, \ \beta := \frac{1}{\Delta_0} \langle \nu_0, g_{\mu_0} \rangle \text{ and } \gamma := \frac{1}{\Delta_0} \langle \nu_0, g_{\nu_0} \rangle.$$

Lemma 2 *(i) We list a couple of formulas that will be used later*

$$\alpha\gamma - \beta^2 = \frac{1}{\Delta_0}, \tag{27}$$

$$\langle \mu_0 * \nu_0, f_1 \rangle = \langle \nu_0 * \mu_0, f_1 \rangle, \tag{28}$$

$$\langle \mu_0 * \nu_0, f_2 \rangle = \langle \nu_0 * \mu_0, f_2 \rangle, \tag{29}$$

$$\langle \mu_0, f_3 \rangle = \langle \nu_0, f_3 \rangle = 0, \tag{30}$$

$$\langle \mu_0 * \nu_0, f_3 \rangle = -\langle \nu_0 * \mu_0, f_3 \rangle = -\Delta_0. \tag{31}$$

(ii) The functions f_1, f_2, and f_3 presented in Definition 8 are odd.

Proof (i) The first equality follows from the definition of Δ. The equalities (28) and (29) follow from (11). For the last statement of (i) we refer to (17).

(ii) We have seen in Proposition 2 that $f_1 := g_{\mu_0}$ and $f_2 := g_{\nu_0}$ are odd functions, so it is left to verify, using the fact that g is central and invariant, and that the functions $g_\mu : \mu \in M(X)$ are odd, that f_3 is odd:

$$\langle \check{\mu}, f_3 \rangle = \left\langle \check{\mu}, \frac{R(\mu_0 * \nu_0) - R(\nu_0 * \mu_0)}{2} g \right\rangle$$

$$
\begin{aligned}
&= \frac{1}{2}(\langle \check{\mu} * \mu_0 * \nu_0, g\rangle - \langle \check{\mu} * \nu_0 * \mu_0, g\rangle)\\
&= \frac{1}{2}(\langle \mu_0 * \nu_0 * \check{\mu}, g\rangle - \langle \nu_0 * \mu_0 * \check{\mu}, g\rangle)\\
&= \frac{1}{2}(\langle \check{\mu}, g_{\mu_0*\nu_0}\rangle + \langle \mu_0 * \nu_0, g\rangle\langle \check{\mu}, g\rangle - \langle \check{\mu}, g_{\nu_0*\mu_0}\rangle - \langle \nu_0 * \mu_0, g\rangle\langle \check{\mu}, g\rangle)\\
&= \frac{1}{2}(-\langle \mu, g_{\mu_0*\nu_0}\rangle + \langle \mu_0 * \nu_0, g\rangle\langle \mu, g\rangle + \langle \mu, g_{\nu_0*\mu_0}\rangle - \langle \nu_0 * \mu_0, g\rangle\langle \mu, g\rangle)\\
&= -\frac{1}{2}(\langle \mu_0 * \nu_0 * \mu, g\rangle - \langle \nu_0 * \mu_0 * \mu, g\rangle)\\
&= -\left\langle \mu, \frac{R(\mu_0 * \nu_0) - R(\nu_0 * \mu_0)}{2} g\right\rangle\\
&= -\langle \mu, f_3\rangle,
\end{aligned}
$$

so f_3 is odd.

The next theorem is one of the important results of this chapter. It was derived for groups in [25].

Theorem 5 *Let $(X, *)$ be a hypergroup and $g \in C_b(X)$ be a non-abelian d'Alembert function, then*

(a) $\dim W(g) = 4$.

(b) A basis of $W(g)_o$ is $\{f_1, f_2, f_3\}$. If $f \in W(g)_o$ then

$$
\begin{aligned}
f = &\left[\gamma\langle \mu_0, f\rangle - \beta\langle \nu_0, f\rangle\right] f_1 + [-\beta\langle \mu_0, f\rangle + \alpha\langle \nu_0, f\rangle] f_2 \\
&- \frac{1}{2\Delta_0}[\langle \mu_0 * \nu_0, f\rangle - \langle \nu_0 * \mu_0, f\rangle] f_3,
\end{aligned}
\tag{32}
$$

where the functions f_1, f_2, f_3 and the constants α, β, γ are introduced in the Definition 8.

(c) $W(g) = span\,\{L(\mu)g : \mu \in M(X)\} = span\,\{R(\mu)g : \mu \in M(X)\}$.

(d) As operators on $W(g)$ we have the identity

$$
L(\mu) + L(\check{\mu}) = 2\langle \mu, g\rangle I. \tag{33}
$$

(e) The matrix-coefficients of subrepresentations of $L/W(g)$ and $R/W(g)$ are in $W(g)$.

(f) If E is an L-invariant subspace of $W(g)$. Then, $\chi_{L/E} = dim(E)g$ where the character $\chi_{L/E}$ of L/E is defined by

$$
\chi_{L/E}(x) = tr(L(x)/E), x \in X.
$$

Proof (a) follows from (b) and Proposition 4(iv). (b) We first show that the formula (32) holds. Let h be the function on X defined by

$$
h = \left[\gamma\langle \mu_0, f\rangle - \beta\langle \nu_0, f\rangle\right] f_1 + [-\beta\langle \mu_0, f\rangle + \alpha\langle \nu_0, f\rangle] f_2 -
$$

$$\frac{1}{2\Delta_0}[\langle\mu_0 * \nu_0, f\rangle - \langle\nu_0 * \mu_0, f\rangle]f_3.$$

Since both f and h belong to $W(g)_0$ it suffices by Lemma 1, to show that $\langle\mu_0, f\rangle = \langle\mu_0, h\rangle$, $\langle\nu_0, f\rangle = \langle\nu_0, h\rangle$ and $\langle\mu_0 * \nu_0, f\rangle = \langle\mu_0 * \nu_0, h\rangle$. Noting that $\langle\mu_0, f_3\rangle = \langle\nu_0, f_3\rangle = 0$, it is easy to see that $\langle\mu_0, f\rangle = \langle\mu_0, h\rangle$ and that $\langle\nu_0, f\rangle = \langle\nu_0, h\rangle$, so it is left to verify that $\langle\mu_0 * \nu_0, h\rangle = \langle\mu_0 * \nu_0, f\rangle$.

Since $\langle\mu_0 * \nu_0, f_3\rangle = -\Delta_0$ by using the formula (23) and (30), we get that

$$\begin{aligned}
&\langle\mu_0 * \nu_0, h\rangle - \langle\mu_0 * \nu_0, f\rangle \\
&= \left[\gamma\langle\mu_0, f\rangle - \beta\langle\nu_0, f\rangle\right]\langle\mu_0 * \nu_0, g_{\mu_0}\rangle + [-\beta\langle\mu_0, f\rangle + \alpha\langle\nu_0, f\rangle]\langle\mu_0 * \nu_0, g_{\nu_0}\rangle \\
&\quad + \frac{1}{2}(\langle\mu_0 * \nu_0, f\rangle - \langle\nu_0 * \mu_0, f\rangle) - \langle\mu_0 * \nu_0, f\rangle \\
&= \left[\gamma\langle\mu_0, f\rangle - \beta\langle\nu_0, f\rangle\right]\langle\mu_0 * \nu_0, g_{\mu_0}\rangle + [-\beta\langle\mu_0, f\rangle + \alpha\langle\nu_0, f\rangle]\langle\mu_0 * \nu_0, g_{\nu_0}\rangle \\
&\quad - \langle\mu_0, f\rangle\langle\nu_0, g\rangle - \langle\nu_0, f\rangle\langle\mu_0, g\rangle \\
&= \langle\mu_0, f\rangle\left\{\gamma\langle\mu_0 * \nu_0, g_{\mu_0}\rangle - \beta\langle\mu_0 * \nu_0, g_{\nu_0}\rangle - \langle\nu_0, g\rangle\right\} \\
&\quad + \langle\nu_0, f\rangle\left\{-\beta\langle\mu_0 * \nu_0, g_{\mu_0}\rangle + \alpha\langle\mu_0 * \nu_0, g_{\nu_0}\rangle - \langle\mu_0, g\rangle\right\}.
\end{aligned}$$

By (11), we find that the coefficient of $\langle\mu_0, f\rangle$ vanishes. Indeed

$$\begin{aligned}
&\Delta_0\left\{\gamma\langle\mu_0 * \nu_0, g_{\mu_0}\rangle - \beta\langle\mu_0 * \nu_0, g_{\nu_0}\rangle\right\} \\
&= \langle\nu_0, g_{\nu_0}\rangle\left[\langle\mu_0, g\rangle\langle\nu_0, g_{\mu_0}\rangle + \langle\mu_0, g_{\mu_0}\rangle\langle\nu_0, g\rangle\right] \\
&\quad - \langle\nu_0, g_{\mu_0}\rangle\left[\langle\nu_0, g\rangle\langle\mu_0, g_{\nu_0}\rangle + \langle\nu_0, g_{\nu_0}\rangle\langle\mu_0, g\rangle\right] - \Delta_0\langle\nu_0, g\rangle \\
&= \left[\langle\nu_0, g_{\mu_0}\rangle\langle\mu_0, g_{\mu_0}\rangle - \langle\nu_0, g_{\mu_0}\rangle^2\right]\langle\nu_0, g\rangle - \Delta_0\langle\nu_0, g\rangle = 0.
\end{aligned}$$

Similar arguments show that the coefficient of $\langle\nu_0, f\rangle$ vanishes. We have now shown that the formula (32) holds, so that $\{f_1, f_2, f_3\}$ spans $W(g)_o$. It remains to show that $\{f_1, f_2, f_3\}$ is a linearly independent set. So assume that $c_1 f_1 + c_2 f_2 + c_3 f_3 = 0$, where $c_1, c_2, c_3 \in \mathbb{C}$ are constants. Using $\langle\mu_0, f_3\rangle = \langle\nu_0, f_3\rangle = 0$, we find that

$$\begin{cases} c_1\langle\mu_0, f_1\rangle + c_2\langle\mu_0, f_2\rangle = 0 \\ c_1\langle\nu_0, f_1\rangle + c_2\langle\nu_0, f_2\rangle = 0 \end{cases},$$

The determinant of this system is

$$\begin{vmatrix} \langle\mu_0, f_1\rangle & \langle\mu_0, f_2\rangle \\ \langle\nu_0, f_1\rangle & \langle\nu_0, f_2\rangle \end{vmatrix} = \begin{vmatrix} \langle\mu_0, g_{\mu_0}\rangle & \langle\mu_0, g_{\nu_0}\rangle \\ \langle\nu_0, g_{\mu_0}\rangle & \langle\nu_0, g_{\nu_0}\rangle \end{vmatrix} =$$

$$\langle\mu_0, g_{\mu_0}\rangle\langle\nu_0, g_{\nu_0}\rangle - \langle\nu_0, g_{\mu_0}\rangle^2 = \Delta_0 \neq 0,$$

so $c_1 = c_2 = 0$. Appliquing at $\mu_0 * \nu_0$, we now find that $0 = c_3\langle\mu_0 * \nu_0, f_3\rangle = c_3(-\Delta_0) = -c_3\Delta_0$, so that also $c_3 = 0$.

(*c*) As mentioned in Proposition 4, $W(g)$ is an invariant under the left representation L of X. Now $g \in W(g)$, so $span\,\{L(\mu)g : \mu \in M(X)\} \subseteq W(g)$. On the other hand

$$\langle \mu, g_\nu \rangle = \langle \nu * \mu, g \rangle - \langle \mu, g \rangle \langle \nu, g \rangle$$
$$= \langle \mu, L(\check{\nu})g \rangle - \langle \nu, g \rangle \langle \mu, L(\delta_e)g \rangle,$$

which means that $g_\nu = L(\check{\nu})g - \langle \nu, g \rangle L(\delta_e)g \in span\{L(\mu)g : \mu \in M(X)\}$ for any $\nu \in M(X)$. As R and L commute, we have $f_3 \in span\{L(\mu)g : \mu \in M(X)\}$. Since $\{g, g_{\mu_0}, g_{\nu_0}, \frac{R(\mu_0 * \nu_0) - R(\nu_0 * \mu_0)}{2} g\}$ is a basis of $W(g)$, then $W(g) \subseteq span\{L(\mu)g : \mu \in M(X)\}$.Thus

$$W(g) = span\{L(\mu)g : \mu \in M(X)\}.$$

The last equality of (c) comes from the fact that g is central, so that $L(\check{\mu})g = R(\mu)g$ for any $\mu \in M(X)$.

(d) Using that g is central we get that

$$\langle \nu, (L(\mu) + L(\check{\mu}))g \rangle = \langle \check{\mu} * \nu, g \rangle + \langle \mu * \nu, g \rangle$$
$$= 2\langle \nu, g \rangle \langle \mu, g \rangle \text{ for all } \mu, \nu \in M(X),$$

which means that $[L(\mu) + L(\check{\mu})]g = 2\langle \mu, g \rangle g$ for all $\mu, \nu \in M(X)$.

For any $f = \sum_{j=1}^{n} c_j R(\mu_j)g \in W(g)$ we get, since L and R commute, that

$$\left(L(\mu) + L(\check{\mu})\right) f = \left(L(\mu) + L(\check{\mu})\right) \sum_{j=1}^{n} c_j R(\mu_j)g$$
$$= \sum_{j=1}^{n} c_j R(\mu_j) \left(L(\mu) + L(\check{\mu})\right) g$$
$$= \sum_{j=1}^{n} c_j R(\mu_j) 2\langle \mu, g \rangle g = 2\langle \mu, g \rangle f,$$

proving the equality (33).

(*e*) Fix a basis of $W(g)$ and let for any $x \in X$

$$M(x) = \begin{pmatrix} a_{11}(x) & a_{12}(x) & a_{13}(x) & a_{14}(x) \\ a_{21}(x) & a_{22}(x) & a_{23}(x) & a_{24}(x) \\ a_{31}(x) & a_{32}(x) & a_{33}(x) & a_{34}(x) \\ a_{41}(x) & a_{42}(x) & a_{43}(x) & a_{44}(x) \end{pmatrix},$$

be the corresponding matrix for $L(x)$. Noting that

$$\langle \mu, M \rangle = \begin{pmatrix} \langle \mu, a_{11} \rangle & \langle \mu, a_{12} \rangle & \langle \mu, a_{13} \rangle & \langle \mu, a_{14} \rangle \\ \langle \mu, a_{21} & \langle \mu, a_{22} \rangle & \langle \mu, a_{23} \rangle & \langle \mu, a_{24} \rangle \\ \langle \mu, a_{31} \rangle & \langle \mu, a_{32} \rangle & \langle \mu, a_{33} \rangle & \langle \mu, a_{34} \rangle \\ \langle \mu, a_{41} \rangle & \langle \mu, a_{42} \rangle & \langle \mu, a_{43} \rangle & \langle \mu, a_{44} \rangle \end{pmatrix},$$

for all $\mu \in M(X)$. The identity (33) translates to

$$\langle \mu, M \rangle + \langle \check{\mu}, M \rangle = 2\langle \mu, g \rangle \mathbb{I}_4,$$

which means, for any $\mu \in M(X)$ and $1 \leq i, j \leq 4$, that we have

$$\langle \mu, a_{ii} \rangle + \langle \check{\mu}, a_{ii} \rangle = 2\langle \mu, g \rangle \text{ and } \langle \mu, a_{ij} \rangle + \langle \check{\mu}, a_{ij} \rangle = 0 \text{ for } i \neq j. \tag{34}$$

Since L is a representation of X, then

$$\langle \mu * \nu, M \rangle = \langle \mu, M \rangle \langle \nu, M \rangle \text{ for all } \mu, \nu \in M(X),$$

this identity translates to

$$\langle \mu * \nu, a_{ij} \rangle = \sum_{k=1}^{4} \langle \mu, a_{ik} \rangle \langle \nu, a_{kj} \rangle, 1 \leq i, j \leq 4 \tag{35}$$

Using the formulas (34) and (35), we get that

$$\begin{aligned}
\langle \mu * \nu, a_{ij} \rangle + \langle \mu * \check{\nu}, a_{ij} \rangle &= \sum_{k=1}^{4} \langle \mu, a_{ik} \rangle \langle \nu, a_{kj} \rangle + \sum_{k=1}^{4} \langle \mu, a_{ik} \rangle \langle \check{\nu}, a_{kj} \rangle \\
&= \sum_{k=1}^{4} \langle \mu, a_{ik} \rangle \left(\langle \nu, a_{kj} \rangle + \langle \check{\nu}, a_{kj} \rangle \right) \\
&= \langle \mu, a_{ij} \rangle \left(\langle \nu, a_{jj} \rangle + \langle \check{\nu}, a_{jj} \rangle \right) \\
&= 2\langle \mu, a_{ij} \rangle \langle \nu, g \rangle,
\end{aligned}$$

which means that the matrix-coefficients of subrepresentations of $L/W(g)$ are in $W(g)$.

(f) Let E be an L-invariant subspace of $W(g)$. The character $\chi_{L/E}$ is a central function and hence (by (e)) a central Wilson function. By Corollary 1, there exists a constant $c \in \mathbb{C}$ such that $\chi_{L/E} = cg$. Evaluating at $e \in X$ we find that $c = cg(e) = \chi_{L/E}(e) = tr(\mathbb{I}_E) = dim(E)$. So $g = \frac{1}{\dim E} tr(\chi_{L/E})$.

Lemma 3 *There are no* 1*-dimensional* L*-invariant subspaces of* $W(g)$ *where* L *is the left regular representation of* X *on* $W(g)$.

Proof We assume to the contrary that $\mathbb{C}f$ is a 1-dimensional L-invariant subspace of $W(g)$. Then, $f \neq 0$ and $L(\mu)f$ has for any $\mu \in M(X)$ the form $L(\mu)f = \langle \mu, \chi \rangle f$, where $\langle \mu, \chi \rangle \in \mathbb{C}$. Since L is a representation of X, then

$$\langle \mu * \nu, \chi \rangle f = L(\mu) \langle \nu, \chi \rangle f = \langle \mu, \chi \rangle \langle \nu, \chi \rangle f \text{ for all } \mu, \nu \in M(X),$$

which implies that

$$\langle \mu * \nu, \chi \rangle = \langle \mu, \chi \rangle \langle \nu, \chi \rangle \text{ for all } \mu, \nu \in M(X),$$

this means that χ is a multiplicative function of X and $\chi(e) = 1$. Evaluating $L(\mu)f = \langle \mu, \chi \rangle f$ at e we get that

$$\langle \check{\mu}, f \rangle = \langle \mu, \chi \rangle f(e).$$

In particular, $f(e) \neq 0$, because otherwise $f = 0$. Since f is a Wilson function then so is the function $\check{\chi} = \frac{f}{f(e)}$, thus, we get

$$\langle \mu * \nu, \check{\chi} \rangle + \langle \mu * \check{\nu}, \check{\chi} \rangle = 2\langle \mu, \check{\chi} \rangle \langle \nu, g \rangle.$$

For $\mu = \delta_e$, we obtain $\langle \nu, g \rangle = \frac{1}{2} \left\{ \langle \nu, \check{\chi} \rangle + \langle \nu, \chi \rangle \right\}$. But this implies the contradiction that g is non-abelian, because χ is multiplicative function and hence abelian.

However, there are 2-dimensional L-invariant vector subspaces of $W(g)$.

Proposition 5 *Fix $\delta \in \mathbb{C}$ such that $\delta^2 = -\Delta_0$. Then*

$$W(g)^{\pm} := \{ f \in W(g) : \frac{R(\mu_0 * \nu_0) - R(\nu_0 * \mu_0)}{2} f = \pm \delta f \},$$

are two 2-dimensional L-invariant subspaces of $W(g)$, and $W(g) = W(g)^+ \oplus W(g)^-$ as direct sum. Furthermore,

(a) $\pi := L/W(g)^+$ is a continuous and irreducible representation of X on $W(g)^+$ with character $\chi_\pi = 2g$.
(b) $\pi(\check{\mu}) = adj(\pi(\mu))$ for all $\mu \in M(X)$, where $adj : \mathcal{L}(W(g)^+) \to \mathcal{L}(W(g)^+)$ the adjugate map defined by

$$adj \begin{pmatrix} a & c \\ b & d \end{pmatrix} = \begin{pmatrix} d & -c \\ -b & a \end{pmatrix}.$$

Proof The statement about the invariance of $W(g)^+$ and $W(g)^-$is easily deduced from the fact that, the left and the right representations commute, L leaves any eigenspace of $\frac{R(\mu_0 * \nu_0) - R(\nu_0 * \mu_0)}{2}$ invariant. We start by proving that

$$\left(\frac{R(\mu_0 * \nu_0) - R(\nu_0 * \mu_0)}{2} \right)^2 = -\Delta_0 I \text{ as operators on } W(g). \tag{36}$$

We know from (33) that

$$\left(\frac{R(\mu_0 * \nu_0) - R(\nu_0 * \mu_0)}{2} \right)^2 g = -\Delta_0 g.$$

Since the left and right representations commute, we get that

$$\left(\frac{R(\mu_0 * \nu_0) - R(\nu_0 * \mu_0)}{2} \right)^2 f = -\Delta_0 f,$$

for any f of the form $f = \sum_{j=1}^{n} c_j L(\mu_j)g$, where $c_j \in \mathbb{C}$ and $\mu_j \in M(X)$, $j = 1, 2\ldots, n$. But these functions f constitute $W(g)$ by Theorem 5(c). This proves (36). Let $T := \frac{R(\mu_0 * \nu_0) - R(\nu_0 * \mu_0)}{2\delta}$, so that $T^2 = I$. Then

$$f = \frac{f + Tf}{2} + \frac{f - Tf}{2},$$

is the desired decomposition of $f \in W(g)$ into elements from $W(g)^+$ and $W(g)^-$. So $W(g) = W(g)^+ \oplus W(g)^-$.

We next prove that $\dim W(g)^+ = \dim W(g)^- = 2$. Let f_3 be the function defined in Definition 8. The computation

$$\begin{aligned}
\frac{R(\mu_0 * \nu_0) - R(\nu_0 * \mu_0)}{2}(f_3 + \delta g) &= \frac{R(\mu_0 * \nu_0) - R(\nu_0 * \mu_0)}{2} f_3 \\
&\quad + \delta \frac{R(\mu_0 * \nu_0) - R(\nu_0 * \mu_0)}{2} g \\
&= \left(\frac{R(\mu_0 * \nu_0) - R(\nu_0 * \mu_0)}{2} \right)^2 g + \delta f_3 \\
&= -\Delta_0 g + \delta f_3 \\
&= \delta \left(f_3 + \frac{-\Delta_0}{\delta} g \right) \\
&= \delta (f_3 + \delta g),
\end{aligned}$$

shows that $f_3 + \delta g \in W(g)^+$. Similarly we find that $f_3 - \delta g \in W(g)^-$. Thus, both $W(g)^+$ and $W(g)^-$ have dimensions, at least one. However, there are no 1-dimensional invariant subspaces (by Lemma 3), so both $W(g)^+$ and $W(g)^-$ must be at least 2-dimensional. But $\dim W(g) = 4$ (Theorem 5(a)), so none of them can have dimension strictly bigger than 2.

(a) $W(g)^+$ is irreducible under L, because a nontrivial invariant subspace of it would be 1-dimensional, and there are no such subspaces according to Lemma 3. By definition of L we see that π is continuous [3]. The statement about the character is immediate from Theorem 5(f).

(b) Fix a basis of $W(g)^+$ and let for any $\mu \in M(X)$

$$\pi(\mu) = \begin{pmatrix} \langle \mu, a_{11} \rangle & \langle \mu, a_{12} \rangle \\ \langle \mu, a_{21} \rangle & \langle \mu, a_{22} \rangle \end{pmatrix} \in M(2, \mathbb{C}),$$

be the corresponding matrix for $\pi(\mu)$. The formula $L(\mu) + L(\mu) = 2\langle \mu, g \rangle I$ on $W(g)$ translates to

$$\begin{pmatrix} \langle \mu, a_{11} \rangle & \langle \mu, a_{12} \rangle \\ \langle \mu, a_{21} \rangle & \langle \mu, a_{22} \rangle \end{pmatrix} + \begin{pmatrix} \langle \check{\mu}, a_{11} \rangle & \langle \check{\mu}, a_{12} \rangle \\ \langle \check{\mu}, a_{21} \rangle & \langle \check{\mu}, a_{22} \rangle \end{pmatrix} = \begin{pmatrix} 2\langle \mu, g \rangle & 0 \\ 0 & 2\langle \mu, g \rangle \end{pmatrix}$$

while the formula $tr(\pi(\mu)) = \chi_\pi(\mu) = 2\langle\mu, g\rangle$ from (a) translates to $\langle\mu, a_{11}\rangle + \langle\mu, a_{22}\rangle = 2\langle\mu, g\rangle$. Using these two formulas, we find for any $\mu \in M(X)$ that

$$\begin{aligned}\langle\check{\mu}, a_{11}\rangle &= 2\langle\mu, g\rangle - \langle\mu, a_{11}\rangle = 2\langle\mu, g\rangle - [\langle\mu, a_{11}\rangle + \langle\mu, a_{22}\rangle] + \langle\mu, a_{22}\rangle \\ &= 2\langle\mu, g\rangle - 2\langle\mu, g\rangle + \langle\mu, a_{22}\rangle = \langle\mu, a_{22}\rangle.\end{aligned}$$

Replacing μ by $\check{\mu}$ we get that $\langle\check{\mu}, a_{22}\rangle = \langle\mu, a_{11}\rangle$. Furthermore, we find that $\langle\check{\mu}, a_{12}\rangle = -\langle\mu, a_{12}\rangle$ and $\langle\check{\mu}, a_{21}\rangle = -\langle\mu, a_{21}\rangle$. Thus,

$$\pi(\check{\mu}) = \begin{pmatrix} \langle\mu, a_{22}\rangle & -\langle\mu, a_{12}\rangle \\ -\langle\mu, a_{21}\rangle & \langle\mu, a_{11}\rangle \end{pmatrix} = adj \begin{pmatrix} \langle\mu, a_{11}\rangle & \langle\mu, a_{12}\rangle \\ \langle\mu, a_{21}\rangle & \langle\mu, a_{22}\rangle \end{pmatrix} = adj(\pi(\mu)).$$

5 The General Case

We return to general d'Alembert functions on hypergroups, i.e., nonzero solution $g \in C_b(X)$ of (4). The first main result of this chapter is the following Davison's structure theorem [10, 25].

Theorem 6 *(a) The d'Alembert functions on $(X, *)$ are the functions of the form*

$$g(x) = \frac{1}{2}tr(\pi(x)),$$

where π range over the 2-dimensional continuous representations of X for which $\pi(\check{x}) = adj o \pi(x)$ for all $x \in X$.

(b) $g = \frac{1}{2} tr o \pi$ is non-abelian if and only if π is irreducible. If g is non-abelian, then π is unique up to equivalence.

(c) If g is abelian, then π can be chosen as a direct sum of two 1-dimensional representations of X, i.e., of two multiplicative functions. If χ is one of these multiplicative functions, then $\check{\chi}$ is the other.

Proof (a) It is easy to verify that any function of the form $g = \frac{1}{2}\chi_\pi$, where π is a 2-dimensional continuous representation of X such that $\pi(\check{x}) = adj o \pi(x)$ for all $x \in X$, is a d'Alembert function on X.

The converse, if g is non-abelian d'Alembert functions on X, then we refer to Proposition 5 for a representation π with the desired properties. If g is an abelian d'Alembert functions on X, we may in the notation of theorem 3 as π choose

$$\pi(x) = \begin{pmatrix} \chi(x) & 0 \\ 0 & \chi(\check{x}) \end{pmatrix} \quad \text{for all } x \in X.$$

(b) Assume first that $g = \frac{1}{2}\chi_\pi$ is non-abelian, we show that π is irreducible. By contradiction, if π were no irreducible, then there would exists an invariant 1-dimensional subspace $\mathbb{C}.f$ of the representation space. So $\pi(\mu)f = \chi(\mu)f$ for all

$\mu \in M(X)$, where $\chi(\mu) \in \mathbb{C}$. Let $\{f, h\}$ be a basis of the representation space with respect to that basis, π takes the form

$$\pi(\mu) = \begin{pmatrix} \chi(\mu) & a(\mu) \\ 0 & \chi_1(\mu) \end{pmatrix} \quad \text{for all } \mu \in M(X),$$

where a, χ_1 are complex valued functions. It follows from π being a representation that χ and $\chi_1 : X \longrightarrow \mathbb{C}$ are multiplicative functions on X. Multiplicative function on X is abelian function on X. Hence, so does $g = \frac{1}{2}\chi_\pi = \frac{1}{2}(\chi + \chi_1)$, i.e., g is abelian. But that contradicts our assumption.

Assume conversely that π is irreducible. We prove that g is non-abelian by contradiction. If g were abelian, then g can, according to Theorem 3, be written in the form $g = \frac{1}{2}(\phi + \check{\phi})$ where ϕ is multiplicative function on X. Thus, $\chi_\pi = \phi + \check{\phi}$. Viewing ϕ and $\check{\phi}$ as 1-dimensional and hence, irreducible representation, we see that the three multiplicative functions χ_π, ϕ, and $\check{\phi}$ are not linearly independent. But multiplicative functions are linearly independent (see [19, Proposition 3]). If ϕ and $\check{\phi}$ are equivalent, they coincide, so χ_π and ϕ are linearly dependent, and thus π and ϕ are equivalent. But they cannot be, being of dimension 2 and 1, respectively. Other possibilities cannot occur; π and ϕ are not equivalent, being of different dimension. Similarly for π and $\check{\phi}$.

The essential uniqueness of π is a consequence of the fact that characters of inequivalent irreducible finite-dimensional representations are linearly independent ([7], Proposition 2, Chap. *VIII*, §13, *no.* 3).

(*c*) We saw in the beginning of the proof that π can be chosen as a direct sum of two 1-dimensional representation of $M(X)$, i.e., of two multiplicative functions. The uniqueness follows from ([19], Proposition 3).

As an immediate consequence of Theorem 6 we have the following Corollary.

Corollary 2 *If g is a d'Alembert function on $(X, *)$ then there is a continuous and multiplicative map $\varphi : X \longrightarrow Mat_2(\mathbb{C})$ with $\check{\varphi} = adj \circ \varphi$ such that*

$$g = \frac{1}{2} tr \circ \varphi,$$

where $Mat_2(\mathbb{C})$ is the space of complex matrix of order 2 and $adj : Mat_2(\mathbb{C}) \longrightarrow Mat_2(\mathbb{C})$ $\begin{pmatrix} a & b \\ c & d \end{pmatrix} \longrightarrow \begin{pmatrix} d & -b \\ -c & a \end{pmatrix}$.

6 Superstability of the D'Alembert Equation (4)

There is a strong stability phenomenon which is known as a superstability. An equation of homomorphism is called superstable if each approximate homomorphism is actually a true homomorphism. This property was first observed by J. Baker, J.

Lawrence, and F. Zorzitto [4] in the following Theorem: *Let V be a vector space. If a function $f : V \longrightarrow \mathbb{R}$ satisfies the inequality*

$$|f(x+y) - f(x)f(y)| \leq \varepsilon$$

for some $\varepsilon > 0$ and for all $x; y \in V$. Then, either f is a bounded function or

$$f(x+y) = f(x)f(y), \quad x, y \in V$$

Later this result was generalized by J. Baker [3] and L. Székelyhidi [29].

Székelyhidy in [27] (Theorem 7.1), dealt with the superstability of exponential (i.e., multiplicative) functions on hypergroups. Precisely, he proved the following result. *Let K be a hypergroup and let $f, g : K \longrightarrow \mathbb{C}$ be continuous functions with the property that the function*

$$y \longmapsto \int_K f d(\delta_x * \delta_y) - f(x)g(y),$$

is bounded for all y in K. Then, either f is bounded, or g is exponential (i.e., multiplicative function).

In present section, we shall extend the investigation given by J. Baker [3], L. Székelyhidi [27, 29], R. Badora [2], and E. Elqorachi and M. Akkouchi [11] to the Eq. (4).

In Theorem 7, the superstability of Eq. (4) will be investigated on any hypergroup.

Lemma 4 *Let $\delta > 0$ be given. Assume that the continuous function $f : X \longrightarrow \mathbb{C}$ satisfies the inequality*

$$\left|\langle \delta_x * \delta_y, f\rangle + \langle \delta_x * \delta_{\check{y}}, f\rangle - 2f(x)f(y)\right| \leq \delta, \ x, y \in X \tag{37}$$

If f is unbounded then it satisfies the d'Alembert's long functional equation

$$\langle \delta_x * \delta_y, f\rangle + \langle \delta_x * \delta_{\check{y}}, f\rangle + \langle \delta_y * \delta_x, f\rangle + \langle \delta_{\check{y}} * \delta_x, f\rangle = 4f(x)f(y). \tag{38}$$

Proof Assume that f is an unbounded function satisfying the inequality (37). For all $x, y, z \in X$, we have

$$\begin{aligned}
&|2f(z)| \left|\langle \delta_x * \delta_y, f\rangle + \langle \delta_x * \delta_{\check{y}}, f\rangle + \langle \delta_y * \delta_x, f\rangle + \langle \delta_{\check{y}} * \delta_x, f\rangle - 4f(x)f(y)\right| \\
&= \big|2f(z)\langle \delta_x * \delta_y, f\rangle + 2f(z)\langle \delta_x * \delta_{\check{y}}, f\rangle + 2f(z)\langle \delta_y * \delta_x, f\rangle \\
&\quad + 2f(z)\langle \delta_{\check{y}} * \delta_x, f\rangle - 8f(z)f(x)f(y)\big| \\
&= \left| \int_X 2f(z)f(t)d(\delta_x * \delta_y)(t) + \int_X 2f(z)f(t)d(\delta_x * \delta_{\check{y}})(t) \right. \\
&\quad \left. + \int_X 2f(z)f(t)d(\delta_y * \delta_x)(t) + \int_X 2f(z)f(t)d(\delta_{\check{y}} * \delta_x)(t) - 8f(z)f(x)f(y) \right|
\end{aligned}$$

$$
\begin{aligned}
&\le \left|\int_X \big(\langle \delta_z * \delta_t, f\rangle + \langle \delta_z * \delta_{\check{t}}, f\rangle - 2f(z)f(t)\big)\, d(\delta_x * \delta_y)(t)\right| \\
&+ \left|\int_X \big(\langle \delta_z * \delta_t, f\rangle + \langle \delta_z * \delta_{\check{t}}, f\rangle - 2f(z)f(t)\big)\, d(\delta_x * \delta_{\check{y}})(t)\right| \\
&+ \left|\int_X \big(\langle \delta_z * \delta_t, f\rangle + \langle \delta_z * \delta_{\check{t}}, f\rangle - 2f(z)f(t)\big)\, d(\delta_y * \delta_x)(t)\right| \\
&+ \left|\int_X \big(\langle \delta_z * \delta_t, f\rangle + \langle \delta_z * \delta_{\check{t}}, f\rangle - 2f(z)f(t)\big)\, d(\delta_{\check{y}} * \delta_x)(t)\right| \\
&+ \left|\int_X \big(\langle \delta_t * \delta_y, f\rangle + \langle \delta_t * \delta_{\check{y}}, f\rangle - 2f(t)f(y)\big)\, d(\delta_z * \delta_x)(t)\right| \\
&+ \left|\int_X (\langle \delta_t * \delta_x, f\rangle + \langle \delta_t * \delta_{\check{x}}, f\rangle - 2f(t)f(x))\, d(\delta_z * \delta_y)(t)\right| \\
&+ \left|\int_X (\langle \delta_t * \delta_x, f\rangle + \langle \delta_t * \delta_{\check{x}}, f\rangle - 2f(t)f(x))\, d(\delta_z * \delta_{\check{y}})(t)\right| \\
&+ \left|\int_X \big(\langle \delta_t * \delta_y, f\rangle + \langle \delta_t * \delta_{\check{y}}, f\rangle - 2f(t)f(x)\big)\, d(\delta_z * \delta_{\check{x}})(t)\right| \\
&+ |2f(y)|\, |\langle \delta_z * \delta_x, f\rangle + \langle \delta_z * \delta_{\check{x}}, f\rangle - 2f(z)f(x)| \\
&+ 2\,|f(x)|\, \big|\langle \delta_z * \delta_y, f\rangle + \langle \delta_z * \delta_{\check{y}}, f\rangle - 2f(z)f(y)\big| .
\end{aligned}
$$

By virtue of inequality (37), we have

$$
\begin{aligned}
&|2f(z)|\, \big|\langle \delta_x * \delta_y, f\rangle + \langle \delta_x * \delta_{\check{y}}, f\rangle + \langle \delta_y * \delta_x, f\rangle + \langle \delta_{\check{y}} * \delta_x, f\rangle - 4f(x)f(y)\big| \\
&\le 8\delta + 2(|f(y)| + |f(x)|)\delta. \qquad (39)
\end{aligned}
$$

If we fix x, y, the right hand side of the above inequality is bounded function of z. Since f is unbounded, from the preceding (39), we conclude that f is a solution of the d'Alembert long equation(38), which ends the proof.

We have the following result on the superstability of the d'Alembert equation which generalizes the Baker's result on the classical d'Alembert functional equation on an abelian group [3] (Theorem 5).

Theorem 7 *Let $\delta > 0$ be given. Assume that the continuous function $f : X \longrightarrow \mathbb{C}$ satisfies the inequality*

$$\left|\langle \delta_x * \delta_y, f\rangle + \langle \delta_x * \delta_{\check{y}}, f\rangle - 2f(x)f(y)\right| \leq \delta,\ x, y \in X \tag{40}$$

then either

$$|f(x)| \leq \frac{1+\sqrt{1+2\delta}}{2},\quad x \in X,$$

or

$$\langle \delta_x * \delta_y, f\rangle + \langle \delta_x * \delta_{\check{y}}, f\rangle = 2f(x)f(y),\ x, y \in X.$$

Proof Assume that f satisfies inequality (40). If f is bounded, let $A = \sup|f|$, then we get for all $x \in X$ that $|2f(x)f(x)| \leq \delta + 2A$, from which we obtain that $2A^2 - 2A - \delta \leq 0$ such that

$$A \leq \frac{1+\sqrt{1+2\delta}}{2},\quad x \in X.$$

Now we consider the case of f unbounded. For all $x, y, z \in X$, we have

$$\begin{aligned}
&|2f(z)|\left|\langle \delta_x * \delta_y, f\rangle + \langle \delta_x * \delta_{\check{y}}, f\rangle - 2f(x)f(y)\right| \\
&= \left|\int_X 2f(z)f(t)d(\delta_x * \delta_y)(t) + \int_X 2f(z)f(t)d(\delta_x * \delta_{\check{y}})(t) - 4f(x)f(y)f(z)\right| \\
&\leq \left|\int_X \left(\langle \delta_t * \delta_z, f\rangle + \langle \delta_t * \delta_{\check{z}}, f\rangle - 2f(t)f(z)\right) d(\delta_x * \delta_y)(t)\right| \\
&\quad + \left|\int_X \left(\langle \delta_t * \delta_z, f\rangle + \langle \delta_t * \delta_{\check{z}}, f\rangle - 2f(t)f(z)\right) d(\delta_x * \delta_{\check{y}})(t)\right| \\
&\quad + \left|\int_X \left(\langle \delta_x * \delta_t, f\rangle + \langle \delta_x * \delta_{\check{t}}, f\rangle - 2f(x)f(t)\right) d(\delta_y * \delta_z)(t)\right| \\
&\quad + \left|\int_X \left(\langle \delta_x * \delta_t, f\rangle + \langle \delta_x * \delta_{\check{t}}, f\rangle - 2f(x)f(t)\right) d(\delta_y * \delta_{\check{z}})(t)\right| \\
&\quad + \left|\int_X \left(\langle \delta_x * \delta_t, f\rangle + \langle \delta_x * \delta_{\check{t}}, f\rangle - 2f(x)f(t)\right) d(\delta_{\check{z}} * \delta_y)(t)\right| \\
&\quad + \left|\int_X \left(\langle \delta_x * \delta_t, f\rangle + \langle \delta_x * \delta_{\check{t}}, f\rangle - 2f(x)f(t)\right) d(\delta_z * \delta_y)(t)\right|
\end{aligned}$$

$$+\left|\int_X \left(\langle\delta_t * \delta_y, f\rangle + \langle\delta_t * \delta_{\check{y}}, f\rangle - 2f(t)f(y)\right) d(\delta_x * \delta_{\check{z}})(t)\right|$$

$$+\left|\int_X \left(\langle\delta_t * \delta_y, f\rangle + \langle\delta_t * \delta_{\check{y}}, f\rangle - 2f(t)f(y)\right) d(\delta_x * \delta_z)(t)\right|$$

$$+ \left|2f(x)\langle\delta_y * \delta_z, f\rangle + 2f(x)\left\langle\delta_y * \delta_{\check{z}}, f\right\rangle + 2f(x)\left\langle\delta_{\check{z}} * \delta_y, f\right\rangle\right.$$

$$\left.+\, 2f(x)\left\langle\delta_z * \delta_y, f\right\rangle - 8f(x)f(y)f(z)\right|$$

$$+ \left|2\langle\delta_x * \delta_z, f\rangle f(y) + 2\langle\delta_x * \delta_{\check{z}}, f\rangle f(y) - 4f(x)f(y)f(z)\right|$$

$$\leq 8\delta + 2\delta\,|f(y)| + |2f(x)|\,\left|\langle\delta_y * \delta_z, f\rangle + \left\langle\delta_y * \delta_{\check{z}}, f\right\rangle\right.$$

$$\left.+ \left\langle\delta_z * \delta_y, f\right\rangle + \left\langle\delta_{\check{z}} * \delta_y, f\right\rangle - 4f(y)f(z)\right|.$$

In virtue of inequality (40), we obtain

$$|2f(z)|\,\left|\langle\delta_x * \delta_y, f\rangle + \langle\delta_x * \delta_{\check{y}}, f\rangle - 2f(x)f(y)\right|$$

$$\leq 8\delta + 2\delta\,|f(y)| + |2f(x)|\,\left|\langle\delta_y * \delta_z, f\rangle + \left\langle\delta_y * \delta_{\check{z}}, f\right\rangle + \left\langle\delta_z * \delta_y, f\right\rangle + \left\langle\delta_{\check{z}} * \delta_y, f\right\rangle\right.$$

$$\left.-4f(y)f(z)\right|.$$

Or f is unbounded then by Lemma 4 it is a solution of (38). We conclude that

$$|2f(z)|\,\left|\langle\delta_x * \delta_y, f\rangle + \langle\delta_x * \delta_{\check{y}}, f\rangle - 2f(x)f(y)\right| \leq 8\delta + 2\delta\,|f(y)|. \qquad (41)$$

Again the right hand side of (41) as a function of z is bounded for all fixed x, y. Since f is unbounded, from the preceding (41), we conclude that f satisfies the Eq. (4), and the proof of the theorem is finished.

Acknowledgement Our sincere regards and gratitude go to Professor Henrik Stetkær for fruitful discussions and for sending us some of his papers on functional equations.

References

1. Aoki, T.: On the stability of the linear transformation in Banach spaces. J. Math. Soc. Jpn. **2**, 44–66 (1950)
2. Badora, R.: On Hyers–Ulam stability of Wilson's functional equation. Aeq. Math. **60**, 211–218 (2000)
3. Baker, J.A.: The stability of the cosine equation. Proc. Am. Math. Soc. **80**, 411±416 (1980)
4. Baker, J.A., Lawrence, J., Zorzitto, F.: *The stability of the equation* $f(x + y) = f(x)f(y)$. Proc. Am. Math. Soc. **74**, 242–246 (1979)
5. Bouikhalene, B.: On Hyers–Ulam stability of generalised Wilson's equations, J. Inequal. Pure Appl. Math. **5**(4), (2004), Article 100.
6. Bloom, W.R., Hayer, H.: Harmonic analysis of probability measures on hypergroups, Berlin-New York, 1995

7. Bourbaki, N.: Eléments de Mathématiques. Livre II. Algèbre, Hermann 1958
8. Corovei, I.: The cosine functional equation on nilpotent groups. Aeq. Math.**15** , 99–106 (1977)
9. Corovei, I.: The functional equation$f(xy) + f(yx) + f(xy^{-1}) + f(y^{-1}x) = 4f(x)f(y)$ *for nilpotent groups.*, (Romanian, English summary) . Bul. Ştiinţ. Instit. Politehn. Cluj-Napoca Ser. Mat.-Fiz.-Mec Apl. **20**, 25–28 (1977)
10. Davison, T.: D'Alembert's functional equation on topological monoids. . Publ. Math. Debr. **75/1–2**, 41–66 (2009)
11. Elqorachi, E., Akkouchi, M.: The superstability of the generalized d'Alembert functional equation. Georgian Math. J. **10**(3), 503–508 (2003)
12. Hyers, D.H.: On the stability of the linear functional equation. . Proc. Natl. Acad. Sci. U.S.A. **27**, 222–224 (1941)
13. Hyers, D.H., Isac, G.I., and Rassias, Th. M: Stability of functional equations in several variables. Progress in nonlinear differential equations and their applications, 34. Birkhauser , Boston (1998)
14. Jewett, R.I.: Spaces with an abstract convolution of measures. Adv. Math. **18**, 1–101 (1975)
15. Kannappan, P.: The functional equation$f(xy) + f(xy^{-1}) = 2f(x)f(y)$ *for groups*. Proc. Am. Math. Soc. **19**, 69–74 (1968)
16. Orosz, Á. and Székelyhidi, L.: Moment function on polynomial hypergroups in several variables. Publ. Math. Debr. **65**(3–4), 429–438 (2004)
17. Orosz, Á. Sine and cosine equation on discrete polynomial hypergroups. Aeq. Math. **72**, 225–233 (2006)
18. Rassias, Th. M.: On the stability of linear mapping in Banach spaces. Proc. Am. Math. Soc. **72**, 297–300 (1978)
19. Roukbi, A., Zeglami, D.: D'Alembert's functional equations on hypergroups Adv. Pure Appl. Math. **2**, 147–166 (2011)
20. Roukbi, A., Zeglami, D., Kabbaj, S.: Hyers–Ulam stability of Winson's functional equation. Math. Sci. Adv. Appl. **22**, 16–26 (2013)
21. Sinopoulos, P.: Functional equations on semigroups. Aeq. Math. **59**, 255–261 (2000)
22. Stetkær, H.: Functional equation on abelian groups with involution. Aeq. Math. **54**, 144–172 (1997)
23. Stetkær, H.: D'Alembert's functional equations on metabelian groups. Aeq. Math. **59**, 306–320 (2000)
24. Stetkær, H.: D'Alembert's and Wilson's functional equations on step 2 nilpotent groups. Aeq. Math. **67**, 241–262 (2004)
25. Stetkær, H.: Trigonometric functionals equations on groups, Manuscript presented at a talk during the first Spring School Mehdia / CNG, Kenitra, Morocco, April 2009
26. Stetkær, H., Functional Equations on Groups, World Scientic, Hackensack, 2013.
27. Székelyhidi, L.: Functional equations on hypergroups, in: Functional equations, Inequalities and applications, Rassias, Th. M. (ed.), Kluwer , Boston, , 2003, 167–181.
28. Székelyhidi, L.: Functional equations on topological Sturm-Liouville hypergroups. Math. Pannonica **17**/2 (2006), 169–182.
29. Székelyhidi, L.: On a theorem of Baker, Lawrence and Zorzitto. Proc. Am. Math. Soc. **84**, 95–96 (1982)
30. Wilson, W.H.: On a certain related functional equations. Proc. Am. Soc. **26**, 300–312 (1919–1920)
31. Zeglami, D., Kabbaj, S., Charifi, A., Roukbi, A.: $\mu-$ trigonometric functional equations and stability problem on hypergroups. Functional Equations in Mathematical Analysis 2011, (26) 337–358. Springer optimization and its applications 52, doi:10007/978-1-46-14-0055-4_26
32. Zeglami, D., Roukbi, A., & Kabbaj, S., Hyers–Ulam stability of generalized Wilson's and d'Alembert's functional equations, Afr. Mat. (2013), DOI 10.1007/s13370-013-0199-6.
33. Zeglami, D., Kabbaj, S., & Roukbi, A. Superstability of a generalization of the cosine equation, British J. Math. Comput. Sci. 4 (2014), no. 5, 719–734.

34. Zeglami, D., The superstability of a variant of Wilson's functional equations on an arbitrary group, Afr. Mat. (2014), DOI 10.1007/s13370-014-0229-z.
35. Zeglami, D., Kabbaj, S., & Charifi, A., On the stability of the generalized mixed trigonometric functional equations, Adv. Pure Appl. Math. 2014, to apper.
36. Zeglami, D., Roukbi, A., Kabbaj, S., Rassias, Th. M., Hyers-Ulam stability of Wilson's functional equation on hypergroups. International J. of Scientific & Innovative Math. Research, (2013) 1(1), 66–80.

Printed in the United States
By Bookmasters